TELECOMMUNICATIONS ENGINEERING

TELECOMMUNICATIONS ENGINEERING

SECOND EDITION

J. DUNLOP
and
D.G. SMITH

Department of Electronic
and Electrical Engineering,
University of Strathclyde,
Glasgow

VNR
International

VAN NOSTRAND REINHOLD (INTERNATIONAL)

First published in 1984 by
Van Nostrand Reinhold (International) Ltd
11 New Fetter Lane, London EC4P 4EE

Reprinted 1984, 1986, 1987 (twice)

Second edition 1989

Phototypeset in 10/12 pt Times by
Thomson Press (India) Ltd, New Delhi

Printed in Hong Kong

ISBN 0–278 00082–7

British Library Cataloguing in Publication Data

Dunlop, J. (John)
Telecommunications engineering.—2nd ed.
1. Telecommunications systems
I. Title II. Smith, D.G. (D. Geoffrey)
621.38

ISBN 0–278–00082–7

CONTENTS

PREFACE

The influence of telecommunications has increased steadily since the introduction of telegraphy, radio and telephony. Now, most of us are directly dependent on one or more of its many facets for the efficient execution of our work, at home, or in our leisure.

Consequently, as a subject for study it has become more and more important, finding its way into a large range of higher education courses, given at a variety of levels. For many students, telecommunications will be presented as an area of which they should be aware. The course they follow will include the essential features and principles of communicating by electromagnetic energy, without developing them to any great depth. For others, however, the subject is of more specialized interest; they will start with an overview course and proceed to specialize in some aspects at a later time. We have written our book with both types of student in mind. We have brought together a broader range of material than is usually found in one text, and we have tried to combine an analytical approach to important concepts with a descriptive account of system design. In several places we have stressed the approximate nature of analysis, and the need to exercise engineering judgement in its application. The intention has been to avoid too much detail, so that the text will stand on its own as a general undergraduate-level introduction, and it will also provide a strong foundation for those who will eventually develop more specialized interests.

We have assumed that the reader is familiar with basic concepts in electronic engineering, electromagnetic theory, probability theory and differential calculus.

Chapter 1 begins with the theoretical description of signals and the channels through which they are transmitted. Emphasis is placed on numerical methods of analysis such as the discrete Fourier transform, and the relationship between the time and frequency domain representations is covered in detail. This chapter also deals with the description and transmission of information-bearing signals.

Chapter 2 is concerned with analogue modulation theory. In this chapter there is a strong link between the theoretical concepts of modulation theory and the practical significance of this theory. The chapter assumes that the reader has a realistic knowledge of electronic circuit techniques.

Chapter 3 is devoted to discrete signals and in particular the coding and transmission of analogue signals in digital format. This chapter also emphasizes the relationship between the theoretical concepts and their practical significance.

Chapters 4 and 5 are concerned with the performance of telecommunications systems in noise. Chapter 4 covers the performance of analogue systems and concentrates on the spectral properties of noise. Chapter 5 covers the performance of digital systems and is based on the statistical properties of noise. This chapter also deals in detail with the practical implication of error correcting codes, a topic which is often ignored by more specialized texts in digital communications.

In Chapter 6 the elements of high-frequency transmission-line theory are discussed, with particular emphasis on loss-less lines. The purpose is to introduce the concepts of impedance, reflection and standing waves, and to show how the designer can influence the behaviour of the line.

Basic antenna analysis, and examples of some commonly used arrays and microwave antennas, are introduced in Chapter 7, while Chapters 8 and 9 describe the essential features of waveguide-based microwave components. A fairly full treatment of the propagation of signals along waveguide is considered from both the descriptive and field-theory analysis points of view.

Telephone system equipment represents the largest part of a country's investment in telecommunications, yet teletraffic theory and basic system design do not always form part of a telecommunications class syllabus. Chapter 10 is a comprehensive chapter on traditional switching systems and the techniques used in their analysis. Care has been taken to limit the theoretical discussion to simple cases, to enable the underlying concepts to be emphasized.

Chapter 11 is devoted to television systems. In a text of this nature such a coverage must be selective. We have endeavoured to cover the main topics in modern colour television systems from the measurement of light to the transmission of teletext information. The three main television systems, NTSC, PAL and SECAM, are covered but the major part of this chapter is devoted to the PAL system.

One of the outstanding major developments in recent years has been the production of optical fibres of extremely low loss, making optical communication systems very attractive, both technically and commercially. Chapter 12 discusses the main features of these systems, without introducing any of the analytical techniques used by specialists. The chapter is intended to give an impression of the exciting future for this new technology.

We cannot claim to have produced a universal text; some omissions will not turn out to be justified, and topics which appear to be of only specialized interest now may suddenly assume a much more general importance. However, we hope that we have provided in one volume a coverage which will find acceptance by many students who are taking an interest in this stimulating and expanding field of engineering.

PREFACE TO THE SECOND EDITION

Since the first edition Telecommunications have changed considerably. In particular, digital systems have become more extensive, and satellite and mobile links are more common. To reflect such changes we have added two new chapters to the text and several have been modified, some considerably.

Chapter 3 has been extended to give a detailed description of the *A* law compression characteristic used in PCM systems and substantial extra material is included on PCM transmission techniques. A section is also included on other digital transmission formats such as differential PCM and delta modulation.

Chapter 10 introduces several elements of switching, both analogue and digital. In the latter case the fundamentals of both time and space switching are discussed. The methods used to analyse the switching performance are covered although the discussion is limited to simple cases in order to highlight the underlying concepts. Some of the material on older signalling systems has been omitted from this edition.

Chapter 13 deals with the topic of packet transmission which is finding increasing use in both wide area and local area networks. This chapter provides the theoretical background for scheduled and random access transmission and draws attention to the limitations of these theoretical descriptions and to the need of using reliable computer models for estimating the performance of practical systems. This chapter also introduces the concept of the 7-layer Open Systems Interconnection reference model and illustrates how some of the OSI layers are incorporated in packet switched systems.

An introduction to satellite communication is given in Chapter 14. So as not to over-complicate the concepts involved, discussion is limited to geo-stationary systems. At the time of writing, the development of direct broadcasting of television programmes by satellite is at a very early stage, but there is little doubt that it will become an increasingly important application.

We are most grateful to the many readers who have made constructive suggestions for improvement to the text and who have indentified several errors that existed in the first edition. We hope that the errors have been corrected and that we have been able to remove a number of areas of difficulty by additional explanation or discussion.

Geoffrey Smith is Professor of Communication Engineering at the University of Strathclyde. His main research interests include performance evaluation of networks, mobile communications, integrated local communications and broadland optical systems.

John Dunlop is a reader at the University of Strathclyde and has been involved in international research projects in digital communications for a number of years. He also has research interests in speech coding, local and wide area networks, cellular radio and underwater communications.

LIST OF SYMBOLS

C	capacitance per unit length of transmission line
A	telephone traffic, in erlangs
a	normalized propagation delay
α	attenuation coefficient on a transmission line
α	traffic offered per free source
B	bandwidth of a signal or channel
B	call congestion
β	phase constant on a transmission line
β	modulation index
c	velocity of light in free space
$C(n)$	discrete spectrum
C_n	nth harmonic in a Fourier series
δ_s	skin depth
$\delta(t-t_0)$	impulse function at t_0
DFT	discrete Fourier transform
E	time congestion
$E(f)$	energy density spectrum
EIRP	effective isotropic radiated power
$E_N(A)$	Erlang's loss function
$E(N, s, \alpha)$	Engset's loss function
ε	permittivity
ε_0	permittivity of free space
ε_r	relative permittivity of dielectric
η_0	free space characteristic impedance
η	Single-sided power spectral density of white noise
F	noise figure of a network
f_0	fundamental frequency of a periodic wave
f_c	cut-off frequency
$f(t), h(t)$	general function of time
$F_v(v)$	cumulative distribution function
G	conductance per unit length of transmission line
G	normalized offered traffic
γ	propagation coefficient on a transmission line
$G(f)$	power spectral density

$G(i)$	probability of any i devices being busy
H	magnetic field
H_{av}	entropy of a message (bits/symbol)
$H(f)$	Fourier transform of $h(t)$
$H(i)$	probability of a particular i devices being busy
h	thickness of dielectric in microstrip line
$h(k)$	discrete signal
$[i]$	probability that a network is in state i
IDFT	inverse discrete Fourier transform
IF	intermediate frequency
$I_0(x)$	modified Bessel function
$J_n(\beta)$	Bessel functions of the first kind
k_c	$2\pi/\lambda_c$
L	inductance per unit length of transmission line
LAN	local area network
λ	likelihood ratio
λ	mean packet arrival rate
λ	wavelength
λ_c	cut-off wavelength
λ_{cmn}	cut-off wavelength of the TE_{mn} or TM_{mn} mode
λ_g	guide wavelength
λ_i	call arrival rate in state i
λ_0	free space wavelength
$L(f)$	frequency domain output of a network
$l(t)$	time domain output of a network
m	depth of modulation
μ	permeability
μ_i	call departure rate in state i
μ_0	permeability of free space
N	number of devices
N	electron density in ionosphere (electrons/m^3)
n	refractive index of glass fibre
N_i	normalized noise power at the input of a network
N_o	normalized noise power at the output of a network
n_0	refractive index of free space
$n(t)$	elemental noise voltage
P	power in a signal
P_e	error probability
$P(f)$	transfer function of a network
p'_{nm}	root of $J'_n(k_c, a)$
ψ	angle of reflection coefficient
$p(t)$	impulse response of a network
$p_v(v)$	probability density function
R	resistance per unit length of transmission line

$R_h(\tau)$	autocorrelation function of $h(t)$
ρ	reflection coefficient at transmission line load
s	mean call holding time
S	normalized throughput
S	number of traffic sources
S	voltage standing wave ratio (VSWR)
S_c	normalized carrier power
S_i	normalized signal power at the input of a network
σ	rms voltage of a random signal
S_o	normalized signal power at the output of a network
SNR	signal to noise ratio (power)
T	period of a periodic wave
θ_i	angle of incidence of radio wave to ionosphere
t_a	mean access delay
t_d	mean packet delivery time
t_p	end to end propagation delay
t_r	token rotation time
t_t	token transmission time
t_s	scan time
t_{sl}	walk time
τ	dummy time variable
T_e	effective noise temperature of a network or antenna
TE_{mn}	transverse electric waveguide mode
TM_{mn}	transverse magnetic waveguide mode
T_s	standard noise temperature (290 K)
V	peak voltage of a waveform
v	transmission line wave velocity
V_1	incident (forward) voltage on a transmission line
V_2	reflected (backward) voltage on a transmission line
v_g	group velocity
$V_n(t)$	bandlimited noise voltage
v_{ph}	phase velocity
W	highest frequency component in a signal
WAN	wide area network width of strip in microstrip line
$x(t)$	amplitude of in-phase noise component
y	mean call arrival rate
$y(t)$	amplitude of quadrature noise component
Z_L	transmission line load impedance
Z_0	characteristic impedance of transmission line
Z_0	characteristic impedance of microstrip line

1 / SIGNALS AND CHANNELS

1.1 Introduction

Telecommunication engineering is concerned with the transmission of information between two distant points. Intuitively we may say that a signal contains information if it tells us something we did not already know. This definition is too imprecise for telecommunications studies, and we shall devote a section of this chapter to a formal description of information. For the present it is sufficient to say that a signal that contains information varies in an unpredictable or random manner. We have thus specified a primary character of the signals in telecommunications systems; they are random in nature.

These random signals can be broadly subdivided into discrete signals that have a fixed number of possible values, and continuous signals that have any value between given limits. Whichever type of signal we deal with, the telecommunication system that it uses can be represented by the generalized model of Fig. 1.1. The central feature of this model is the transmission medium or channel. Some examples of channels are coaxial cables, radio links, optical fibres and ultrasonic transmission through solids and liquids. It is clear from these examples that the characteristics of channels can vary widely. The common feature of all channels, however, is that they modify or distort the waveform of the transmitted signal. In some cases the distortion can be so severe that the signal becomes totally unrecognizable.

In many instances it is possible to minimize distortion by careful choice of the transmitted signal waveform. To do this the telecommunications engineer must be able to define and analyse the properties of both the signals and the channels over which they are transmitted. In this chapter we shall concentrate on the techniques used in signal and linear systems analysis, although we should point out that many telecommunications systems do have non-linear characteristics.

1.2 The Frequency and Time Domains

The analysis of linear systems is relatively straightforward if the applied signals are sinusoidal. We have already indicated that the signals encountered in

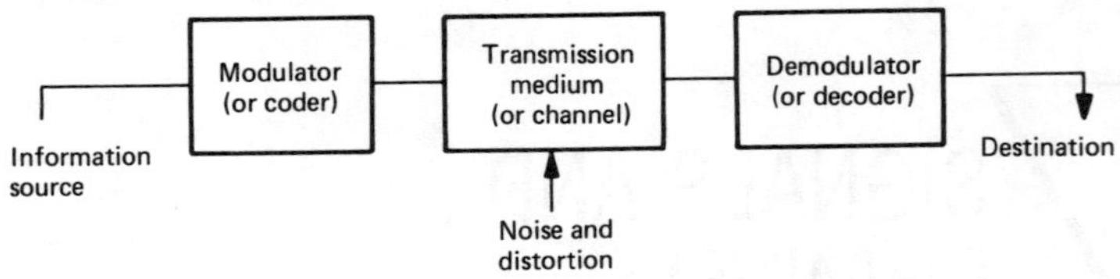

Fig. 1.1 Basic elements of a telecommunications system.

telecommunications systems are random in nature and, as such, are non-deterministic. It is often possible to approximate such signals by periodic functions that themselves can be decomposed into a sum of sinusoidal components. The signal waveforms are functions of time and the variation of signal amplitude with time is known as the 'time domain representation' of the signal. Alternatively, if a signal is decomposed into a sum of sinusoidal components, the amplitude and phase of these components can be expressed as a function of frequency. This leads us to the 'frequency domain representation' of the signal.

The relationship between frequency domain and time domain is an extremely important one and is specified by Fourier's theorem. The response of a linear system to a signal can be determined in the time domain by using the principle of convolution, and in the frequency domain by applying the principle of superposition to the responses produced by the individual sinusoidal components. We will consider the frequency domain first, as this makes use of the theorems of linear network analysis which will be familiar to readers with an electronics background. Time domain analysis is considered in detail in Section 1.11. Frequency domain analysis will be introduced using traditional Fourier methods and we will then develop the discrete Fourier transform (DFT) which is now an essential tool in computer aided analysis of modern telecommunications systems.

1.3 Continuous Fourier Analysis

Fourier's theorem states that any single-valued periodic function, which has a repetition interval T, can be represented by an infinite series of sine and cosine terms which are harmonics of $f_0 = 1/T$. The theorem is given by Eqn (1.1).

$$h(t) = \frac{a_0}{T} + \frac{2}{T}\sum_{n=1}^{\infty}(a_n \cos 2\pi n f_0 t + b_n \sin 2\pi n f_0 t) \tag{1.1}$$

where $f_0 = 1/T$ is the fundamental frequency. The response of a linear system to a waveform $h(t)$ that is not a simple harmonic function is found by summing the responses produced by the individual sinusoidal components of which $h(t)$ is composed. The term a_0/T is known as the dc component and is the mean value of $h(t)$.

$$\frac{a_0}{T} = \frac{1}{T}\int_{-T/2}^{T/2} h(t)\,\mathrm{d}t$$

i.e.

$$a_0 = \int_{-T/2}^{T/2} h(t)\,\mathrm{d}t \tag{1.2}$$

The amplitudes of the sine and cosine terms are given by

$$a_n = \int_{-T/2}^{T/2} h(t)\cos(2\pi n f_0 t)\,\mathrm{d}t$$

$$b_n = \int_{-T/2}^{T/2} h(t)\sin(2\pi n f_0 t)\,\mathrm{d}t \tag{1.3}$$

The Fourier series thus contains an infinite number of sine and cosine terms. This can be reduced to a more compact form as follows; let

$$x(t) = a_n \cos(2\pi n f_0 t) + b_n \sin(2\pi n f_0 t)$$

$$\cos\phi_n = \frac{a_n}{\sqrt{(a_n^2 + b_n^2)}}$$

$$\sin\phi_n = \frac{-b_n}{\sqrt{(a_n^2 + b_n^2)}}$$

Hence $\phi_n = \arctan[-b_n/a_n]$ and

$$x(t) = (a_n^2 + b_n^2)^{1/2}[\cos(2\pi n f_0 t)\cos\phi_n - \sin(2\pi n f_0 t)\sin\phi_n]$$

i.e.

$$x(t) = (a_n^2 + b_n^2)^{1/2}\cos(2\pi n f_0 t + \phi_n)$$

Hence the Fourier series can be modified to

$$h(t) = \frac{a_0}{T} + \frac{2}{T}\sum_{n=1}^{\infty} C_n \cos(2\pi n f_0 t + \phi_n)$$

where

$$C_n = (a_n^2 + b_n^2)^{1/2} \quad \text{and} \quad \phi_n = \arctan\left[\frac{-b_n}{a_n}\right]. \tag{1.4}$$

A graph of C_n against frequency is known as the *amplitude spectrum* of $h(t)$ and a graph of ϕ_n against frequency is known as the *phase spectrum* of $h(t)$.

Note that if the voltage developed across a 1 Ω resistance is

$$v(t) = \frac{2C_n}{T}\cos(2\pi n f_0 t + \phi_n)$$

the average power dissipated in the resistance is

$$P = \frac{2}{T^2} C_n^2 = \frac{2}{T^2}(a_n^2 + b_n^2) \tag{1.5}$$

A graph of C_n^2 against frequency is known as the *power spectrum* of $h(t)$. The total power developed in a 1 Ω resistance by $h(t)$ is thus given by

$$P_T = \frac{a_0^2}{T^2} + \frac{2}{T^2} \sum_{n=1}^{\infty} C_n^2 \tag{1.6}$$

The mean square value of $h(t)$ is given by

$$\sigma^2 = \frac{1}{T} \int_{-T/2}^{T/2} |h(t)|^2 \, dt$$

This is effectively the power dissipated when a voltage equal to $h(t)$ is developed across a resistance of 1Ω. The frequency and time domain representations of $h(t)$ are thus related by Eqn (1.7). This equation is formally known as Parseval's theorem.

$$\frac{1}{T} \int_{-T/2}^{T/2} |h(t)|^2 \, dt = \frac{a_0^2}{T^2} + \frac{2}{T^2} \cdot \sum_{n=1}^{\infty} C_n^2 \tag{1.7}$$

EXAMPLE: Find the amplitude and power spectrum of the periodic rectangular pulse train of Fig. 1.2.

The zero frequency (mean) value is a_0/T where

$$a_0 = \int_{-T/2}^{T/2} h(t) \, dt = \int_{-t_{1/2}}^{t_{1/2}} A \, dt = At_1$$

$$a_n = \int_{-T/2}^{T/2} h(t) \cos(2\pi n f_0 t) \, dt = \int_{-t_{1/2}}^{t_{1/2}} A \cos(2\pi n f_0 t) \, dt$$

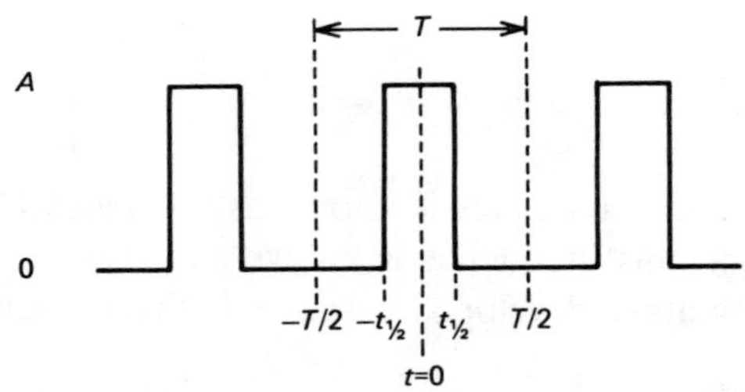

Fig. 1.2 Rectangular periodic pulse train.

i.e.

$$a_n = \frac{A}{2\pi n f_0} [\sin(\pi n f_0 t_1) - \sin(-\pi n f_0 t_1)]$$

hence

$$a_n = \frac{A}{\pi n f} \sin(\pi n f_0 t_1)$$

similarly

$$b_n = \frac{A}{2\pi n f_0} [\cos(\pi n f_0 t_1) - \cos(-\pi n f_0 t_1)] = 0$$

Hence in this example $C_n = a_n$.

The amplitude spectrum is $C_n = At_1(\sin \pi n f_0 t_1)/\pi n f_0 t_1$, which is often written $C_n = At_1 \operatorname{sinc}(\pi n f_0 t_1)$.

The amplitude spectrum is plotted in Fig. 1.3 and it should be noted that the envelope of this spectrum is a sinc function that has unity value when $\pi f t_1 = 0$ and zero value when $\pi f t_1 = m\pi$, i.e. when $f = m/t_1$. In this particular example $\phi_n = 0$ indicating all harmonics are in phase. The power spectrum of $h(t)$ is simply the square of the amplitude spectrum.

A convenient alternative form of Eqn (1.1) can be developed by writing the sine and cosine terms in exponential notation i.e.

$$\cos(2\pi n f_0 t) = [\exp(j2\pi n f_0 t) - \exp(-j2\pi n f_0 t)]/2$$
$$\sin(2\pi n f_0 t) = [\exp(j2\pi n f_0 t) - \exp(-j2\pi n f_0 t)]/2j$$

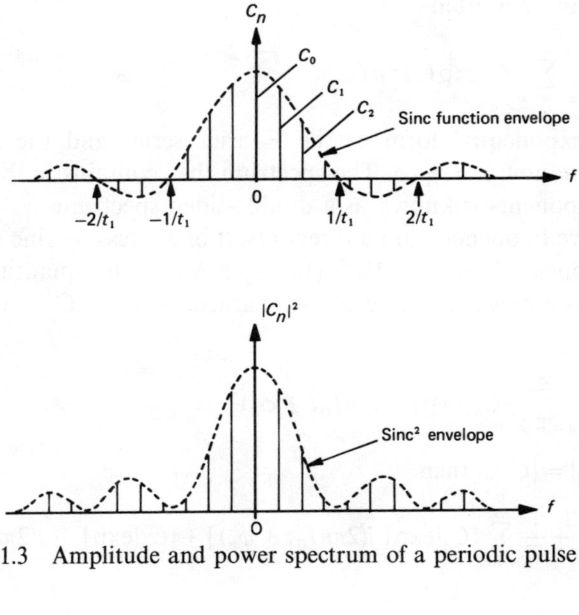

Fig. 1.3 Amplitude and power spectrum of a periodic pulse train.

Substitution in Eqn (1.1) gives

$$h(t)=\frac{a_0}{T}+\frac{1}{T}\sum_{n=1}^{\infty}(a_n-jb_n)\exp(j2\pi nf_0t)+(a_n+jb_n)\exp(-j2\pi nf_0t)$$

If $C_n=(a_n-jb_n)$, its complex conjugate is $C_n{}^*=(a_n+jb_n)$, and

$$(a_n-jb_n)=\int_{-T/2}^{T/2}h(t)[\cos(2\pi nf_0t)-j\sin(2\pi nf_0t)]\,dt$$

i.e.

$$C_n=\int_{-T/2}^{T/2}h(t)\exp(-j2\pi nf_0t)\,dt \tag{1.8}$$

Similarly

$$C_n{}^*=\int_{-T/2}^{T/2}h(t)\exp(j2\pi nf_0t)\,dt$$

i.e. $C_n{}^*=C_{-n}$.

Hence

$$h(t)=\frac{a_0}{T}+\frac{1}{T}\sum_{n=1}^{\infty}[C_n\exp(j2\pi nf_0t)+C_{-n}\exp(-j2\pi nf_0t)]$$

and

$$C_0=\int_{-T/2}^{T/2}h(t)\exp(j^0)\,dt=a_0$$

so that $h(t)$ can be written

$$h(t)=\frac{1}{T}\sum_{n=-\infty}^{\infty}C_n\exp(j2\pi nf_0t) \tag{1.9}$$

This is the exponential form of the Fourier series and the limits of the summation are now $n=\pm\infty$. The spectrum that contains both positive and negative components is known as a double-sided spectrum.

The negative frequencies are a direct result of expressing sine and cosine in complex exponential form. In Eqn. (1.8) C_n is a complex quantity and can be separated into a magnitude and phase characteristic, i.e. $C_n=|C_n|\exp(j\phi_n)$; i.e.

$$h(t)=\frac{1}{T}\sum_{n=-\infty}^{\infty}|C_n|\exp[j(2\pi nf_0t+\phi_n)]$$

But since $|C_n|=|C_{-n}|$ then

$$h(t)=\frac{C_0}{T}+\frac{1}{T}\sum_{n=1}^{\infty}|C_n|\exp[j(2\pi nf_0t+\phi_n)]+|C_n|\exp[-j(2\pi nf_0t+\phi_n)]$$

i.e.

$$h(t)=\frac{C_0}{T}+\frac{2}{T}\sum_{n=1}^{\infty}|C_n|\cos(2\pi n f_0 t+\phi_n)$$

Eqns (1.9) and (1.4) are therefore equivalent, but some care is required in interpreting Eqn (1.9). The harmonic amplitude C_n/T is exactly half the value given by Eqn (1.4), but it is defined for both negative and positive values of n. The correct amplitude is obtained by summing the equal coefficients which are obtained for negative and positive values of n. This is quite reasonable because only one frequency component actually exists.

The power of any frequency is derived from Eqn (1.9) in a similar way. Since C_n is a complex quantity the power at any value of n is $(C_n/T)\cdot(C_n{}^*/T)$,

i.e.

$$(C_n/T)\cdot(C_n{}^*/T)=(a_n-jb_n)(a_n+jb_n)/T^2=(a_n{}^2+b_b{}^2)/T$$

Both negative and positive values of n will contribute an equal amount of power; the total power at any frequency is thus $2(a_n{}^2+b_n{}^2)/T^2$. This of course agrees with Eqn (1.5), since physically only a single component exists at any one frequency. We can write Parseval's theorem for the exponential series as

$$\frac{1}{T}\int_{-T/2}^{T/2}|h(t)|^2\,\mathrm{d}t=\frac{1}{T^2}\sum_{n=-\infty}^{\infty}|C_n|^2 \tag{1.10}$$

Note that only a single value of n appears at $n=0$.

EXAMPLE: We will now evaluate the amplitude spectrum of the waveform in Fig. 1.2 using the exponential series

$$h(t)=\frac{1}{T}\sum_{n=-\infty}^{\infty}C_n\exp(j2\pi n f_0 t)\quad\text{where } C_n=\int_{-t_{1/2}}^{t_{1/2}}A\exp(-j2\pi n f_0 t)\,\mathrm{d}t$$

i.e.

$$C_n=\left[\frac{-A}{j2\pi n f_0}\exp(-j2\pi n f_0 t)\right]_{-t_{1/2}}^{t_{1/2}}$$

$$=\frac{A}{\pi n f_0}\left[\frac{\exp(-j\pi n f_0 t_1)-\exp(-j\pi n f_0 t_1)}{2j}\right]$$

i.e.

$$C_n=\frac{A}{\pi n f_0}\sin(\pi n f_0 t_1)$$

or

$$C_n=At_1\operatorname{sinc}(\pi n f_0 t_1)$$

This is identical to the equation obtained from the cosine series.

1.4 Odd and Even Functions

The waveform $h(t)$ is defined as an even function if $h(t) = h(-t)$; it has the property of being symmetrical about the $t = 0$ axis. If $h(t) = -h(-t)$ the waveform is an Odd function and has skew symmetry about the $t = 0$ axis. A function that has no symmetry about the $t = 0$ axis is neither odd nor even; the sawtooth waveform of Fig. 1.4(c) is an example of such a waveform.

Odd and even functions have properties that may be used to simplify Fourier analysis, e.g.

If

$$h(t) = \frac{a_0}{T} + \frac{2}{T}\sum_{n=1}^{\infty} [a_n \cos(2\pi n f_0 t) + b_n(\sin 2\pi n f_0 t)]$$

then

$$h(-t) = \frac{a_0}{T} + \frac{2}{T}\sum_{n=1}^{\infty} [a_n \cos(2\pi n f_0 t) - b_n \sin(2\pi n f_0 t)]$$

If $h(t) = h(-t)$ this can only be true if $b_n = 0$, i.e. the Fourier series of an even function has cosine terms only. Alternatively, all phase angles ϕ_n in Eqn (1.4) are 0 or $\pm\pi$ and all values of C_n in Eqn (1.8) are real. If $h(t) = -h(-t)$ then $a_n = 0$, i.e. the Fourier series of an odd function contains only sine terms. Alternatively, all phase angles ϕ_n in Eqn (1.4) are $\pm\pi/2$ and all values of C_n in Eqn (1.8) are imaginary. If $h(t)$ has no symmetry about $t = 0$ the Fourier series contains both sines and cosines, the phase angles ϕ_n of Eqn (1.8) are

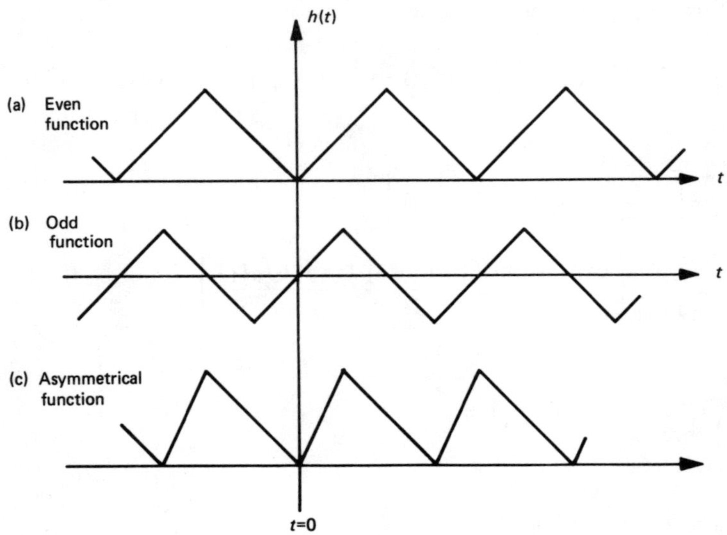

Fig. 1.4 Examples of periodic functions.

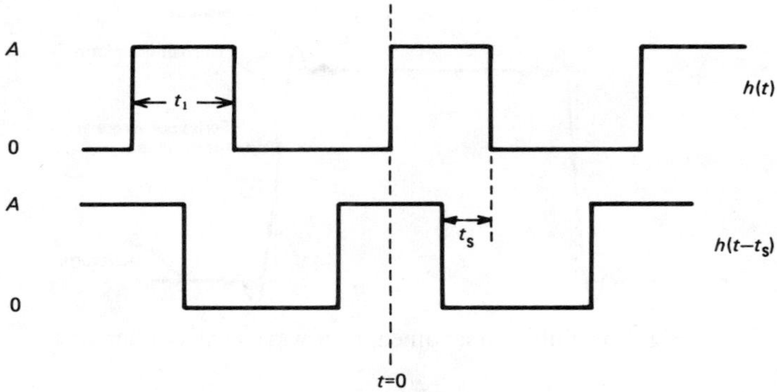

Fig. 1.5 Time shifting.

given by arctan $(-b_n/a_n)$, and all values of C_n in Eqn (1.8) are complex.

Many waveforms that are not symmetrical about $t=0$ can be made either odd or even by shifting the waveform relative to the $t=0$ axis. The shifting process is illustrated in Fig. 1.5. It is of interest to examine the effect of such a shift on the Fourier series. We shall consider the Fourier series of the shifted waveform $h(t-t_s)$. Let $(t-t_s)=t_x$; the amplitude spectrum is thus given by

$$C_n=\int_{-t_{1/2}}^{t_{1/2}} A\exp(-j2\pi nf_0t_x)\,dt_x$$

This evaluates to $C_n=At_1\,\mathrm{sinc}\,(\pi nf_0t_1)$; hence shifting the time axis does not affect the amplitude spectrum. It does, however, affect the phase spectrum:

$$h(t-t_s)=\frac{1}{T}\sum_{n=-\infty}^{\infty} C_n\exp(j2\pi nf_0t)\exp(-j2\pi nf_0t_s)$$

This effectively adds a phase shift of $\phi_s=-2\pi nf_0t_s$ to each component in the series. Interpreted in another way, a time delay of t_s is equivalent to a phase shift of $2\pi f_0t_s$ in the fundamental, $4\pi f_0t_s$ in the second harmonic, etc.

1.5 Waveform Synthesis

This can be regarded as the inverse of Fourier analysis. In effect the Fourier series indicates that any periodic waveform can be synthesized by adding an infinite number of cosine waves with specific amplitudes and phases. In most practical cases a very good approximation to a given periodic waveform can be obtained by truncating the series to only a few terms. As the number of terms in the series is increased the mean square error between the synthesized waveform and the desired waveform decreases. A difficulty does arise in the vicinity of a discontinuity, however. As the number of terms in the series tends

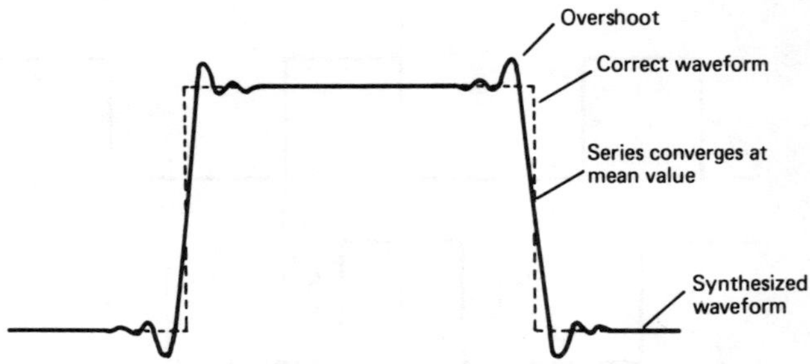

Fig. 1.6 Gibb's phenomenon in waveform synthesis.

to infinity the mean value of the synthesized waveform approaches the mean value of the desired waveform at the discontinuity. The amplitude of the synthesized waveform on either side of the discontinuity is subject to error, which is not reduced when the number of terms in the series is increased. This error is known as Gibb's phenomenon, and is illustrated for a synthesized rectangular wave in Fig. 1.6.

This is in fact a convergence property of the Fourier series. The Fourier series converges to the mean value of $h(t)$ at discontinuities in the waveform of $h(t)$. The conditions required for the convergence of the series are

(i) $h(t)$ must have a finite number of maxima and minima in the interval T;
(ii) $h(t)$ must have a finite number of discontinuities in the interval T;
(iii) $h(t)$ must satisfy the inequality $\int_0^T |h(t)|\,dt < \infty$.

1.6 The Fourier Integral

The Fourier series representation of $h(t)$ is only valid when $h(t)$ is periodic. We have already indicated that information-bearing signals change in a random fashion and do not therefore belong to this category. The amplitude spectra of non-periodic signals are obtained from the Fourier integral. The Fourier integral may be developed from the Fourier series by allowing the period T to approach infinity. In Fig. 1.2 allowing $T \to \infty$ means that $h(t)$ becomes a single pulse of width t_1 seconds.

$$h(t) = \frac{1}{T} \sum_{n=-\infty}^{\infty} C_n \exp(j2\pi n f_0 t) \qquad \text{where } f_0 = 1/T$$

and

$$C_n = \int_{-T/2}^{T/2} h(t) \exp(-j2\pi n f_0 t)\,dt$$

If we let Δf be the spacing between harmonics in the Fourier series then $\Delta f = (n+1)f_0 - n f_0 = 1/T$.

The Fourier series may thus be written

$$h(t) = \sum_{n=-\infty}^{\infty} C_n \exp(j2\pi n f_0 t)\Delta f$$

As $T \to \infty$ then $\Delta f \to 0$ and the discrete harmonics in the series merge, and an amplitude spectrum that is a continuous function of frequency results, i.e. $\lim_{T\to\infty} C_n = H(f)$. The harmonic number n now has all possible values and the summation of the series can thus be replaced by an integral, i.e, nf_0 is

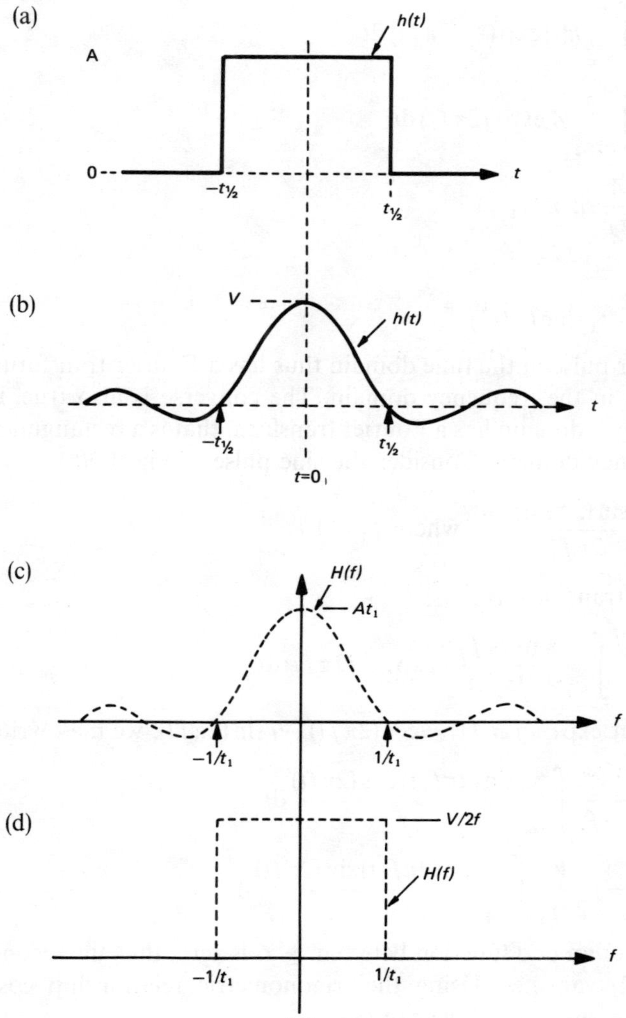

Fig. 1.7 Relationship between frequency and time domain.

replaced by a continuous function f and

$$h(t)=\int_{-\infty}^{\infty} H(f)\exp(j2\pi ft)\,\mathrm{d}f$$

$$H(f)=\int_{-\infty}^{\infty} h(t)\exp(-j2\pi ft)\,\mathrm{d}t \tag{1.11}$$

These two integrals are known as the *Fourier transform pair*. To illustrate the use of the Fourier transform, assume $h(t)$ is a single pulse of amplitude A and duration t_1 seconds.

$$H(f)=\int_{-\infty}^{\infty} h(t)\exp(-j2\pi ft)\,\mathrm{d}t$$

$$=\int_{-t_{1/2}}^{t_{1/2}} A\exp(j2\pi ft)\,\mathrm{d}t$$

$$=\frac{A}{\pi f}\sin \pi f t_1$$

i.e.

$$H(f)=At_1\,\mathrm{sinc}\,(\pi f t_1)$$

A rectangular pulse in the time domain thus has a Fourier transform that is a sinc function in the frequency domain. The converse is also true, i.e. a sinc pulse in the time domain has a Fourier transform that is a rectangular function in the frequency domain. Consider the sinc pulse of Fig. 1.7(b).

$$h(t)=V\frac{\sin(2\pi f_1 t)}{2\pi f_1 t} \qquad \text{where } f_1=1/t_1$$

The Fourier transform is

$$H(f)=V\int_{-\infty}^{\infty}\frac{\sin(2\pi f_1 t)}{2\pi f_1 t}\exp(-j2\pi ft)\,\mathrm{d}t$$

Recalling that $\exp(-j2\pi ft)=\cos(2\pi ft)-j\sin(2\pi ft)$, we may write

$$H(f)=\frac{V}{2\pi f_1}\int_{-\infty}^{\infty}\frac{\sin(2\pi f_1 t)\cos(2\pi ft)}{t}\,\mathrm{d}t - j\frac{V}{2\pi f_1}\int_{-\infty}^{\infty}\frac{\sin(2\pi f_1 t)\sin(2\pi ft)}{t}\,\mathrm{d}t \tag{1.12}$$

The integral of an odd function between $\pm\infty$ is zero; thus the second integral of Eqn (1.12) vanishes. Using the trigonometric relationship $\cos\phi\sin\theta=\frac{1}{2}[\sin(\phi+\theta)-\sin(\phi-\theta)]$ we can say

$$H(f)=\frac{V}{2\pi f_1}\int_{0}^{\infty}\frac{\sin[2\pi(f+f_1)t]}{t}\,\mathrm{d}t-\frac{V}{2\pi f_1}\int_{0}^{\infty}\frac{\sin[2\pi(f-f_1)t]}{t}\,\mathrm{d}t \tag{1.13}$$

At this point we make use of the standard integral

$$\int_0^\infty \frac{\sin ax}{x}\,\mathrm{d}x = \pi/2 \qquad \text{for } a>0$$

$$0 \qquad \text{for } a=0$$

$$-\pi/2 \qquad \text{for } a<0$$

Re-writing Eqn (1.13) as

$$H(f)=\frac{V}{2\pi f_1}(I_1-I_2)$$

there are three frequency ranges of interest:

$$-\infty<f<f_1 \text{ gives } I_1=-\pi/2, \quad I_2=-\pi/2$$

$$-f_1<f<f_1 \text{ gives } I_1=\pi/2, \quad I_2=-\pi/2$$

$$f_1<f<\infty \text{ gives } I_1=\pi/2, \quad I_2=\pi/2$$

Hence $H(f)=V/2f_1$ for $-f_1<f<f_1=0$ else. The resulting $H(f)$ is shown in Fig. 1.7(d).

Comparing Fig. 1.7(a) with Fig. 1.3 shows another important relationship. The envelope of the amplitude spectrum of a single pulse is identical to the envelope of the amplitude spectrum of a periodic pulse train of the same pulse width. This relationship is not restricted to rectangular pulses and is useful in determining the spectral envelope of signals composed of randomly occurring pulses, such as are encountered in digital communications systems.

The fact that all frequencies are present in the amplitude spectrum of a non-periodic signal requires careful interpretation when considering the power dissipated by such signals.

1.7 Power and Energy Density Spectrum

The power density spectrum of a non-periodic signal is developed in a similar way to the amplitude spectrum. If we assume that $h(t)$ is a periodic function we may write Eqn (1.10) as

$$\frac{1}{T}\int_{-T/2}^{T/2}|h(t)|^2\,\mathrm{d}t=\frac{1}{T}\sum_{n=-\infty}^{\infty}|C_n|^2\,\Delta f$$

As $T\to\infty$ for non-periodic signals, this equation can be written in the limit as

$$\frac{1}{T}\int_{-\infty}^{\infty}|h(t)|^2\,\mathrm{d}t=\frac{1}{T}\int_{-\infty}^{\infty}|H(f)|^2\,\mathrm{d}f \tag{1.14}$$

The power spectrum of a non-periodic signal is then defined as

$$G(f)=\frac{|H(f)|^2}{T} \tag{1.15}$$

The power spectral density is a measure of the distribution of power as a function of frequency. It is a useful concept for random signals, such as noise, that have a finite power and are eternal; T is then the period of measurement. When T is large the power is independent of the value of T. If signals exist for a finite time only the power spectrum approaches zero as $T \to \infty$. When dealing with such signals the concept of energy density spectrum is more meaningful. Equation (1.14) can also be written

$$\int_{-\infty}^{\infty} |h(t)|^2 \, dt = \int_{-\infty}^{\infty} |H(f)|^2 \, df \tag{1.16}$$

This is Parseval's theorem for non-periodic signals. The LHS of the equation represents the total energy dissipated in a 1 Ω resistance by a voltage equal in amplitude to $h(t)$. It is clear, therefore, that $|H(f)|^2$ is an energy density, that is, a measure of the distribution of the energy of $h(t)$ with frequency. The double-sided (defined for $\pm f$) energy density spectrum of $h(t)$ is

$$E(f) = |H(f)|^2 \tag{1.17}$$

The total energy within the frequency range f_1 to f_2 is

$$E = \int_{-f_2}^{-f_1} E(f) \, df + \int_{f_1}^{f_2} E(f) \, df \qquad \text{joules} \tag{1.18}$$

i.e., half the energy is contributed by the negative components. In particular, as $f_2 \to f_1$, the total energy $\to 0$. Thus although the energy density spectrum of a non-periodic signal is continuous, the energy at a specific frequency is zero.

1.8 Signal Transmission Through Linear Systems

We noted in Section 1.1 that all communications channels have the common feature of modifying or distorting the waveforms of signals transmitted through them. The amount of distortion produced by a channel with a given transfer function (attenuation and phase shift as a function of frequency) is readily calculated using Fourier transform techniques.

If we assume that $P(f)$ is the channel transfer function (often a voltage ratio in electrical networks) we can obtain the amplitude spectrum of the signal at the channel output by multiplying the amplitude spectrum of the input signal by the network transfer function, i.e.

$$L(f) = H(f) \cdot P(f) \tag{1.19}$$

We can then obtain the output signal $l(t)$ by taking the Fourier transform of Eqn (1.19), i.e.

$$l(t) = \int_{-\infty}^{\infty} H(f) \cdot P(f) \exp(j2\pi f t) \, dt \tag{1.20}$$

Note that $P(f) = |P(f)| \exp(-j2\pi ft)$ where $|P(f)|$ represents attenuation as a function of frequency (i.e. the frequency response of the channel) and $\phi(f)$ represents the phase shift produced. Both $|P(f)|$ and $\phi(f)$ produce signal distortion. Phase distortion is normally neglected when speech and music signals are transmitted over a channel, but it assumes special significance for digital transmission. This topic is covered further in Chapter 3.

When considering signal transmission through networks, we are often concerned with the loss of signal power or energy that occurs during transmission. If $G(f)$ is the power spectral density of a signal and $P(f)$ is the transfer function of a channel the power spectral density at the output is

$$G_0(f) = G_i(f) \cdot |P(f)|^2 \tag{1.21}$$

The total power in a given frequency range f_1 to f_2 at the channel output is

$$W = \int_{-f_2}^{-f_1} G_0(f)\,\mathrm{d}f + \int_{f_1}^{f_2} G_0(f)\,\mathrm{d}f \qquad \text{watts}$$

i.e.

$$W = \int_{-f_2}^{-f_1} G_i(f)|P(f)|^2\,\mathrm{d}f + \int_{f_1}^{f_2} G_i(f)|P(f)|^2\,\mathrm{d}f \tag{1.22}$$

In most practical cases Eqn (1.22) can only be solved by numerical integration. In such cases the use of the discrete Fourier transform, which is discussed in Section 1.10, is particularly useful. To illustrate the application of Eqn (1.22) we will consider a specific example in which the integrals can be evaluated in closed form.

EXAMPLE: A sinc pulse of amplitude V and zero crossings at intervals of $\pm nt_1/2$ is passed through a low-pass RC network of the type shown in Fig. 1.8. If the value of t_1 for the pulse is 2 ms, find the cut-off frequency of the filter in order that 60% of the pulse energy is transmitted.

We calculate the incident energy of the pulse using Parseval's theorem, i.e.

$$E = \int_{-\infty}^{\infty} |h(t)|^2\,\mathrm{d}t = \int_{-\infty}^{\infty} |H(f)|^2\,\mathrm{d}f$$

We have shown that for the sinc pulse $H(f) = V/2f_1$ for $-f_1 < f < f_1$ and zero otherwise.

In this example $f_1 = 1/t_1$ and the energy is

$$E_i = \int_{-f_1}^{f_1} \left(\frac{V}{2f_1}\right)^2 \mathrm{d}f = \frac{V^2}{2f_1} \qquad \text{joules}$$

The network transfer function is $P(f) = 1/[1 + j(f/f_c)]$ where $f_c = 1/2\pi RC$ is the network cut-off frequency. We note from Fig. 1.8 that the network response extends to negative frequencies because $H(f)$ is a double-sided spectrum. For this network $P(f) = f_c^2/(f_c^2 + f^2)$.

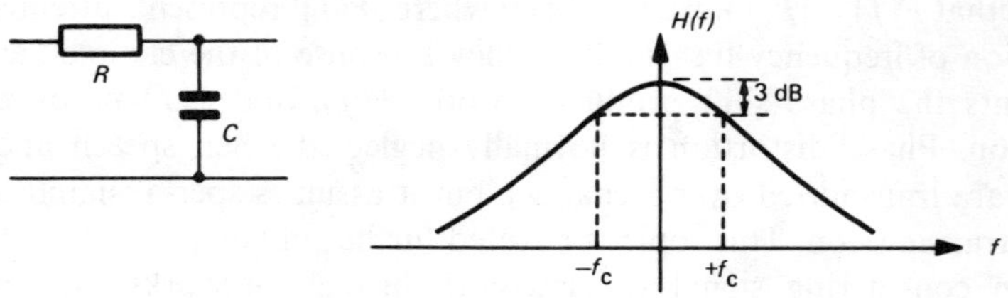

Fig. 1.8 Low-pass RC network.

The energy in the transmitted signal is then

$$E_0 = \int_{-f_1}^{f_1} \left(\frac{V}{2f_1}\right)^2 \frac{f_c^{\,2}}{f_c^{\,2}+f^2}\,\mathrm{d}f$$

$$= 2f_c^{\,2}\left(\frac{V}{2f_1}\right)^2 \frac{1}{f_c}\tan^{-1}\left(\frac{f_1}{f_c}\right)$$

But $E_0 = 0.6E_i$. Hence

$$\frac{V^2}{2f_1^{\,2}} f_c \tan^{-1}\left(\frac{f_1}{f_c}\right) = \frac{0.6\,V^2}{2f_1}$$

or

$$\tan^{-1}\left(\frac{f_1}{f_c}\right) = 0.6\frac{f_1}{f_c}$$

The solution to this equation is $f_1/f_c \simeq 1{\cdot}755$, i.e.

$$f_c = \frac{1}{1.755t_1} = 285\ \mathrm{Hz}$$

1.9 The Impulse Function

If we consider the rectangular pulse of Fig. 1.7(a) and let $A = 1/t_1$, the pulse area becomes unity, i.e. area $= (1/t_1)\cdot t_1 = 1$. If t_1 is allowed to approach zero, then in order to preserve unit area the pulse amplitude A is allowed to approach infinity. Such a pulse cannot be produced practically, but it is extremely useful for analytical purposes and is known as the unit impulse. The unit impulse function is formally defined by Eqn (1.23):

$$\int_a^b \delta(t-t_0)\,\mathrm{d}t = 1 \qquad \text{for } a < t_0 < b, \qquad = 0 \text{ else} \tag{1.23}$$

The impulse exists only at time t_0 and has zero value for all other values of t. The amplitude of the impulse at $t = t_0$ is undefined; instead the impulse is defined in terms of its area (or weight) at time $t = t_0$. If any continuous function $h(t)$ is multiplied by an impulse with unit weight at time $t = t_0$ the resulting function is given by

$$\int_a^b h(t)\delta(t - t_0)\,\mathrm{d}t = h(t_0) = 1 \qquad \text{for } a < t_0 < b, \qquad = 0 \text{ else} \tag{1.24}$$

Hence multiplying $h(t)$ by an impulse function at $t = t_0$ and performing the integration of Eqn (1.24) is equivalent to taking an instantaneous sample of $h(t)$ at $t = t_0$. The impulse function is defined only in integral form and expressions such as $h(t_0) = h(t)\cdot\delta(t - t_0)$ are strictly meaningless. However, it is common practice to express the integral equation (1.24) in this form, the process of integration being implicit. The Fourier transform of an impulse function is particularly important. The Fourier transform of the unit impulse defined by Eqn (1.23) is

$$\Delta(f) = \int_{-\infty}^{\infty} \delta(t - t_0)\exp(-j2\pi f t)\,\mathrm{d}t = 1\exp(-j2\pi f t_0) \tag{1.25}$$

This means that $|\Delta(f)|$ has unity value for all values of f. The function $\exp(-j2\pi f t)$ represents the phase of each component in $\Delta(f)$, i.e. $\phi = 2\pi f t_0$.

If, instead of a single impulse, we consider a periodic train of impulses separated by a period T, the amplitude spectrum is obtained from the Fourier series. The amplitude of the nth harmonic is then

$$C_n = \int_{-\infty}^{\infty} h(t)\exp(-j2\pi n f_0 t)\,\mathrm{d}t \qquad \text{where } h(t) = \delta(t - nT)$$

i.e.

$$C_n = \int_{-\infty}^{\infty} \delta(t - nT)\exp(-j2\pi n f_0 t)\,\mathrm{d}t = 1\exp(-j2\pi) \tag{1.26}$$

Hence each component in the Fourier series has unity value and a phase of 2π radians. This periodic train of impulses is used to obtain regularly spaced samples of a continuous wave form $h(t)$ and is of fundamental importance in the digital transmission of analogue signals.

Now that we have defined the impulse (or delta) function we can show that the Fourier integral can also be used to define the amplitude spectrum of a periodic signal and is therefore a general transform. If

$$h(t) = \frac{1}{T}\sum_{n=-\infty}^{\infty} C_n \exp(j2\pi n f_0 t)$$

the Fourier transform of $h(t)$ is

$$H(f)=\frac{1}{T}\int_{-\infty}^{\infty}\sum_{n=-\infty}^{\infty} C_n \exp(j2\pi(f-nf_0)t)\,dt$$
$$=\frac{1}{T}\sum_{n=-\infty}^{\infty} C_n \int_{-\infty}^{\infty} \exp[2j\pi(f-nf_0)t]\,dt$$
$$=\frac{1}{T}\sum_{n=-\infty}^{\infty} C_n\delta(f-nf_0) \tag{1.27}$$

The Fourier transform of a periodic signal is thus a set of impulses located at harmonics of the fundamental frequency $f_0 = 1/T$.

1.10 The Discrete Fourier Transform (DFT)

We pointed out in Section 1.8 the extensive use is made of computer-aided analysis in the study of modern telecommunication systems. Computers cannot handle continuous signals but can process signals that are defined at discrete intervals of time. The DFT is an extension of the continuous Fourier transform designed specifically to operate on signals that have been sampled at regular intervals of time. The sampling process may be regarded as multiplying the continuous signal by a periodic series of impulses. We have shown [Eqn (1.27)] that the spectrum of such a period signal is a series of harmonics all of equal amplitude. When such a spectrum is multiplied by the spectrum of a continuous signal, each component in the continuous signal will form sum and difference frequencies with the harmonics of the periodic impulse train. If we assume that the impulses are separated by an interval T_s and that the maximum frequency component of the continuous signal is W Hz, the amplitude spectrum of the sampled signal will take the form of Fig. 1.9. It will be noted from Fig. 1.9 that provided the sampling frequency $f_s(=1/T_s)$ is at least $2W$, there will be no overlap (aliasing) between the signal spectrum and the first lower side-band of the sampled signal spectrum. The original signal is defined by its amplitude spectrum which is preserved in the sampled version provided that $f_s \geqslant 2W$. This is in fact a statement of the 'sampling theorem' that we consider in more detail in Chapter 3.

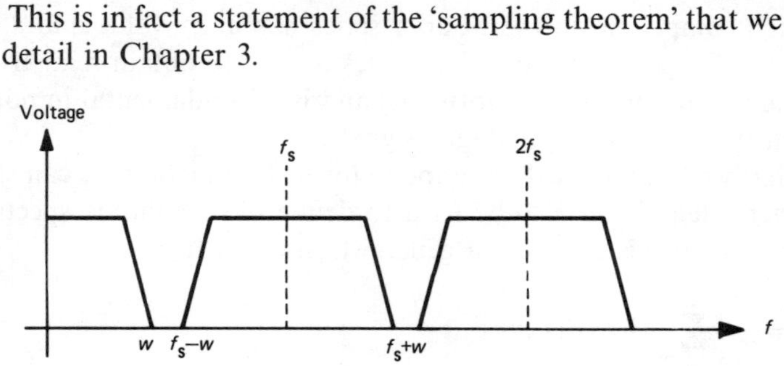

Fig. 1.9 Spectrum of a sampled signal.

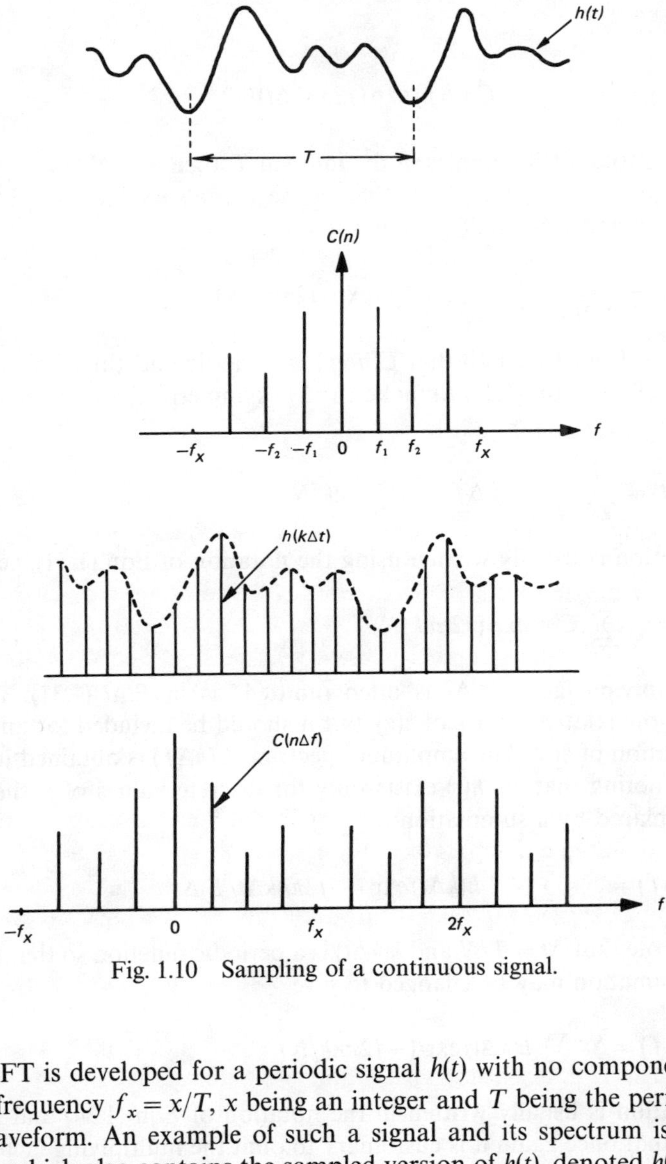

Fig. 1.10 Sampling of a continuous signal.

The DFT is developed for a periodic signal $h(t)$ with no components at or above a frequency $f_x = x/T$, x being an integer and T being the period of the signal waveform. An example of such a signal and its spectrum is given in Fig. 1.10, which also contains the sampled version of $h(t)$, denoted $h(k\Delta t)$, and its amplitude spectrum $C(n\Delta f)$. The sampling frequency ($f_s = 1/\Delta T$) is chosen to equal $2f_x$ which avoids aliasing.

The Fourier series for $h(k\Delta t)$ is

$$h(k\Delta t) = \frac{1}{T} \sum_{n=-(x-1)}^{x-1} C_n \exp(j2\pi nk\Delta t/T) \tag{1.28}$$

but over the range $-(x-1) \leqslant n \leqslant (x-1)$ the coefficients C_n are identical to

$C(n\Delta f)$; hence

$$h(k\Delta t)=\frac{1}{T}\sum_{n=-(x-1)}^{(x-1)} C(n\Delta f)\exp(j2\pi nk\Delta t)/T) \qquad (1.29)$$

If there is a total of N samples in the interval T, then $T = N\Delta t$ and the range of k is $0, \pm 1, \pm 2 \ldots \pm[(N/2)-1]$. Since we can also write $t = 1/2f_x = T/2x$ then $N = 2x$ and Eqn (1.29) becomes

$$h(k\Delta t)=\frac{1}{N\Delta t}\sum_{n=-N/2+1}^{N/2+1} C(n\Delta f)\exp(j2\pi nk/N) \qquad (1.30)$$

We observe from Fig. 1.10 that $C(n\Delta f)$ is periodic and thus we can change the range of n in Eqn (1.30) to make $C(n\Delta f)$ symmetrical about the frequency f_x; thus

$$h(k\Delta t)=\frac{1}{N\Delta t}\sum_{n=0}^{n-1} C(n\Delta f)\exp(j2\pi nk/N)$$

This equation is usually written using the notation of Eqn (1.31), i.e.

$$h(k)=\frac{1}{N}\sum_{n=0}^{n-1} C(n)\exp(j2\pi nk/N) \qquad (1.31)$$

The multiplying factor $1/\Delta t$ is often omitted, as in Eqn (1.31). This does not affect the relative values of $h(k)$, but it should be included for an accurate representation of $h(k)$. The amplitude spectrum $C(n\Delta f)$ is obtained using Eqn (1.8) and noting that, as $h(t)$ exists only for discrete values of t, the integral can be replaced by a summation:

$$C(n\Delta f)=\sum_{k=-(N/2-1)}^{N/2-1} h(k\Delta t)\exp(-j2\pi nk\Delta t/T)\Delta t \qquad (1.32)$$

Here we note that $\Delta t = T/N$ and $h(k\Delta t)$ is a periodic function so that the limits of the summation may be changed to give

$$C(n\Delta f)=\Delta t\sum_{k=0}^{N-1} h(k\Delta t)\exp(-j2\pi nk/N) \qquad (1.33)$$

This equation is usually written in the notation of Eqn (1.34) and it should be noted that once again it is customary to omit the multiplying factor, which in this case is Δt:

$$C(n)=\sum_{k=0}^{N-1} h(k)\exp(-j2\pi nk/N) \qquad (1.34)$$

Equation (1.34) is known as the Discrete Fourier Transform (DFT) of $h(t)$, and Eqn (1.31) is known as the Inverse Discrete Fourier Transform (IDFT) of $C(n\Delta f)$. It should be noted that in both Eqns (1.31) and (1.34) there is no

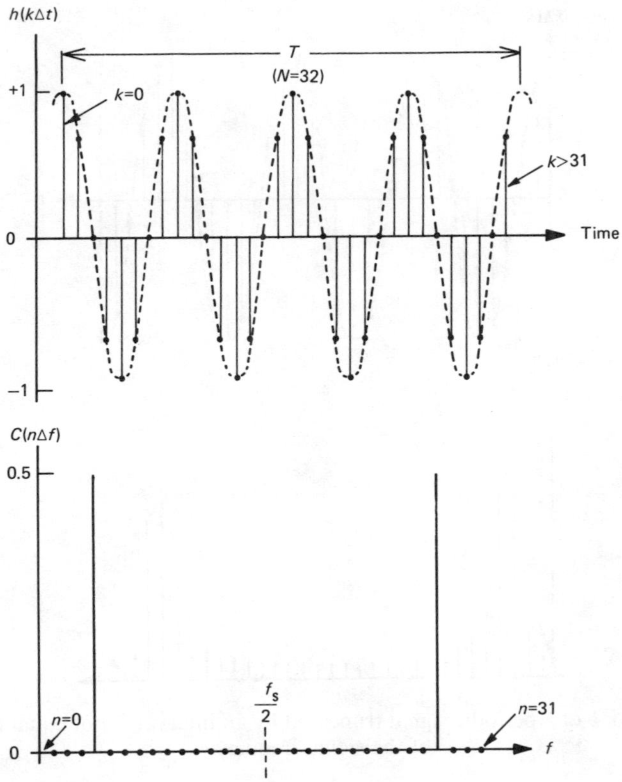

Fig. 1.11 DFT of a periodic-band-limited waveform.

explicit frequency or time scale as the coefficients k, n and N simply have numerical values.

Some care is required in the use of the DFT because, as we have shown, it is valid only for the special case of a band-limited periodic signal. A waveform of this type and its DFT is shown in Fig. 1.11. In this figure $h(t)$ is a single tone with four complete cycles in the interval T. The DFT has a component at a value of $n=4$ (i.e. the fourth harmonic of the fundamental frequency $f_0 = 1/T$) and a second component at a value of $n=28$. This second component is the equivalent of $n=-4$ resulting from the change of range in Eqn (1.30). We note, therefore, that in the special case of a band-limited periodic function the DFT produces the correct spectrum of $h(t)$. In all other cases the DFT will only produce an approximation to the amplitude spectrum of $h(t)$.

Consider next the DFT of the waveform of Fig. 1.12. In this case the interval T contains 3.5 cycles of $h(t)$. The DFT requires the signal to be periodic with period T, and this means that discontinuities must now exist at the extremities of the interval T.

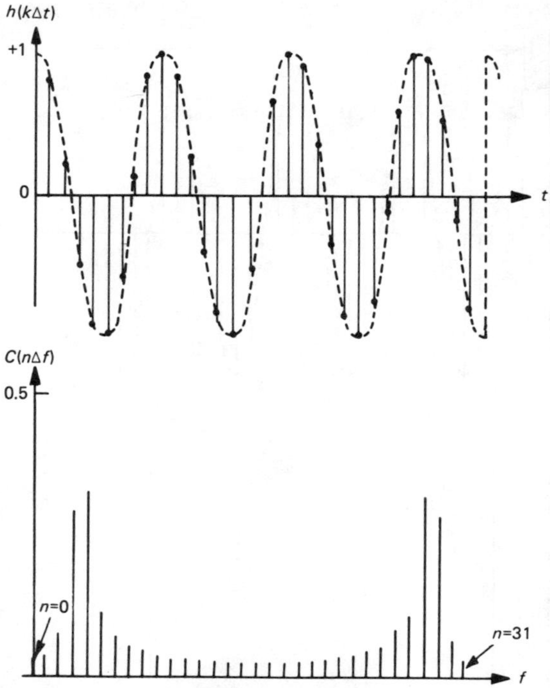

Fig. 1.12 DFT of a periodic signal truncated by an interval T not equal to a multiple of the signal frequency.

In other words, the periodic signal is no longer band-limited, and a form of distortion known as leakage is introduced into the spectrum. This form of distortion is considered in more detail in Section 1.11 after the concept of convolution has been introduced. It suffices here to note that sampling a non-band-limited signal produces a discrete spectrum of the form shown in Fig. 1.12. This is clearly only an approximation to the original spectrum; the approximation can be made more accurate by increasing the interval of observation T (for non-periodic signals) or by making T equal to a multiple of the actual period for periodic signals. In addition, increasing the sampling frequency always reduces aliasing which is produced by sampling non-band-limited signals.

The DFT can be calculated directly from Eqn (1.34) but it will be noticed that each coefficient $C(n)$ requires N complex multiplications and additions. There are N spectral coefficients so that a total of N^2 complex multiplications and additions will be required in the complete DFT calculation. Multiplication is a relatively slow process in a general purpose digital processor, and for this reason the DFT is usually calculated by using the 'Fast Fourier Transform' algorithm. This is an algorithm designed to reduce the number of multiplications required to evaluate the DFT. The algorithm achieves a reduction from N^2 to $N \log_2 N$ multiplications by dividing Eqn (1.34) into the sum of several smaller sequences.[1] This can be very significant when N is a very large number. (It should be noted that special purpose processors are now available which can perform multiplication in one machine cycle.)

1.11 Time Domain Analysis

The time domain and frequency domain are uniquely linked by the Fourier transform and consequently the frequency domain analysis of the previous sections can also be undertaken in the time domain. To illustrate this point, consider Eqn (1.19) which relates the spectrum at a network output to the product of the input spectrum and the network transfer function. If the input to the network $h(t)$ is a unit impulse we have shown in Eqn (1.25) that the spectrum $\Delta(f)$ has unity value for all f. Hence the spectrum at the network output is simply $L(f) = P(f)$ where $P(f)$ is the network transfer function. The response in the time domain is the Fourier transform of $P(f)$, and is known as the impulse response, i.e.

$$p(t) = \int_{-\infty}^{\infty} P(f) \exp(j2\pi f t) \, \mathrm{d}f \tag{1.35}$$

Having defined impulse response we now make use of Eqn (1.24), which states that the value of a signal $h(t)$ at any time t_0 is obtained by multiplying $h(t)$ by a unit impulse centred at t_0. The signal $h(t)$ can thus be regarded as an infinite number of impulses, the weight of each impulse being equal to the instantaneous value of $h(t)$. Each of these impulses will produce an impulse response and the network output is then obtained by the superposition of the individual impulse responses. The response of a linear network $l(t)$ to an input signal $h(t)$ is given in terms of the network impulse response by Eqn (1.36):

$$l(t) = \int_{-\infty}^{\infty} h(\tau) p(t-\tau) \, \mathrm{d}\tau \tag{1.36}$$

In this equation τ is a dummy time variable and both $h(\tau)$ and $p(\tau)$ are continuous functions. Equation (1.36) therefore states that the output of a linear network at time t is given by the sum of all values of the input $h(\tau)$ weighted by the appropriate value of $p(\tau)$ at time t. The integral in Eqn (1.36) is known as the convolution integral and the equation is often written as

$$l(\tau) = h(t) * p(t) \tag{1.37}$$

where the symbol $*$ denotes convolution. Comparing Eqn (1.37) with Eqn (1.19) we note the important relationship that multiplication in the frequency domain is equivalent to convolution in the time domain. The converse is also true; that is, multiplication in the time domain is equivalent to convolution in the frequency domain. We will now consider some examples of convolution.

The first example concerns frequency domain convolution. If we consider the waveform of Fig. 1.12 we note that in selecting a time window of T sec. we are in effect multiplying the continuous signal $h(t)$ by a rectangular pulse of unity amplitude and duration T. This is equivalent to convolving the amplitude spectrum of $h(t)$ with the spectrum of the rectangular window function which, as we have already seen, is a sinc function. The spectrum of $h(t)$ is actually a delta function at $\pm f_0$ since only a single frequency is present. The convolution integral is thus

$$\int_{-\infty}^{\infty} H(f)\delta(f-f_0)\,\mathrm{d}f + \int_{-\infty}^{\infty} H(f)\delta(f+f_0)\,\mathrm{d}f = H(-f_0) + H(f_0) \tag{1.38}$$

The original spectrum centred at $f=0$ is thus transferred to frequencies $\pm f_0$. The procedure is illustrated in Fig. 1.13 and it is interesting to compare this spectrum with the DFT of Fig. 1.12. The effect of truncating the signal $h(t)$ in the time domain causes a spreading of the spectrum (leakage) in the frequency domain.

We next consider the transmission of a rectangular pulse through a low-pass RC network of the form shown in Fig. 1.8. The impulse response of this network is $p(t) = (1/RC)\exp(-t/RC)$. This may be proved as follows:

$$P(f) = \int_{-\infty}^{\infty} p(t)\exp(-j2\pi ft)\,\mathrm{d}t$$

Since $p(t)$ is the impulse response of a real network it has a value of zero for $t<0$; hence

$$P(f) = \frac{1}{RC}\int_{0}^{\infty} \exp(-t/RC)\exp(-j2\pi ft)\,\mathrm{d}t$$

$$= \frac{-1}{RC(j2\pi f + 1/RC)}[\exp[-(j2\pi f + 1/RC)t]]_0^{\infty}$$

i.e.

$$P(f) = \frac{1}{1 + j2\pi fRC} = \frac{1}{1 + j(f/f_c)}$$

which agrees with the expression obtained by network analysis.

Before proceeding further we shall consider the physical interpretation of Eqn (1.36). In this equation t represents the present instant in time and we note that for particularly realizable filters $p(t-\tau)$ must be zero for $t<\tau$; in other words, the impulse response must be zero for all time before an impulse occurs. The impulse response of an ideal filter exists for all values of τ and is therefore unrealizable (see Section 3.3). If we confine our interest to practical networks, then it is clear that the impulse response $p(t-\tau)$ scans the signal

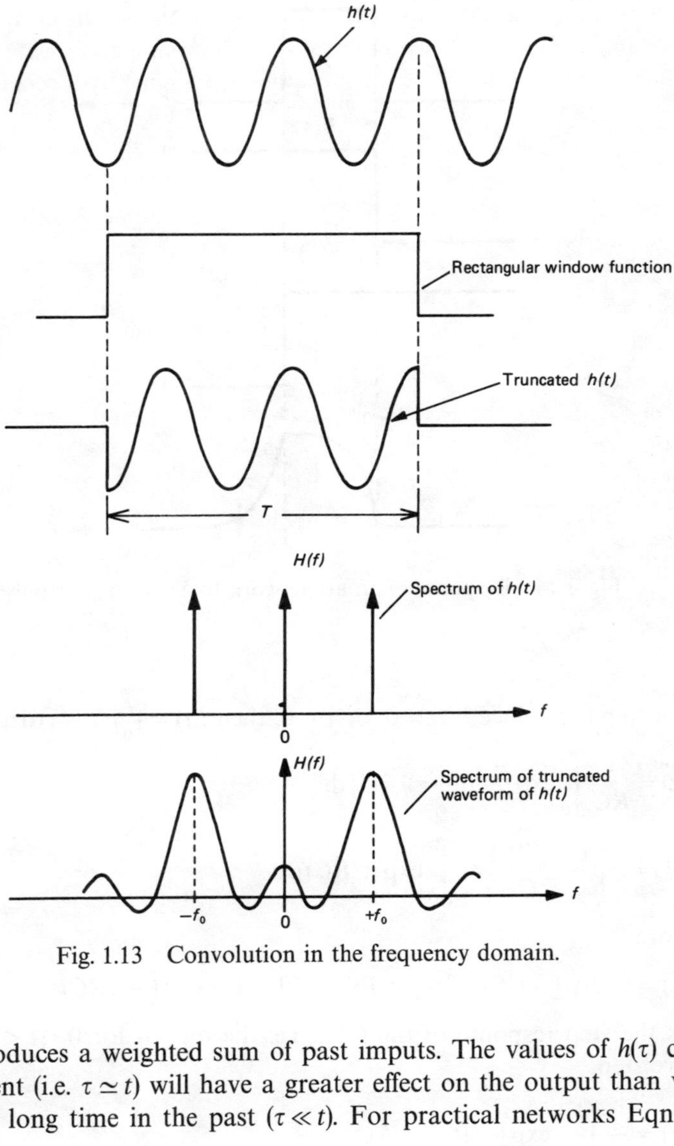

Fig. 1.13 Convolution in the frequency domain.

$h(\tau)$ and produces a weighted sum of past imputs. The values of $h(\tau)$ closest to the present (i.e. $\tau \simeq t$) will have a greater effect on the output than values occurring a long time in the past ($\tau \ll t$). For practical networks Eqn (1.36) becomes

$$l(t) = \int_{-\infty}^{t} h(t)p(t-\tau)\,\mathrm{d}\tau \tag{1.39}$$

We can now consider the response of the RC network to a rectangular pulse of width t_1. This is split into a positive step of unit amplitude at $\tau = 0$ followed by a negative step of unit amplitude at $\tau = t_1$, as shown in Fig. 1.14. Since the system is linear the output is obtained by superposition. Considering the

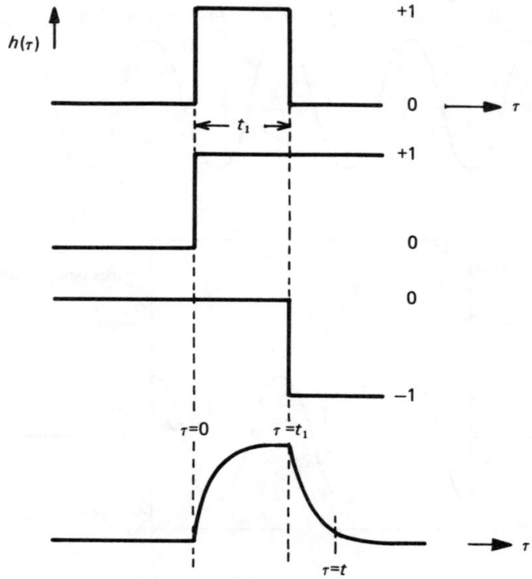

Fig. 1.14 Response of an RC network to a rectangular pulse.

positive step first, since $h(\tau)=0$ for $\tau<0$, then $l(t)=\int_0^t p(t-\tau)\,d\tau$, i.e.

$$l(t)=\frac{1}{RC}\int_0^t \exp[-(t-\tau)/RC]\,dt$$

$$=\exp\frac{(-t/RC)}{RC}d\tau\int_0^t \exp(\tau/RC)\,d\tau$$

Therefore

$$l(t)=-\exp(-t/RC)[\exp(t/RC)-1]=1-\exp(-t/RC)$$

This is the step response of the filter, i.e. the output for $0<t<t_1$. For the negative step

$$l(t)=-\int_{-t_1}^t \exp[-(t-\tau)/RC]\,d\tau$$

i.e.

$$l(t)=\exp[(t_1-t)/RC]-1$$

The output of the filter at times in excess of $t=t_1$ is therefore

$$l(t)=\exp[(t_1-t)/RC]-\exp(-t/RC)$$

As we have already pointed out, time domain analysis is equivalent to frequency domain analysis and it is not possible to give a general rule as to which technique is more appropriate to particular situations. Time domain analysis is frequently used in digital systems, especially in specifying network characteristics to minimize signal distortion.

1.12 Correlation Functions

These functions have particular application in the time domain specification of signals that vary in an unpredictable manner (i.e. information-bearing signals, noise, etc.). The autocorrelation function of a waveform $h(t)$ is defined as

$$R_h(\tau) = \lim_{T\to\infty} \frac{1}{T}\int_{-T/2}^{T/2} h(t)h(t+\tau)\,\mathrm{d}t \tag{1.40}$$

The autocorrelation function is the average of the product $h(t)h(t+\tau)$ and will clearly depend on the value of τ. The function $h(t+\tau)$ is a replica of $h(t)$ delayed by an interval τ. The numerical value of $R_h(\tau)$ is a measure of the similarity (or correlation) of $h(t)$ and $h(t+\tau)$. If there is no similarity between $h(t)$ and $h(t+\tau)$ and each has zero mean value then $R_h(\tau)$ is zero. Completely random signals, such as white noise, have an autocorrelation function equal to zero. The maximum value of $R_h(\tau)$ for any signal will occur when $\tau = 0$. In these circumstances

$$R_h(0) = \lim_{T\to\infty} \frac{1}{T}\int_{-T/2}^{T/2} h^2(t)\,\mathrm{d}t$$

which is the mean square value of $h(t)$.

The autocorrelation function is widely used in signal analysis for recognizing signals in the presence of noise and also for estimating the power spectral density of random signals. We will derive as an example the autocorrelation function of a periodic pulse train and a random binary data signal. Consider the periodic waveform of Fig. 1.15: it is only necessary to average the signal over one period and thus

$$R_h(\tau) = \frac{1}{T}\int_{-T/2}^{T/2} h(t)h(t+\tau)\,\mathrm{d}t$$

The waveform of $h(t)h(t+\tau)$ is shown in Fig. 1.15. It is a periodic pulse train of amplitude A^2, pulse duration $t_1 - |\tau|$ and period T. The autocorrelation function is thus

$$R_h(\tau) = \frac{A^2}{T}(t_1 - |\tau|) \qquad \text{for } -t_1 < \tau < t_1$$

but because $h(t+T) = h(t)$ the autocorrelation function is periodic. The value of $R_h(\tau)$ is plotted as a function of τ in Fig. 1.15.

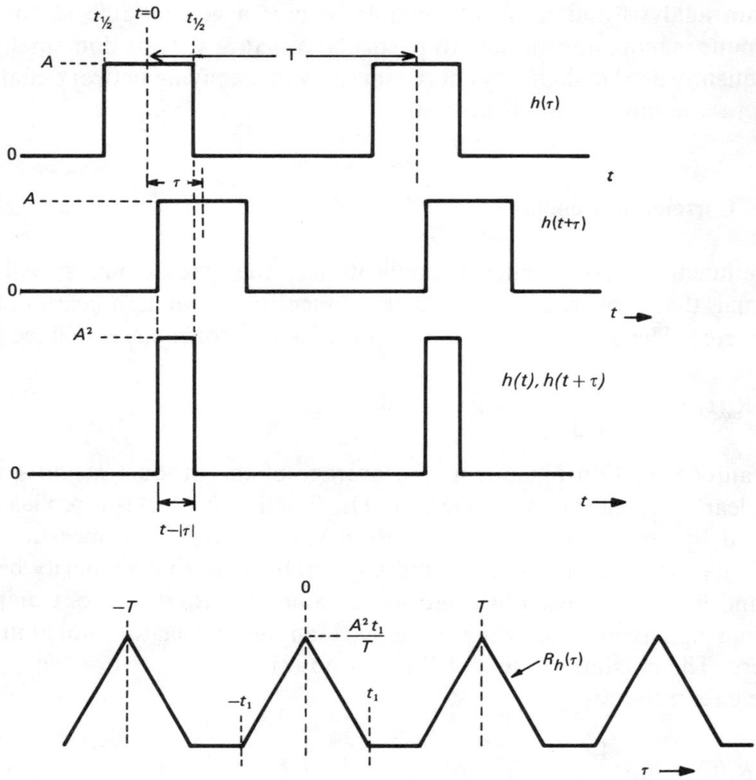

Fig. 1.15 Autocorrelation function of a periodic pulse train.

We note that if $h(t)$ is a single pulse the autocorrelation function is modified to

$$R_h(\tau) = \frac{1}{t_1} \int_{-t_{1/2}}^{t_{1/2}} h(t)h(t+\tau)\,\mathrm{d}t$$

which evaluates to

$$R_h(\tau) = A^2\left(1 - \frac{|\tau|}{t_1}\right) \qquad \text{for } -t_1 < \tau < t_1$$

In this case $R_h(\tau)$ is a single triangular pulse of peak amplitude A^2. The autocorrelation function of a random binary pulse train is obtained in a similar way. Assume such a signal is composed of pulses of amplitude A volts and 0 volts and each of the duration t_1, both of equal probability. During any time interval T there will be an equal number of pulses of amplitude A volts and 0 volts. The average value of $h(t)h(t+\tau)$ must therefore be

$$R_h(\tau) = \frac{A^2}{2}\left(1 - \frac{|\tau|}{t_1}\right) \qquad \text{for } -t_1 < \tau < t_1$$

and this is the autocorrelation function of a random binary pulse train.

Equation (1.40) is very similar to the convolution integral of Eqn (1.36). Remembering that convolution in the time domain is equivalent to multiplication in the frequency domain, the Fourier transform of $R_h(\tau)$ is equal to the Fourier transform of $h(t)$ multiplied by the Fourier transform of $h(t+\tau)$. The amplitude spectrum of $h(t)$ is identical to the amplitude spectrum of $h(t+\tau)$, i.e.

$$\int_{-\infty}^{\infty} R_h(\tau)\exp(j2\pi f\tau)\,\mathrm{d}\tau = |H(f)|^2 \tag{1.41}$$

Thus the power spectral density of any signal is the Fourier transform of its autocorrelation function. The power spectral density of the random data signal is thus

$$G(f) = \int_{-\infty}^{\infty} \frac{A^2}{2}\left(1+\frac{|\tau|}{t_1}\right)\exp(j2\pi f\tau)\,\mathrm{d}\tau$$

i.e.

$$G(f) = \int_{-\infty}^{\infty} \frac{A^2}{t_1}\left(1+\frac{|\tau|}{t_1}\right)[\cos(2\pi f\tau) + j\sin(2\pi f\tau)]\,\mathrm{d}\tau \tag{1.42}$$

Since $R_h(\tau)$ is an even function of τ the imaginary terms in Eqn (1.42) vanish and

$$G(f) = \int_{-\infty}^{\infty} \frac{A^2}{2}\left(1+\frac{|\tau|}{t_1}\right)\cos(2\pi f\tau)\,\mathrm{d}\tau$$

which evaluates to $G(f) = (At_1)^2\,\mathrm{sinc}^2(\pi f t_1)$.

The autocorrelation function is a measure of the degree of similarity between $h(t)$ and a delayed version of the same waveform. The cross-relation function is a measure of the degree of similarity between two different waveforms $h(t)$ and $g(t)$. The cross-correlation function is defined as

$$R_{hg}(\tau) = \lim_{T\to\infty}\frac{1}{T}\int_{-T/2}^{T/2} h(t)g(t+\tau)\,\mathrm{d}t \tag{1.43}$$

This function finds specific application in the detection of signals at low signal-to-noise ratios. Correlation detection is considered in detail in Chapter 5.

It should be clear from the previous two sections that both time domain and frequency domain techniques are important tools in the analysis of telecommunications systems. They should be regarded as complementary, as it is not possible to give a general rule as to which technique is more appropriate to a particular situation.

1.13 Information Content of Signals

In previous sections we have considered signals in terms of waveforms and spectra. In this section we consider the information content of signals and show how it is related to the information capacity of communication channels.

Information is conveyed by a signal that changes in an unpredictable fashion. It is important to have some method of evaluating the information content of a signal because this will determine whether or not the signal can be transmitted over a particular channel. The information content of a signal is measured in bits, which as we shall show later is not necessarily related to the number of binary digits required to transmit it. The information capacity of a communication channel is limited by bandwidth, which determines the maximum signalling speed, and by noise, which determines the number of distinguishable signal levels.

We will consider the specific example of a teleprinter which is restricted to transmitting the four signals *ABCD*. If one of these symbols is transmitted there are four (4^1) possible messages which are *A* or *B* or *C* or *D*. If two symbols are sent there are 16 (4^2) possible messages, viz.

	AA	or	*AB*	or	*AC*	or	*AD*
or	*BA*	or	*BB*	or	*BC*	or	*BD*
or	*CA*	or	*CB*	or	*CC*	or	*CD*
or	*DA*	or	*DB*	or	*DC*	or	*DD*

If P symbols are sent the number of possible messages is 4^P. If the teleprinter can transmit n different symbols, the number of different messages that could be transmitted when P symbols are sent is n^P. Obviously the greater the number of possible messages the less predictable is any particular message. Intuitively we would argue that the more unpredictable a particular message the more information it contains. It is reasonable to assume that the information content is a function of the *unpredictability* of a message. In algebraic form the information content H is

$$H \propto f(n^P) \tag{1.44}$$

Equation (1.44) can be made a function of time by assuming that one symbol is transmitted every t_1 seconds. The total number of symbols transmitted in T seconds is thus T/t_1 and the information content of such a message of T seconds duration would be

$$H \propto f(n^{T/t_1}) \tag{1.45}$$

It is reasonable to assume that a similar message of duration $2T$ seconds would contain twice as much information as a message of duration T seconds; in other words $f(n^{T/t_1})$ should be linearly related to T. This defines the function f as a logarithm, i.e.

$$H \propto \log_x (n^{T/t_1})$$

or

$$H = K\frac{T}{t_1}\log_x n \qquad \text{bits} \tag{1.46}$$

We are still required to define the numerical values of K and x. The constant of proportionality is taken as unity and the base of the logarithm is specified by defining the unit of information. To illustrate this idea consider the simplest possible system, i.e. a teleprinter that can send only two possible symbols, A or B. The simplest possible message will occur when only one of the two possible symbols is sent. The information content of such a message is defined as 1 bit.

In such a system, T/t_1 symbols are sent in T seconds and the information transmitted is T/t_1 bits, i.e.

$$H = \frac{T}{t_1}\log_x n \qquad \text{where } n = 2$$

or

$$\frac{T}{t_1}\log_x 2 = \frac{T}{t_1}$$

Hence

$$x = 2$$

The information transmitted by a teleprinter that can send n different symbols is

$$H = \frac{T}{t_1}\log_2 n \qquad \text{bits} \tag{1.47}$$

The information rate is $(1/t_1)\log_2 n$ bits/s and the information per symbol is $\log_2 n$ bits. In arriving at this result we have made the implicit assumption that each of the n different symbols has equal probability of being sent. When the probability is not equal our definition of information informs us that symbols that occur least frequently contain a greater amount of information than symbols that occur very frequently. Before considering probability in detail it is important to note that the symbol example chosen is not restricted to alphabetic characters.

Consider the example of a voltage pulse, and assume that each pulse can have any one of eight different voltages. A typical signal is shown in Fig. 1.16. If we assume that each of the eight levels is equi-probable the information per pulse is

$$H = \log_2 8 = 3 \text{ bits/pulse}$$

It is not always convenient to use base 2 logarithms so, making use of the relationship

$$\log_2 n = \frac{\log_{10} n}{\log_{10} 2}$$

then

$$H = 3.32\log_{10} n \qquad \text{bits/pulse} \tag{1.48}$$

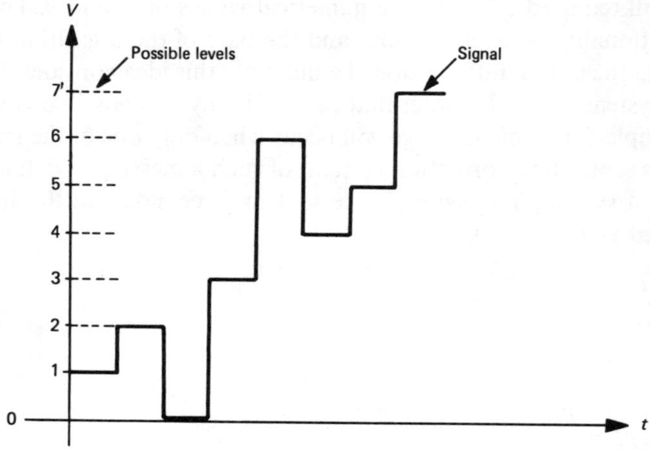

Fig. 1.16 Information-bearing signal.

The information capacity of a channel must be greater than the information rate of the transmitted signal in order for reliable communication to occur. We show in Section 3.7 that an ideal low-pass channel of bandwidth B can transmit pulses at a maximum rate of $2B$ per second. If we use the pulse analogy and assume that m different pulse levels can be distinguished at the channel output, the maximum rate at which information can be transmitted over the channel becomes

$$C = 2B\log_2 m \qquad \text{bits/sec} \tag{1.49}$$

This is known as Hartley's law. If we know the channel capacity we can determine the speed (i.e. number of symbols/s) at which a particular message can be sent. In other words, Hartley's law determines the time required to transmit a given amount of information over a particular channel. No indication is given, however, as to how the value of m is determined. This in fact depends on the signal-to-noise ratio (SNR) at the channel output. Before considering this further we will introduce probability into information theory.

To illustrate the significance of probability, we will assume that the n equiprobable symbols sent by a teleprinter are divided into two groups. The first group contains n_1 symbols and the second group contains n_2 symbols. The receiver is interested in determining which group a symbol belongs to rather than the value of the symbol itself, i.e. the receiver can receive one of two possible messages. The probability of the first message is $P_1 = n_1/n$ and the probability of the second message is $P_2 = n_2/n$, where probability is defined as

$$P = \frac{\text{number of times an event occurs}}{\text{number of possibilities}}$$

We require, from the point of view of the receiver, to find the information transmitted by this system. The information when all symbols are equi-probable is $\log_2 n$ bits/symbol. The total information transmitted by n symbols is then $n \log_2 n$ bits. If we now assume an observer within each group the total information within group 1 due to the n equi-probable symbols within the group is $n_1 \log_2 n_1$. Similarly the total information within group 2 is $n_2 \log_2 n_2$. Our receiver is not interested in the information contained within each group, so that the information content as far as this reciever is concerned must be equal to

(total information in system) – (information within the groups)
i.e. useful information $H = n \log_2 n - n_1 \log_2 n_1 - n_2 \log_2 n_2$ (1.50)

The average information is thus $H_{av} = H/n$ or

$$H_{av} = \left(\frac{n_1 + n_2}{n}\right) \log_2 n - \frac{n_1}{n} \log_2 n_1 - \frac{n_2}{n} \log_2 n_2$$

or

$$H_{av} = \frac{n_1}{n} (\log_2 n - \log_2 n_1) + \frac{n_2}{n} (\log_2 n - \log_2 n_2)$$

i.e.

$$H_{av} = \frac{-n_1}{n} \log_2 \frac{n_1}{n} - \frac{n_2}{n} \log_2 \frac{n_2}{n}$$

But n_1/n and n_2/n are probabilities; hence

$$H_{av} = -P_1 \log_2 P_1 - P_2 \log_2 P_2 \tag{1.51}$$

The more unpredictable an event the more information it contains; for instance let $P_1 = 0.8$, which means that $P_2 = 0.2$ since $P_1 + P_2 = 1$.

The information associated with the first event is

$$H_1 = -\log_2 0.8 = -3.32 \log_{10} 0.8 = 0.32 \text{ bits}$$

The information associated with the second event is

$$H_2 = -\log_2 0.2 = -3.32 \log_{10} 0.2 = 2.32 \text{ bits}$$

This agrees with our concept of information. The average information in this case would be

$$H_{av} = (0.8 \times 0.32) + (0.2 \times 2.32) = 0.72 \text{ bits/symbol}$$

which is considerably less than the information transmitted by the symbol with lower probability. It is of interest to determine the maximum value of H_{av}, and to do this we eliminate P_2 from Eqn (1.51), i.e.

$$H_{av} = -P_1 \log_2 P_1 + (P_1 - 1) \log_2 (1 - P_1) \tag{1.52}$$

To find the maximum value of H_{av} we differentiate with respect to P_1 and set the result equal to zero:

$$\frac{dH_{av}}{dP_1} = -P_1\frac{1}{P_1} - \log_2 P_1 + (P_1 - 1)\frac{-1}{1-P_1} + \log_2(1-P_1)$$

i.e.

$$\frac{dH_{av}}{dP_1} = \log_2(1-P_1) - \log_2(P_1)$$

which is zero when $(1-P_1) = P_2$ or $P_1 = P_2 = 0.5$. The average information is a maximum when the symbols are equi-probable.

In developing Eqn (1.50), the original n symbols were divided into two separate groups. This idea can be extended for any number of groups up to a maximum of n. When the number of groups equals the number of symbols, we are in effect saying that each individual symbol has its own probability of occurrence and the average information is

$$H_{av} = -\sum_{i=1}^{n} P_i \log_2 P_i \tag{1.53}$$

Equation (1.49) is similar to an equation in statistical mechanics that defines a quantity known as 'entropy'. For this reason H_{av} is usually known as the entropy of a message. In particular if all symbols are equi-probable $P_i = 1/n$ and Eqn (1.49) becomes

$$H_{av} = \sum_{i=1}^{n} P_i \log_2 n = \log_2 n \qquad \text{since } \sum_{i=1}^{n} P_i = 1$$

Extension of the analysis for maximum entropy produces the same result as for the two-symbol case; that is, the entropy of a message is a maximum when all symbols are equi-probable. In any other situation the entropy will be less than the maximum and the message is said to contain 'redundancy'.

When all symbols are equi-probable, the average information is a maximum and it is not possible to make other than a pure guess at what the next symbol will be after a number have been received. In certain circumstances this can be a serious problem because if an error occurs during transmission the receiver will not be aware of it. When all symbols are not equi-probable, it becomes feasible to predict what the next symbol in a received sequence should be. The redundancy in such a message is defined as

$$R = \frac{H_{av(max)} - H_{av}}{H_{av(max)}} \times 100\% \tag{1.54}$$

The significance of redundancy in a message will be illustrated by reference to the English language. If we assumed that all letters in the English alphabet were equi-probable, the average information per letter would be

$\log_2 26 = 4.7$ bits. If the relative frequencies of occurrence of individual letters are taken into account (E has a probability of 0.1073, Z has a probability of 0.006) the figure works out as $H_{av} = 4.15$ bits/letter. This gives a redundancy of 11.7%. The redundancy is actually much higher than this because of the interdependence between letters, words and groups of words within English text. For example, if the letter Q occurs in a message it is almost certain that the next letter will be U. The U contains no information because it can be guessed with almost 100% certainty. If the letters IN have been received, the probability that the next letter will be G is much higher than the probability that it will be Z. There are many examples of this interdependence, which can be extended to words and sentences. The overall effect of redundancy is twofold; it reduces the rate of transmission of information but at the same time it allows the receiver to detect, and sometimes correct, errors.

Consider the received message

Thi ship wilp arrive on September 28

It is clear that we can detect and correct errors in the alphabetic section of the message. There is no way that we can detect an error in the date (unless it is a number greater than 30), however. The numerical part of message thus contains no redundancy. This is very important because data transmission occurs as a sequence of binary numbers that has no inherent redundancy. It is important to detect occasional errors when they occur, and in data systems a form of redundancy known as 'parity checking' is often employed. The binary digits are divided into groups of 7, e.g., 1000001, and an extra digit is added to make the total number of 1s in the group of 8 either even or odd depending on the system. The receiver then checks each group of 8 digits to determine whether an odd or even parity has been preserved. If the parity check is not valid an error is detected. (See also Chapter 3, p. 112.)

1.14 Information Transmission

The information capacity of a communications channel is specified by Hartley's law [Eqn (1.49)], but this equation does not tell us how to evaluate the number of detectable levels, m. All signals in telecommunications are subject to corruption by noise, and we can make a qualitative statement to the effect that the difference between detectable levels must be greater than the noise present during transmission. If this were not the case, signal plus noise could produce a false level indication. Noise is a random disturbance that may be analysed either in terms of its statistical properties or in terms of its spectral properties. From either description we are able to define a mean square value for the noise that is equivalent to the power developed by the noise voltage in a resistance of $1\,\Omega$.

The relationship between the number of messages that can be transmitted and noise power was obtained by Shannon[2] in 1948 using the mathematics of n-dimensional space. The mathematics of n-dimensional space is a theoretical extension of the familiar mathematics of two- and three-dimensional space. An n-dimensional space is termed a 'hyperspace' and is defined by a set of n mutually perpendicular axes. If q is a point in this hyperspace, its distance from the origin (the point of intersection of the n mutually perpendicular axes) is d where

$$d^2 = x_1{}^2 + x_2{}^2 + x_3{}^2 + \cdots\cdots x_n{}^2 \tag{1.55}$$

x_n being the perpendicular distance from the point to the nth axis. When dealing wih hyperspace, the 'volume' of an n-dimensional figure is defined as the product of the lengths of its sides. If we are considering an n-dimensional cube (hypercube) in which all the sides have length L units, the volume is given by L^n. (Note that when $n = 2$ the hypercube is actually a square, and the 'volume' is interpreted as an area.)

A circle is a two-dimensional figure whose volume (i.e., area) is πr^2. If two concentric circles are drawn, one with a radius of 1 and the other with a radius of $\frac{1}{2}$, then one-quarter of the total area is enclosed within the inner circle which has half the total radius. The volume of a three-dimensional sphere is $\frac{4}{3}\pi r^3$. This means that a sphere of radius $\frac{1}{2}$ would contain only one-eighth of the volume of a sphere of radius 1. An n-dimensional sphere (hypersphere) has a volume proportional to r^n where r is its radius. The volume of such a hypersphere of radius $\frac{1}{2}$ will thus be equal to $(\frac{1}{2})^n$ of the volume of a hypersphere of radius 1. In other words, the larger the value of n the smaller is the percentage of total volume contained within the hypersphere of radius $\frac{1}{2}$. When $n = 7$ only 1/128 of the total volume lies within the hypersphere of radius $\frac{1}{2}$. For a hypersphere for which $n = 100$ it turns out that only 0.004% of the total volume of a hypersphere of radius 1 lies within a hypersphere of radius 0·99. This leads to the important conclusion that for a sphere of n dimensions (where $n \gg 1$) practically all of the volume lies very close to the surface of the sphere. This property is fundamental to Shannon's derivation of the information capacity of a channel.

Shannon postulates a signal source that can send out messages of duration T seconds; the source is limited to a bandwidth of B Hz and has a statistically stationary output (the properties of the output averaged over a long period are constant). We have already indicated in Section 1.10 that any signal band limited to B Hz can be represented by $2B$ samples/seconds. The energy in each of these samples is proportional to the square of the sample amplitude and the total signal energy will therefore be proportional to the sum of these mean square values. If the signal amplitudes are denoted by x_1, x_2, etc., then the total normalized signal energy is

$$E = x_1{}^2 + x_2{}^2 + \cdots + x_n{}^2 \tag{1.56}$$

If we compare Eqns (1.55) and (1.56) we see that the signal energy E is equivalent to the square of the distance of a point in a hyperspace from the origin of the hyperspace.

When a message is long in duration the average power becomes independent of the message duration and is given by the signal energy E divided by the duration of the message T. A message of duration T seconds can be represented by $2BT$ samples, the amplitude of each sample being equivalent to the perpendicular distance of a point from one of the axes in an n-dimensional space. If the average energy per sample is S, the total energy in the signal of duration T is $2BTS$ joules. This total energy is equivalent to the square of the distance of a point q from the origin in an n-dimensional space. Alternatively the message of duration T can be represented as a point in an n-dimensional space. Each different message will be represented by a different point. As the number of samples is made larger and larger, the points representing the different messages of duration T will be very close to the surface of a hypersphere of radius $(2BTS)^{1/2}$, because this is where most of the enclosed volume is contained.

Received signals are always accompanied by noise, which in an ideal case occupies the same bandwidth as the signal. This means that the noise signal can also be represented by $2BT$ samples. If the average energy per noise sample is N joules the total noise energy is $2BTN$ joules. We have stated that each message is represented by a point in the hypersphere of radius $(2BTS)^{1/2}$ and therefore the sum of message and noise will be a point whose distance is $(2BTN)^{1/2}$ from each point representing the message alone. This means that all possible message + noise combinations are represented by points that are very close to the surface of a hypersphere of radius $(2BTN)^{1/2}$ centred on the point representing the noise alone. The receiver actually receives message + noise with a total energy of $2BT(S+N)$ and the points representing each combination must lie within a hypersphere of radius $[2BT(S+N)]^{1/2}$.

Because each message must be a distance $(2BTN)^{1/2}$ from points representing message + noise, we can represent the system as a hypersphere of radius $[2BT(S+N)]^{1/2}$ filled with non-overlapping hyperspheres of radius $(2BTN)^{1/2}$. The centres of the small hyperspheres represent a distinguishable message that could have been sent. The total number of distinguishable messages is given by the number of non-overlapping hyperspheres of radius $(2BTN)^{1/2}$ which can exist within a hypersphere of radius $[2BT(S+N)]^{1/2}$. This will be equal to the ratio of the volumes of the two hyperspheres. The physical picture is given in Fig. 1.17. If the number of dimensions is $2BT$, the ratio of the volumes of the two hyperspheres is

$$\left[\frac{(2BT(S+N))^{1/2}}{(2BTN)^{1/2}}\right]^{2BT} = \left[\frac{S+N}{N}\right]^{BT} \tag{1.57}$$

This gives the number of distinguishable messages that can be sent. Assuming

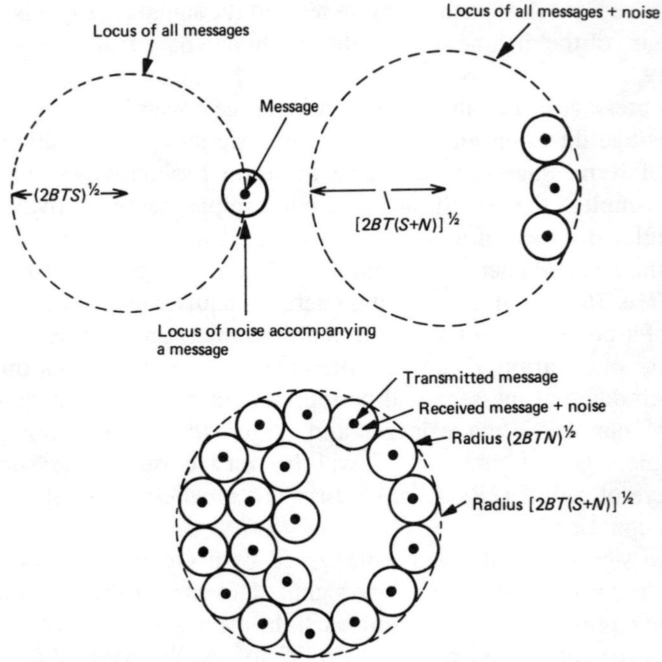

Fig. 1.17 Multidimensional representation of signal + noise.

each message to be equi-probable, the information transmitted is

$$\log_2\left[\frac{S+N}{N}\right]^{BT} = BT\log_2\left(1+\frac{S}{N}\right) \quad \text{bits} \tag{1.58}$$

The channel capacity is thus

$$C = B\log_2\left(1+\frac{S}{N}\right) \quad \text{bits/second}$$

The ratio of signal energy to noise energy is identical to the corresponding power ratio so that Shannon's law for channel capacity is

$$C = B\log_2\left(1+\frac{S_p}{N_p}\right) \quad \text{b/s} \tag{1.59}$$

This law is a fundamental law of telecommunications, and states that if a channel has a bandwidth B and the mean SNR is S_p/N_p the maximum rate at which information may be transmitted is C bits/second. In other words there is a theoretical limit on the amount of information that can be transmitted over any telecommunications channel and, as we shall show in later chapters, all practical systems fall short of this theoretical limit by varying degrees. As an illustration of the theoretical application of Shannon's law we will assume that a link of effective bandwidth 3 kHz is to be used for pulse transmission. This link is to be used for six identical channels, the data rate being 1 kb/s per channel. The SNR on the link is $2/\log_{10} D$ where D is the length of the link in kilometres. We are required to determine the maximum distance over which reliable communication is possible.

The total data rate is $6 \times 1000 = 6$ kb/s. Hence from Shannon's law

$$6 \times 10^3 = 1 \times 10^3 \log_2\left(1 + \frac{2}{\log_{10} D}\right)$$

which reduces to

$$1 + \frac{2}{\log_{10} D} = 4.0$$

or

$$D = 4.36\,\text{km}$$

This example does not consider the way in which the information is actually transmitted. In order to achieve the theoretical channel capacity Shannon postulated that each message should be represented by a large number of samples (data points). The receiver, in effect, compares the received message, corrupted by noise, with each possible uncorrupted message that could have been transmitted. These uncorrupted messages are the centres of the hyperspheres of radius $(2BTN)^{1/2}$. The receiver then decides that the message which was actually transmitted is the one (in the n-dimensional space) closest to the point representing the message + noise. As the number of data points describing each message increases, the closer is any individual message to the surface of the hypersphere of radius $(2BTS)^{1/2}$. Alternatively, the larger the number of data points, the more accurate is Shannon's law.

The practical drawback is that the larger the number of data points the longer is the time taken to transmit them over a link of fixed bandwidth. The receiver cannot decide which message was transmitted until all data points have been received. Thus in attempting to realize a communications system that obeys Shannon's law, long delays would be introduced between transmission and reception (the information rate is not affected by a delay).

Shannon's law would not seem to be a practical proposition, but it is extremely useful as a standard for comparing the performance of various telecommunications systems. Of particular importance is the possibility of exchanging bandwidth and SNR in order to achieve a given information transmission rate. We will now examine this possibility in some detail. It will be noted from Eqn (1.59) that the value C can remain fixed when the bandwidth B is changed, provided that the SNR is modified accordingly. This means that SNR and bandwidth can be exchanged without affecting the channel capacity. Figure 1.18 shows bandwidth plotted against SNR for a constant channel capacity of 1 b/s.

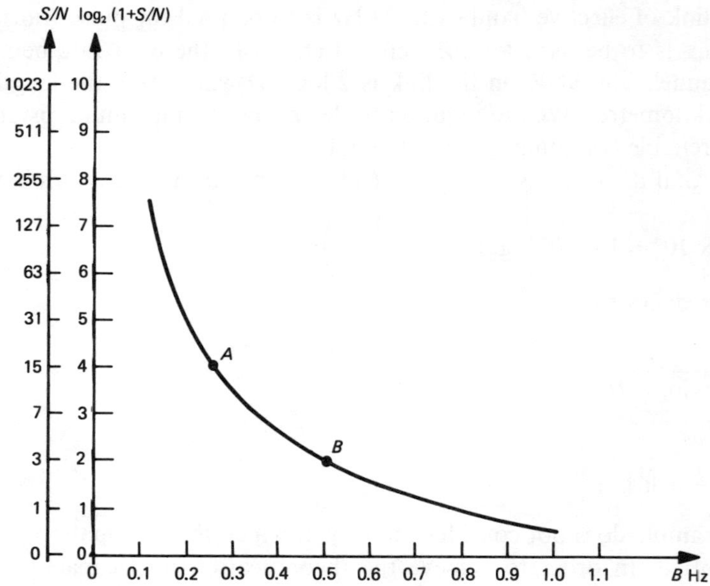

Fig. 1.18 Shannon's law for $C = 1$ b/s.

Point A on this graph shows that it is possible to transmit 1 b/s in a bandwidth of 0.25 Hz provided that the signal power is 15 times the noise power. If the bandwidth is now doubled (point B on the graph) the same rate of transmission is possible with an SNR of 3. If we assume that the noise power is constant then evidently much less signal power is required at point B. As a simple physical example assume that 1 b/s is being transmitted at point B on the curve using binary pulses. If the bandwidth is decreased then clearly fewer binary pulses can be transmitted per second. To maintain the same information rate, the binary pulses must be replaced by pulses with more than two levels, which consequently increases the mean signal power. For large SNRs, Shannon's law approximates to

$$C = B\log_2(S_p/N_p) \tag{1.60}$$

This equation states that in order to maintain a given rate of transmission of information, a linear change in bandwidth must be accompanied by an exponential change in SNR. Thus very large increases in signal power are required to compensate for relatively small reductions in the channel bandwidth B. The assumption that noise power remains constant is not in fact justified. In telecommunications systems the noise power is linearly related to bandwidth, i.e.

$$N_p = \eta B \tag{1.61}$$

η is known as the noise power spectral density. This means that when the bandwidth is doubled the noise power increases by 3 dB; hence the reduction of required signal power that accompanies an increase in bandwidth is partially offset by the extra noise power. At low SNRs, the saving in received signal power will be cancelled by the extra noise. We can write Shannon's law as

$$C = 1.45B\ln(1 + S_p/N_p)$$

If S_p/N_p is very small, $\ln(1 + S_p/N_p) \simeq S_p/N_p$.
Hence $C = 1.45B \cdot S_p/N_p$. But $N_p = \eta B$; thus

$$C = 1.45S_p/\eta \text{ b/s} \tag{1.62}$$

The channel capacity now becomes independent of the bandwdith and a further increase in bandwidth has no effect. Hence there is a limit to the amount of information that can be transmitted with a fixed signal power regardless of the bandwidth. Once again this is a theoretical limit and all practical systems fall short of this limit. In the following chapters we consider both analogue and digital communications systems and compare their relative performance in terms of the theoretical limits imposed by Shannon's law.

1.15 Conclusion

In this chapter we have considered, in a general way, methods of describing both signal and channel characteristics met in telecommunications systems. We have introduced the idea that the channel characteristic intimately influences the signal waveform of the transmitted information. In defining information, which is the basic entity to be transmitted, we were able to show once again an intimate relationship between channel characteristics and the rate at which this information can be transmitted.

In the following chapters we will be considering specific telecommunications systems in detail and showing that the relationships outlined in this chapter are common to all systems. Of particular importance to the telecommunications engineer is the effect of the omnipresent noise which corrupts the information-bearing signal in addition to any distortion produced by channel characteristics. The effect of noise is quite different in analogue and digital systems, and we will show that this is one reason why digital systems have a performance much closer to the theoretical performance of Shannon's law than their analogue counterparts.

References

1. Brigham, O., *The Fast Fourier Transform*, Prentice-Hall, London and New Jersey, 1974.
2. Shannon, C.E., 'A mathematical theory of communication', *Bell Systems Technical Journal*, **27**, 379–423 (1948).

Problems

1.1 Each of the pulse trains shown in the figure represents a voltage across a

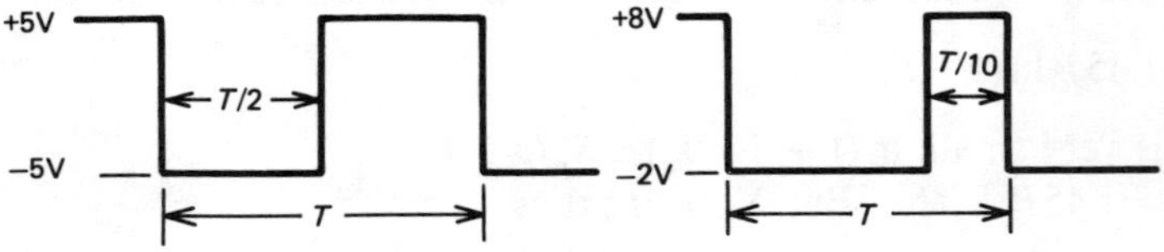

1Ω resistance. Find the total average power dissipated by each voltage and also the percentage of the total average power contributed by the first harmonic in each case.

Answer: 25 W; 10 W; 19.2%.

1.2 Assuming the integrator shown in figure is ideal (the output voltage increases at a rate of $-1/RC$ volts/second when the input is 1 volt), sketch the output voltage

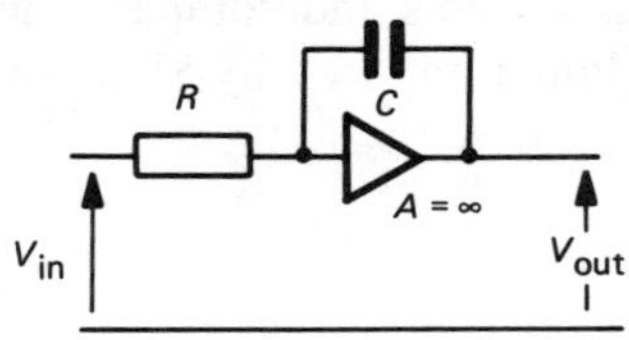

when the input is a square wave of amplitude $+A$ volts or $-A$ volts and period T.
If $1/RC = 2/T$, plot the amplitude spectrum of the input and output signals.

Answer: $(AT/2)\,\mathrm{sinc}\,(\pi nfT/2)$; $AT\,\mathrm{sinc}^2\,(\pi nfT/2)$.

1.3 Using the principle of superposition, or otherwise, obtain an expression for the amplitude of the nth harmonic of the exponential Fourier series of the waveform shown in the figure.

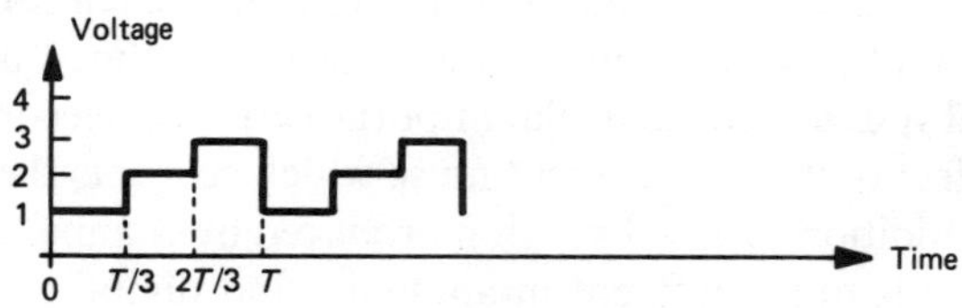

Answer: $[\cos(2n\pi/3) - 1]/\pi nf$.

1.4 Plot the single-sided and double-sided power spectral densities for a pulsed waveform with period 0.1 ms and pulse width of 0.01 ms. The amplitude of the narrow pulse is 10 V and the amplitude of the wide pulse is 0 V.
What fraction of the total average power is contained within a bandwidth extending from 0 to 100 kHz?

Answer: 91%.

1.5 A rectangular pulse train with amplitude 0 V or 10 V is applied to the input of the RC network shown in the figure, the pulse repitition rate being 20,000 per second.

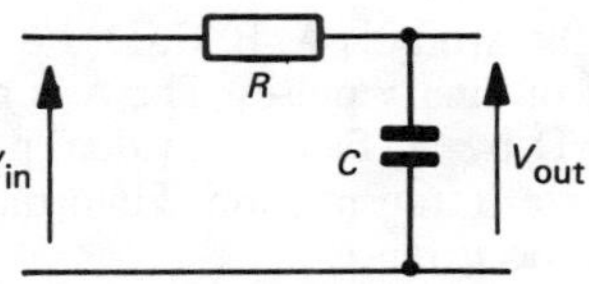

If the half-power frequency of the network is 500 kHz and the pulse train has a duty cycle of 0.01, sketch the signal power spectrum at the network output. How much power is contained in the second harmonic?

Answer: 127 μW.

1.6 Find the Fourier transform of the function $V(t)$ which is zero for negative values of its argument and is equal to $\exp(-t)$ for positive values. Find the transform when $\exp(-t)$ is replaced by $t*\exp(-t)$.

Answer: $1/(1+j2\pi f)$; $1/(1+j2\pi f)^2$.

1.7 Show that the response of a linear system with a transfer function $H(jf)$ to a unit impulse function is

$$v(t) = 2\int_0^{\infty} m\cos(2\pi ft + \phi)\,\mathrm{d}f$$

where $m = |H|$ and $\phi = \arg(H)$. (*Hint*: observe that m is an even function and ϕ is an odd function.)

1.8 Show that the autocorrelation function of a periodic waveform has the same period as the waveform. A voltage waveform is odd with period $2p$ and has the value A V over half its period and 0 V over the remainder of its period. Show that for $0 < t < P$ its autocorrelation function is given by

$$R(\tau) = (1 - 2\tau/p)A^2$$

1.9 A voltage waveform $e(t)$ has an arithmetic mean given by

$$\bar{e} = \lim_{T\to\infty} 1/2T \int_{-T}^{T} e(t)\,\mathrm{d}t$$

and $R(\tau)$ is its autocorrelation function. Deduce that the autocorrelation function of $e(t) + C$ (where C is a constant) is

$$R(\tau) + 2C\bar{e} + C^2$$

1.10 A system can send out a group of four pulses, each of 1 ms width and with equal probability of having an amplitude of 0, 1, 2 or 3 V. The four pulses are always followed by a pulse of amplitude -1 V to separate the groups. What is the average rate of information transmitted by this system?

Answer: 1600 b/s.

1.11 The probabilities of the previous question are altered such that the 0 V level occurs one-half of the time on average, the 1 V level occurs one-quarter of the time on average, the remaining levels occurring one-eighth of the time each. Find the average rate of transmission of information and determine the redundancy.

Answer: 1400 b/s; 12.5%.

1.12 An alphabet consists of the symbols A, B, C, D. For transmission, each symbol is coded into a sequence of binary pulses. The A is represented by 00, the B by 01, the C by 10, and the D by 11. Each individual pulse interval is 5 ms.

Calculate the average rate of transmission of information if the different symbols have equal probability of occurrence.

Find the average rate of transmission of information when the probability of occurrence of each symbol is $P(A) = 1/5$, $P(B) = 1/4$, $P(C) = 1/4$, $P(D) = 3/10$.

Answer: 200 b/s; 198 b/s.

1.13 A telemetering system can transmit eight different characters which are coded for this purpose into pulses of varying duration. The width of each coded pulse is inversely proportional to the probability of the character it represents. The transmitted pulses have durations of 1, 2, 3, 4, 5, 6, 7 and 8 ms, respectively. Find the average rate of information transmitted by this system.

If the eight characters are coded into 3-digit binary words find the necessary digit rate to maintain the same transmitted information rate.

Answer: 888.8 b/s; 1019.4 digits/second. & a B.W of 9KHz

1.14 A space vehicle at a distance of 381 000 kilometres from the Earth's surface is equipped with a transmitter with a power of 6 W. The attenuation of the signal between the transmitter and a receiver on the Earth is given by $10 + 7\log_{10}(X)$ dB, where X is the distance measured in kilometres. The noise power at the receiver input is 0·1 μW. If the receiver requires an input signal-to-noise ratio 12 dB above the value given by Shannon's law, find the maximum rate of transmission of information.

Answer: 50.41 kb/s.

2 / ANALOGUE MODULATION THEORY

By definition, an information-bearing signal is non-deterministic, i.e. it changes in an unpredictable manner. Such a signal cannot be defined in terms of a specific amplitude and phase spectrum, but it is usually possible to specify its power spectrum.

The characteristics of the channel over which the signal is to be transmitted may be specified in terms of a frequency and phase response. For efficient transmission to occur, the parameters of the signal must match the characteristics of the channel. When this match does not occur, the signal must be modified or processed. The processing is termed modulation, and the need for it may be made clearer by considering two specific examples, viz. frequency multiplexing and electromagnetic radiation from an antenna (aerial).

Frequency multiplexing is commonly used in long-distance telephone transmission, in which many narrowband voice channels are accommodated in a wideband coaxial cable. The bandwidth of such a cable is typically 4 MHz, and the bandwidth of each voice channel is about 3 kHz. The 4 MHz bandwidth is divided up into intervals of 4 kHz, and one voice channel is transmitted in each interval. Hence each voice channel must be processed (modulated) in order to shift its amplitude spectrum into the appropriate frequency slot. This form of processing is termed frequency division multiplexing (FDM), and is discussed in more detail in Section 2.19.

In Chapter 7 we show that, for efficient radiation of electromagnetic energy to occur from an antenna, the wavelength of the radiated signal must be comparable with the physical dimensions of the antenna. For audio-frequency signals antennas of several hundred kilometres length would be required – clearly a practical impossibility. For convenient antenna dimensions the radiated signal must be of a very high frequency. In this particular example the high-frequency signal would be varied (modulated) in some way to obtain efficient transmission of the low-frequency information.

This chapter is concerned only with continuous wave (i.e. sinusoidal) modulation and assumes a noise-free environment. It should be stressed at this point that real communication is always accompanied by noise, but it is desirable to consider the effects of noise after the basic ideas of modulation have been presented. The general expression for a sinusoidal carrier is

$$f_c(t) = A\cos(2\pi f_c t + \phi) \tag{2.1}$$

The three parameters A, f_c and ϕ may be varied for the purpose of transmitting information giving respectively amplitude, frequency and phase modulation.

2.1 Double Sideband Amplitude Modulation (DSB-AM)

With this type of modulation the carrier amplitude is made proportional to the instantaneous amplitude of the modulating signal. Figure 2.1 shows that the original modulating signal is reproduced as an envelope variation of the carrier. Let $A = K + f_m(t)$ where K is the unmodulated carrier amplitude and $f_m(t) = a\cos(2\pi f_m t)$ is the modulating signal. The modulated carrier, assuming $\phi = 0$, is

$$f_c(t) = [K + a\cos(2\pi f_m t)]\cos(2\pi f_c t)$$

i.e.

$$f_c(t) = K[1 + m\cos(2\pi f_m t)]\cos(2\pi f_c t) \tag{2.2}$$

m is the depth of modulation and is defined as

$$m = \frac{\text{modulating signal amplitude}}{\text{unmodulated carrier amplitude}} = \frac{a}{K}$$

For an undistorted envelope, $m \leqslant 1$. If this condition is not observed the envelope becomes distorted, and a carrier phase reversal occurs as shown in Fig. 2.2

Equation (2.2) may be expanded to yield

$$f_c(t) = K\left[\cos(2\pi f_c t) + \frac{m}{2}\cos\{2\pi(f_c - f_m)t\} + \frac{m}{2}\cos\{2\pi(f_c + f_m)t\}\right] \tag{2.3}$$

The amplitude spectrum of the modulated carrier clearly consists of three components: the carrier frequency f_c and the lower and upper side frequencies

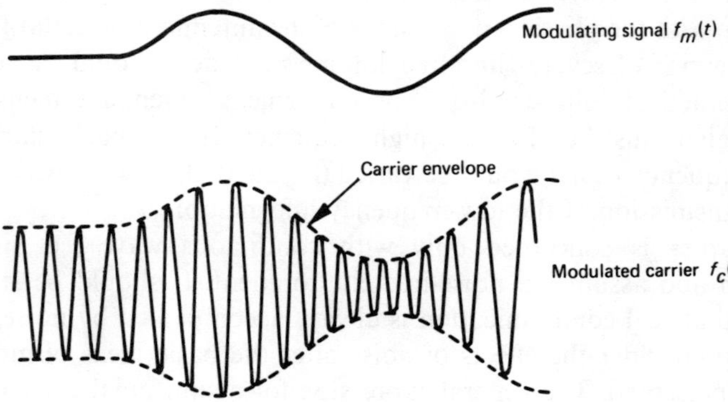

Fig. 2.1 DSB-AM.

$(f_c - f_m)$ and $(f_c + f_m)$. If $f_m(t)$ is a multitone signal, e.g.

$$f_m(t) = a_1 \cos(2\pi f_1 t) + a_2 \cos(2\pi f_2 t) + a_3 \cos(2\pi f_3 t) \tag{2.4}$$

the modulated carrier becomes

$$f_c(t) = K[1 + m_1 \cos(2\pi f_1 t) + m_2 \cos(2\pi f_2 t) + m_3 \cos(2\pi f_3 t)] \cos 2\pi f_c t \tag{2.5}$$

The depth of modulation is $m = m_1 + m_2 + m_3$ and, once again for an undistorted envelope, $m \leqslant 1$. The amplitude spectrum of $f_c(t)$ now contains a band of frequencies, termed sidebands, above and below the carrier. If the modulating signal is expressed in terms of a two-sided amplitude spectrum the process of full amplitude modulation (AM) (which is an alternative name

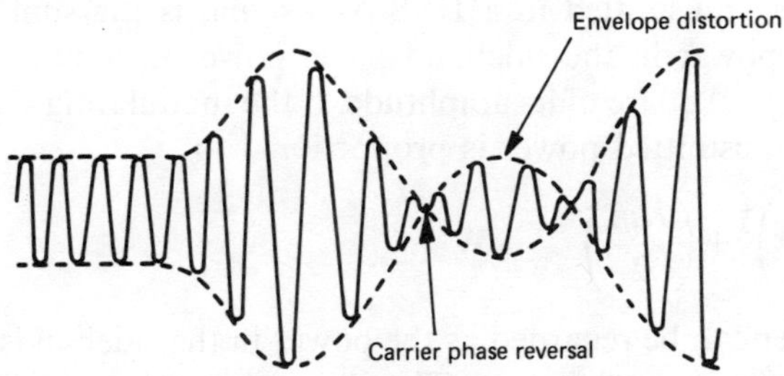

Fig. 2.2 Effect of overmodulation.

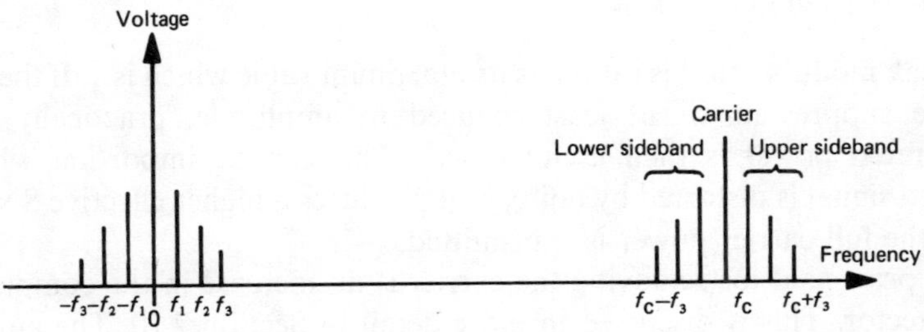

Figure 2.3 Amplitude spectrum of DSB-AM.

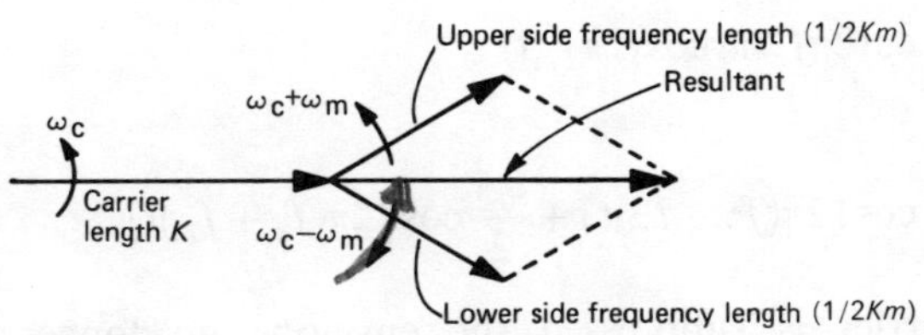

Fig. 2.4 Phasor representation of DSB-AM.

for this type of modulation) reproduces the spectrum of $f_m(t)$ centred at frequencies $\pm f_c$. This process is illustrated in Fig. 2.3.

The phasor representation of Eqn (2.3) is a single component of length K rotating with angular velocity $\omega_c(=2\pi f_c)$ rad/s representing the carrier added to two phasors of length $Km/2$ rotating in opposite directions with angular velocity ω_m rad/s relative to the carrier. This is illustrated in Fig. 2.4. The resultant is a single phasor rotating with angular velocity ω_c rad/s with an amplitude varying between the limits $K(1-m)$ and $K(1+m)$.

2.2 Double Sideband Suppressed Carrier Amplitude Modulation (DSB-SC-AM)

The total power transmitted in a DSB-AM signal is the sum of the carrier power and the power in the sidebands. The power in a sinusoidal signal is proportional to the square of its amplitude. If the modulating signal is a single tone the total transmitted power is proportional to

$$K^2 + \left(\frac{Km}{2}\right)^2 + \left(\frac{Km}{2}\right)^2 \quad \text{W}$$

The useful power can be regarded as the power in the sidebands as the carrier component carries no information. The ratio of useful power to total power is therefore

$$\frac{m^2}{2}:\left(1+\frac{m^2}{2}\right)$$

For peak modulation this ratio has its maximum value which is $\frac{1}{3}$. If the carrier can be suppressed, or at least reduced in amplitude, practically all the transmitted power is then useful power. This can be important when the received signal is distorted by noise, as it produces a higher effective SNR than when the full carrier power is transmitted.

The price paid for removing the carrier is an increase in the complexity of the detector. This is discussed in more detail in Section 2.10. The amplitude spectrum of DSB-SC-AM may be derived by assuming a carrier $A\cos(2\pi f_c t)$ and a modulating signal given by $a\cos(2\pi f_m t)$. The modulated signal is simply the product of these two components, i.e.

$$f_c(t) = A\cos(2\pi f_c t)a\cos(2\pi f_m t)$$

i.e.

$$f_c(t) = \frac{aA}{2}\cos[2\pi(f_c - f_m)t] + \frac{aA}{2}\cos[2\pi(f_c + f_m)t] \tag{2.6}$$

When the carrier is suppressed the envelope no longer represents the modulating signal. The modulated carrier and the phasor representation of

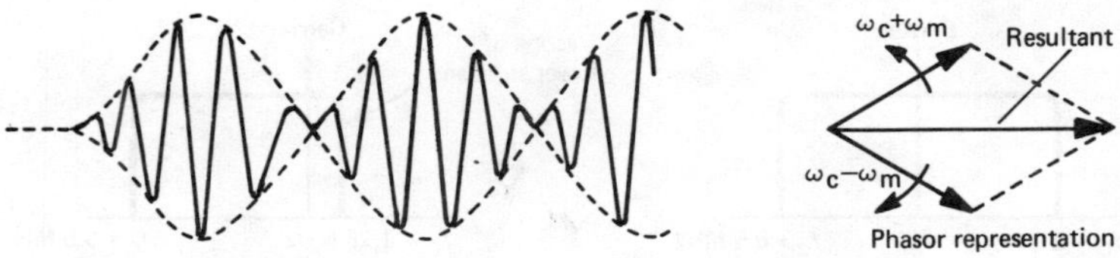

Fig. 2.5 SCDSB-AM waveform and phasor representation.

Eqn (2.6) are illustrated in Fig. 2.5. For this specific case the resultant phasor is a single component with angular velocity ω_c rad/s varying in amplitude between the limits $\pm(aA)$.

2.3 Single Sideband Amplitude Modulation (SSB-AM)

The reason for suppressing the carrier component was that it contained no information. It can also be observed that the signal information transmitted in the lower sideband is the same as is transmitted in the upper sideband. If either one is suppressed, therefore, it is possible to transmit the same information, but the bandwidth requirement is halved. The SSB signal that is transmitted can be either sideband, e.g.

$$f_c(t) = \frac{aA}{2}\cos\left[2\pi(f_c + f_m)t\right] \tag{2.7}$$

It should be noted that Eqn (2.7) seems to infer that a single sinusoidal component is transmitted. This is a special case since it is applicable to a single modulating tone. When the modulating signal is multitone the SSB signal becomes a band of frequencies. The resultant phasor is then the resultant of several phasors each rotating with different angular velocities and lengths. The price paid for reducing the signal bandwidth is an increase in the complexity of the receiver, which is dealth with in Section 2.11.

It becomes clear that, since SSB-AM has half the bandwidth of DSB-AM, twice as many independent information-bearing signals can be transmitted over a channel of fixed bandwidth when SSB-AM is used to produce frequency multiplexing.

2.4 Vestigial Sideband Amplitude Modulation (VSB-AM)

This is used for wideband modulating signals, such as television, where the bandwidth of the modulating signal can extend up to 5.5 MHz (for a 625-line system). The required bandwidth for a DSB-AM transmission would therefore be 11 MHz. This is regarded as excessive both from the point of view of

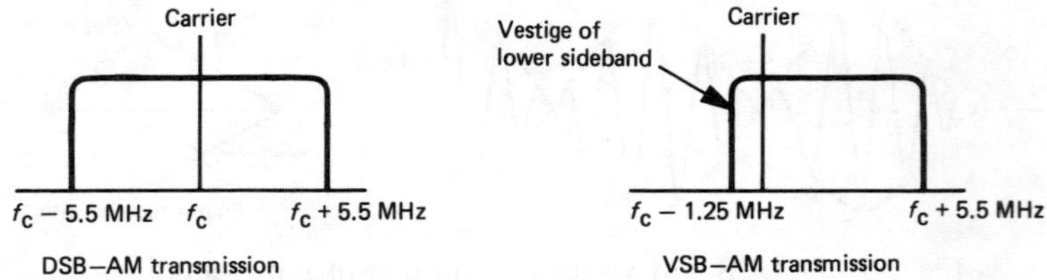

Fig. 2.6 Amplitude spectrum of a television signal.

transmission bandwidth occupation and of cost. It is generally accepted that the wider the bandwidth of a receiver, the greater the cost.

Since the amplitude spectrum of a video waveform has a dc component it would be extremely difficult to produce SSB-AM for a television signal. As a compromise, part (i.e. the vestige) of one sideband and the whole of the other is transmitted. A typical example is shown in Fig. 2.6.

The VSB-AM signal thus has both lower power and less bandwidth than DSB-AM and higher power and greater bandwidth than SSB-AM. The VSB-AM signal, however, does permit a much simpler receiver than an SSB-AM signal.

2.5 DSB-AM Modulators

DSB-AM is produced by multiplying together the carrier and the modulating signal. The multiplication is achieved by using a network with a non-linear characteristic. There are basically two types of non-linear networks, one in which the characteristic is continuous and the other in which the characteristic is noncontinuous, e.g. a switch.

Non-linear networks are not true multipliers because other components, which have to be filtered off, are produced. The diode modulator is an example of a modulator with a continuous non-linear characteristic of the form shown in Fig. 2.7.

The input/output characteristic of the circuit of Fig. 2.7 can be written in

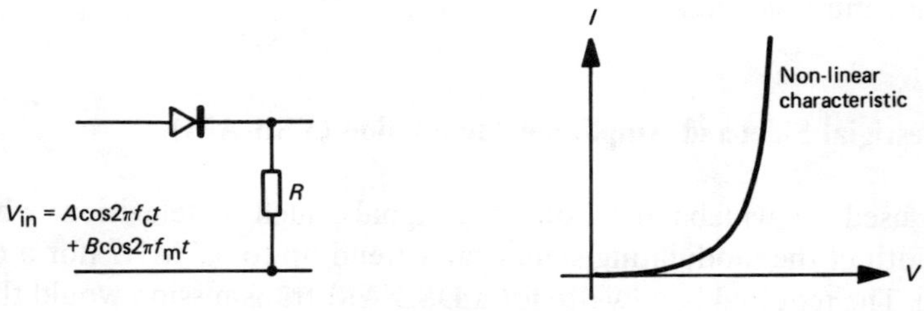

Fig. 2.7 Diode modulator.

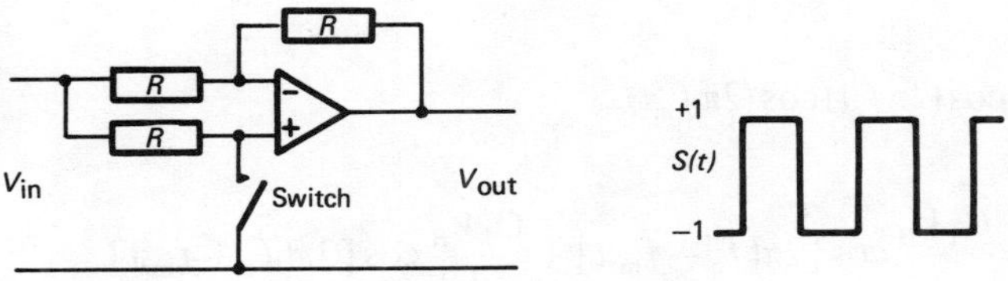

Fig. 2.8 Switching modulator.

terms of a power series, i.e.

$$V_{out} = aV_{in} + bV_{in}^{2} + cV_{in}^{2} + \cdots \tag{2.8}$$

If

$$V_{in} = [A\cos(2\pi f_c t) + B\cos(2\pi f_m t)]$$

the output is given by

$$V_{out} = a[A\cos(2\pi f_c t) + B\cos(2\pi f_m t)] + b[A^2\cos^2(2\pi f_c t) + B^2\cos^2(2\pi f_m t) + 2AB\cos(2\pi f_c t)\cos(2\pi f_m t)] + \cdots$$

i.e.

$$V_{out} = aA\cos(2\pi f_c t)[1 + K_1\cos(2\pi f_m t)] + \text{other terms} \tag{2.9}$$

where $K_1 = 2B/a$. If the "other terms" are filtered off, Eqn (2.9) has the same form as Eqn (2.2). Hence the diode produces the required modulation.

Considering the circuit of Fig. 2.8, it can be shown that when the switch is open the circuit has a voltage gain of +1 and when the switch is closed the circuit has a voltage gain of −1. If the switch is replaced by a semiconductor device that is pulsed at the carrier frequency the output of the amplifier is V_{in} multiplied by a square wave of amplitude ±1. The Fourier series of the square wave is

$$S(t) = C_1\cos(2\pi f_c t) + C_3\cos(6\pi f_c t) + C_5\cos(10\pi f_c t) + \cdots$$

If the input is $V_{in} = V_A + V_B\cos(2\pi f_m t)$ then the output will contain a term

$$C_1 V_A\cos(2\pi f_c t)[1 + K_2\cos(2\pi f_m t)] \tag{2.10}$$

where $K_2 = V_B/(C_1 V_A)$, which again has the same form as Eqn (2.2). Hence the switching modulator also produces DSB-AM. Filtering is required in this case also, to remove the unwanted frequency components.

2.6 DSB-SC-AM Modulators

The switching modulator may also be used to produce DSB-SC modulation. If the input to the circuit of Fig. 2.8 is $V_B\cos(2\pi f_m t)$ the output will contain a term

$$C_1 V_B \cos(2\pi f_c t)\cos(2\pi f_m t)$$

i.e.

$$V_{out} = \frac{C_1 V_B}{2}\cos[2\pi(f_c - f_m)t] + \frac{C_1 V_B}{2}\cos[2\pi(f_c + f_m)t] + \text{higher-frequency terms} \qquad (2.11)$$

This equation, when compared with Eqn (2.3), will be seen to contain no carrier frequency f_c.

2.7 SSB-AM Modulators

SSB-AM can be produced from DSB-SC-AM by filtering off one of the sidebands. The filtering process is relatively difficult to accomplish at high frequency where the sideband separation would be a very small fraction of the filter centre frequency. The problem is eased considerably if the initial modulation takes place at a low carrier frequency. The selected sideband can then be shifted to the required frequency range by a second modulation. The sidebands of the second modulation are widely spaced, and less exact filtering is necessary.

The two-stage modulation process is shown in Fig. 2.9; a typical first modulating frequency would be in the region of 100 kHz. Most modern high-power SSB-AM transmitters working the HF band (3–30 MHz) generate signals in this way. An interesting alternative method is based on the Hilbert transformation. The Hilbert transform of a signal $f(t)$ is the sum of the individual frequency components that have all been shifted in phase by 90°. The SSB-AM signal for a single modulating tone is

$$f_c(t) = A\cos[2\pi(f_c + f_m)t]$$

i.e.

$$f_c(t) = A[\cos(2\pi f_c t)\cos(2\pi f_m t) - \sin(2\pi f_c t)\sin(2\pi f_m t)] \qquad (2.12)$$

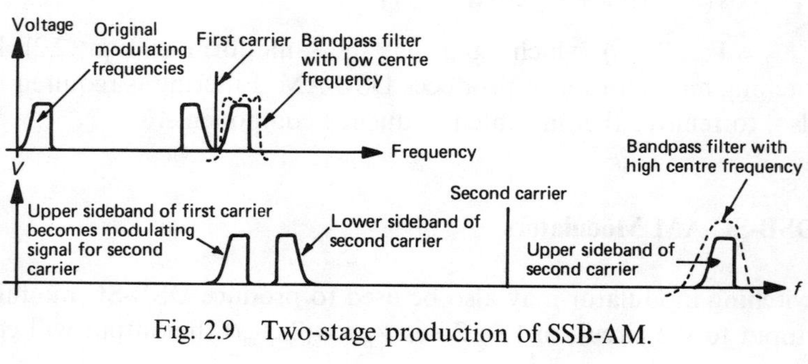

Fig. 2.9 Two-stage production of SSB-AM.

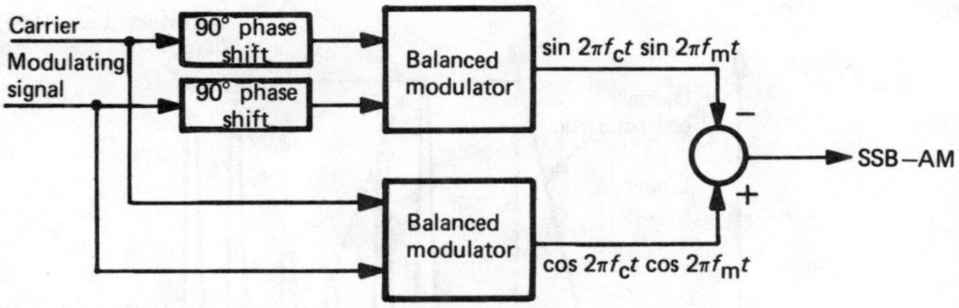

Fig. 2.10 Hilbert transform production of SSB-AM.

Equation (2.12) reveals that $f_c(t)$ is obtained by subtracting the outputs of two DSB-SC-AM modulators, often called balanced modulators. There is a quadrature phase relationship between the inputs of the first modulator relative to the inputs of the second modulator. A block diagram of this type of SSB-AM modulator is given in Fig. 2.10, and it should be noted that sideband filters are not required. The modulator does, however, require the provision of wideband phase-shifting networks.

2.8 VSB-AM Modulators

VSB-AM modulation is produced by filtering the corresponding DSB-AM signal. In a television transmission system the required filtering is usually achieved by a series of cascaded high-frequency tuned amplifiers. The centre frequency of each amplifier in the chain is chosen so that the overall cascaded frequency response is asymmetrically positioned relative to the carrier frequency.

2.9 DSB-AM Detection

The detection (i.e. demodulation) of DSB-AM modulation can be considered broadly under two headings; these are noncoherent detection and coherent (synchronous) detection. Traditionally broadcast receivers have been of the superheterodyne type (see later) and have made use of envelope detection. With the advent of the integrated phase locked loop (PLL), this position is likely to change radically in the future, and coherent detectors will probably become standard. The envelope (non-coherent) detector, as its name suggests, physically reproduces the envelope of the modulated carrier. This detector is basically a half-wave rectifier, and commonly makes use of a silicon diode whose typical current voltage relationship is shown in Fig. 2.11.

The diode acts as a rectifier and its effect on the envelope of a DSB-AM signal is shown in Fig. 2.11. For large carrier amplitudes the envelope is reproduced by the linear part of the characteristic (provided the depth of

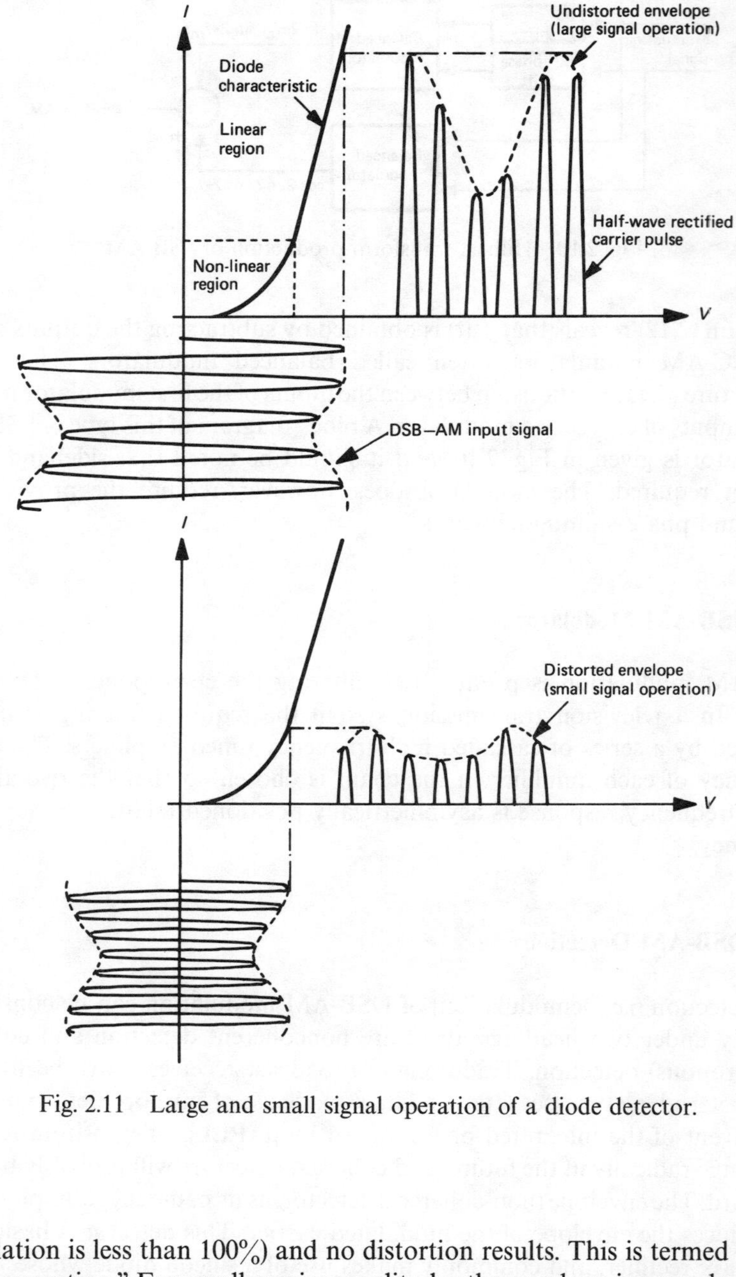

Fig. 2.11 Large and small signal operation of a diode detector.

modulation is less than 100%) and no distortion results. This is termed "large signal operation." For small carrier amplitudes the envelope is reproduced by the non-linear part of the characteristic and distortion of the envelope occurs. Assuming the input to the detector is

$$V_{\text{in}} = K[1 + m\cos(2\pi f_m t)]\cos(2\pi f_c t)$$

when operating in the large signal mode, the output is given by

$$i = P[1 + m\cos(2\pi f_m t)]F(t) \tag{2.13}$$

where $F(t)$ is a half-wave rectified sinusoid of carrier frequency. Representing $F(t)$ by its Fourier series the output is

$$i = P[1 + m\cos(2\pi f_m t)][C_0 + C_1 \cos(2\pi f_c t) + C_2 \cos(4\pi f_c t) + C_3 \cos(6\pi f_c t) + \cdots] \tag{2.14}$$

i.e.

$$i = PC_0 + PC_0 \cos(2\pi f_m t) + \text{unwanted terms} \tag{2.15}$$

The unwanted terms can be filtered off leaving the original modulating signal. To ensure that the envelope remains in the linear region the depth of modulation must be less than 100%. When operating in the small signal mode the diode current is given by

$$i = aV_{in} + bV_{in}{}^2 - cV_{in}{}^3 + \cdots \tag{2.16}$$

Letting

$$V_{in} = K[\cos(2\pi f_c t) + \tfrac{1}{2}m\cos\{2\pi(f_c - f_m)t\} + \tfrac{1}{2}m\cos\{2\pi(f_c + f_m)t\}]$$

The square term gives a contribution to the output current of

$$bK^2\{\cos(2\pi f_c t) + \tfrac{1}{2}m\cos[2\pi(f_c - f_m)t] + \tfrac{1}{2}m\cos[2\pi(f_c + f_m)t]\}^2$$

i.e.

$$b\{K^2 m\cos(2\pi f_m t) + \tfrac{1}{4}K^2 m^2 \cos(4\pi f_m t)\} + \text{unwanted terms} \tag{2.17}$$

The unwanted terms are outside the modulating signal bandwidth and can be filtered off. The output current thus contains the required modulating frequency together with a second harmonic distortion term. For the lower modulating frequencies the second harmonic will be within the modulating signal bandwidth and cannot be filtered off. (When the modulating signal is multitone, cross-modulation between the individual components will also occur.) The ratio of fundamental to second harmonic is

$$K^2 m : K^2\frac{m^2}{4} \qquad \text{i.e.} \qquad 1 : \frac{m^2}{4}$$

Thus, the relative distortion is proportional to the depth of modulation, m. (For a depth of modulation of 30%, the second harmonic distortion is 7%.)

To minimize distortion, the receiver should be designed so that the signal voltage is as large as possible. For modulating signals derived from speech or music, the peak modulation is restricted to a depth of 80%; this restricts the average depth of modulation to about 30%, which limits distortion under both large and small signal operation.

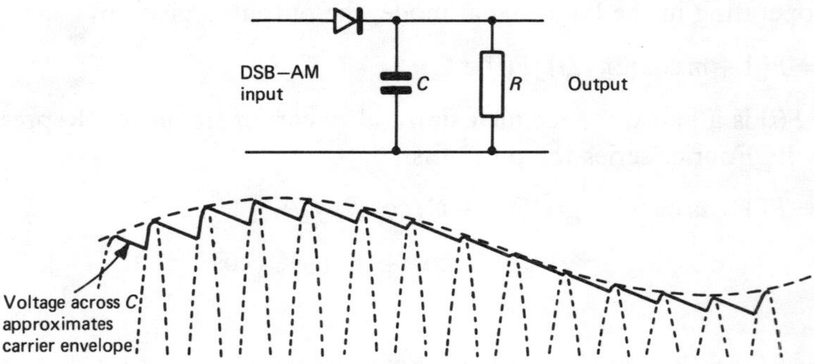

Fig. 2.12 Physical action of envelope detector.

The "unwanted terms" which appear in Eqns (2.15) and (2.17) have frequencies very much higher than the modulating frequencies and can be removed by a simple RC filter as shown in Fig. 2.12. The physical action of this detector is identical to the action of a half-wave rectifier with capacitive smoothing. The RC time constant is chosen such that the capacitor can follow the highest modulating frequencies in the envelope of the rectified carrier without losing excessive charge between carrier pulses.

In the foregoing analysis we assumed that a single modulated carrier appeared at the input of the envelope detector. In a broadcast environment there will of course be many different modulated carriers present in the antenna circuits of all receivers. A primary function of the receiver, therefore, will be to select one of these carriers and reject all others. The obvious way to achieve this would be to precede the detector with a band-pass filter designed to pass the required carrier and its sidebands and to reject the rest. In modern receivers the necessary frequency response is obtained by use of ceramic resonators. These circuits are available as hybrid units, and a typical example with its frequency response is shown in Fig. 2.13(a).

The circuits can be designed to have a flat response over the required bandwidth and a very large attenuation on either side. One problem, however, is that the centre frequency has a fixed value. It is not possible to vary this centre frequency, in order to receive other modulated carriers, and at the same time to maintain the required attenuation on either side of the passband. The superheterodyne receiver was designed specifically to overcome such a problem by transferring the sidebands of the selected carrier frequency to a constant intermediate frequency (IF). A standard IF of 470 kHz has been adopted for receivers working in the medium wave band (540 to 1640 kHz).

A block diagram of the superheterodyne receiver is shown in Fig. 2.13(b) and by examining the functions of each of the blocks the derivation of the term superheterodyne will become apparent. The sidebands of the selected carrier are transferred to a frequency of 470 kHz by multiplying (heterodyning)

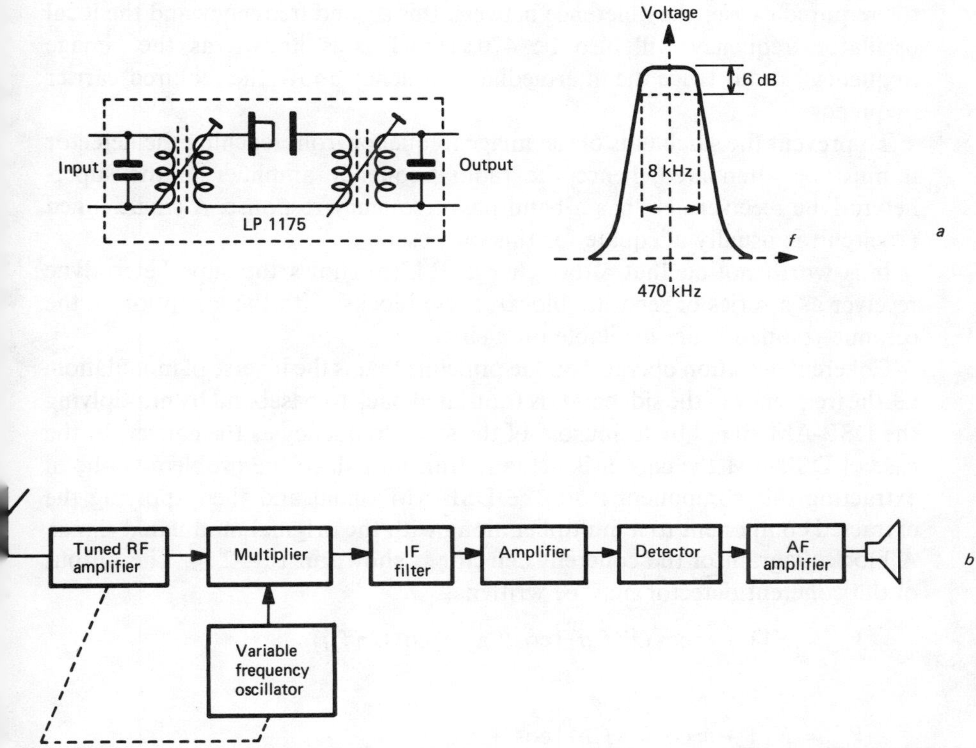

Fig. 2.13 The superheterodyne principle. (a) Ceramic resonator circuit and frequency response; (b) the superheterodyne receiver.

the carrier with the output of a local oscillator. The output of the multiplier will contain sum and difference frequencies and the difference frequency between the local oscillator and selected carrier is made equal to 470 kHz. To produce this figure, the local oscillator must have a frequency of 470 kHz above or below the required carrier. If the local oscillator is above the carrier frequency the required frequency range of the oscillator is 1010 to 2110 kHz for a receiver operating in the medium waveband. This is a frequency ratio of approximately 2:1 (i.e. 1 octave). If the local oscillator is below the carrier frequency the corresponding range would be 70 to 1170 kHz, which is a ratio of 17:1. The local oscillator is designed to work above the required carrier frequency as it is then required to have a range of about 1 octave. Hence the derivation of the term superheterodyne.

It is not the purpose of this text to discuss in detail the design of superheterodyne receivers,[1] but there is one important point worth noting that is a disadvantage of the superheterodyne principle. When the local oscillator is 470 kHz above the required carrier the sidebands will be transferred to 470 kHz. However, if there is a second carrier at a frequency of 940 kHz above the required carrier the difference between this second frequency and the local oscillator frequency will also be 470 kHz. This is known as the "image frequency" and is twice the intermediate frequency above the required carrier frequency.

To prevent the sidebands of the image frequency from reaching the detector it must be attenuated; hence the radio frequency amplifier in the superheterodyne receiver also has a band-pass frequency response. A single tuned LC circuit is usually adequate for this purpose.

It is worth noting that although Fig. 2.13(b) shows the superheterodyne receiver as a series of separate blocks, these blocks, with the exception of the ceramic resonator, are available on a single chip.

Coherent detection operates on the principle that is the inverse of modulation, i.e. the frequency of the sidebands is translated back to baseband by multiplying the DSB-AM signal by a sinusoid of the same frequency as the carrier. In the case of DSB-AM the carrier is actually transmitted, so the problem is one of extracting this component from the DSB-AM signal and then applying the extracted component to a multiplier along with the original modulated signal. A block diagram of the coherent detector is shown in Fig. 2.14. The output of the coherent detector may be written

$$V_{out} = K[1 + m\cos(2\pi f_m t)]\cos(2\pi f_c t)\cos(2\pi f_c t)$$

i.e.

$$V_{out} = K[1 + m\cos(2\pi f_m t)]\cos^2(2\pi f_c t)$$

but

$$\cos^2 2\pi f_c t = \tfrac{1}{2}[1 + \cos(4\pi f_c t)]$$

Therefore

$$V_{out} = \tfrac{1}{2}Km\cos(2\pi f_m t) + \text{unwanted terms} \tag{2.18}$$

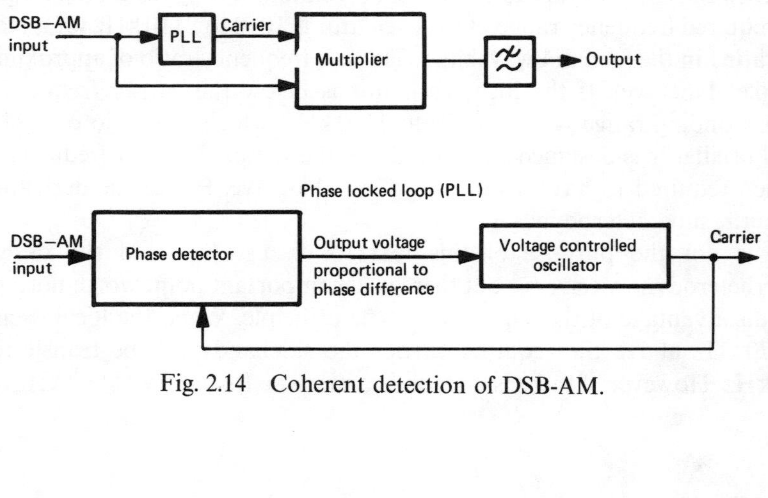

Fig. 2.14 Coherent detection of DSB-AM.

The unwanted terms will be sidebands of $2f_c$ and are easily filtered off. A convenient method of extracting the carrier is to use a phase locked loop as shown in Fig. 2.14.

The basic operation of the loop may be described as follows. The input to the phase detector is the DSB-AM signal and a sinusoid from a voltage controlled oscillator (VCO). The output of the multiplier is a signal proportional to the phase difference between the carrier of the DSB-AM signal and the VCO output. This error signal is fed via a low-pass filter to the VCO input. Its effect is to cause the output frequency of the VCO to change in order to minimize the phase error. Hence the PLL is essentially a negative feedback system. When the loop is locked, the frequency of the VCO output is equal in frequency to the incoming carrier.

The PLL is a very important circuit as it also finds use as a frequency modulation (FM) detector. This is described more fully in Section 2.18.

2.10 DSB-SC-AM Detection

This type of modulation requires coherent detection, but since there is no carrier transmitted the required component must be generated by a local oscillator. Special circuits are required to ensure that the local oscillator is locked in phase to the incoming signal. If the incoming signal is represented by

$$V_{in} = K\cos(2\pi f_c t)\cos(2\pi f_m t)$$

and the output of the local oscillator is $\cos(2\pi f_c t + \phi)$, the output of the coherent detector will be

$$V_{out} = \tfrac{1}{2}K\cos(2\pi f_m t)[\cos(4\pi f_c t + \phi) + \cos\phi] \qquad (2.19)$$

After filtering off the components centred at $2f_c$ this gives a term $\frac{1}{2}K\cos(2\pi f_m t)\cos\phi$. When $\phi = 0$ the output is a maximum and is proportional to the original modulating signal; when $\phi = 90^\circ$ the output is zero. A circuit for demodulating a DSB-SC-AM signal, which makes use of this property, is shown in Fig. 2.15.

If the local oscillator has the same phase as the missing carrier the "in phase" channel will have the correct output and the "quadrature channel" will have zero output. The output of the multiplier is a voltage required to maintain the desired value of the VCO phase. If there is a phase error between the VCO output and the missing carrier, the output of the "in phase" channel will drop and the output of the "quadrature channel" will become non-zero.

The multiplier now produces an output signal that, when applied to the VCO, will cause the phase error between the missing carrier and the VCO output to tend to zero. Hence the circuit automatically produces a locally generated sinusoid of the correct phase. This circuit was initially developed for data communications and is known as the Costas loop.

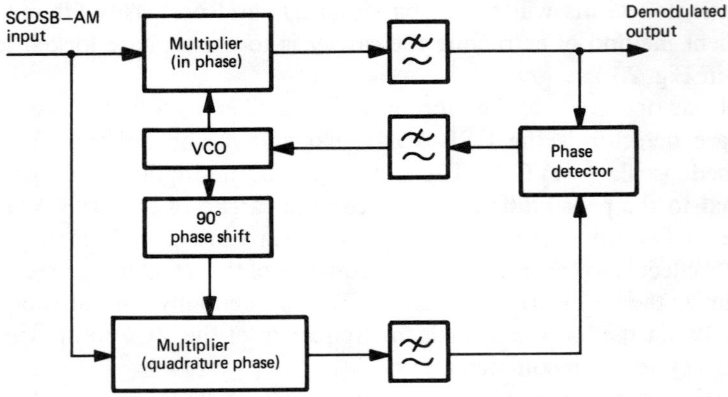

Fig. 2.15 The Costas loop.

2.11 SSB-AM Detection

SSB-AM signals require coherent detection and , once again, a local oscillator is required as no carrier is transmitted.

Assuming the SSB-AM signal is $K\cos 2\pi(f_c + f_m)t$ and the local oscillator signal is $\cos(2\pi f_c t + \phi)$ the output of the coherent detector will be

$$V_{out} = \tfrac{1}{2}K\cos[2\pi(2f_c + f_m)t + \phi] + \tfrac{1}{2}K\cos(2\pi f_m t - \phi) \tag{2.20}$$

When the term at frequency $(2f_c + f_m)$ is filtered off the remaining component can be written

$$\tfrac{1}{2}K\cos(2\pi f_m t)\cos\phi + \sin(2\pi f_m t)\sin\phi \tag{2.21}$$

If $\phi = 0$ the output is thus $\frac{1}{2}K\cos(2\pi f_m t)$ and when $\phi = 90°$ the output is $\frac{1}{2}K\sin(2\pi f_m t)$. Hence in contrast to DSB-SC-AM there is also an output (shifted in phase by 90°) when $\phi = 90°$. For human listeners, whose hearing is relatively insensitive to phase distortion, the requirement of phase coherence can be relaxed. It is, however, necessary to have frequency coherence between the missing carrier and the local oscillator output. Modern point-to-point and mobile SSB-AM communications systems use crystal controlled oscillators and frequency synthesizers to achieve the required frequency stability. This is typically 1 part in 10^6 (i.e. a stability of 1 Hz at a frequency of 1 MHz).

2.12 VSB-AM Detection

VSB-AM transmission is virtually exclusive to television systems, and a typical television detection system will therefore be described in this section. Television receivers use a conventional diode detector to demodulate the video carrier but some pre-processing of the VSB-AM signal occurs in the amplifiers that precede the detector. To illustrate the detection procedure it will be assumed that a VSB-AM signal has been produced from a two-tone modulating signal, i.e.

$$f_m(t) = q_1 \cos(2\pi f_1 t) + q_2 \cos(2\pi f_2 t)$$

The DSB-AM signal from which the vestigial sideband signal is derived will be

$$f_c(t) = K[\cos(2\pi f_c t) + \tfrac{1}{2}m_1 \cos\{2\pi(f_c - f_1)t\} + \tfrac{1}{2}m_2 \cos\{2\pi(f_c - f_2)t\} + \tfrac{1}{2}m_1 \cos\{2\pi(f_c + f_1)t\} + \tfrac{1}{2}m_2 \cos\{2\pi(f_c + f_2)t\}] \qquad (2.22)$$

To produce a VSB-AM signal the component at a frequency of $(f_c - f_2)$ in the lower sideband is assumed suppressed.

It has been shown in Section 2.9 that, if the diode detector is working in the linear region, the input signal is effectively multiplied by the carrier and its harmonics [Eqn (2.14)]. The detector output will therefore contain components

$$K[\cos^2(2\pi f_c t) + \tfrac{1}{4}m_1 \cos\{2\pi(2f_c - f_1)t\} + \tfrac{1}{4}m_1 \cos(-2\pi f_1 t) + \tfrac{1}{4}m_1 \cos\{2\pi(2f_c + f_1)t\} + \tfrac{1}{4}m_1 \cos(2\pi f_1 t) + \tfrac{1}{4}m_2 \cos 2\pi\{(2f_c + f_2)t\} + \tfrac{1}{4}m_2 \cos(2\pi f_2 t)] + \text{unwanted terms} \qquad (2.23)$$

The unwanted terms can be filtered off, and it then becomes clear that the detector output contains components

$$\tfrac{1}{4}K[2m_1 \cos(2\pi f_1 t) + m_2 \cos(2\pi f_2 t)]$$

i.e.

$$\tfrac{1}{4}[2a_1 \cos(2\pi f_1 t) + a_2 \cos(2\pi f_2 t)]$$

The amplitude of the component transmitted in both sidebands is doubled relative to the amplitude of the component transmitted in the single sideband. To compensate for this, the gains of the amplifiers preceding the detector are designed to be asymmetric about the carrier frequency. The characteristic is shown in Fig. 2.16. The response is adjusted in such a way that the components

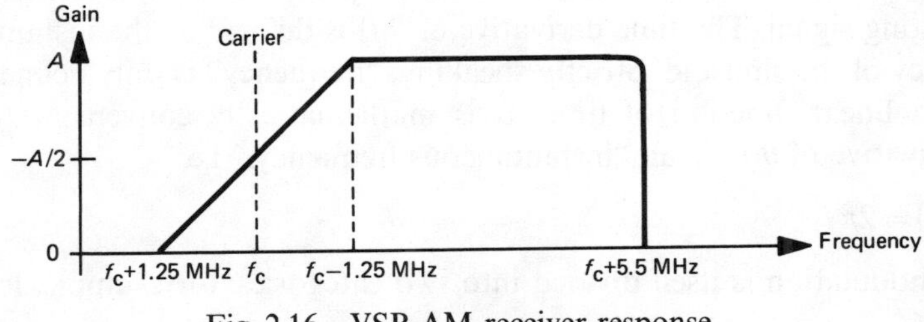

Fig. 2.16 VSB-AM receiver response.

present in both sidebands add to give the same output that would occur if only one sideband were transmitted.

2.13 Economic Factors Affecting the Choice of AM Systems

It is apparent that the special amplifier responses would not be required if the preprocessing was done at the transmitter, which would have the added advantage of reducing the transmitted power. The present standards were chosen on an economic basis. If the processing were done at the transmitter, the television receiver amplifier (of which there are many) would require a flat response over a bandwidth of 6.75 MHz. When the processing is done in the receiver, the required amplifier bandwidths are reduced to 4.25 MHz. This represents a considerable saving in the production costs of a typical television receiver.

The economic argument also decides to a large extent the standards adopted in other forms of telecommunication. A local broadcast transmission system, for example, will have a single high-power transmitter with many receivers. The unit cost of each receiver is lowest when DSB-AM is used, which dictates the use of DSB-AM in this particular situation. The conditions in a mobile communications system will be somewhat different; there will usually be a limited number of receivers, and transmitted power will be at a premium. In this environment SSB-AM with its lower transmitted power becomes attractive. SSB-AM is also used in frequency multiplexed systems over coaxial cables. The obvious advantage here is that twice as many channels may be transmitted as with DSB-AM for a given cable bandwidth.

2.14 Angle Modulation

This is an alternative to amplitude modulation and, as its name suggests, information is transmitted by varying the phase angle of a sinusoidal signal. The sinusoidal signal can be conveniently written as

$$f_c(t) = A\cos\theta(t)$$

where θ is a phase angle that is made proportional to a function of the modulating signal. The time derivative of $\theta(t)$ is defined as the instantaneous frequency of the sinusoid. Strictly speaking, "frequency" is only defined when $\theta(t)$ is a linear function of time. It is mathematically convenient to write the derivative of $\theta(t)$ as an "instantaneous frequency," i.e.

$$\dot{\theta}(t) = 2\pi f_i$$

Angle modulation is itself divided into two categories; for example, let

$$\theta(t) = 2\pi f_c t + \phi + K_1 f_m(t) \tag{2.24}$$

This is termed phase modulation because $\theta(t)$ varies linearly with the amplitude of the modulating signal. Frequency modulation is produced if $\dot{\theta}(t)$ varies linearly with the amplitude of the modulating signal, e.g., let

$$\dot{\theta}(t) = 2\pi f_c + 2\pi K_2 f_m(t) = 2\pi f_i \tag{2.25}$$

Therefore

$$\theta(t) = 2\pi f_c t + \phi + 2\pi K_2 \int_0^t f_m(t)\,dt \tag{2.26}$$

In this case $\theta(t)$ is linearly proportional to the amplitude of the integral of the modulating signal.

2.15 Phase Modulating (PM)

It is shown in Section 2.16 that angle modulation may be considered as a non-linear process in terms of the relationship between time and frequency domains. This means that it becomes very difficult to derive expressions for angle-modulated waves unless very simple modulating waveforms are considered. In this context a single modulating tone may be regarded as such a signal. The modulating signal is given by

$$f_m(t) = a\cos(2\pi f_m t)$$

The PM carrier produced by this signal has the form

$$f_c(t) = A\cos[2\pi f_c t + \Delta\theta\cos(2\pi f_m t)] \tag{2.27}$$

where $\Delta\theta = K_1 a$ is the phase shift produced when the modulating signal has its maximum positive value, i.e.

$$f_c(t) = A\cos(2\pi f_c t)\cos[\Delta\theta\cos(2\pi f_m t)] - A\sin(2\pi f_c t)\sin[\Delta\theta\cos(2\pi f_m t)]$$

If $\Delta\theta$ is restricted to a low value ($\Delta\theta \ll 1$) then $\cos[\Delta\theta\cos(2\pi f_m t)] \simeq 1$ and $\sin[\Delta\theta\cos(2\pi f_m t)] \simeq \Delta\theta\cos(2\pi f_m t)$. The modulated carrier may then be approximated as

$$f_c(t) = A[\cos(2\pi f_c t) - \Delta\theta\sin(2\pi f_c t)\cos(2\pi f_m t)]$$

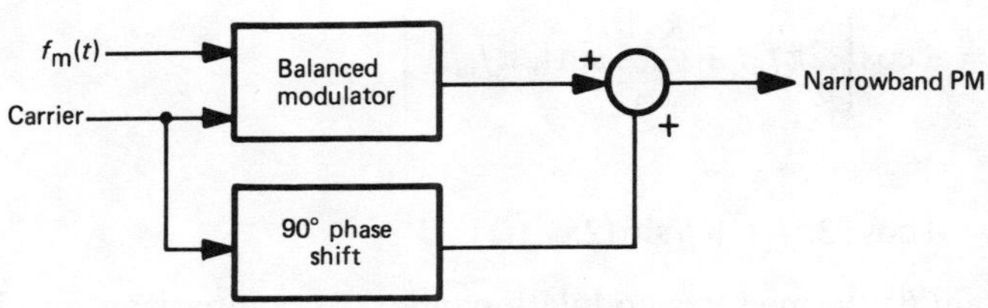

Fig. 2.17 PM.

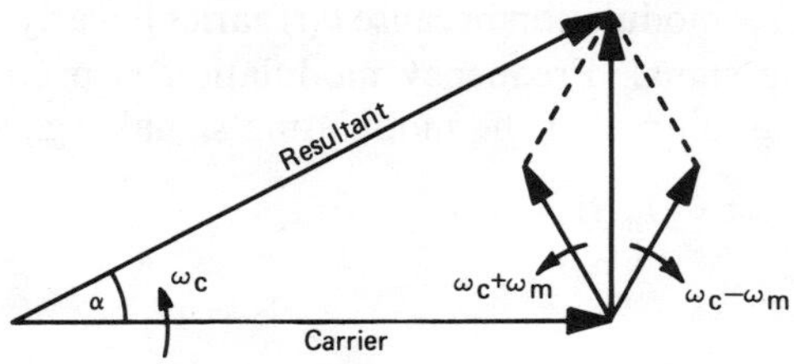

Fig. 2.18 Phasor representation of narrowband PM.

This can be expanded to yield

$$f_c(t) = A[\cos(2\pi f_c t) + \tfrac{1}{2}\Delta\theta \sin\{2\pi(f_c - f_m)t\} + \tfrac{1}{2}\Delta\theta \sin\{2\pi(f_c + f_m)t\}] \quad (2.28)$$

When this equation is compared with the expression for DSB-AM [Eqn (2.3)] it will be seen that, provided the maximum phase shift is restricted to low values, the PM signal is equivalent to DSB-AM in which the carrier has been shifted in phase by 90° relative to the sidebands. PM may thus be produced from DSB-SC-AM by reintroducing the suppressed carrier in phase quadrature. This process is illustrated in block schematic form in Fig. 2.17. The phasor representation of narrowband PM ($\Delta\theta \ll 1$) is given in Fig. 2.18. For small values of $\Delta\theta$ the resultant phasor has an almost constant amplitude but the phase angle α is a function of time.

2.16 Frequency Modulation (FM)

The expression for a frequency modulated carrier is developed in a similar fashion to the expression for a PM carrier. The modulating signal is again assumed to be a single tone, and the phase of the carrier is now related to the time integral of $f_m(t)$, i.e.

$$f_c(t) = A\cos\left[2\pi f_c t + 2\pi K_2 \int_0^t a\cos(2\pi f_m t)\,dt\right] \quad (2.29)$$

Therefore

$$f_c(t) = A\cos\left[2\pi f_c t + \frac{K_2 a}{f_m}\sin(2\pi f_m t)\right]$$

i.e.

$$f_c(t) = A\cos[2\pi f_c t + \beta\sin(2\pi f_m t)] \quad (2.30)$$

The constant β is termed the modulation index. By comparison with Eqn (2.27), it is clear that $K_2 a$ represents the maximum change in carrier frequency that is

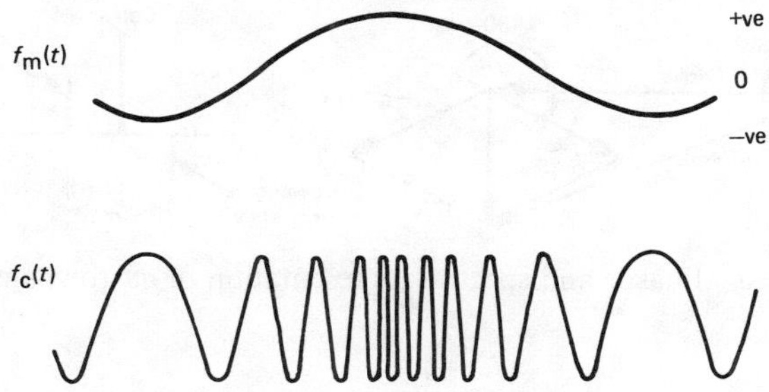

Fig. 2.19 FM carrier.

produced by the modulating signal. Letting $K_2 a = \Delta f$ the modulation index is defined as $\beta = \Delta f / f_m$. It is important to note that β depends both on the carrier deviation, which is linearly proportional to the signal amplitude, and also on the frequency of the modulating signal. The phase shift $\Delta\theta$ that occurs in PM is independent of the modulating signal frequency. The FM waveform that results from a single modulating tone is shown in Fig. 2.19.

The amplitude spectrum of the FM carrier is obtained by expanding the trigonometric function of Eqn (2.30).

$$f_c(t) = A\cos(2\pi f_c t)\cos[\beta\sin(2\pi f_m t)] - A\sin(2\pi f_c t)\sin[\beta\sin(2\pi f_m t)] \tag{2.31}$$

It is convenient to consider Eqn (2.31) for two specific types of FM – narrowband FM and wideband FM. The reason for this distinction is made clearer by considering the series expansion of the trigonometric functions $\cos[\beta\sin(2\pi f_m t)]$ and $\sin[\beta\sin(2\pi f_m t)]$.

$$\cos[\beta\sin(2\pi f_m t)] = 1 - \frac{\beta^2\sin^2(2\pi f_m t)}{2!} + \frac{\beta^4\sin^4(2\pi f_m t)}{4!} + \frac{\beta^6\sin^6(2\pi f_m t)}{6!} + \cdots$$

$$\sin[\beta\sin(2\pi f_m t)] = \beta\sin(2\pi f_m t) - \frac{\beta^3\sin^3(2\pi f_m t)}{3!} + \frac{\beta^5\sin^5(2\pi f_m t)}{5!} + \cdots$$

Narrowband FM is produced when $\beta \ll 1$. The series expansions for this case can be approximated by

$$\cos[\beta\sin(2\pi f_m t)] \simeq 1 \qquad \text{and} \qquad \sin[\beta\sin(2\pi f_m t)] \simeq \beta\sin(2\pi f_m t)$$

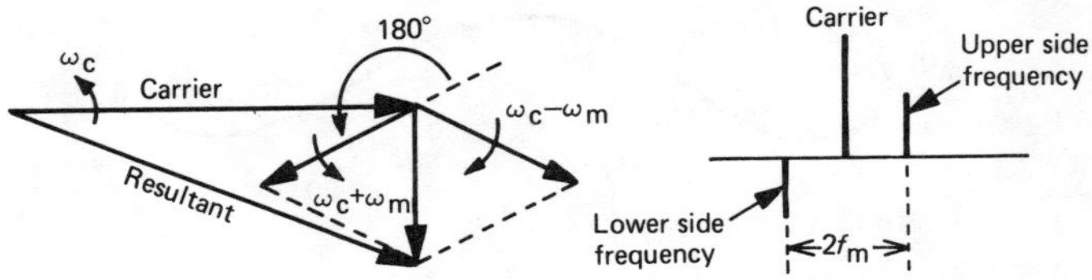

Fig. 2.20 Phasor and spectral representation of narrowband FM.

Using these approximations, Eqn (2.31) can be rewritten as

$$f_c(t) = A\left[\cos(2\pi f_c t) - \frac{\beta}{2}\cos\{2\pi(f_c - f_m)t\} + \frac{\beta}{2}\cos 2\pi\{(f_c + f_m)t\}\right] \tag{2.32}$$

This expression shows that narrowband FM is equivalent to DSB-AM with a phase shift of 180° in the lower sideband. The bandwidth in this case is identical to the bandwidth of a DSB-AM signal and is exactly twice the bandwidth of the modulating signal. The resultant phasor, which is illustrated in Fig. 2.20, is similar to the resultant for PM, i.e. it has almost constant amplitude and a phase angle that is a function of time.

It is evident that as the value of β increases more terms in the series expansions of $\cos[\beta\sin(2\pi f_m t)]$ and $\sin[\beta\sin(2\pi f_m t)]$ become significant and cannot be ignored. This illustrates the non-linear relationship between the time domain and frequency domain, which is a feature of angle-modulated signals. As an example, consider the series expansions of $\cos\beta\sin(2\pi f_m t)$ and $\sin\beta\sin(2\pi f_m t)$ when $\beta = 0.5$. The new approximations are

$$\cos[\beta\sin(2\pi f_m t)] \simeq 1 - \tfrac{1}{2}\beta^2\sin^2(2\pi f_m t) = 1 - \tfrac{1}{4}\beta^2[1 - \cos(4\pi f_m t)]$$

$$\sin[\beta\sin(2\pi f_m t)] \simeq \beta\sin(2\pi f_m t)$$

Equation (2.31) may be rewritten for this example as

$$\begin{aligned} f_c(t) = A\{&(1 - \tfrac{1}{4}\beta^2)\cos 2\pi f_c t \\ &- \tfrac{1}{2}\beta[\cos\{2\pi(f_c - f_m)t\} - \cos\{2\pi(f_c + f_m)t\}] \\ &+ \tfrac{1}{8}\beta^2[\cos\{2\pi(f_c - 2f_m)t\} + \cos\{2\pi(f_c + 2f_m)t\}] \end{aligned} \tag{2.33}$$

A simple increase in the value of β (produced by an increase in the amplitude of the modulating signal or a decrease in its frequency) produces a decrease in the amplitude of the component at frequency f_c and results in two additional frequency components spaced at $\pm 2f_m$ relative to f_c. The bandwidth thus effectively doubles, when compared with the previous case, simply as a result of increasing the amplitude of the modulating signal. This non-linear relationship between time and frequency domains becomes apparent when Fig. 2.21 is compared with Fig. 2.20.

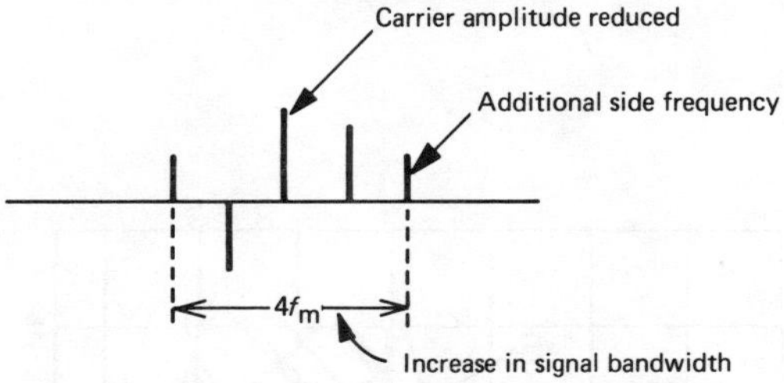

Fig. 2.21 Non-linearity of the FM process.

Further increases in the value of β will clearly increase the number of significant terms in the series expansion of the two trigonometric functions. Both these functions are in fact periodic and can therefore be expressed in terms of a Fourier series, i.e.

$$\cos[\beta \sin(2\pi f_m t)] = C_0 + C_2 \cos(4\pi f_m t) + C_4 \cos(8\pi f_m t) + \cdots$$

$$\sin[\beta \sin(2\pi f_m t)] = C_1 \sin(2\pi f_m t) + C_3 \sin(6\pi f_m t) + \cdots$$

The Fourier coefficients C_n are themselves infinite series and may only be evaluated numerically for specific values of n and β. These coefficients are termed Bessel functions of the first kind, i.e. $C_n = J_n(\beta)$. Graphs of $J_n(\beta)$ for several values of n are given in Fig. 2.22. Using the Fourier series representation, Eqn (2.31) may be rewritten as

$$\begin{aligned} f_c(t) = J_0(\beta)\cos(2\pi f_c t) &- J_1(\beta)[\cos\{2\pi(f_c - f_m)t\} - \cos\{2\pi(f_c + f_m)t\}] \\ &+ J_2(\beta)[\cos\{2\pi(f_c - 2f_m)t\} + \cos\{2\pi(f_c + 2f_m)t\}] \\ &- J_3(\beta)[\cos\{2\pi(f_c - 3f_m)t\} - \cos\{2\pi(f_c + 3f_m)t\}] \\ &+ \cdots \end{aligned} \qquad (2.34)$$

The amplitudes of the carrier and sidebands, of which there is an infinite number, depend on the value of β which fixes the appropriate Bessel function value. Tables of Bessel functions for a range of n and β are given in Appendix 1.

The bandwidth of the wideband FM signal is apparently infinite, but in practice some simplifying approximations are possible. With reference to Fig. 2.22, it can be observed that as the value of n is increased then β must be made progressively larger before the appropriate value of $J_n(\beta)$ becomes non-zero. Extrapolating this trend it is found that for large values of n the value of $J_n(\beta)$ is approximately zero when $\beta < n$, e.g. when $\beta = 100$ then $J_{101}(\beta)$, $J_{102}(\beta)$, etc., are approximately zero since $\beta < n$. In this example, therefore, there are 100 values of $J_n(\beta)$ (excluding the carrier $J_0(\beta)$) that have a non-zero value. Hence this leads to the approximation that when β is large there are β values of $J_n(\beta)$ that are non-zero. Under these circumstances there will be β pairs of sidebands, each spaced by f_m, in the amplitude spectrum of the modulated signal.

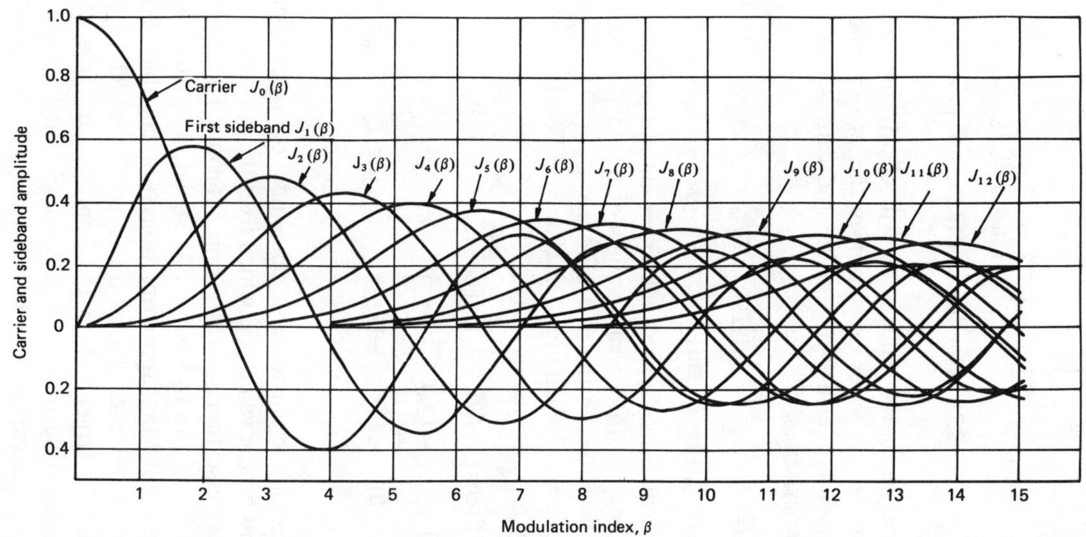

Fig. 2.22 Bessel functions of order n for various values of β.

The signal bandwidth thus becomes

$$B = 2\beta f_m = 2\Delta f$$

i.e. for large values of β the signal bandwidth is approximately twice the carrier deviation. An accurate figure for the bandwidth of an FM signal for either very large or very small values of β is given by Carson's rule, which states

$$B = 2f_m(1 + \beta)$$

i.e.

$$\text{when} \quad \beta \gg 1 \quad B \simeq 2f_m\beta = 2\Delta f$$

$$\text{when} \quad \beta \ll 1 \quad B \simeq 2f_m$$

For other values of β the signal bandwidth must be computed from Bessel function tables. The spectrum of the FM signal is then defined as containing all sideband components with an amplitude $\geqslant 1\%$ of the unmodulated carrier amplitude.

In the UK, commercial FM broadcasting stations restrict the maximum carrier deviation to ± 75 kHz. A single modulating tone of frequency 15 kHz producing peak frequency deviation of the carrier would yield a modulation index $\beta = 5$. The number of sideband terms with an amplitude of 0.01 or greater is obtained from the Bessel function tables as 8. The bandwidth is then

$$B = 2 \times 8 \times f_m = 240\,\text{kHz}$$

If the peak frequency deviation is produced by a 5 kHz modulating tone the modulation index will have a value of $\beta = 15$. The number of significant sidebands increases to 19 and the bandwidth becomes

$$B = 2 \times 19 \times f_m = 190\,\text{kHz}$$

Although the larger value of β produces more significant sidebands, the sidebands are actually closer together and the bandwidth of the FM signal is less.

It will be apparent from the foregoing analysis that FM and PM are very closely related, so much so that an "instantaneous frequency" and "modulation index" can also be defined for PM. It is instructive to compare PM and FM from the point of view of these parameters, as the comparison gives some insight into why FM is used almost exclusively in preference to PM. For a single modulating tone the instantaneous frequency in an FM wave has been defined as

$$2\pi f_i = 2\pi f_c + 2\pi K_2 a \cos(2\pi f_m t)$$

i.e.

$$f_i = f_c + \Delta f \cos(2\pi f_m t)$$

which gives

$$f_c(t) = A\cos\left[2\pi f_c t + \frac{\Delta f}{f_m}\sin(2\pi f_m t)\right]$$

For PM,

$$\theta(t) = 2\pi f_c t + \Delta\theta\cos(2\pi f_m t)$$

Therefore

$$\dot{\theta}(t) = 2\pi f_i = 2\pi f_c - \Delta\theta 2\pi f_m \sin(2\pi f_m t)$$

i.e.

$$\dot{\theta}(t) = 2\pi f_c - 2\pi\Delta f\sin(2\pi f_m t)$$

The PM carrier may therefore be written in terms of a frequency deviation $\Delta f = \Delta\theta f_m$, i.e.

$$f_c(t) = A\cos\left[2\pi f_c t + \frac{\Delta f}{f_m}\cos(2\pi f_m t)\right] \tag{2.35}$$

If a modulation index β_p is defined for PM, Eqn (2.35) can be written

$$f_c(t) = A\cos\left[2\pi f_c t + \beta_p\cos(2\pi f_m t)\right]$$

which is similar to Eqn (2.30) except that β_p does not depend on the value of f_m.

Having shown that a PM carrier can be considered in this way, it is then possible to obtain the bandwidth of a wideband PM signal by reference to Bessel function tables. If a frequency deviation $\Delta f = \pm 75$ kHz is produced in a PM carrier by a modulating tone of 15 kHz there will be 8 significant sidebands and, as with the FM carrier, the bandwidth of the signal will be approximately 240 kHz. However, since β_p does not depend on the value of f_m there will be 8 sidebands for all possible values of f_m. A modulating tone of 50 Hz would result in a bandwidth of $2 \times 8 \times 50 = 800$ Hz. The value for FM would be approximately 150 kHz. Thus it becomes apparent that PM makes less efficient use of bandwidth than frequency modulation, since the bandwidth of any receiver must be equal to the maximum bandwidth of the received signal. It is shown in Chapter 4 that SNRs in modulated systems are directly related to the bandwidth of the modulate wave. Hence it may be concluded that the signal-to-noise performance of FM will be superior to PM, and for this reason FM is usually chosen in preference to PM. (There is one notable exception to this rule, namely the transmission of wideband data signals, which is dealt with in Chapter 3.)

Because of the inherent non-linearity of the FM process, it is not possible to derive the bandwidth for a multi-tone modulating signal. (The bandwidth of an FM wave has been derived by Black[2] for a two-tone modulating signal.)

One observation that can be made is that if the modulating signal is multitone then no single component can produce peak frequency deviation, or overmodulation would result. Under these circumstances, practical measurements have shown that a reasonably accurate measure of the signal bandwidth is given by a modified form of Carson's rule, i.e., $B = 2(\Delta f + D)$, where D is the bandwidth of the modulating signal.

2.17 Frequency Modulators

FM may be produced directly by varying the frequency of a voltage-controlled oscillator or indirectly from PM. Figure 2.23 illustrates two methods of directly producing frequency modulation. The first circuit is an RC oscillator and the second is a parallel resonant LC circuit.

The RC circuit has the advantage of requiring no inductance. The integrator charges a capacitor C from a current proportional to V_{in}. When the capacitor charges to the switching voltage of the comparator the comparator output goes low, producing a rapid discharge of C. This causes the comparator output to return high and C begins to charge once again. The charging time of the capacitor is inversely proportional to charging current and hence the frequency of oscillation is linearly related to V_{in}. The comparator output will be a variable-frequency pulse train; a sinusoidal FM waveform is obtained by appropriate filtering.

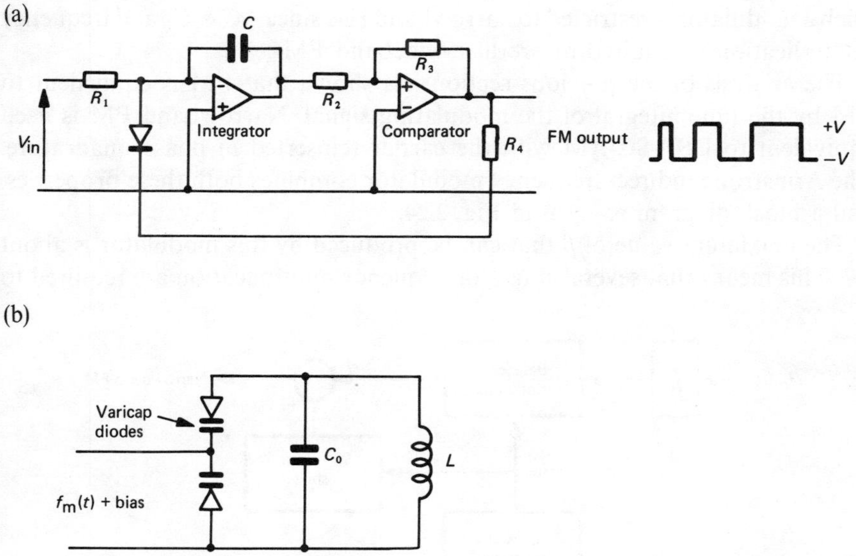

Fig. 2.23 Direct FM. (a) RC voltage controlled oscillator, $V_{in} = f_m(t) + \text{bias}$; (b) Varicap diode tuning of LC tank circuit.

The LC circuit forms the tank circuit of an LC oscillator. A common method of altering the resonant frequency of such a tuned circuit is to use the dependence of capacitance of a reverse biased P-N junction on the reverse bias voltage. Varactor (varicap) diodes are specifically designed for this purpose and Fig. 2.23 shows a typical arrangement of these diodes in a parallel resonant circuit. The total capacitance in this circuit will be $C = C_0 + \Delta C$ where $\Delta C = Kf_m(t)$. The resonant frequency of the circuit is given by

$$2\pi f_r = \frac{1}{(LC)^{1/2}} = \frac{1}{(LC_0)^{1/2}} \frac{1}{[1 + (\Delta C/C_0)]^{1/2}}$$

If the change in capacitance is small then $\Delta C \ll C_0$ and the binomial expansion of $[1 + (\Delta C/C_0)]^{-1/2}$ is approximated by $(1 - \Delta C/2C_0)$, i.e.

$$2\pi f_r = \frac{1}{(LC_0)^{1/2}}\left(1 - \frac{\Delta C}{C_0}\right)$$

Therefore

$$2\pi f_r = 2\pi f_c\{1 - [Kf_m(t)/2C_0]\}$$

or

$$f_r = f_c[1 - K_c f_m(t)] \tag{2.36}$$

The resonant frequency of the tuned circuit is thus directly related to the amplitude of the modulating signal. The tuned circuit is used as the frequency determining network in a feedback oscillator thus producing FM directly. Such a modulator is restricted to narrowband FM since $\Delta C \ll C_0$ and frequency multiplication is required to produce wideband FM.

The analysis of the previous section has shown that FM is equivalent to PM by the time integral of the modulating signal. Narrowband PM is itself equivalent to DSB-SC-AM with the carrier reinserted in phase quadrature. The Armstrong indirect frequency modulator combines both these properties, and a block diagram is given in Fig. 2.24.

The maximum value of β that can be produced by this modulator is about 0.2. This means that several stages of frequency multiplication are required to

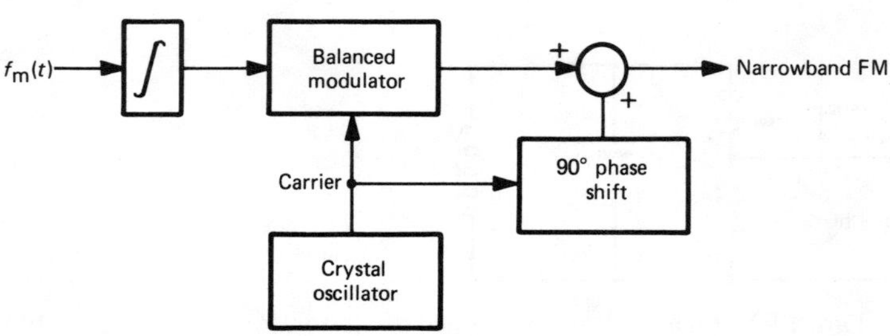

Fig. 2.24 Armstrong FM.

produce wideband FM. The advantage of the Armstrong modulator is that the carrier is produced by a stable crystal oscillator. It should be noted that when the frequency of the carrier is multiplied by n (by means of a non-linear device) the frequency deviation, and hence the value of β, is also multiplied by the same factor.

2.18 Demodulation of a Frequency Modulated Wave

The primary function of a frequency modulation detector is to produce a voltage proportional to the instantaneous frequency f_i of the modulated wave. The FM waveform is given by

$$f_c(t) = A\cos\left(2\pi f_c t + K\int_0^t f_m(t)\,dt\right)$$

If this waveform is differentiated, the resultant waveform is

$$f_c(t) = -A[2\pi f_c + K f_m(t)]\sin\left(2\pi f_c t + K\int_0^t f_m(t)\,dt\right)$$

This is a frequency modulated wave that now has an envelope of magnitude proportional to the amplitude of the modulating signal $f_m(t)$. The modulating signal may then be recovered by envelope detection. The envelope detector will, of course, ignore the frequency variations of the carrier. Traditionally FM detectors have relied upon the properties of tuned circuits to perform the required differentiation. This is demonstrated in Fig. 2.25, where the resonant frequency of the tuned circuit is chosen such that the carrier frequency f_c is on the slope of the circuit response.

The linearity of a single tuned circuit is limited to a relatively small frequency range. This range may be extended by introducing a second tuned circuit with

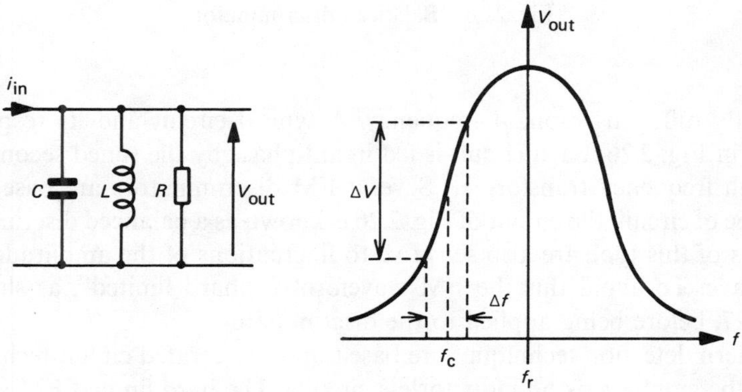

Fig. 2.25 Detection of FM with a single tuned circuit.

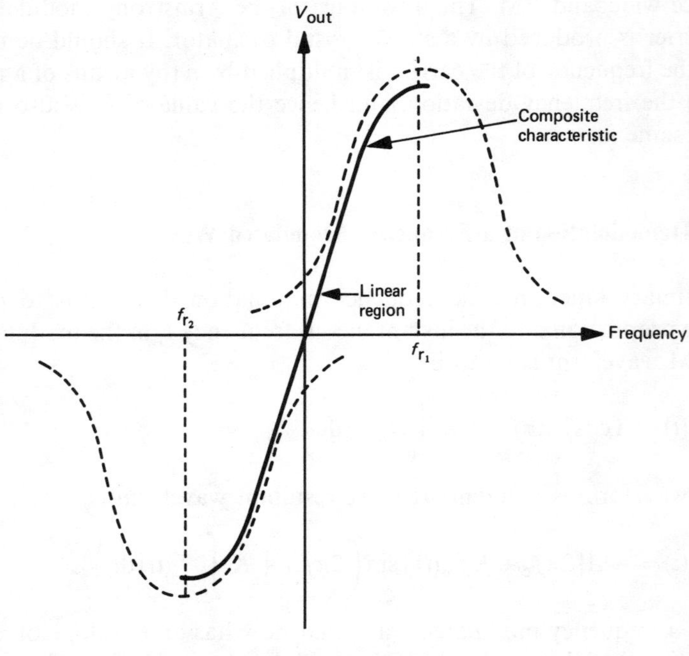

Extended range frequency-to-voltage conversion

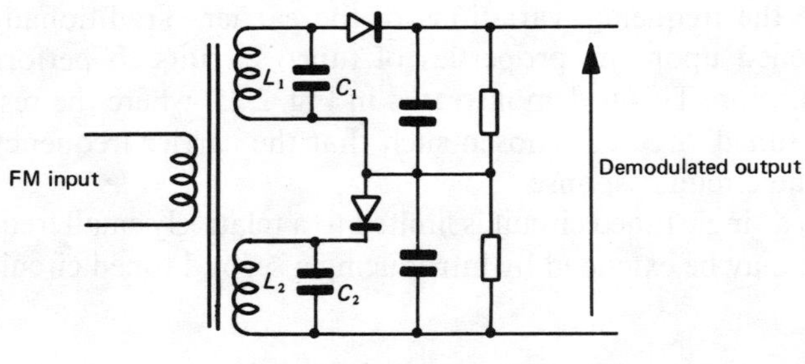

Fig. 2.26 Balanced discriminator.

a slightly different resonant frequency. A typical circuit and its response is shown in Fig. 2.26. Each circuit is fed in antiphase by the tuned secondary of the high frequency transformer. Several FM discriminators are based upon this type of circuit; the circuit of Fig. 2.26 is known as a balanced discriminator. Circuits of this type are also sensitive to fluctuations of the amplitude of the FM wave. To avoid this the FM waveform is "hard limited", as shown in Fig. 2.27, before being applied to the discriminator.

Modern detection techniques are based upon integrated circuit technology, where the emphasis is on inductorless circuits. The hard limited FM signal is in fact a variable frequency pulse waveform. The instantaneous frequency is

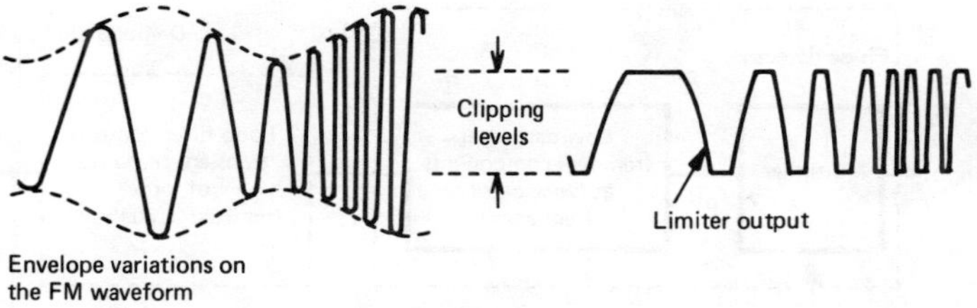

Fig. 2.27 Action of the limiter.

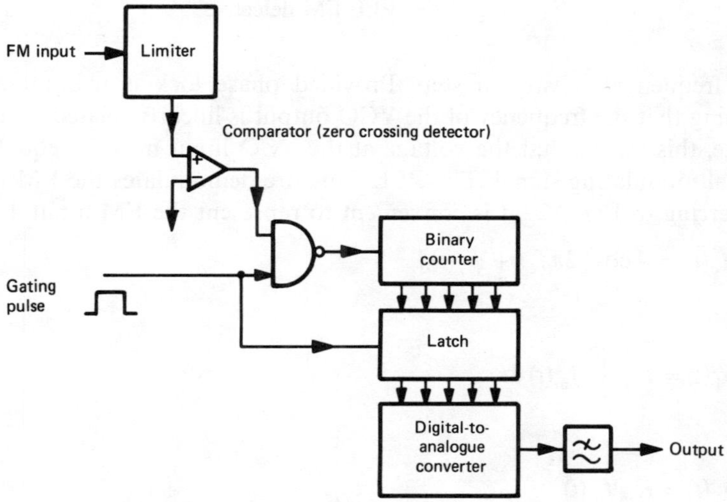

Fig. 2.28 Digital (zero crossing) FM detector.

preserved in the zero crossings of this pulsed signal. A completely digital circuit based upon a zero crossing detector is shown in Fig. 2.28. The number of zero crossings in a fixed interval are gated to the input of a binary counter. The counter outputs are then used as inputs to a digital-to-analogue converter (DAC). The analogue output voltage is thus proportional to the number of zero crossings that occur during the gating interval. Thus the output of the DAC is proportional to the original modulating signal $f_m(t)$.

Probably the most important integrated inductorless FM detector is the phase locked loop (PLL). This circuit may be used in connection with the detection of AM and is therefore an attractive proposition (see Fig. 2.13). When used as an FM detector the output of the PLL is actually the input voltage of the voltage controlled oscillator. This configuration is shown in Fig. 2.29.

It was stated in Section 2.10 that the ouput of the multiplier causes the VCO to minimize the phase error between the input signal and the VCO output. If the frequency of the input signal varies, the multiplier output will cause the

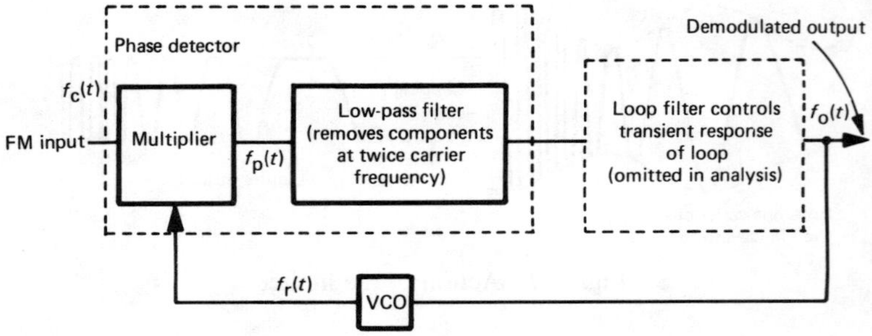

Fig. 2.29 PLL FM detector.

VCO frequency to vary in step. Provided phase lock is maintained, and assuming that the frequency of the VCO output is linearly related to its input voltage, this means that the voltage at the VCO input must be equal to the original modulating signal. The PLL therefore demodulates the FM input.

Referring to Fig. 2.29 it is convenient to represent the FM input as

$$f_c(t) = A\cos[2\pi f_c t + \phi_1(t)]$$

where

$$\phi_1(t) = K_m \int_0^t f_m(t)\,dt$$

i.e.

$$\dot{\phi}_1(t) = K_m f_m(t)$$

The VCO output can be written

$$f_r(t) = B\sin[2\pi f_c t + \phi_2(t)]$$

The multiplier output will be the product of $f_c(t)$ and $f_r(t)$; thus

$$f_p(t) = \tfrac{1}{2}K_d AB[\sin\{\phi_2(t) - \phi_1(t)\} + \sin\{4\pi f_c t + \phi_1(t) + \phi_2(t)\}]$$

where K_d is the gain of the multiplier and has units of volts per radian. The component at the frequency of $2f_c$ in the output of the multiplier is blocked by the low-pass filter so that the input to the VCO is

$$f_0(t) = \tfrac{1}{2}K_d AB\sin\{\phi_2(t) - \phi_1(t)\}$$

When the loop is locked the phase error will be very small, i.e.

$$\phi_2(t) \simeq \phi_1(t)$$

Thus

$$f_0(t) = \tfrac{1}{2}K_d AB\{\phi_2(t) - \phi_1(t)\} \tag{2.37}$$

It should be noted that when the loop is locked there is actually a phase difference of 90° between $f_c(t)$ and $f_r(t)$. The output frequency of the VCO is $2\pi f_c + \dot{\phi}_2(t)$ where $\dot{\phi}_2(t) = K_0 f_0(t)$, i.e.

$$\phi_2(t) = K_0 \int_0^t f_0(t)\,dt$$

Equation (2.37) can thus be written

$$f_0(t) = \tfrac{1}{2} K_d AB \left[K_0 \int_0^t f_0(t)\,dt - \phi_1(t) \right] \tag{2.38}$$

This equation upon differentiation yields

$$\dot{\phi}_1(t) = K_0 \left[\frac{2\dot{f}_0(t)}{K_0 K_d AB} - f_0(t) \right] \tag{2.39}$$

It is of interest at this point to examine the dimensions of the various constants in Eqn (2.39). K_d relates the frequency of the VCO output to the voltage appearing at its input and has dimensions of hertz per volt, or alternatively radians per second per volt. Hence the product $K_0 K_d$ has dimensions of $\sec^{-1}$, which is a frequency. Thus if the frequency of $f_0(t) \ll K_0 K_d$ then $\dot{f}_0(t)/(K_0 K_d) \ll 1$ and $\dot{\phi}_1(t) \simeq -K_0 f_0(t)$.

But

$$\dot{\phi}_1(t) = K_m f_m(t)$$

Therefore

$$f_0(t) = -\frac{K_m}{K_0} f_m(t)$$

The input to the VCO is thus linearly proportional to the modulating signal, which means, in effect, that the PLL demodulates an FM waveform.

An insight into the physical action of the PLL can be gained by considering the loop when the input is an unmodulated carrier. The phase of $f_c(t)$ will be a linear function of time, i.e. $2\pi f_c t$ can be regarded as a ramp function. To maintain a small error at the phase detector output the phase of the VCO output must also be a ramp of the same slope. If the frequency of $f_c(t)$ increases the slope of the "phase ramp" also increases. The error between the phase ramp of $f_c(t)$ and the phase ramp of the VCO output then begins to increase producing an increase in the phase detector output. This increase modifies the frequency of the VCO output and a new state of equilibrium is reached when the two slopes are again equal. There will now be a larger output from the phase detector indicating a greater, but constant, phase error. The slopes of the phases of $f_c(t)$ and the VCO, and thus the two frequencies, will, however, be the same.

The analysis presented here deals only with the steady-state response of the PLL assuming that the loop is locked. If this is not the case the PLL will behave as a negative feedback system with the usual problems of transients and stability. A discussion of the transient properties of the PLL is outside the scope of this work, but a full treatment is given by Stremler.[3]

2.19 Frequency Division Multiplex (FDM) Transmission

This is a form of transmission used extensively in telephone systems for the simultaneous transmission of several separate telephone circuits over a wideband link. Each of the telephone signals has a bandwidth limited to 3.4 kHz, and therefore many can be transmitted over, say, a coaxial cable with a bandwidth of several megahertz.

Each signal modulates a sinusoidal carrier of different frequency, SSB modulation being used. The signals are therefore effectively stacked one above the other throughout the transmission bandwidth. To facilitate separation of the signals at the far end of the link, adjacent signals are separated by a guard band f_g (Fig. 2.30). Each signal in an FDM transmission system thus occupies part of transmission bandwidth for the whole of the transmission time.

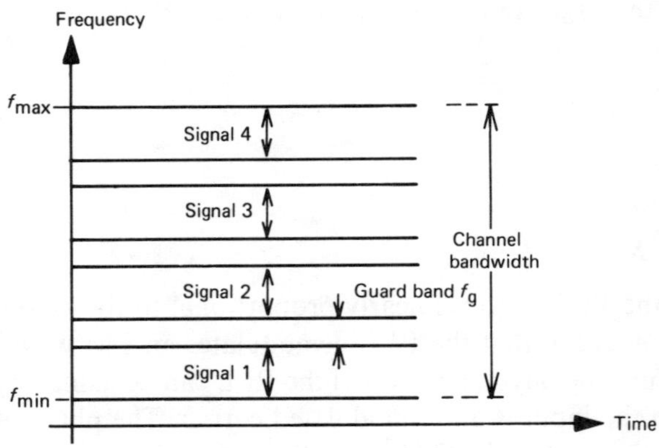

Fig. 2.30 FDM transmission.

2.20 Conclusion

The economic argument for use of the various AM systems was given in Section 2.13. FDM of telephone circuits was based on SSB-AM because here the emphasis is on packing as many circuits as possible into a finite bandwidth.

The arguments for the use of FM are less well defined. We show in Chapter 4 that for a fixed transmitter power FM has a superior signal-to-noise performance over all types of AM. Consequently, FM has been adopted as the standard for high-quality sound broadcast transmission in several countries

FM is also an extremely important technique used in the transmission of binary data signals over telephone circuits.

References

1. Scroggie, M.G., *Foundations of Wireless*, 8th edn, Illife, London, 1971.
2. Black, H.S., *Modulation Theory*, Van Nostrand Reinhold, Wokingham, UK, 1953.
3. Stremler, F.G., *Introduction to Communication Systems*, Addison-Wesley, London, 1976.

Problems

2.1 A transmitter radiates a DSB-AM signal with a total power of 5 kW at a depth of modulation of 60%. Calculate the power transmitted in the carrier and also in each sideband.

Answer: 4.24 kW; 0.38 kW.

2.2 A carrier wave represented by $10\cos(2\pi 10^6 t)$ V is amplitude modulated by a second wave represented by $3\cos(2\pi 10^3 t)$ V. Calculate
(a) the depth of modulation;
(b) the upper and lower side frequencies;
(c) the amplitude of the side frequencies;
(d) the fraction of the power transmitted in the sidebands.

Answer: (a) 30%; (b) 1.001 MHz, 0.999 MHz; (c) 1.5 V; (d) 4.3%.

2.3 A DSB-AM transmitter produces a total output of 24 kW when modulated to a depth of 100%. Determine the power output when
(a) the carrier is unmodulated;
(b) the carrier is modulated to a depth of 60%, one sideband is suppressed and the carrier component is attenuated by 26 dB.

Answer: (a) 16 kW; (b) 1.48 kW.

2.4 A DSB-AM receiver uses a square law detector. What is the maximum depth of modulation that may be used if the second harmonic distortion of the modulating signal, produced by the detector, is restricted to 10% of the fundamental?

Answer: 40%.

2.5 A carrier of 5 V rms and frequency 1 MHz is added to a modulating signal of 2 V rms and frequency 1 kHz. The composite signal is applied to a biased diode rectifier in which the relationship between current and voltage over the range ± 10 V is $i = (5 + v + 0.05v^2)\mu A$ where v is the instantaneous voltage. Find the depth of modulation of the resulting DSB-AM signal and the frequency of each component in the diode current.

Answer: 28.3%; 0 Hz, 1 kHz, 1 MHz, 2 MHz, 0.999 MHz, 1·001 MHz.

2.6 Narrowband FM is produced indirectly by varying the phase of a carrier of frequency 13 MHz, the maximum phase shift being 0.5 rad. Show that wideband

FM may be produced by multiplying the frequency of the modulated carrier. If the modulating signal has a frequency of 1.5 kHz find the frequency deviation of the carrier after a frequency multiplication of 15.

Answer: 11.25 kHz.

2.7 A frequency modulated wave has a total bandwidth of 165 kHz when the modulating signal is a single tone of frequency 10 kHz. Using Bessel function tables, or otherwise, find the maximum carrier frequency deviation produced by the modulating signal.

Answer: 50 kHz.

2.8 In a direct FM transmitter an inductance of 10 μH is tuned by a capacitor whose capacitance is a function of the amplitude of the modulating signal. When the modulating signal is zero the effective capacitance is 1000 pF. An input signal of $4.5 \cos(3\pi 10^3 t)$ V produces a maximum change in capacitance of 6 pF. Assuming the resultant FM signal is multiplied in frequency by 5, calculate the bandwidth of the eventual output.

Answer: 60 kHz.

2.9 A single tone of frequency 7.5 kHz forms the modulating signal for both a DSB-AM and an FM transmission. When modulated the peak frequency deviation of the FM signal is 60 kHz. Assuming that the same total power is transmitted for each of the modulated signals, find the depth of modulation for the DSB-AM signal when the amplitude of the first pair of sidebands of the FM wave equals the amplitude of the sidebands of the DSB-AM wave.

Answer: 50%.

3 / DISCRETE SIGNALS

The signals considered in Chapter 2 were continuous functions of time. There are many advantages that result from the conversion of analogue signals into a binary coded format. Two such advantages that are easily identified are that the transmission and processing of binary signals are generally much easier to achieve than the transmission and processing of analogue signals.

It is not possible to code a continuous analogue signal into binary format because there is an infinite number of values of the continuous signal. Instead the continuous signal is coded at fixed instants of time. These instants are known as sampling instants, and it is important to determine the effect of the sampling action on the properties of the original continuous signal. The rules governing the sampling of continuous signals are specified by the sampling theorem.

3.1 Sampling of Continuous Signals

The sampling theorem states that if a signal has a maximum frequency of W Hz it is completely defined by samples which occur at intervals of $1/2W$ sec. The sampling theorem can be proved by assuming that $h(t)$ is a non-periodic signal band limited to W Hz. The amplitude spectrum is given by

$$H(f) = \int_{-\infty}^{\infty} h(t) \exp(-j2\pi ft)\, dt \tag{3.1}$$

Since $H(f)$ is band limited to $\pm W$ Hz it is convenient to make $H(f)$ a periodic function of frequency (Fig. 3.1). The value of $H(f)$ in the region $-W$ to $+W$ can be expressed in terms of a Fourier series.

The well-known time domain Fourier series is

$$h(t) = 1/T \sum_{n=-\infty}^{\infty} C_n(\exp j2\pi nt/T)$$

and the corresponding series in the frequency domain is

$$H(f) = 1/2W \sum_{n=-\infty}^{\infty} X_n \exp(j2\pi nf/2W) \tag{3.2}$$

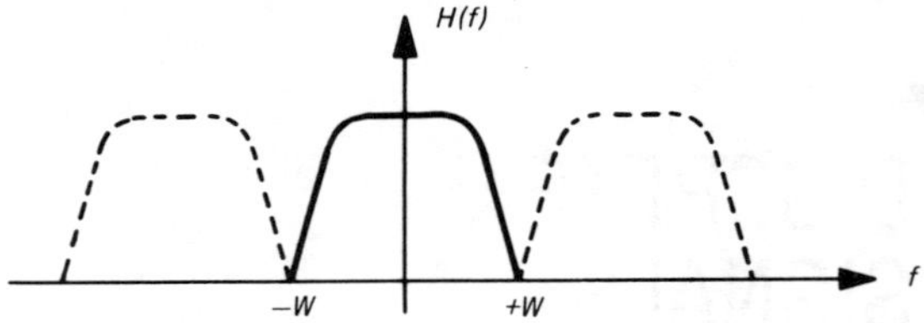

Fig. 3.1 Representation of $H(f)$ as a periodic function.

The values of the Fourier coefficients are given by

$$X_n = \int_{-W}^{W} H(f)\exp(-j\pi nf/W)\,\mathrm{d}f$$

But $H(f)$ is the Fourier transform of $h(t)$:

$$h(t) = \int_{-W}^{W} H(f)\exp(j2\pi ft)\,\mathrm{d}f \tag{3.3}$$

Hence if $t = -n/2W$

$$h(-n/2W) = \int_{-W}^{W} H(f)\exp(-j\pi nf/W)\,\mathrm{d}f$$

i.e.

$$h(-n/2W) = X_n \tag{3.4}$$

The values $h(-n/2W)$ are samples of $h(t)$ taken at equally spaced instants of time, the time between samples being $1/2W$ sec. These samples define X_n, which in turn completely defines $H(f)$. Since $H(f)$ is the Fourier transform of $h(t)$, then $H(f)$ defines $h(t)$ for all values of t. Hence $h(-n/2W)$ completely defines $h(t)$ for all values of t.

The physical interpretation of the sampling process is shown in Fig. 3.2. The continuous signal $h(t)$ is multiplied by a periodic pulse train $S(t)$ in which the pulse width is much less than the pulse period.

Since the sampling pulse train is periodic it can be expanded in a Fourier

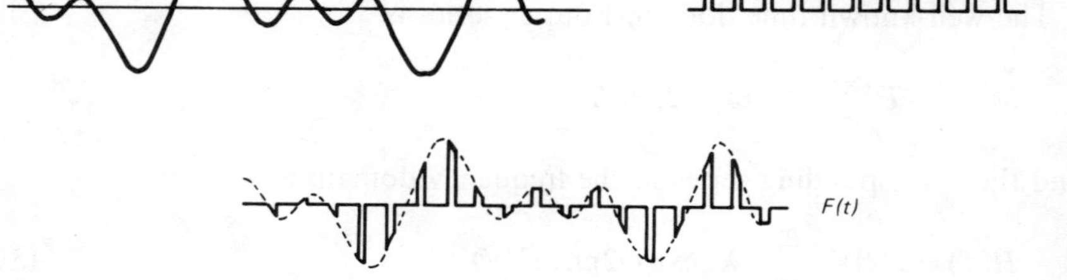

Fig. 3.2 The sampling process.

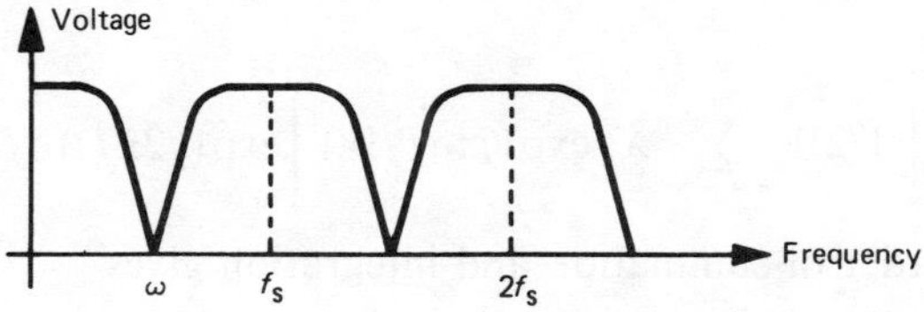

Fig. 3.3 Amplitude spectrum of a sampled signal.

series as

$$S(t) = a_0 + a_1 \cos \omega_s t + a_2 \cos 2\omega_s t + \cdots$$

where $\omega_s = 2\pi/T_s$. If the continuous signal $h(t)$ is assumed to be a single tone $\cos \omega_m t$ then

$$\begin{aligned} F(t) = {} & a_0 \cos \omega_m t + (a_1/2) \cos (\omega_s - \omega_m)t + (a_1/2) \cos (\omega_s + \omega_m)t \\ & + (a_2/2) \cos (2\omega_s - \omega_m)t + (a_2/2) \cos (2\omega_s + \omega_m)t \\ & + \cdots \end{aligned} \tag{3.5}$$

The spectrum of $F(t)$ thus contains the original spectrum of $h(t)$ and upper and lower sidebands centred at harmonics of the sampling frequency. The amplitude spectrum for $F(t)$ when $h(t)$ is a multitone signal band limited to W Hz is shown in Fig. 3.3. It can be seen from this figure that if the sampling frequency $f_s = 2W$, the sidebands just fail to overlap. If $f_s < 2W$, overlap (aliasing) occurs and distortion of the spectrum of $h(t)$ results.

3.2 Reconstruction of the Continuous Signal

In order to reproduce a continuous signal from the samples some form of interpolation is required. The output from a sample-and-hold circuit is shown in Fig. 3.4. This circuit holds the level of the last sample until a new sample arrives and then assumes the value of the new sample. It can be seen from the figure that there is a considerable error between the output of the sample-and-hold circuit and the original value of $h(t)$. The exact interpolation function is obtained once again by considering Eqn (3.3).

If the Fourier series representation of $H(f)$ [Eqn (3.2)] is substituted in

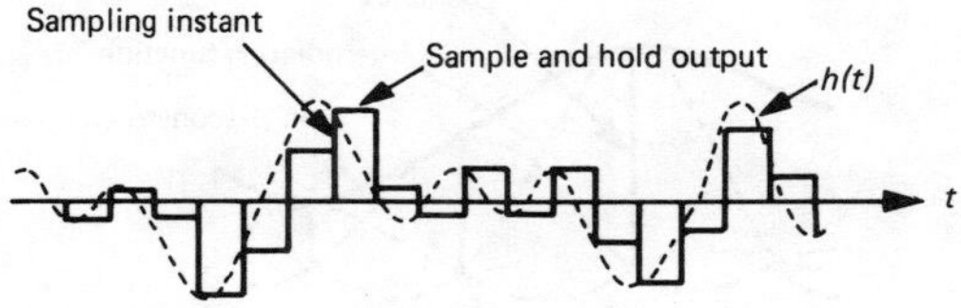

Fig. 3.4 Sample and hold interpolation.

Eqn (3.3), then

$$h(t) = \int_{-W}^{W} \left[1/2W \sum_{n=-\infty}^{\infty} X_n \exp(j\pi n f/W) \right] \exp(j2\pi f t)\,\mathrm{d}f$$

Changing the order of summation and integration gives

$$h(t) = \sum_{n=-\infty}^{\infty} X_n/2W \int_{-W}^{W} \exp[j2\pi f(t + n/2W)]\,\mathrm{d}f$$

i.e.

$$h(t) = \sum_{n=-\infty}^{\infty} X_n \operatorname{sinc}[2\pi W(t + n/2W)]$$

but

$$X_n = h(-n/2W)$$

Therefore

$$h(t) = \sum_{n=-\infty}^{\infty} h(-n/2W) \operatorname{sinc}[2\pi W(t + n/2W)]$$

or alternatively

$$h(t) = \sum_{n=-\infty}^{\infty} h(n/2W) \operatorname{sinc}[2\pi W(t - n/2W)] \tag{3.6}$$

The values $h(n/2W)$ are the samples of $h(t)$ and $\operatorname{sinc}[2\pi W(t - n/2W)]$ is the required interpolation function. This function is centred (has unity value) at intervals of time spaced at $1/2W$. The function has zeros at instants of time equal to $(n + p)/2W$ where p has integer values, except zero, between $\pm\infty$. The value of $h(t)$ at a sampling instant, therefore, is equal to the amplitude of the sample

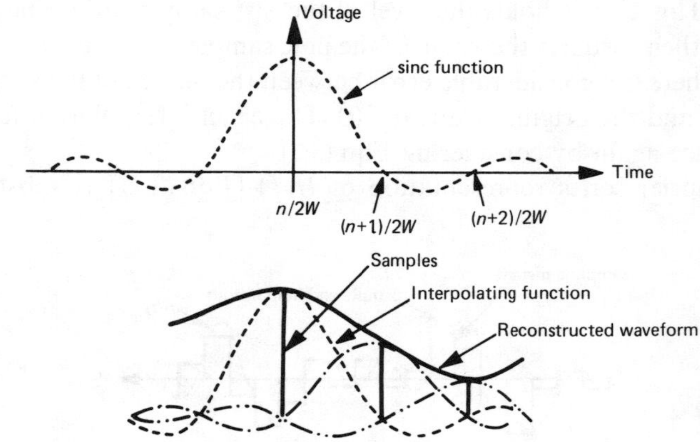

Fig. 3.5 Sinc function interpolation.

only, since the weights of the other samples are zero. The value of $h(t)$ between the sampling instants is given by the summation of the corresponding sinc functions and this process is shown in Fig. 3.5.

3.3 Low-pass Filtering of a Sampled Signal

The previous section has shown that the required interpolation function for perfect signal reconstruction is a sinc function. This function is the impulse response of an ideal low-pass filter. If $P(f)$ is the transfer function of a network and $G(f)$ is the spectrum of an input signal, the spectrum of the signal at the network output $L(f)$ is given by

$$L(f) = P(f)G(f)$$

If the input to the network is a very narrow pulse (which approximates to a unit impulse), $G(f) \simeq 1$ and the spectrum of the output is then $L(f) = P(f)$.

The response in the time domain is obtained by taking the Fourier transform of $P(f)$. If the network is an ideal low-pass filter with cut-off frequency f_c the transfer function is given by

$$P(f) = |P(f)| \exp(-j2\pi n f t_0)$$

where

$$|P(f)| = 1 \qquad \text{for } |f| \leqslant f_c, \qquad = 0 \text{ else}$$

The factor $\exp(-j2\pi f t_0)$ is the linear phase characteristic of the ideal filter. The impulse response $p(t)$ is

$$p(t) = \int_{-f_c}^{f_c} \exp[-j2\pi f(t_0 - t)]\,df$$

i.e.

$$p(t) = 2f_c \frac{\sin[2\pi f_c(t - t_0)]}{2\pi f_c(t - t_0)} \tag{3.7}$$

This function has the required form when $f_c = W$.

In the frequency domain the effect of the ideal low-pass filter is to remove completely all spectral components above W Hz.

In practice the ideal low-pass filter is not physically realizable. This may be seen from the fact that the impulse response exists for $t < 0$ which means that an output exists before the impulse is applied, which is physically impossible. The ideal filter may be approximated by a physically realizable network, the approximation becoming more exact as the complexity of the network is increased. One such approximation with a sharp cut-off characteristic is the Chebyshev approximation. The impulse response of a fifth-order Chebyshev

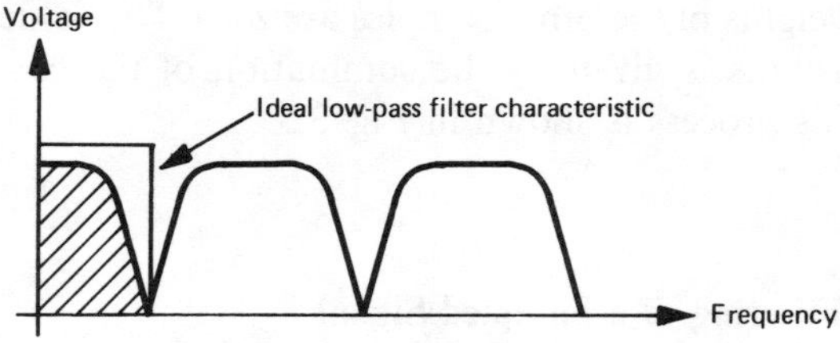

Fig. 3.6 Ideal filtering of a sampled signal.

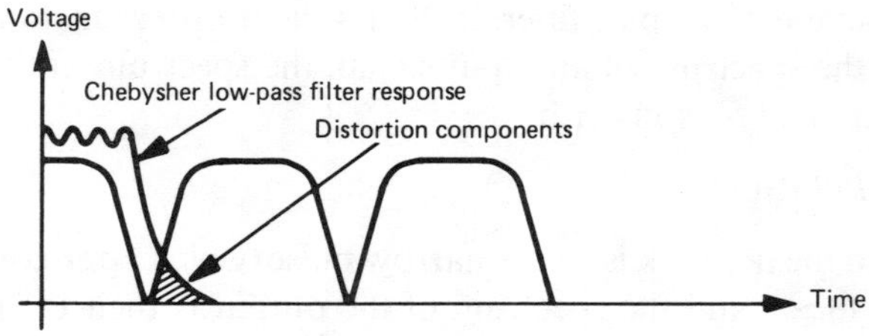

Fig. 3.7 Practical filtering of a sampled signal.

low-pass filter is

$$p(t) = Ae^{-at} + Be^{-bt}(\cos\theta t + \sin\theta t) + Ce^{-ct}(\cos\phi t + \sin\phi t) \tag{3.8}$$

This impulse response will clearly produce some distortion, which is most easily evaluated in the frequency domain as shown in Fig. 3.7.

Since the filter has a finite transmission above W Hz some distortion components outside the signal bandwidth will appear in the filter output. To minimize this distortion the sampling frequency is increased above $2W$ Hz. A practical telephone channel has an upper cut-off frequency of 3.4 kHz and the sampling rate employed is 8 kHz, that is 2.35 times the maximum signal frequency.

3.4 Time Division Multiplex (TDM) Transmission

When a signal is sampled by narrow pulses there are large intervals between the samples in which no signal exists. It is possible during these intervals to transmit the samples of other signals. This process is shown in Fig. 3.8 and is called "time division multiplex" (TDM) transmission. Since each sampled signal gives rise to a continuous signal after filtering, TDM transmission allows simultaneous transmission of several signals over a single wideband link. It is therefore an alternative to FDM transmission, described in Section 2.19.

The switches at transmitter and receiver (which would be solid state devices) are synchronized and perform the sampling and interlacing. The samples

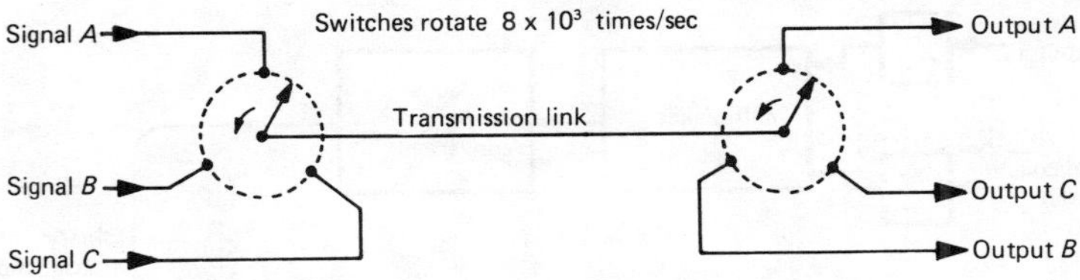

Fig. 3.8 TDM transmission.

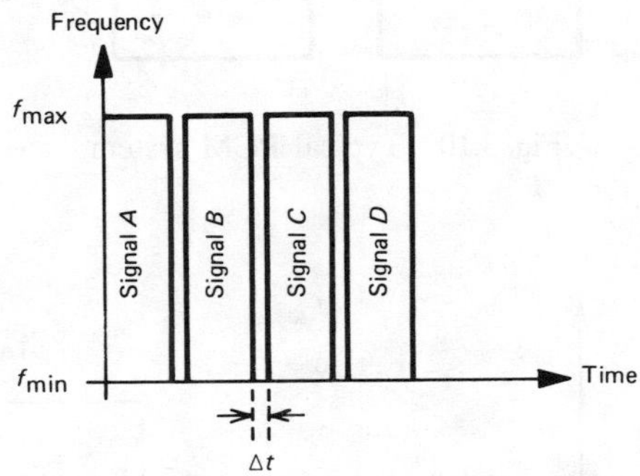

Fig. 3.9 Spectrum/time diagram for TDM.

themselves are very narrow and consequently have a large bandwidth. When transmitted over a link with a fixed bandwidth the samples spread and can overlap adjacent samples. To minimize this, guard bands Δt are allowed between adjacent signals. In the case of TDM transmission each signal therefore occupies the whole of the transmission bandwidth for part of the transmission time (Fig. 3.9).

3.5 Pulse Code Modulation (PCM)

Although it is possible to transmit the samples directly in a TDM system there is a considerable advantage in coding each sample into a binary word before transmission. This is because binary signals have a much greater immunity to noise than analogue signals. A widely used form of digital transmission is pulse code modulation, the essentials of which are shown in Fig. 3.10.

In this system, 30 speech channels are each limited in frequency to 3.4 kHz and sampled at 8 kHz. The sampled signals are converted into binary form for transmission. In addition to the coded signals other signals are sent over the link for synchronization and identification. Once each sample has been converted into a binary code this effectively means that it has been quantized into one of a fixed number of levels. The greater the number of quantization levels the greater is the accuracy of the quantized representation, but also the greater is the

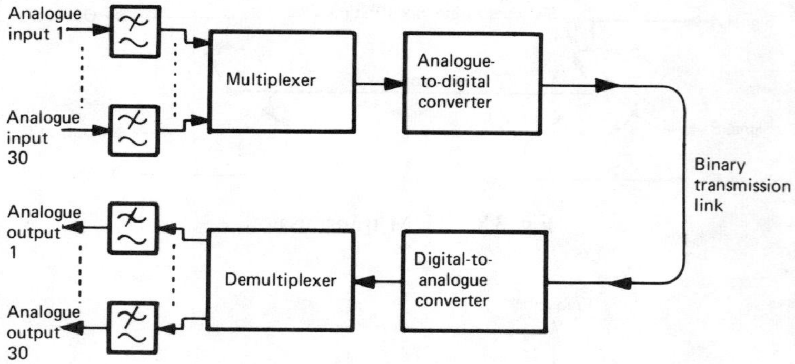

Fig. 3.10 Typical PCM system.

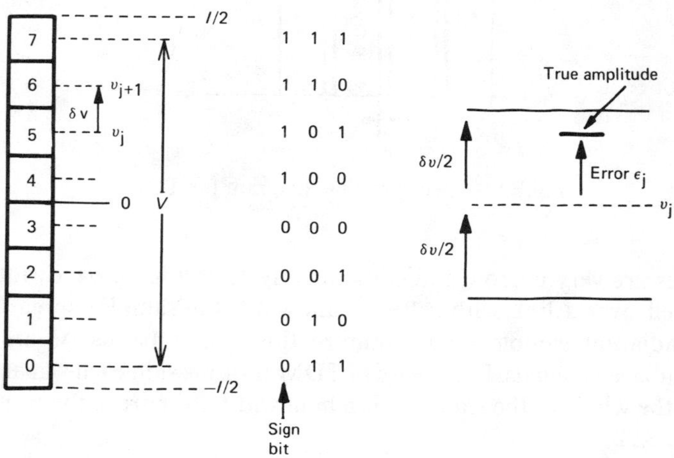

Fig. 3.11 8 level (3 bit) uniform quantization.

number of binary digits (bits) required to represent the sample. Since more bits require a higher transmission bandwidth a balance must be struck between accuracy and bandwidth.

The quantization process is shown in Fig. 3.11 and it is clear that once quantized the precise amplitude of the original sample can never be restored. This gives rise to an error in the recovered analogue signal, known as quantization noise. The quantization noise for an eight-level PCM system can be calculated by reference to Fig. 3.11. This shows that the peak-to-peak input is I V. If this value is quantized into M equally spaced levels the spacing between each level is $\delta v = I/M$ V. The amplitudes reproduced after decoding are usually the mid-point of each quantizing interval, which gives a maximum decoded peak-to-peak value of $V = (M-1)\delta v$ V. If a level of v_j is reproduced by the decoder the true amplitude will be anywhere in the range $v_j \pm \delta v/2$; hence a

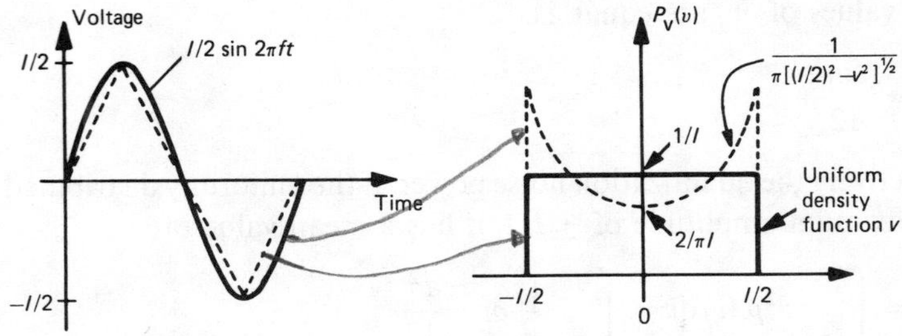

Fig. 3.12 Sinusoidal and triangular waveforms and their amplitude density functions.

maximum error of $\pm\delta v/2$ will be present on any decoded output. This error is random in nature and is called quantization noise. In order to calculate the magnitude of this quantization noise we need to know the amplitude probability density function of the signal to be coded. From this knowledge we can calculate the probability that the signal will be in any quantizing interval and the mean square error for each of the quantizing steps.

The simplest case to evaluate is a signal with a uniformly distributed amplitude density function. Such a signal is a triangular wave of amplitude $\pm I/2$. The density function for a triangular wave and a sine wave are shown in Fig. 3.12. A signal with a uniform probability density has equal probability of being in any of the quantizing intervals and also has equal probability of having any particular amplitude within a given quantizing interval. The error involved is shown in Fig. 3.11 and has a mean square value given by

$$\bar{\varepsilon}_j^2 = \frac{1}{\delta v_j}\int_{-\delta v_j/2}^{\delta v_j/2} \varepsilon_j^2 \, d\varepsilon = (\delta v_j)^2/12 \tag{3.9}$$

The total mean square error throughout the range of the coder is

$$\bar{\sigma}^2 = \sum_j \bar{\varepsilon}_j^2 P(v_j)$$

where $P(v_j)$ is the probability that the signal will be in the jth quantization interval, i.e.

$$\bar{\sigma}^2 = \frac{1}{12}\sum_j \delta v_j^2 p(v_j) \tag{3.10}$$

But $\sum \delta v_j^2 P(v_j)$ is the mean square value of δv_j, i.e.

$$\bar{\sigma}^2 = \frac{\overline{(\delta v_j)^2}}{12} \tag{3.11}$$

If the signal has a uniform amplitude distribution and the quantization is linear, then all values of δv_j are equal, i.e.

$$\bar{\sigma}^2 = \frac{(\delta v)^2}{12} \tag{3.12}$$

$\bar{\sigma}^2$ is effectively the quantization noise power. If the uniformly distributed signal has a maximum amplitude of $\pm I/2$, it has a mean value of

$$\bar{S}^2 = \int_{-I/2}^{I/2} v^2 p_v(v)\,\mathrm{d}v = \int_{-I/2}^{1/2} \frac{v^2}{I}\delta v = \frac{I^2}{12} \tag{3.13}$$

The mean SNR is thus

$$\text{SNR} = \frac{\bar{S}^2}{\bar{\sigma}^2} = \frac{I^2}{(\delta v)^2} = M^2$$

In a binary system the number of bits (m) required to code M levels must satisfy the relationship $M = 2^m$; hence

$$\text{SNR} = 10\log_{10}[2^{2m}] \quad \text{dB}$$

i.e.

$$\text{SNR} = 20m\log_{10}2 \quad \text{dB}$$

or

$$\text{mean SNR} = 6m \quad \text{dB} \tag{3.14}$$

The maximum signal power is $(I/2)^2$, which gives a maximum SNR of $3\,M^2$, i.e.

$$\text{maximum SNR} = (4.8 + 6m) \quad \text{dB} \tag{3.15}$$

This figure is applicable only to signals with a uniform amplitude distribution that have been uniformly quantized. The corresponding figure for a sine wave can be evaluated by noting that the probability density function of $v = (I/2)\sin(2\pi f t)$ is

$$p_v(v) = 1/(\pi[(I/2)^2 - v^2]^{1/2})$$

The individual values of $\bar{\varepsilon}_j^2$ can be obtained from Eqn (3.9) using the appropriate density function for each interval. Finally, the mean square quantization noise is obtained from Eqn (3.10). The mean SNR for a sine wave of amplitude $\pm I/2$ is approximately

$$\text{mean SNR} = (1.8 + 6m) \quad \text{dB} \tag{3.16}$$

Equation (3.16) shows that the SNR of the decoded signal, in dB, increases linearly with m which is itself linearly related to transmission bandwidth. This means that the SNR, measured as a ratio, increases exponentially with bandwidth. This is close to the theoretical relationship between SNR and bandwidth given by Shannon's law, but in this particular case the noise is quantization noise rather than fluctuation noise.

In deriving Eqn (3.16) it was assumed that the input waveform fully occupied the range of the coder, which would correspond to the loudest talker in a telephone system. It follows that the SNR for the quietest talker would be considerably lower than this value. PCM encoders are designed to make the SNR constant over as wide a range of input amplitudes as possible and this is achieved by making the quantization step size a function of the input waveform amplitude. This is known as non-linear (or non-uniform) quantization. Essentially the step size of a non-linear coder is reduced near zero and increases towards the maximum input level. Non-linear coding is often achieved in practice by first COMpressing the signal then coding the compressed signal in a linear coder and finally exPANDING the decoded signal with the inverse of the compression characteristic. The combined process is known as COMPANDING. The process is illustrated in Fig. 3.13, the compression law used in Europe is known as the '*A*' Law.

The relationship between output and input for the *A* law compression characteristic is

$$V_{out} = \frac{1 + \ln(AV_{in})}{1 + \ln(A)} \qquad \text{for } 1/A < |V_{in}| < 1$$

$$V_{out} = \frac{AV_{in}}{1 + \ln(A)} \qquad \text{for } 0 < |V_{in}| < 1/A \tag{3.17}$$

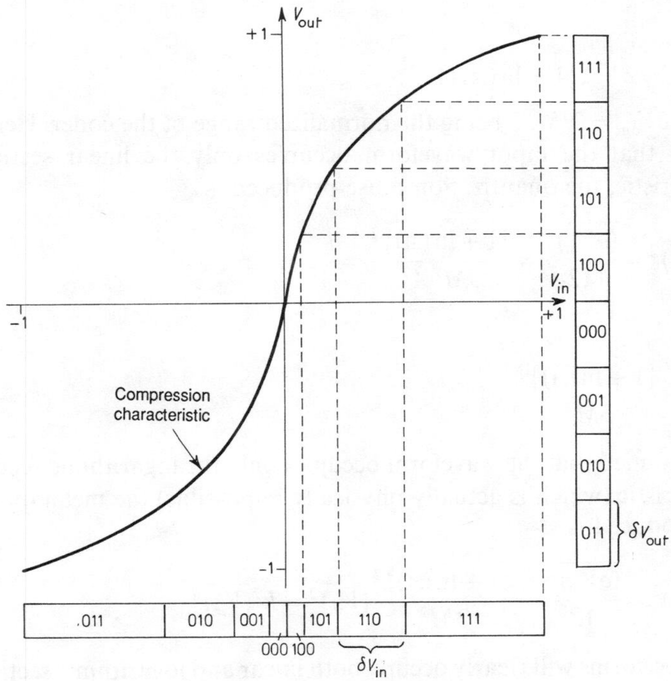

Fig. 3.13 Non-linear quantization.

The A law characteristic has a normalized range of ± 1 and is divided into a linear and a non-linear section. The linear section ensures that $V_{out} = 0$ when $V_{in} = 0$; the logarithmic section gives $V_{out} = 1$ when $V_{in} = 1$. The characteristic is continuous at $V_{in} = 1/A$. A is known as the compression characteristic and has a value related to the required dynamic range of the system. The larger the value of A the greater the compression achieved.

The quantization noise produced with this characteristic is evaluated by considering the linear quantization of the compressed signal in terms of the equivalent non-linear quantization of the un-compressed input waveform. If it is assumed that the number of quantization levels M is large, then for the linear section of the characteristic

$$\delta V_{out} = \frac{A\,\delta V_{in}}{1 + \ln(A)}$$

i.e.

$$\delta V_{in} = \frac{\delta V_{out}(1 + \ln(A))}{A}$$

For the logarithmic part

$$\delta V_{out} = \frac{\delta V_{in}}{\{1 + \ln(A) V_{in}\}} \qquad \left(\text{since } \frac{\mathrm{d}}{\mathrm{d}x}\{\ln(ax+b)\} = \frac{a}{ax+b}\right)$$

i.e.

$$\delta V_{in} = \delta V_{out}\{1 + \ln(A)\}\, V_{in}$$

Further $\delta V_{out} = 2/M$, 2 being the normalized range of the coder. Hence, if it is assumed that the input waveform occupies only the linear section of the characteristic, the quantization noise produced is

$$\overline{(\sigma_{lin})^2} = \frac{\overline{(\delta V_{in})^2}}{12} = \frac{\{1 + \ln(A)\}^2}{3M^2A^2} = \frac{k^2}{A^2} \tag{3.18}$$

where

$$k^2 = \frac{\{1 + \ln(A)\}^2}{3M^2}$$

If it is assumed that the waveform occupies only the logarithmic section of the characteristic (which is actually physically impossible) the mean quantization noise produced is

$$\overline{(\sigma_{log})^2} = \frac{\overline{(\delta V_{in})^2}}{12} = \frac{\{1 + \ln(A)\}^2}{3M^2}\overline{(V_{in})^2} = k^2\overline{(V_{in})^2} \tag{3.19}$$

Input waveforms will clearly occupy both linear and logarithmic sections of the characteristic. In order to determine the mean quantization noise it is necessary to know the probability that the waveform will be in the linear section and the mean square value of the input voltage when the waveform is in the logarithmic section. This requires a knowledge of the amplitude distribution function of the input waveform. The particular interest here is the amplitude distribution of speech waveforms. The amplitude probability density function of a typical speech waveform may be approximated by

$$p_v(v) = \frac{1}{\sqrt{2}\sigma_s} \exp\left[\frac{-\sqrt{2}|v|}{\sigma_s}\right] \tag{3.20}$$

σ_s is the rms speech voltage (normalized)

The probability that the speech voltage will be within the linear section of the compression characteristic is then

$$P_A = \int_{-1/A}^{1/A} p_v(v)\,dv$$

$$= 2\int_0^{1/A} \frac{1}{\sqrt{2}\sigma_s} \exp(-\sqrt{2}|v|/\sigma_s)\,dv$$

$$= 1 - \exp(-\sqrt{2}/A\sigma_s)$$

The mean square voltage when the waveform is in the logarithmic section is

$$\overline{(v_{\log})^2} = \int_{1/A}^{1} v^2 p_v(v)\,dv$$

$$= 2\int_{1/A}^{1} \frac{v^2}{\sqrt{2}\sigma_s} \exp(-\sqrt{2}|v|/\sigma_s)\,dv$$

Integrating by parts and assuming that $A \gg 1$ gives

$$\overline{(v_{\log})^2} = \left(\frac{1}{A^2} + \frac{2\sigma_s}{A} + (\sigma_s)^2\right) \exp(-\sqrt{2}/A\sigma_s)$$

The total quantization noise is then

$$(\sigma_s)^2 = \overline{(\sigma_{\text{lin}})^2} + \overline{(\sigma_{\log})^2} = k^2\left(\frac{1}{A^2} P_A + \overline{(v_{\log})^2}\right)$$

The signal to quatization noise power is then

$$\text{SNR} = \frac{(\sigma_s)^2}{(\sigma_n)^2}$$

i.e.

$$\text{SNR} = \frac{(\sigma_s)^2}{k^2\left[\frac{1}{A^2}\{1+\alpha\} + \left\{\frac{1}{A^2} + \frac{\sqrt{2}\sigma_s}{A} + (\sigma_s)^2\right\}\alpha\right]} \tag{3.21}$$

where

$$\alpha = \exp(-\sqrt{2}/A\sigma_s)$$

When $\sigma_s = 1/A$ then $\sigma_s = \overline{(v_{\log})^2}$
If $\sigma_s = 1/A$ in Eqn (3.21) then
Under these circumstances

$$(\sigma_n)^2 = k^2\overline{(v_{\log})^2}$$

Hence

$$\text{SNR} = \frac{1}{k^2} = \frac{3M^2}{\{1 + \ln(A)\}} \tag{3.22}$$

If $\sigma_s = 1/A$ in Eqn (3.21) then

$$\text{SNR} = \frac{1/A^2}{\dfrac{k^2}{A^2}\{(0.76) + (1 + \sqrt{2} + 1)(0.24)\}} = \frac{1}{1.58k^2}$$

But

$$10\log_{10}(1/1.58) = -2\,\text{dB}$$

i.e. the SNR is constant for large values of σ_s, and drops by 2 dB at $\sigma_s = 1/A$. This characteristic is illustrated in Fig. 3.14.

The system is thus designed so that the rms voltage produced by the quietest talker is equal to $1/A$, which will produce an effectively constant SNR for all users. The SNR is a function of both A and M which are in turn chosen to give an acceptable performance for a specified dynamic range. The question which then arises is: what is a realistic dynamic range? The answer to this question is provided by measurements on a sample of the population. Such measurements reveal that 98% of the population will have a rms voice amplitude within ± 13 dB of the median talker, and 99.8% of the population will have a rms amplitude

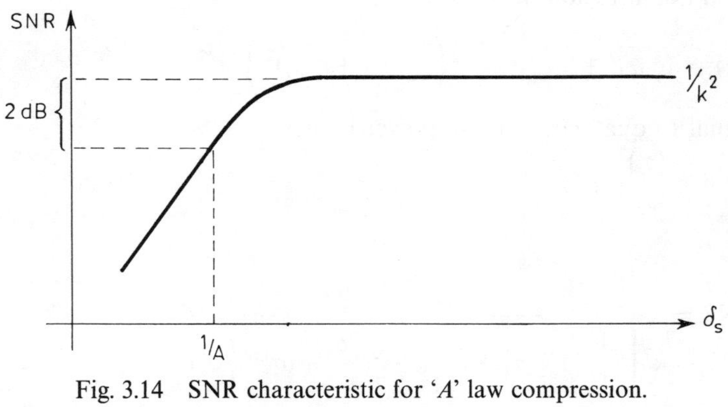

Fig. 3.14 SNR characteristic for 'A' law compression.

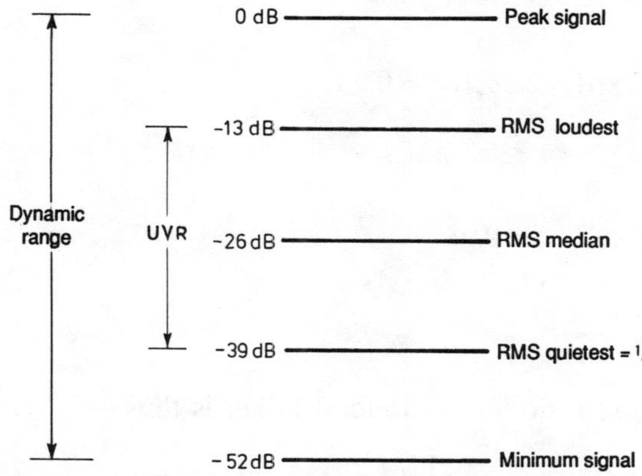

Fig. 3.15 Dynamic range of a non-linear PCM system.

within ± 17 dB of the median talker. The ratio of the rms output of the loudest talker to the quietest talker is known as the useful volume range (UVR) and will thus be between 26 dB and 34 dB, depending on which statistic is used.

The required dynamic range of the coder will actually exceed this value by approximately 26 dB. The reason for this is that the loudness of all talkers varies about their individual rms values. Measurements have shown that individual talkers have an output amplitude within ± 13 dB of their rms values for 99% of the time. The total dynamic range is thus UVR + 26 dB as shown in Fig. 3.15.

In Fig. 3.15 the peak signal has been normalized to 0 dB. Hence it may be seen that the required value of the compression coefficient is

$$A = \text{UVR} + 13\,\text{dB}$$

Thus if UVR = 30 dB, $20\log_{10}(1/A) = -43$

i.e.

$$A = 141$$

The CCITT recommended value for A is 87.6 which gives a UVR of 26 dB. With this value of A and an 8 bit code ($M = 255$) the signal-to-noise ratio is 38 dB. The UVR may be increased at the expense of SNR, e.g. for a UVR of 30 dB ($A = 141$) and an 8 bit code SNR = 36 dB. It is interesting, at this point, to consider the number of quantization levels in a linear coder required to produce the same UVR and signal-to-noise ratio as the CCITT recommendation. To illustrate this point is will be assumed that the probability of the speech waveform being within the limits of the linear coder ($\pm I/2$) is 0.98. From this assumption

$$\int_{-I/2}^{I/2} p_v(v)\,dv = 0.98$$

Hence

$$2\int_0^{I/2} \frac{1}{\sqrt{2}\sigma_s} \exp(\sqrt{2}v/\sigma_s)\,dv = 0.98$$

or

$$1 - \exp(-I/\sqrt{2}\sigma_s) = 0.98$$

Hence

$$\sigma_s = 0.18I$$

The signal-to-noise ratio for the loudest talker is thus

$$\text{SNR} = (\sigma_s)^2 \frac{12}{\delta v^2} = (0.18I)^2 \frac{12M^2}{I^2}$$

i.e.

$$\text{SNR} = 0.388M^2$$

If the SNR for the quietest talker is to be 38 dB it follows that the SNR for the loudest talker will be $38 + 26 = 64$ dB i.e.

$$10\log_{10}(0.388) + 20\log_{10}(M) = 64$$

letting $M = 2^m$ gives $m = 11.3$

Thus 12 bits are required for a linear coder with the same performance. Companding thus produces an effective saving in bandwidth of the order of 33%.

In the USA companding is carried out using the 'μ' Law characteristic which is given by

$$V_{out} = \ln(1 + \mu V_{in}) \qquad \text{for } 0 < |V_{in}| < 1 \tag{3.23}$$

A typical figure for μ is 255. It should be noted that when V_{in} is low $\ln(1 + \mu V_{in}) \approx \mu V_{in}$, hence this characteristic is also approximately linear for low input voltages.

3.6 PCM Line Signals

Although it is possible to encode signals at their source this is not in widespread use. TDM transmission with PCM encoding is used extensively between telephone exchanges. This transmission makes use of standard telephone lines that were designed specifically for voice communications. The standards of transmission adopted have been dictated largely by the characteristics of the cables "already in the ground". To minimize signal distortion over the audio frequency band these cables are artificially loaded by inductance at intervals of approximately 2 km. This produces very high attenuation of frequencies above about 4 kHz, which makes the line totally unsuitable for digital signals. In addition to the loss of response at high frequency the 2 km lengths of line are transformer coupled and therefore have no dc path. The high frequency response of the section can be considerably improved by simply removing the loading coil. A typical frequency response for a section of loaded and unloaded cable is shown in Fig. 3.16.

In practical PCM systems the loading coils are replaced by regenerative repeaters which effectively isolate each 2 km section of line. The regenerative repeater, in fact, produces a binary signal free from noise at the start of each section. This is shown in Fig. 3.17.

If the repeater input is above the threshold at a timing instant a binary 1 is transmitted to the next section. If the input is below the threshold a binary 0 is

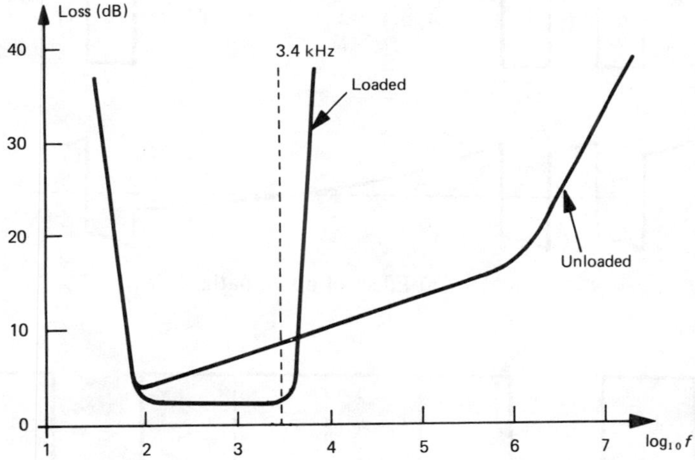

Fig. 3.16 Loss of a loaded and unloaded cable.

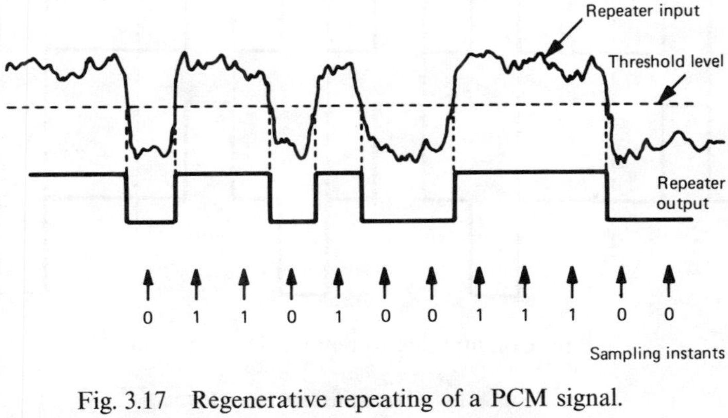

Fig. 3.17 Regenerative repeating of a PCM signal.

transmitted. If the SNR at each repeater input is adequate few decision errors occur and the binary signal is repeated free from noise. This is a distinct advantage of digital transmission over analogue transmission. In the latter case, amplifiers are required at intervals to compensate for signal attenuation. These amplifiers boost the noise as well as the signal. The absence of a dc path presents a serious problem for PCM signals. Its effect on a long sequence of binary 1s is to cause a gradual droop of signal level below the decision threshold, and this will of course produce decision errors. This is illustrated in Fig. 3.18a.

The solution is to remove the dc component in the PCM signal. The first stage in the process is to convert the full-width non-return-to-zero (NRZ) pulses into half-width return-to-zero pulses (RZ). This effectively doubles the signal

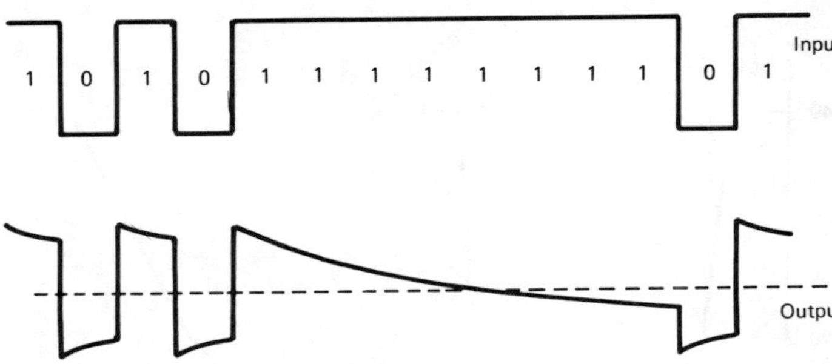

(a) Effect of no dc path.

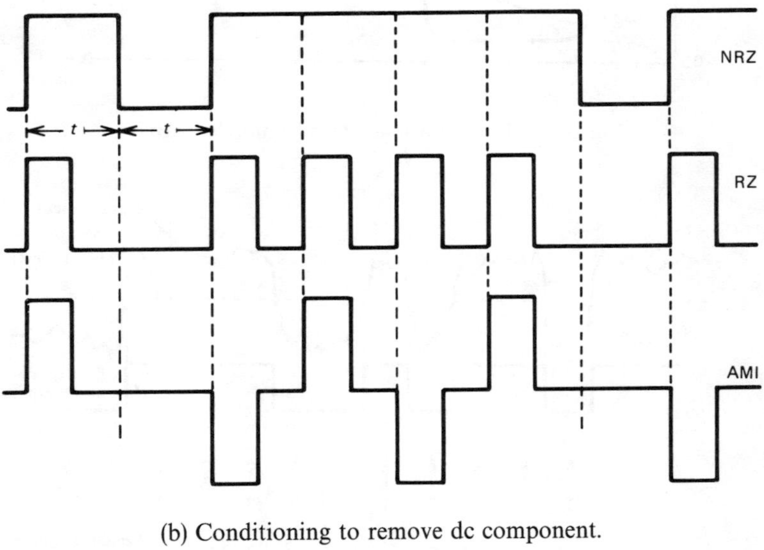

(b) Conditioning to remove dc component.

Fig. 3.18 Transmission waveforms.

bandwidth, but is necessary for synchronization purposes (Section 3.8). The dc component in the RZ waveform is removed by inverting alternate binary 1s, the process being called "alternate mark inversion" (AMI). The original 2-level PCM signal is actually transmitted as a 3-level signal with zero dc component. The bandwidth of a PCM signal is calculated on the basis of AMI. The processing to produce AMI is illustrated in Fig. 3.18b.

3.7 Bandwidth Requirements for PCM Transmission

The PCM signal consists of a random sequence of binary 1 s and 0 s. The transmitted waveform will have maximum bandwidth when the number of transitions per unit time is also a maximum. This corresponds, in the case of NRZ pulses, to an alternating sequence of binary 1 s and 0 s. The maximum bandwidth in the case of RZ pulses and AMI occurs for a sequence of binary 1 s. The transmitted waveforms and their respective amplitude spectra are given in Fig. 3.19.

The CCITT[2] recommendation specifies PCM transmission in terms of time division multiplexing of 32 channels. Each channel is sampled 8×10^3 times per second with each sample represented as an 8 bit code. This gives a total frame bit rate of 2.048 Mb/s. It may be seen from Fig. 3.19 that the minimum bandwidth requirement for AMI is approximately half the bit rate. This is an 'idealized' figure based on the premise that the three levels of the AMI signal could be extracted from a sine wave with a period of twice the bit interval. This bandwidth is insufficient in practice owing to distortion of the pulse waveform, which produces inter-symbol interference. To illustrate this point assume that the line (or channel) has an ideal low pass response with cut off frequency f_c.

The impulse response of such a channel is a sinc function with zeros at intervals $t = n/2f_c$. Hence in theory pulses transmitted at a rate of $2f_c$ per second could be received free of interference from adjacent pulses. It was stated in Chapter 1 that an ideal response cannot be realized in practice; however, the raised cosine response is an approximation to the ideal filter which can be synthesized in practice. The raised cosine response is shown in Fig. 3.20 and it should be observed that the impulse response of such a characteristic also has zeros at intervals of $n/2f_c$.

The raised cosine response is given by

$$H(f) = 0.5[1 + \cos(\pi f/2f_c)] \qquad \text{for } |f| < 2f_c \tag{3.24}$$

The impulse response of the network is given by the Fourier transform of equation (3.24) and is

$$h(t) = 2f_c \operatorname{sinc}(2\pi f_c t)\cdot\cos(2\pi f_c t)/[1 - (4f_c t)^2] \tag{3.25}$$

Thus in order to transmit pulses at a rate of $2f_c$ per second a raised cosine response with cut off frequency $2f_c$ is required. (The raised cosine response is a

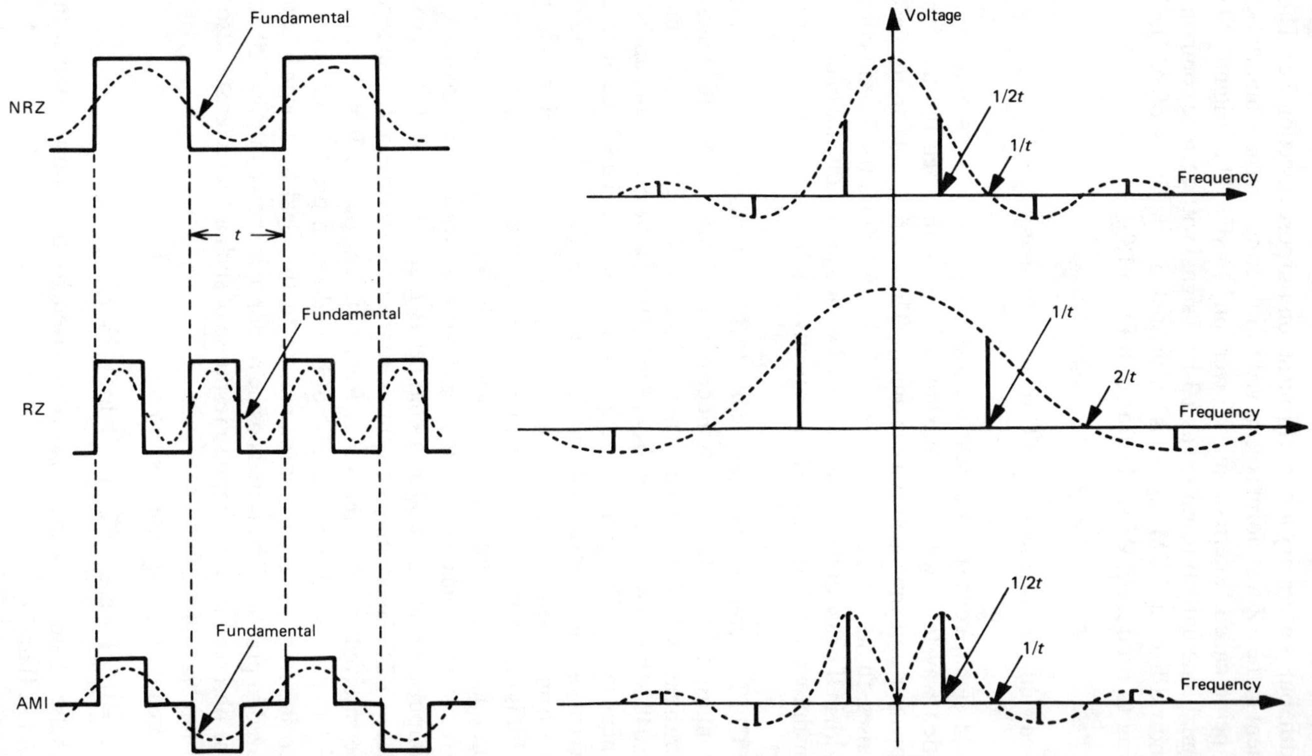

Fig. 3.19 PCM transmission waveforms and their spectra.

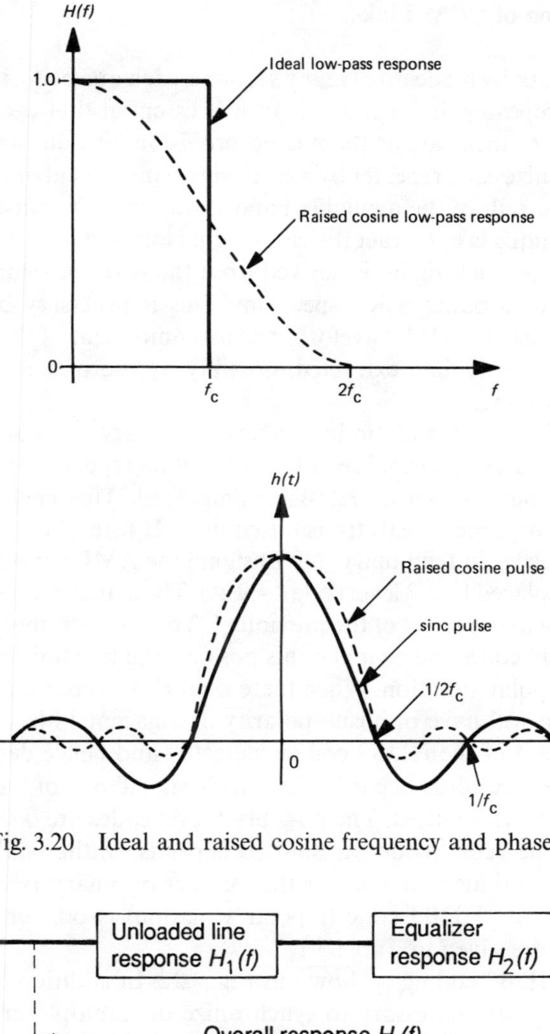

Fig. 3.20 Ideal and raised cosine frequency and phase response.

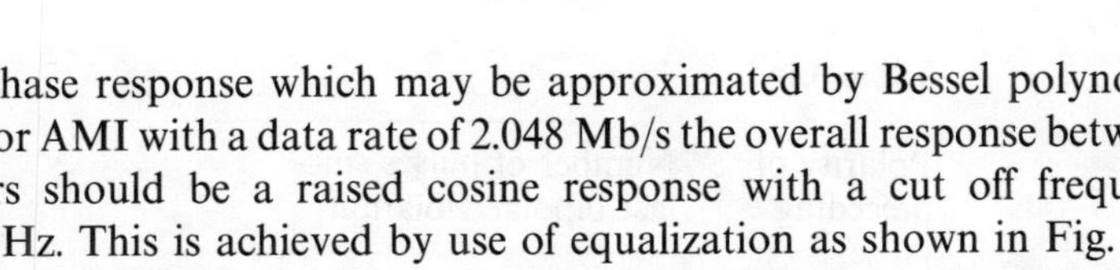

Fig. 3.21 Equalization of line response.

linear phase response which may be approximated by Bessel polynomials[3].) Hence for AMI with a data rate of 2.048 Mb/s the overall response between two repeaters should be a raised cosine response with a cut off frequency of 2.048 MHz. This is achieved by use of equalization as shown in Fig. 3.21.

The overall response $H(f)=H_1(f)H_2(f)$ is designed to be a linear phase approximation of the ideal low pass filter, i.e. to have a raised cosine amplitude response. The equalizer response $H_2(f)$, which may be realized with digital filters, is thus designed to produce this response.

3.8 Synchronization of PCM Links

A PCM link is essentially made up of many sections, each one being terminated by a regenerative repeater. In such a system it is essential that each repeater operates at the same clock rate as there is no provision for data storage. It is possible to synchronize each repeater by a clock signal inserted at one end of the link, but this would reduce the available bandwidth for data transmission. A more efficient technique is to extract the clock signal for each repeater from the data signal itself. The clock signal is derived from the AMI waveform by full-wave rectification to produce a RZ spectrum. This is necessary because, as illustrated in Fig. 3.19, the AMI waveform has no component at the data rate. The component at t^{-1} is then extracted, possibly by use of a phase locked loop, for repeater timing.

A practical problem arises with timing if a series of binary 0s occurs, as would be the case during pauses in normal speech. To maintain repeater timing during such situations a code known as HDB3 is employed. This code limits the maximum number of successive 0s transmitted in AMI format to three. When four successive 0s occur in the binary (NRZ) signal the AMI waveform, which would be zero, is replaced by a 3 level code (– 0 +). The actual code substituted depends upon the AMI polarity of the previous 1. The receiver must be able to recognize the HDB3 code, and to make this possible the transmitter produces what is termed a bipolar violation. When there is no HDB3 code adjacent 1s in the AMI waveform will have opposite polarity. If adjacent 1s have the same polarity the inclusion of the HDB3 code is detected and hence decoded.

When four successive zeros occur in the binary signal, one of four possible HDB3 line signals is transmitted. The possible 3 level codes are 000 +, 000 –, – 00 –, + 00 +. The actual code transmitted depends on the polarity of the preceding binary 1 and also on whether the number of binary 1s which have occurred since the last HDB3 code (bipolar violation) is odd or even. The substitutions used are given in Table 3.1.

An example of HDB3 coding is shown in Fig. 3.22. In addition to repeater synchronization it is also necessary to synchronize the multiplexers.

In the CCITT specification two of the 32 PCM channels are reserved for signalling and synchronization. The channels are numbered 0 to 31, the 32

Table 3.1

Polarity of preceding pulse	Number of pulses since last bipolar violation	
	ODD	EVEN
+	000+	–00–
–	000–	+00+

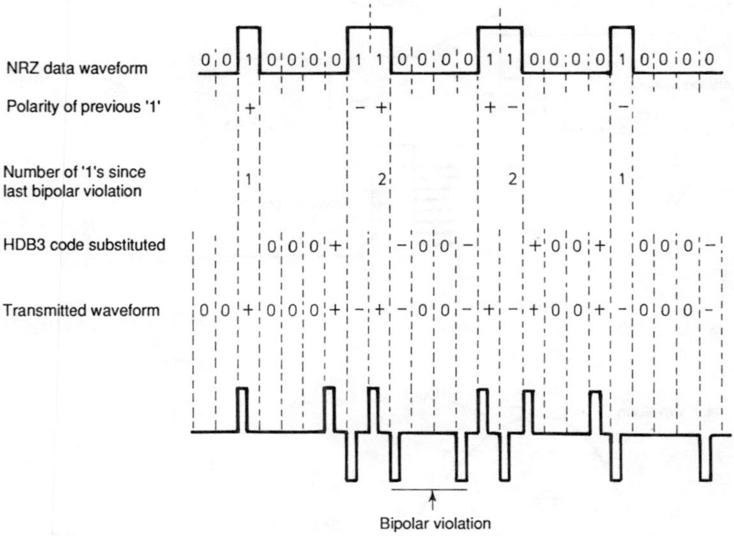

Fig. 3.22 HDB3 coding of line signals.

channels being called a 'frame'. Frame synchronization is achieved by transmitting a fixed code word in channel 0 on alternate frames. Circuits at the receiver search for this code word and its absence in alternate frames and derive from it a synchronizing signal for the demultiplexer. In this way each 8 bit word in the received frame is routed to its correct destination.

3.9 Delta Modulation

This is an alternative binary transmission system using a single digit binary code. Delta modulation does not have the widespread application of standard PCM, it is however used in some rural telephone networks[4] and in digital recording of analogue signals. The fundamental delta modulator, or tracking coder is shown in Fig. 3.23.

The analogue input is compared with the output of a DAC the input of which is derived from an up/down counter. If the amplitude of the analogue input exceeds the output of the DAC the comparator output will be high. This sets the up/down counter to increment on the next clock pulse. If the output of the DAC exceeds the amplitude of the input the comparator output will be low. This sets the up/down counter to decrement on the next clock pulse. The output of the DAC is thus a staircase approximation of the analogue input. The demodulator will consist of the elements in the feedback loop of the modulator. It will be noted from Fig. 3.23 that there are two kinds of distortion produced by this system. Slope overload distortion occurs when the transition from one step to the next

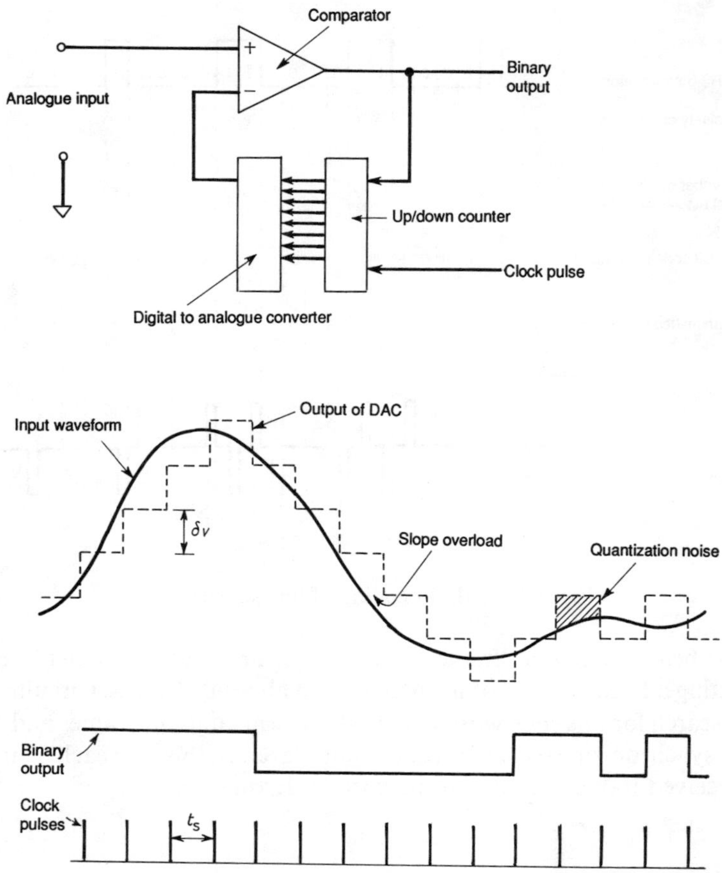

Fig. 3.23 The idealized delta modulator.

fails to cross the input waveform. Quantization distortion occurs due to the finite step size δv.

If t_s is the clocking interval the condition required to prevent slope overload is

$$\dot{h}(t)t_s < \delta v \tag{3.26}$$

For sinusoidal waveforms $h(t) = A\cos 2\pi f_m t$ and thus $\dot{h}(t)_{max} = A2\pi f_m$, the condition to prevent slope overload is then $A2\pi f_m < \delta v f_s$. Alternatively

$$A_{max} = \frac{\delta v f_s}{2\pi f_m} \tag{3.27}$$

where f_s is the clocking frequency. It may be noted that either δv or f_s can be increased to avoid overload, but with some penalty. Increasing f_s increases the

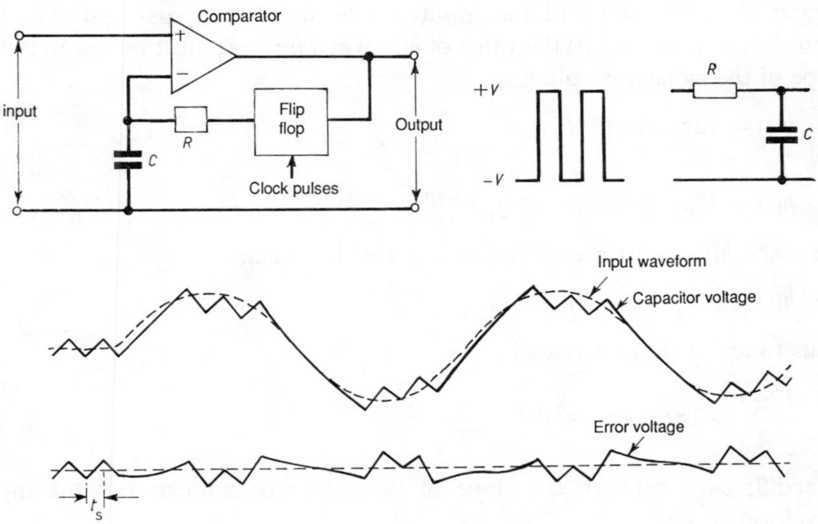

Fig. 3.24 Practical RC delta modulator and waveforms.

transmitted bit rate (and hence bandwidth requirement), increasing δv increases the quantization error.

In practice a much simpler circuit than that of Fig. 3.23 is used, which has the additional advantage that the effect of any digit errors decreases to zero after a given interval. (This would not be the case for the circuit of Fig. 3.23.) The practical circuit is based on a simple RC integrator and is shown in Fig. 3.24. The circuit will function if the output of the flip-flop is $\pm V$ volts (which may be easily achieved with a CMOS device).

The voltage across the capacitor will be a series of positive and negative exponential decays since the capacitor charges from either $+V$ or $-V$. The capacitor voltage for a typical input waveform is illustrated in Fig. 3.24. The error voltage in this figure is the difference between the input voltage and the capacitor voltage. This is approximately triangular when the input signal is zero and this is known as the idling voltage.

Considering a single RC network in which V is the charging voltage and v is the instantaneous voltage across the capacitor, then

$$\frac{V-v}{R} = C\frac{\mathrm{d}v}{\mathrm{d}t} \text{ and integrating gives } \mathrm{e}^{-t/RC} = \frac{V-v}{V-v_i}$$

where v_i is the initial capacitor voltage. Letting $v_1 = -V$ (i.e. logic 0) the capacitor voltage is

$$v = V(1 - 2\,\mathrm{e}^{-t/RC})$$

To avoid overload the slope of the capacitor voltage must be equal to or greater than the slope of the input waveform. If it is assumed that the input $h(t) = A\cos(2\pi f_m t)$ the value of $\dot{h}(t)$, when $h(t) = v$, must be less than the slope of the capacitor voltage.

$$\dot{h}(t) = A2\pi f_m \sin(2\pi f_m t),$$

or

$$h(t) = A2\pi f_m[1 - \cos^2(2\pi f_m t)]^{1/2}$$

but when $h(t) = v$ then $\cos(2\pi f_m t) = v/A$ which means

$$\dot{h}(t) = 2\pi f_m(A^2 - v^2)^{1/2}$$

Thus to avoid slope overload

$$\frac{V - v}{A} > 2\pi f_m(A^2 - v^2)^{1/2} \tag{3.28}$$

The difference between the slope of the capacitor voltage and the input waveform is

$$D = \frac{(V - v)}{RC} - 2\pi f_m(A^2 - v^2)^{1/2}$$

If overload is to be avoided then D must be positive and in the limit $D \rightarrow 0$. D changes from a positive to negative value when $\mathrm{d}D/\mathrm{d}v = 0$.
i.e.

$$\frac{\mathrm{d}D}{\mathrm{d}v} = -\frac{1}{RC} + \frac{2\pi f_m}{2}(A^2 - v^2)^{-1/2}2v = 0$$

Thus

$$\frac{1}{RC} = \frac{2\pi f_m v}{(A^2 - v^2)^{1/2}}$$

From which

$$v = \frac{A}{[1 + (f_m/f_0)^2]^{1/2}} \qquad \text{where } f_0 = \frac{1}{2\pi RC}$$

Substituting this into equation 3.28 gives

$$A_{max} = \frac{V}{[1 + (f_m/f_0)^2]^{1/2}} \tag{3.29}$$

The overload characteristic has a frequency response equivalent to that of a single lag characteristic. The optimum value of f_0 is chosen with reference to the amplitude spectrum of normal speech. In practice it is desirable to work as close to overload as possible, and a suitable choice for f_0 is 150 Hz as illustrated in Fig. 3.25. This allows maximum SNR to be achieved at the expense of some overload in the mid-frequency range.

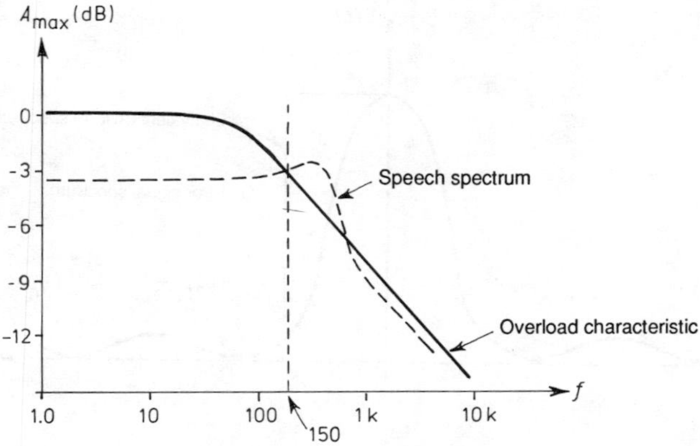

Fig. 3.25 Overload characteristic of RC delta modulator.

Dynamic Range and Quantization Noise

The demodulator in the practical realization is simply the RC network. The quantization noise produced will then be the difference between the input waveform and the voltage across the capacitor of the RC network. When the input signal is zero the error signal is an approximate triangular waveform of period $1/2f_s$ and peak to peak amplitude δv. The slope of this waveform at $v = 0$ is $dv/dt = V/RC$ hence $\delta v = V/RCf_s$. This error waveform is shown in Fig. 3.25 and clearly the coder will not deviate from this idling pattern unless the signal exceeds the value $\delta v/2$. The minimum signal level corresponding to the threshold of coding is thus

$$A_{min} = \frac{V}{2RCf_s} \tag{3.30}$$

The dynamic range of the delta modulator is thus

$$\frac{A_{max}}{A_{min}} = \frac{V}{[1+(f_m/f_0)^2]^{1/2}} \frac{2RCf_s}{V}$$

i.e.

$$\text{dynamic range} = \frac{2RCf_s}{[1+(f_m/f_0)^2]^{1/2}} \tag{3.31}$$

The peak to peak error is

$$\bar{\varepsilon}^2 = \frac{2}{t_s}\int_0^t \left(\delta v \frac{t}{t_s}\right)^2 dt = \frac{(\delta v)^2}{12}$$

This is the mean square noise output when there is zero input. The error when the input is a sine wave is shown in Fig. 3.25 and it has been shown[5] that the

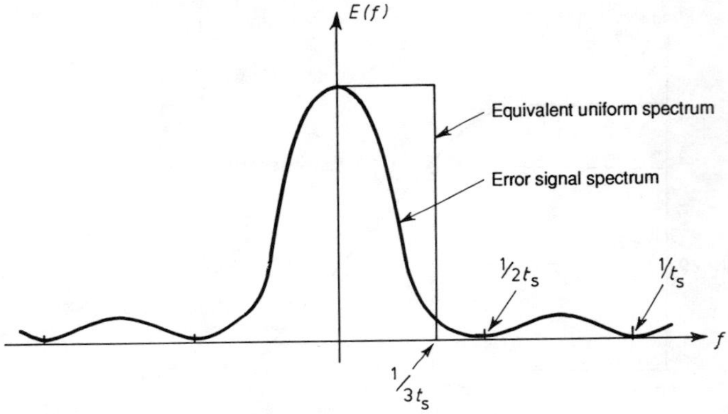

Fig. 3.26 Error signal power spectrum.

mean square error in this case is numerically equal to

$$\bar{\varepsilon}^2 = \frac{(\delta v)^2}{6} \tag{3.32}$$

This is constant from the threshold of coding to overload. The error waveform is actually random but approximately triangular in shape. The power spectrum may thus be approximated by the spectrum of a single triangular pulse (via the autocorrelation function), which is a sinc^2 function with spectral zeros occurring at frequency intervals of $n/2f_s$. The total area under the sinc^2 function is equal to the area under a rectangular figure of the same zero frequency amplitude extending to a frequency of $1/3t_s$, as shown in Fig. 3.26. The total power in the error waveform can thus be represented as $G(f)/3t_s$ where $G(f)$ is a uniform power spectral density, hence

$$\bar{\varepsilon}^2 = G(f)/3t_s \tag{3.33}$$

If the detector output is low pass filtered, the cut off frequency of the filter being f_1, then provided $f_1 < 1/3t_s$ the output noise power will be

$$N_q = G(f)f_1 = \frac{3\bar{\varepsilon}^2 f_1}{f_s} = \frac{(\delta v)^2 f_1}{2f_s}$$

But $\delta v = 2\pi V f_0/f_s$, thus

$$N_q = 2\pi V^2 f_1 (f_0)^2 \tag{3.34}$$

The maximum signal power which the system can handle before overload is

$$S = \frac{(A_{\max})^2}{2} = \frac{V^2}{2[1+(f_m/f_0)^2]}$$

hence

$$\mathrm{SNR}_{\mathrm{max}} = \frac{V^2(f_0)^2}{2[(f_0)^2 + (f_m)^2]} \frac{(f_s)^3}{2\pi^2 V^2 f_1 (f_0)^2}$$

Assuming, on average, that $f_m \gg f_0$ then

$$\mathrm{SNR}_{\mathrm{max}} = \frac{(f_s)^3}{4\pi^2 f_1 (f_m)^2} \tag{3.35}$$

This is an approximate result for single tone inputs. With speech waveforms it is found that overload is minimized if the instantaneous speech amplitude is limited to the amplitude of a single tone of frequency 800 Hz, which could be transmitted without overload. Thus if $f_m = 800$ Hz, $f_1 = 3.4 \times 10^3$ and $f_s = 64 \times 10^3$ (which is equivalent to 8 bit linearly quantized PCM the SNR is 35 dB. The corresponding figure for 8 bit linearly quantized PCM, obtained from Eqn (3.16) is 49.8 dB. It is usually assumed that the minimum acceptable signal to quantization noise for telephone transmission is about 26 dB. This would give delta modulation a dynamic range of only 9 dB as compared with a figure of 23.8 dB for linearly quantized PCM. (Both of these figures are far short of companded PCM with a UVR of 26 dB at a SNR of 38 dB.)

The dynamic range of delta modulation may also be increased by a form of companding. In the case of the delta modulator the slope of the voltage across the integrating capacitor is varied with input signal amplitude. This gives rise to continuously variable slope delta modulation (CVSDM). Essentially the output of the delta modulator is monitored and when several successive pulses have the value '1' or '0' (indicating that slope overload is occurring) the amplitude of the charging voltage V is increased thereby increasing the slope of the integrator output. When the output returns to 010101 overload is removed and the step size may be progressively reduced. The analysis of several forms of CVSDM is covered in Steele[6].

3.10 Differential PCM

A number of analogue waveforms such as speech and video exhibit the property of predictability. This means that the change in value between one sample and the next is small because the rate of change of the analogue waveform is usually low compared with the sampling frequency. Alternatively waveforms such as speech and video have 'instantaneous frequencies' considerably lower than the maximum frequency component on which the sampling frequency is based. The next sample in a waveform can thus be predicted from a knowledge of previous samples. There will be some prediction error, but the peak to peak value of the error will be considerably less than the peak to peak value of the original waveform. Differential pulse code modulation (DPCM) capitalizes on this fact by coding and transmitting the prediction error. The prediction error requires

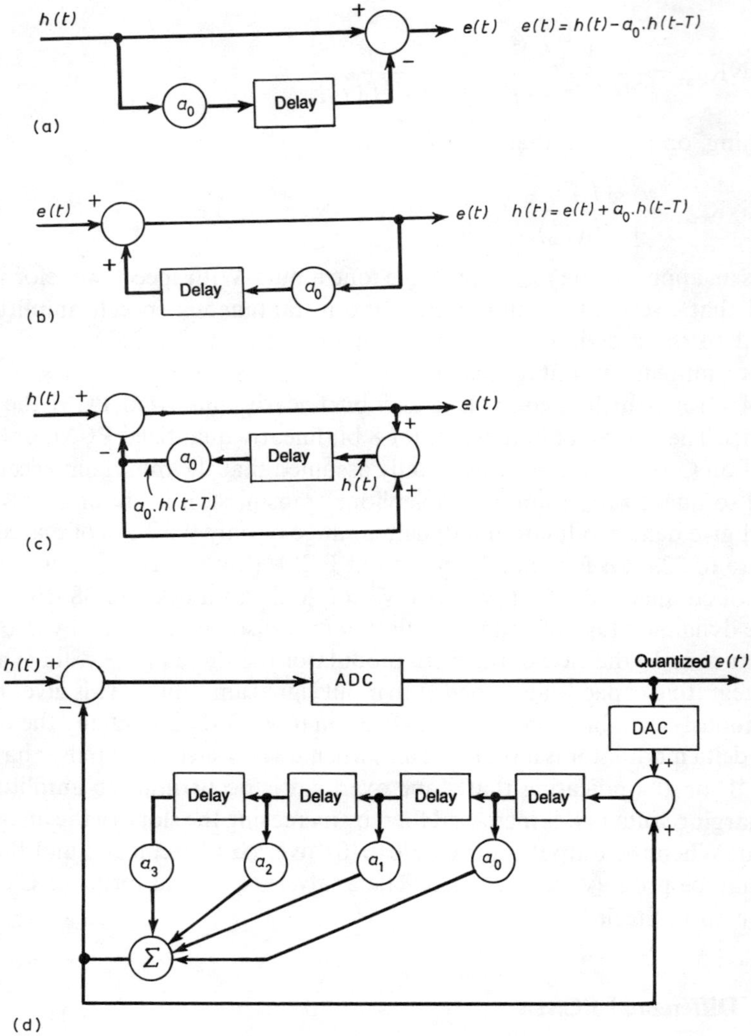

Fig. 3.27 Differential PCM.

fewer quantization levels for a given SNR and hence the required transmission bandwidth is less. A simple predictive coder and decoder are shown in Fig. 3.27.

In Fig. 3.27a, $h(t)$ represents an input sample and $e(t)$ represents the difference between $h(t)$ and the previous sample weighted by the coefficient a_0 ($\leqslant 1$). T is a delay equal to the sampling period. From this figure

$$e(t) = h(t) - a_0 h(t - T)$$

Considering now the circuit of Fig. 3.27b it is evident that

$$h(t) = e(t) + a_0 h(t - T) \tag{3.36}$$

This circuit is known as a predictor because the current sample is predicted from the previous sample and the error signal $e(t)$. The predictor is incorporated into the transmitter as shown in Fig. 3.27c and output $e(t)$ is known as the prediction error. There is an advantage in incorporating the predictor in the transmitter as the feedback loop minimizes quantization error when an analogue to digital converter is employed. The receiver is simply the feedback loop of the transmitter which, in this case, is the circuit of Fig. 3.27b.

In practice the estimate of the predictor circuit is based on estimates of the previous four quantized samples as shown in Fig. 3.27d. This figure also contains an analogue to digital converter in the forward path and a digital to analogue converter in the feedback path. As stated previously, the feedback action minimizes quantization error. The SNR for differential PCM is between 5 and 10 dB higher than for PCM without differential coding. This SNR may be increased further by use of adaptive DPCM (ADPCM). The quantizer step size (i.e. the ADC and DAC) is adapted according to the amplitude of the prediction error. Alternatively, with this type of coder, using a 4 bit quantizer, it is possible to transmit speech with the same quality as 64 kb/s '*A*' law compression at a bit rate of 32 kb/s.

3.11 Data Communications

PCM and delta modulation have been optimized for the transmission of coded voice signals over trunk routes. Data communications deals primarily with the transmission of digital signals between machines. The most common example of this is the connection of a visual display unit (VDU) and keyboard to a distant computer via the public telephone network. It is the most rapidly expanding area of telecommunications; the most significant development in recent years has been the combination of the domestic telephone and television receiver to produce a viewdata terminal. This aspect is covered in detail in Chapter 12.

The bulk of data communications at present is via the public telephone network, and this will be considered in some detail. Plans are well advanced, however, for alternative data communication networks using packet switching techniques. These networks are described in Chapter 13.

When considering data communications, one of the basic parameters that must be defined is the signalling speed. The unit of signalling speed is known as the *baud* after the telegraph engineer Baudot. The signalling speed in bauds is in effect the rate at which pulses are transmitted over the communications link. These pulses need not be binary, which means that the data rate, which is usually expressed in bits/s, does not necessarily equal the signalling speed.

Unlike PCM, which uses unloaded lines for transmission, data signals are transmitted over normal telephone circuits with their severely restricted frequency response. Special problems linked to this response therefore arise in data transmission, and we shall consider these problems in some detail. The

Table 3.2 International 7-bit code

Character	Binary	Octal	Hexadecimal
A	1000001	101	41
B	1000010	102	42
C	1000011	103	43
1	0110001	061	31
2	0110010	062	32
3	0110011	063	33

characteristics of the data signal produced by a VDU keyboard have two well-defined properties:

(1) low data rate, limited by human typing speed, and
(2) sparodic output with long periods of no output at all.

To cope with the second of these characteristics, asynchronous communications is used, but it should be noted that dedicated high-speed data links use synchronous communication.

For transmission purposes, each symbol on a keyboard is represented by a unique binary code. The international standard code represents each character by a 7-bit word, i.e.

$$b_6 b_5 b_4 b_3 b_2 b_1 b_0$$

where b_n is a binary digit. Some examples of the 7-bit code are listed in Table 3.2. The 7-bit word has an additional digit, b_7, called a parity check bit, added for error-detection purposes. The parity bit is chosen so that the number of 1s in each 8-bit word is even. If an odd number of 1s occurs at the receiver, the receiver is aware that an error has occurred.

The asynchronous transmission system requires extra bits to allow the receiver at either end of the link to determine the beginning and end of each symbol. The transmission of each symbol is preceded by a level change from 1 to 0. The 0 level has a duration of 1 digit interval. The symbol is then transmitted serially, least significant digit (b_0) first, followed by the parity check bit (b_7). The end of each symbol is signalled by a binary 1 which lasts for two digit periods. The idle state (no signal) is thus equivalent to binary 1. A typical digit sequence (10 bits) is illustrated in Fig. 3.28. All timing is initiated by the falling edge of the start bit and the following digits are sampled at their mid-points. This means that timing clocks do not have to be closely matched as synchronization occurs on each start bit. A human operated VDU will produce a maximum of about 10 characters/s, which is, of course, very slow compared with the speed of operation of the computer to which it may be connected. If the transmitted pulses are binary, each pulse has a duration of 9.1 ms, which is equivalent to a signalling speed of 110 baud.

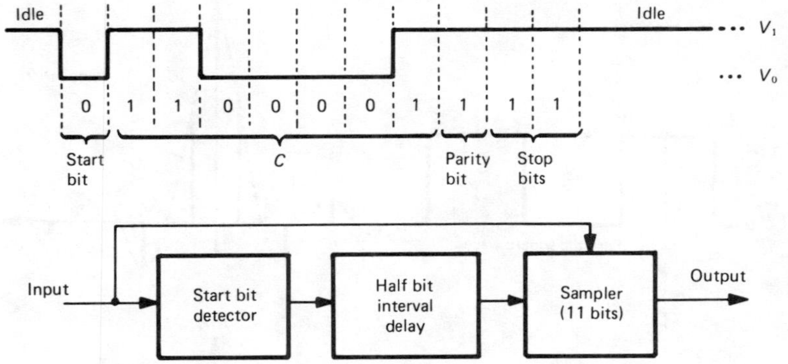

Fig. 3.28 Asynchronous transmission.

3.12 Spectral Properties of Data Signals

We must know something of the spectral properties of data signals before we can specify the most appropriate form of data transmission over telephone circuits. A typical data signal will consist of a random sequence of pulses of binary 1s and 0s. The power spectral density of such a signal was derived in Section 1.12 from its autocorrelation function. A random binary signal with pulse amplitudes of 0 or A V and pulse duration t_1 seconds has an amplitude spectrum given by

$$H(f) = At_1 \operatorname{sinc}(\pi f t_1)$$

It can be seen from Fig. 3.29 that most of the energy in the spectral envelope is confined to frequencies below $f = 1/t_1$ Hz. The bandwidth of the data signal is therefore usually approximated by the reciprocal of the pulse width.

The data spectrum, which has a component at zero frequency must be modified for transmission over a telephone circuit which usually has a bandwidth from 300 Hz to 3.4 kHz. Further, since two-way signalling is required over a single circuit, it is necessary to differentiate between transmitted and received data signals. Both these requirements are met by modulating the data signal on to an audio frequency tone. The three possible forms of modulation are AM, FM and PM.

3.13 Amplitude Shift Keying (ASK)

This is the name given to AM when used to transmit data signals. It is not normally used on telephone lines because the large variations in circuit attenuation which can occur make it difficult to fix a threshold for deciding

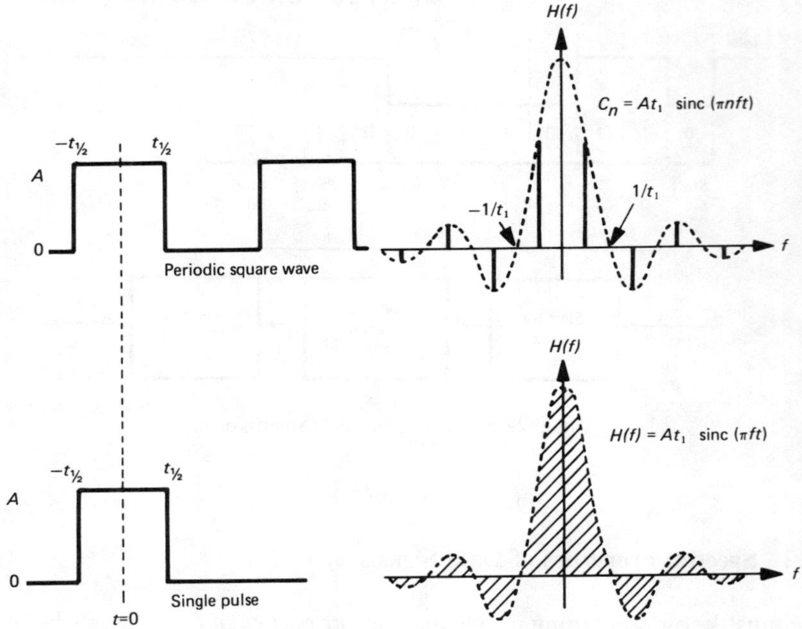

Fig. 3.29 Amplitude spectrum of data signals.

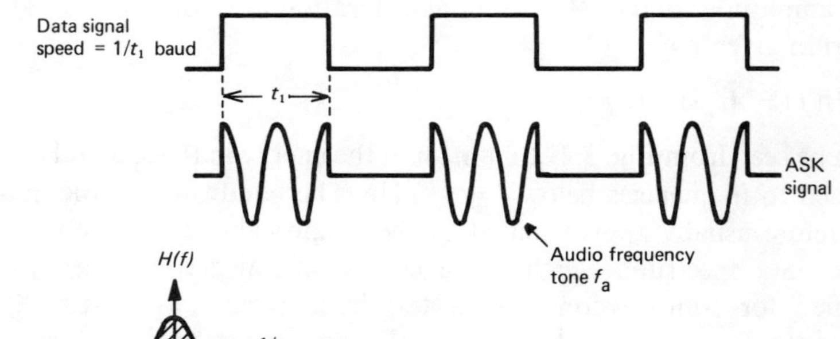

Fig. 3.30 ASK amplitude spectrum.

between binary 1 and 0. We shall, however, consider ASK in some detail because it is convenient to represent FM as the sum of two ASK signals.

The ASK signal is generated by multiplying the data signal by an audio tone. This effectively shifts the data spectrum to a centre frequency equal to that of the audio tone. The process is shown in Fig. 3.30. The bandwidth of the modulated signal is twice the bandwidth of the original data signal. This means that the original 110 baud signalling rate requires a transmission bandwidth of 220 Hz using ASK:

$$\text{ASK} \equiv \text{DSBAM} \equiv (\text{carrier} + \text{upper and lower sidebands})$$

3.14 Frequency Shift Keying (FSK)

This is the binary equivalent of FM. In this case a binary 0 is transmitted as an audio frequency tone f_0 and a binary 1 is transmitted as a tone f_1. Hence the binary signal effectively modulates the frequency of a "carrier".

Although, strictly speaking, FSK is FM, it is more convenient to consider FSK as the sum of two ASK waveforms with different carrier frequencies. The spectrum of the FSK wave is thus the sum of the spectra of the two ASK waves.

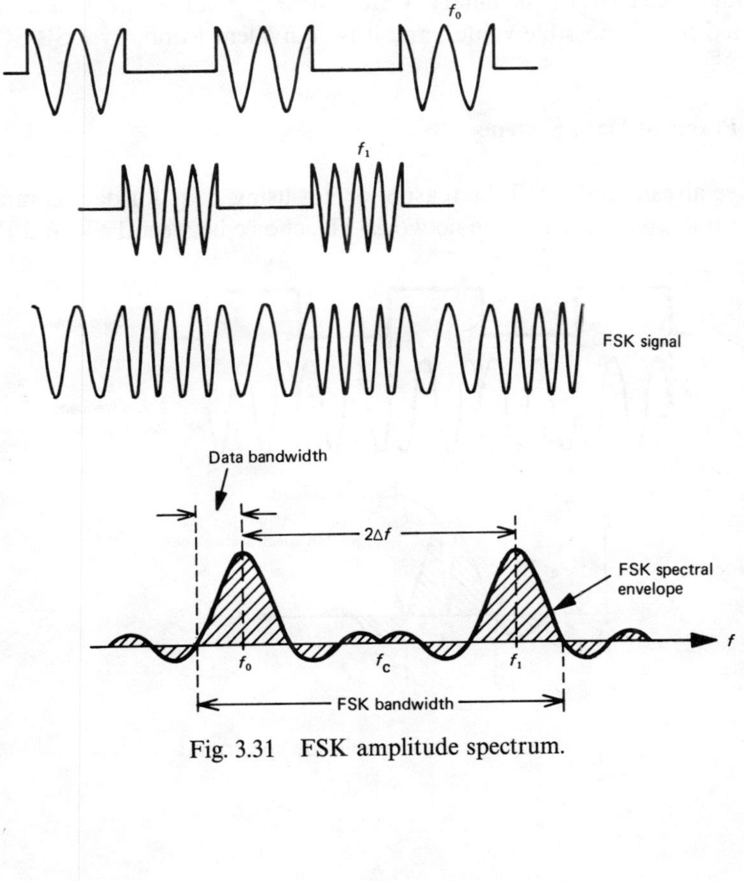

Fig. 3.31 FSK amplitude spectrum.

This spectrum is shown in Fig. 3.31. Using the FM analogy, it is possible to define a "carrier frequency" $f_c = f_0 + (f_1 - f_0)/2$ and a "carrier deviation" $\Delta f = (f_1 - f_0)/2$. The modulation index β is defined as $\beta = \Delta f/B$, where $B = 1/t_1$ is the bandwidth of the data signal. Using these definitions the bandwidth of the FSK signal is

$$B_{FSK} = 2B(1 + \beta) \tag{3.37}$$

This is similar to Carson's rule for continuous FM. Unlike analogue FM there is no advantage in increasing Δf beyond the value $\Delta f = B$ since the receiver only needs to differentiate between the two tones f_0 and f_1.

3.15 Phase Shift Keying (PSK)

This is the binary equivalent of PM, the binary information being transmitted either as zero phase shift or a phase shift of π radians. This is equivalent to multiplying the audio tone by either $+1$ or -1. The bandwidth is thus the same as for ASK. Since there is no dc component in the modulating signal, the carrier in the PSK spectrum will be suppressed. The equivalent modulating signal and the PSK spectral envelope are shown in Fig. 3.32. This form of PSK is sometimes referred to as binary PSK (BPSK) because the phase shift is restricted to two possible values, and it is equivalent to binary DSB-SC-AM.

3.16 Practical Data Systems

We have already indicated the reason for not using ASK for data communications on the public telephone network. The choice between FSK and PSK is

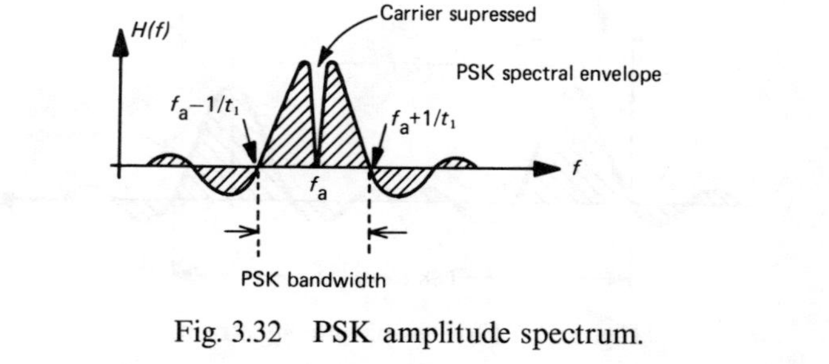

Fig. 3.32 PSK amplitude spectrum.

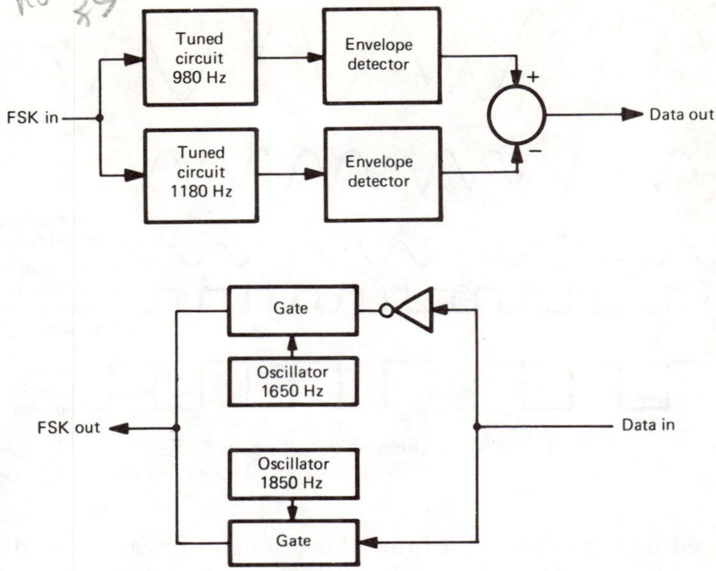

Fig. 3.33 200 baud modem.

determined by the data rate. At low data rates FSK is employed for two-way (duplex) communication. A typical system operating at a signalling rate of 200 baud uses two tones of 980 Hz and 1180 Hz for binary 1 and 0 in one direction and 1650 Hz and 1850 Hz for binary 1 and 0 in the reverse direction. The incoming FSK is separated into two tones using bandpass filters. Envelope detection is then used to reproduce the binary signal. The combined modulator/demodulator (modem) is illustrated in schematic form in Fig. 3.33.

As the data rate is increased higher carrier frequencies are required; otherwise, each data interval would contain very few cycles of carrier, which would make detection extremely difficult. There is a limit on carrier frequency imposed by the upper cut-off frequency of the telephone line. For this reason FSK is limited to signalling speeds up to 600 baud. At speeds above this PSK is employed. This type of modulation makes more efficient use of bandwidth but requires more sophisticated coherent detectors. The reference signal for coherent detection is derived from the PSK signal itself. Since the carrier is suppressed in the PSK spectrum the received waveform is first rectified to produce a component at twice carrier frequency. This component is then limited and divided by two to produce the required reference signal. The required signal processing is illustrated in Fig. 3.34.

In the public telephone network any connection between transmitter and receiver will be made via several different paths which will contain several stages of frequency multiplexing and demultiplexing. Imperfections in the various stages of modulation result in random, slowly varying, phase shifts which are

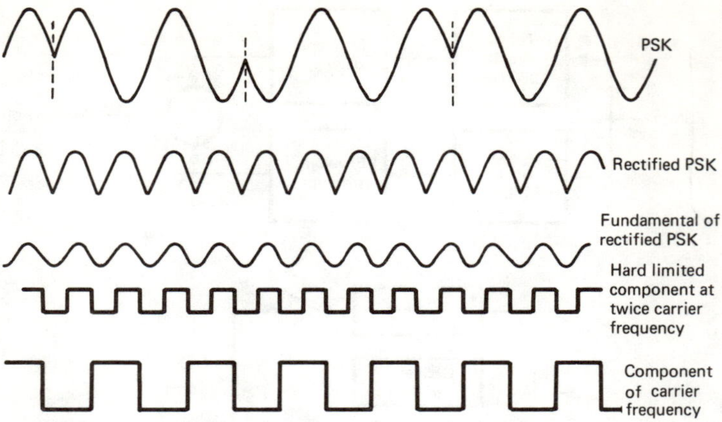

Fig. 3.34 Coherent detection of PSK.

introduced into the PSK waveform. This results in phase ambiguity at the receiver and can produce data inversion.[6] The problem is greatly reduced if differential encoding is employed.

3.17 Differential Phase Shift Keying (DPSK)

DPSK has the advantage of using the phase of the previous bit interval as the reference for the present bit interval. In order to make this possible, a binary 0 is transmitted as the same phase as the previous digit and a binary 1 is transmitted as a change of phase. The relationship between PSK and DPSK is shown in Fig. 3.35. The receiver compares the phase of the current digit with the phase of the previous digit. If they are the same the current digit is interpreted as a 0; otherwise it is interpreted as a 1. DPSK can be produced by pre-coding the data signal which then modulates the carrier as in standard PSK. If A_n is the present input to the encoder (A_n is binary) and C_{n-1} is the previous output the truth

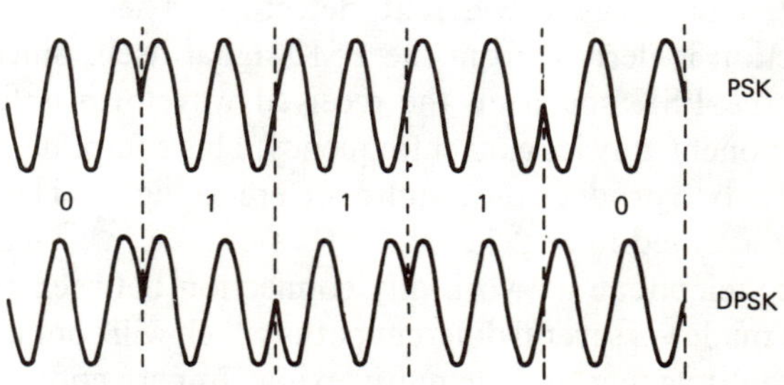

Fig. 3.35 Relationship between PSK and DPSK.

table for the encoder is

A_n	C_{n-1}	C_n
0	0	0
0	1	1
1	0	1
1	1	0

which will be recognized as the exclusive-OR operation

$$C_n = A_n \oplus C_{n-1}$$

We have already noted that pulses can be transmitted at a rate of $1/t_1$ without mutual interference over a channel of cut-off frequency t_1 Hz, provided that the channel has a raised cosine frequency response. This applies to the unmodulated signal. A signalling rate of 1200 baud thus requires a raised cosine channel of bandwidth 1200 Hz. The PSK signal will require a bandpass channel with a raised cosine characteristic with a bandwidth of 2400 Hz. A typical PSK signal system would operate at a carrier frequency of 1.8 kHz and a signalling rate of 1200 baud. The bandwidth occupied by this waveform extends from 600 Hz to 3 kHz. Hence a data rate of 1.2 kbits/s is an upper limit for BPSK.

3.18 Advanced Modulation Methods

The data rate of BPSK is sometimes expressed as 1 bit/baud. Since the baud rate is fixed by the channel characteristics the data rate can only be increased by increasing the number of levels per pulse beyond two. If each pulse has four levels the data rate becomes 2 bits/baud and each level can produce a unique phase shift. Thus it is possible to transmit data at a rate of 2.4 kbits/s without any increase in bandwidth. It is convenient when considering multiphase PSK to represent the transmitted signal in terms of the sum of two quadrature audio frequency tones, i.e.

$$f_c(t) = a\cos 2\pi ft + b\sin 2\pi ft \tag{3.38}$$

Each of the levels in a 4-level pulse can be represented by two binary digits called dibits. Thus it is not actually necessary in practice to produce a 4-level pulse; instead the binary signal can be grouped into dibits and each dibit can be used to produce a unique phase shift in multiples of $\pi/2$. When interpreted in this way each dibit represents one pulse, i.e. the signalling rate in bauds equals half the bit rate. Table 3.3 lists the possible dibits and the values of a and b in Eqn (3.38) necessary to produce the required phase shifts. The resulting quaternary PSK can be represented on a signal space diagram of the type shown in Fig. 3.36.

The data signal is recovered from the QPSK waveform by using two coherent detectors supplied with locally generated carriers in phase quadrature. The data

Table 3.3

Dibit	Phase shift	In-phase component	Quadrature component
		a	b
00	$\pi/4$	$+1$	$+1$
01	$3\pi/4$	-1	$+1$
11	$-3\pi/4$	-1	-1
10	$-\pi/4$	$+1$	-1

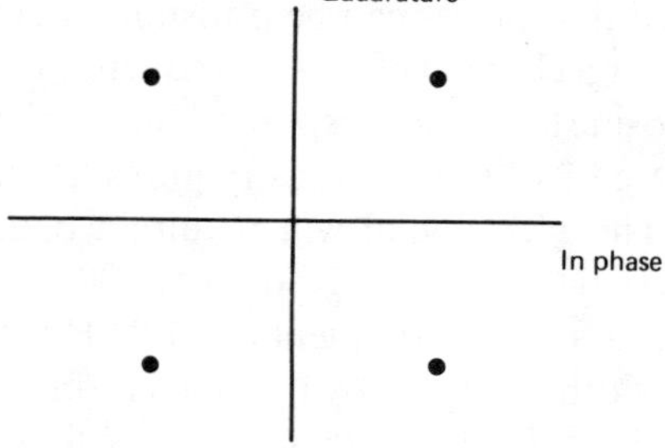

Fig. 3.36 Signal space diagram for QPSK.

rate can be increased further by increasing the number of levels of each pulse beyond four. For example, if the number of levels is increased to 16, it is possible to transmit data at a rate of 4 bits/baud. The QPSK signal was characterized by the fact that the coefficients a and b of Eqn (3.38) always have the same magnitude, thereby producing a resultant of constant amplitude and varying phase. It is also possible for a and b to have different values and the resulting signal is in fact quadrature amplitude modulation. The detection of QAM is covered in Section 11.11 in connection with the transmission of chrominance signals in the PAL colour television system.

The signal space diagram for QAM with 16-level pulses is shown in Fig. 3.37. Each individual level is represented by a unique combination of a and b in Eqn (3.38). It is possible with this system to transmit data at a rate of 4.8 kbits/s over a raised cosine bandpass channel, with a bandwidth of 2.4 kHz.

At these high data rates, intersymbol interference is a severe problem and elaborate equalization networks (transversal digital filters) are always employed. The lines used are not part of the public telephone network and are maintained to close tolerances in respect of loss and bandwidth. Such lines are often known as leased lines. The sophisticated equalization used on these lines means that signalling rates can approach the Nyquist rate, i.e. pulses can be transmitted at rates approaching 2400 baud. This means that it is possible to transmit data rates up to 9.6 kbits/s using 16-level QAM.

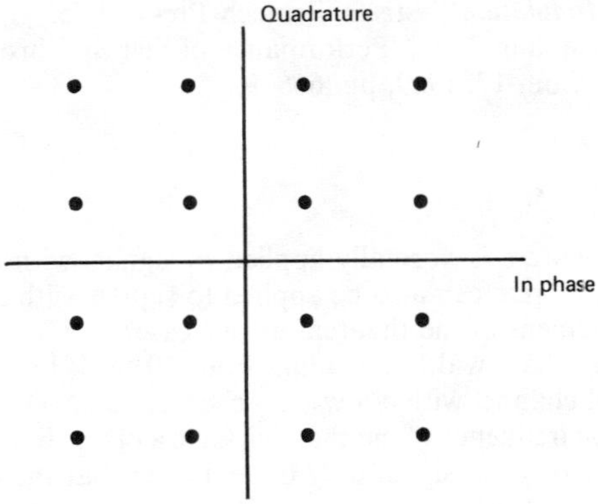

Fig. 3.37 Signal space diagram for 16-level QAM.

The leased lines referred to in the previous paragraph are basically high grade voice channels. Much higher data rates are possible on specially designed wideband links. These wideband links operate in a synchronous fashion and modern networks are adopting packet-switching techniques to maximize the efficiency of usage of these links.

3.19 Conclusions

This chapter has introduced the basic concepts of digital communications and stressed the advantages of transmission of information in digital format. This is a huge growth area in telecommunications systems engineering and is likely to remain so for the foreseeable future. All digital telephone networks, including cellular mobile telephones, are progressively being installed. These networks will gradually replace analogue systems and will provide enhanced services and increased reliability. A detailed treatment of packetized transmission is given in Chapter 13.

References

1. Schwartz, M., *Information Transmission Modulation and Noise*, 3rd edn, McGraw-Hill, 1980.
2. CCITT, *Orange Book*, Vol. III–2, Recommendations G711, G712, G732.
3. Henderson, K.W. and Kautz, W.H., 'Transient Responses of Conventional Filters', *IRE Transactions on Circuit Theory*, Vol. 5, 1958, pp. 333–47.
4. Johnson, F.B., 'Calculating Delta Modulator Performance', *Trans. IEEE*, Vol. AU–16, No. 1, 1968, pp. 121–9.
5. Betts, J.A., *Signal Processing Modulation and Noise*, The English Universities Press, 1972, Ch. 7.

6. Steele, R., *Delta Modulation Systems*, Pentech Press, 1975.
7. Oberst, J.P. and Schilling, D.L., 'Performance of Self Synchronized PSK Systems', *Trans. IEEE*, Vol. Com-17, 1969, pp. 666–9.

Problems

3.1 The sampling theorem is normally applied to signals with a low-pass spectrum. Show that this theorem can also be applied to signals with a bandpass spectrum. What is the statement of the theorem in this case?

A signal with a bandwidth extending from 30 to 34 kHz is to be transmitted through an ideal channel with a low-pass characteristic. Determine the minimum theoretical cut-off frequency of the channel. Give a block diagram of the system and describe how the original signal may be reproduced at the receiver.

Answer: 4 kHz.

3.2 A signal is used to amplitude-modulate a periodic pulse train in which the pulse width is much less than the pulse duration. The sampled signal is to be reconstructed by a sample-and-hold circuit. Sketch the output of this circuit and, by considering its frequency response, find the ratio of the signal frequency amplitude to the amplitude of the lowest frequency distortion component. The signal is a single tone of frequency 2.8 kHz sampled at 8 kHz.

Answer: 1.86:1.

3.3 Explain the difference between the actual bandwidth of a pulse and the bandwidth required for pulse transmission.

Twenty-four speech signals each with a bandwidth of 0 to 4.5 kHz are to be transmitted over a line with a raised cosine frequency response by TDM. Calculate the minimum theoretical bandwidth of the line. Would this bandwidth be adequate in a practical system?

Answer: 216 kHz

3.4 An equalized line with a raised cosine has an effective bandwidth of 200 kHz. Six speech channels are to be transmitted over this line using time multiplexed PCM. Assuming linear quantization and a sampling frequency of 8 kHz for each signal, what will be the average signal power-to-noise ratio at the decoder output. Assume quantization noise only is to be considered.

Answer: 26 dB.

3.5 A single information channel carries voice frequencies in the range 50 Hz to 4.3 kHz. The channel is sampled at a 9 kHz rate and the resulting pulses may be transmitted either directly by pulse amplitude modulation (PAM) or by PCM.

Calculate the minimum bandwidth required for the PAM transmission assuming a line with a raised cosine response.

If the pulses are linearly quantized into eight levels and are transmitted as binary digits, find the bandwidth required to transmit the digital signal and compare it with the analogue figure. If the number of levels of quantization is increased to 128 what is the new bandwidth required? Calculate the increase in SNR at the decoder output, assuming the peak-to-peak voltage swing at the quantizer is 2 V.

Answer: 8 kHz; 24 kHz; 56 kHz; 24 dB.

3.6 A PCM system employing uniform quantization and generating a 7-digit code is capable of handling analogue signals of 5 V peak-to-peak. Calculate the mean signal-to-quantizing noise ratio when the analogue waveform has a probability density function given by

$$P(v) = K\exp(-|v|) \qquad -2.5 < v < 2.5, \qquad = 0 \text{ else}$$

Assume uniform signal distribution within a given quantization interval.

Answer: 38.9 dB.

3.7 Derive an expression for the amplitude spectrum of a single triangular pulse of base width t s and amplitude A V. Hence estimate its bandwidth.

Answer: $2/t$ Hz.

3.8 A data signal consists of a series of binary pulses occuring at a rate of 100 digits/s. This signal is to be transmitted over a telephone line, binary 1 being sent as a 1.5 kHz tone and binary 0 as a 2.8 kHz tone. What is the bandwidth of the transmitted signal?

If the digit rate is increased to 1000 bits/s what are the required upper and lower cut-off frequencies of the line in order that it may transmit this signal?

Answer: 1.5 kHz; 500 Hz; 3.8 kHz.

3.9 If the transmission of question 3.8 is by DPSK, what is the maximum data rate that can be transmitted over the telephone line? What is the optimum carrier frequency in this case?

Answer: 1650 bits/s; 2.35 kHz.

3.10 Derive an expression for the amplitude spectrum of an FSK transmission when the digit stream is a series of alternate 1 s and 0 s.

3.11 A sinusoidal signal is switched periodically from 10 MHz to 11 MHz at a rate of 5000 times/s. Sketch the resulting waveform and identify the modulating signal if the switched sine wave is regarded as FSK.

Find the approximate transmission bandwidth of the FSK signal and compare this with the bandwidth required if the modulating signal is approximated by a sine wave producing the same carrier deviation.

Answer: 1.02 MHz; 1.0 MHz.

4 NOISE IN ANALOGUE COMMUNICATIONS SYSTEMS

4.1 Introduction

We have considered, in the previous chapters, various ways of transmitting information from one location to another. We concern ourselves now with the performance of these systems in a noisy environment. This comparative analysis will provide the insight required to determine the suitability, or otherwise, of using a particular form of transmission in a specific environment. The relative performance of various systems in noise is, of course, only one of the factors taken into account when choosing a particular method of information transmission. The economic considerations, e.g. cost, complexity, maintenance, are sometimes of paramount importance, but are outside the scope of this text.

Noise is defined as any spurious signal that tends to mask or obscure the information in the transmitted signal. The ratio of signal power to noise power at any point in a telecommunications system is known as the SNR, and the fundamental relationship between signal bandwidth and SNR is given by Shannon's law (see Section 1.14). In this chapter we will consider the performance of systems using analogue transmission and then discuss the significance of Shannon's law for the examples chosen.

4.2 Physical Sources of Noise

Noise is usually divided into naturally occurring noise and man-made noise. Man-made noise comes from various sources, the most important types being ignition interference, produced whenever sparks occur at electrical contacts, and crosstalk which is produced by inductive or capacitive coupling between one or more communications channels. Man-made noise can, in theory, be eliminated, although the cost of such elimination is often quite uneconomic. Natural noise is produced by many different phenomena, some examples being lightning discharges, thermal radiation and cosmic radiation. Natural noise cannot be eliminated and communication systems must perform efficiently in the presence of this type of noise, and often in the presence of man-made noise also.

The two most important types of natural noise are thermal noise and shot noise. Thermal noise is produced by random motion of charged particles in resistive materials and by thermal radiation from objects surrounding a telecommunications system, particularly one with an antenna. Shot noise is produced in semiconductor devices and results from the fact that currents flow across p–n junctions in finite quanta rather than continuously.

Thermal noise, caused by random motion of charged particles in resistive materials, produces a mean square noise voltage with a magnitude directly related to the temperature of the resistive material, which is given by

$$\bar{v}^2 = 4kTR\Delta f \quad \mathrm{V}^2 \tag{4.1}$$

In this expression k is Boltzmann's constant (1.38×10^{-23} J/K), T is the absolute temperature in kelvins, R is the resistance of the material in ohms, and Δf is the bandwidth of the measurement. The derivation of Eqn (4.1) is given by King.[1] Any noisy resistance can be represented by an equivalent circuit consisting of a noise-free resistance in series with a voltage source whose mean square amplitude is given by Eqn (4.1). This allows the determination of the noise properties of networks using conventional network theorems. We will consider one such calculation, of particular importance in telecommunications, to illustrate the procedures. We want to calculate the noise power delivered by a resistance to a matched load, i.e. a load with the same resistive value. The circuit is shown in Fig. 4.1.

In this circuit R_1 is the noise source and R_2 is the load. Both source and load will produce a mean square noise voltage which will give rise to a mean square current i^2. The power delivered by R_1 to R_2 is

$$P_n = \frac{\bar{v}_1{}^2 R_2}{(R_1 + R_2)^2} \tag{4.2}$$

If $R_1 = R_2$ (i.e. the source and the load are matched) the power delivered by R_1 to R_2 is

$$P_n = \frac{\overline{v_1{}^2}}{4R_2} = \frac{4kT\Delta f R_2}{4R_2} = kT\Delta f \quad \mathrm{W}$$

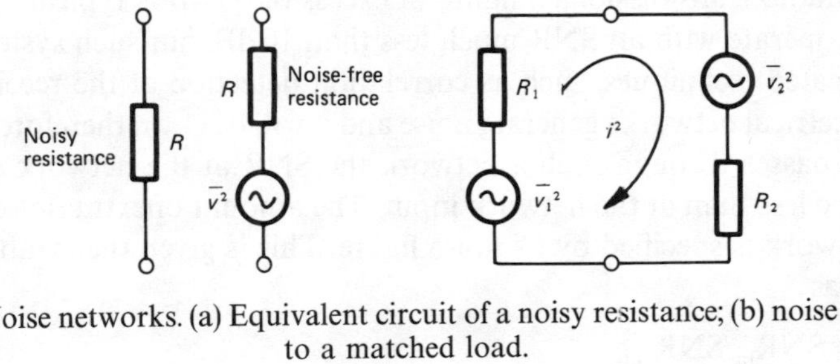

Fig. 4.1 Noise networks. (a) Equivalent circuit of a noisy resistance; (b) noise delivered to a matched load.

This is often written

$$P_n = \eta \Delta f \tag{4.3}$$

The constant η is the single-sided noise power spectral density and is independent of the range of frequencies Δf. If Δf is measured over both negative and positive frequencies the double-sided noise spectral density is $\eta/2$ W/Hz. This agrees with the Fourier analysis developed in Chapter 1, in which we indicated that half the total power was contributed each by the negative and positive components. When the noise spectral density is independent of frequency the noise is said to be white, by analogy with white light. When white noise is filtered by a frequency-selective network the resulting noise is known as coloured noise.

When thermal noise is derived from an antenna, the temperature of the noise source is not necessarily equal to the antenna temperature. The expression for the noise power delivered by an antenna to a matched load is $P_n = kT_e\Delta f$ W. The effective noise temperature of the antenna T_e is related to the temperature of the radiating bodies surrounding the antenna and is determined by measurement. If the radiating bodies have the same temperature as the antenna, this is the value used for T_e. Antennas pointing into space generally receive far less radiation than antennas directed towards bodies on the Earth's surface and consequently have a much lower effective noise temperature.

4.3 Noise Properties of Networks

All telecommunications systems are characterized by the fact that received signals are always accompanied by noise. The effectiveness of such a system is measured in terms of the ratio of signal power to noise at the system output. The SNR at any point in a telecommunications link is usually expressed in decibels:

$$\mathrm{SNR} = 10\log_{10}[S_p/N_p] \tag{4.4}$$

The minimum acceptable SNR for reliable communication is normally considered to be about 10 dB. Many systems operate at much higher ratios than this: the minimum SNR for telephone circuits is around 26 dB and for high-quality audio transmissions a figure in excess of 60 dB is typical. Some space systems operate with an SNR much less than 10 dB, but such systems require sophisticated techniques, such as correlation detection at the receiver.

All electrical networks generate noise and it will be clear, therefore, that when a signal passes through such a network the SNR at the network output will always be less than at the network input. The amount of extra noise generated by a network is specified by its noise figure. This is given the symbol F and is defined as

$$F = \mathrm{SNR}_{in}/\mathrm{SNR}_{out} \tag{4.5}$$

A noiseless network has a noise figure of unity and it therefore follows that real networks always have a noise figure with a numerical value greater than unity. The noise figure of any network is derived in terms of the schematic diagram of Fig. 4.2. The input signal power is S_i and the network power gain is A_p (not necessarily restricted to values greater than unity); the signal power at the network output is thus $S_0 = A_p S_i$. If the input noise power is N_i and the noise power generated within the network is N_a the total noise power at the network output is $N_0 = A_p N_i + N_a$. The output SNR is

$$\mathrm{SNR}_0 = A_p S_i/(A_p N_i + N_a)$$

i.e.

$$F = \frac{S_i}{N_i}\frac{(A_p N_i + N_a)}{A_p S_i} = \frac{A_p N_i + N_a}{A_p N_i} \tag{4.6}$$

The factor $A_p N_i$ is the output noise power of a noise-free network; thus

$$F = \frac{\text{total output noise power}}{\text{output noise power if network was noise free}} \tag{4.7}$$

We see from the definition given as Eqn (4.7) that F is not constant but is related to the noise power at the network input. The value of F is standardized by fixing the input noise power as that produced by a matched source at a standard temperature of 290 K. Use of noise figures in network calculations is thus only valid if the input noise power is $kT_s\Delta f$, where $T_s = 290$ K.

The usefulness of noise figures is demonstrated by considering the cascaded networks of Fig. 4.3. Using the definition of Eqn (4.7) we can write the overall noise figure for the cascaded network as

$$F = \frac{(A_{p1}N_i + N_{a1})A_{p2} + N_{a2}}{A_{p1}A_{p2}N_i}$$

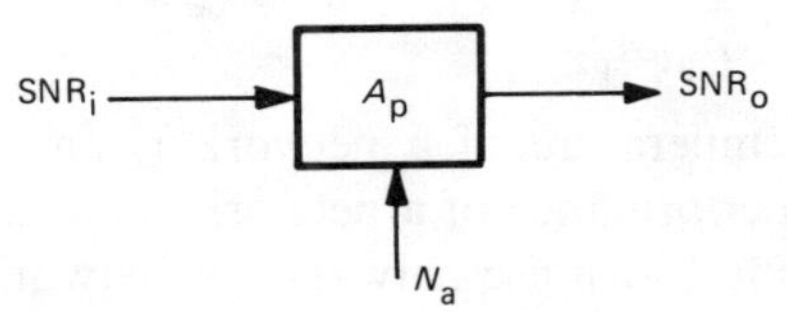

Fig. 4.2 Noise figure of a network.

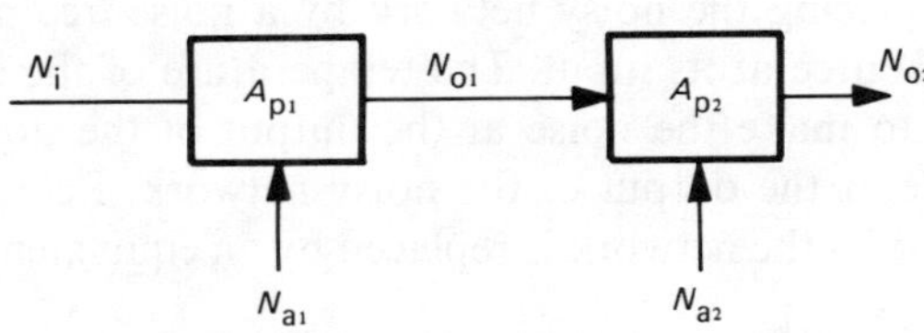

Fig. 4.3 Cascaded noisy networks.

We assume that the networks are matched, i.e. the output resistance of the first network is equal to the input resistance of the second network, and that the input noise N_i is the noise produced by a matched source at 290 K. The noise figure of the first network is therefore

$$F_1 = (A_{p1}N_i + N_{a1})/A_{p1}N_i$$

In defining the noise figure of the second network the input power is also N_i; hence

$$F_2 = (A_{p2}N_i + N_{a2})/A_{p2}N_i$$

Thus the overall noise figure may be written

$$F = \frac{A_{p1}N_i + N_{a1}}{A_{p1}N_i} + \frac{N_{a2}}{A_{p1}A_{p2}N_i}$$

But

$$F_2 - 1 = \left(1 + \frac{N_{a2}}{A_{p2}N_i}\right) - 1$$

Therefore

$$F = F_1 + \frac{F_2 - 1}{A_{p1}} \tag{4.8}$$

We see from Eqn (4.8) that if $A_{p1} \gg 1$, which is usually the case, the major contribution to the overall noise figure is produced by the first network. Evidently it becomes extremely important to ensure that the first network in any cascaded system has as low a noise figure as possible. Equation (4.8) can be expanded to include any number of cascaded networks. The equation for three networks in cascade is

$$F = F_1 + \frac{F_2 - 1}{A_{p1}} + \frac{F_3 - 1}{A_{p1}A_{p2}} \tag{4.9}$$

The effective noise temperature of a network is an alternative method of describing the noise performance of a network. This alternative is especially useful when considering low noise networks or networks in which the input noise is not produced by a matched source at 290 K. In the latter case the use of noise figures is not valid. The effective noise temperature of a network is determined by replacing the noisy network by a noise-free network with an equivalent noise source at its input. The temperature of the equivalent noise source is chosen to make the noise at the output of the noise-free network equal to the noise at the output of the noisy network. Referring to Fig. 4.2, the noise produced by the network is replaced by an equivalent noise source of

$$N_a = kT_e \Delta f A_p$$

The factor $(kT_e\Delta f)$ is the noise delivered by an equivalent matched source at a temperature T_e. The temperature T_e is known as the "effective noise temperature" of the network. If the value of $T_e \ll 290$ K the network itself contributes very little extra noise. The relationship between noise figure and effective noise temperature is

$$F = \frac{A_p N_i + N_a}{A_p N_i} = \frac{A_p k\Delta f(T_s + T_e)}{A_p k\Delta f\, T_s}$$

T_s is the standard temperature equal to 290 K. Hence

$$F = 1 + T_e/T_s \qquad \text{or} \qquad T_e = (F-1)T_s \tag{4.10}$$

The maser microwave amplifier is an example of a very low noise network; the effective noise temperature of such a device would be between 10 and 30 K, which is equivalent to a noise figure of between 1.03 and 1.11.

It is often convenient to represent cascaded networks in terms of effective noise temperature; substituting Eqn (4.10) into Eqn (4.9) gives

$$T_e = T_1 + \frac{T_2}{A_{p1}} + \frac{T_3}{A_{p1}A_{p2}} \tag{4.11}$$

This equation is particularly useful when considering the noise performance of a cascaded system in which the first element is an antenna with an effective noise temperature not equal to 290 K. The cascaded system, excluding the antenna, is replaced by a noise-free system with an equivalent matched noise source at the input. The antenna noise is then included by adding the effective noise temperature of the antenna to the equivalent noise temperature of the noise source, and using the sum as the noise temperature of the matched source.

We indicated when referring to Fig. 4.2 that the power gain A_p was not necessarily greater than unity; when A_p is less than unity the network is passive and is usually characterized in terms of insertion loss rather than power gain. The insertion loss of a passive network is the reciprocal of power gain:

$$\text{Insertion loss } L = \frac{\text{input power}}{\text{output power}} \tag{4.12}$$

When a passive network is matched at both the input and output the insertion loss has the same numerical value as the network noise figure. The noise power delivered to the network by a matched source is $kT_s\Delta f$ W. The noise delivered by the network to its load will be the sum of the input noise power multiplied by the network power again (which is less than unity for passive networks) and the noise power generated within the network. This latter component is represented, as in the determination of effective noise temperature, by an equivalent noise source at the input of the network which is then assumed noise free. The total noise power delivered to this noise-free network is $k\Delta f(T_s + T_e)$. The noise power delivered to the load is $N_0 = A_p k\Delta f(T_s + T_e)$. However, the network is a matched source for its load, and such a source delivers a noise power $N_0 = k\Delta f T_s$. Thus these two values must be equal or

$$A_p k\Delta f(T_s + T_e) = k\Delta f T_s$$

i.e.

$$T_e = T_s(1 - A_p)/A_p = T_s(L - 1) \tag{4.13}$$

But $T_e = (F - 1)T_s$. Thus for a matched passive network the insertion loss L has the same numerical value as the network noise figure, i.e. $F = L$. Both F and L are usually measured in decibels.

EXAMPLE: To illustrate the significance of the analysis presented in this section we will consider a typical system involving a domestic television receiver. The receiver that has a video bandwidth of 5.5 MHz is coupled via a 70 Ω coaxial cable with an insertion loss of 6 dB to an antenna with an effective noise temperature of 290 K. The noise figure of the receiver, referred to a matched source of 70 Ω at 290 K, is 6 dB. We are requested to find the SNR at the receiver output when the open-circuit signal voltage at the antenna terminals is 1 mV rms.

Each component in this system will cause a degradation of the SNR. The maximum value of SNR will occur at the input to the coaxial feeder. We begin by determining the SNR at this point and to do this we consider the antenna as both a matched signal and noise source as shown in Fig. 4.4. We assume that the antenna has a source (see Chapter 7) resistance of 70 Ω, and is therefore matched to the 70 Ω feeder. We also assume that the receiver has an input resistance of 70 Ω.

Since the antenna open-circuit signal voltage is 1 mV rms the voltage across the 70 Ω load will be half this value, i.e., 0.5 mV rms. The signal power delivered to the coaxial feeder will be $S_p = (0.5 \times 10^{-3})^2/70$. The noise power delivered by the antenna, which acts as a matched noise source, is $N_p = k\Delta f T_s$. The SNR at the feeder input is thus

$$\text{SNR} = 10\log_{10} S_p/N_p = 52.5 \quad \text{dB}$$

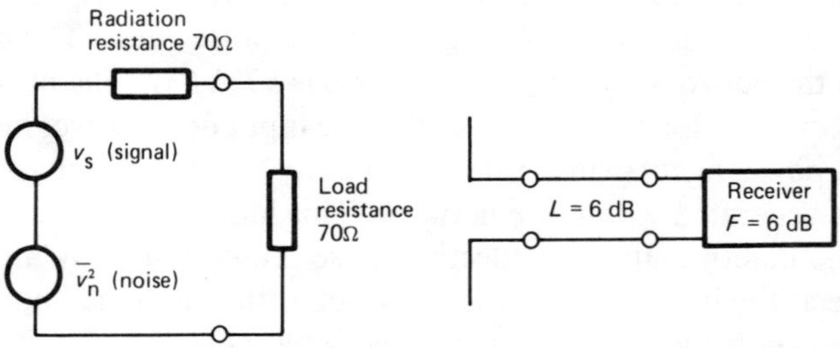

Fig. 4.4 Matched antenna.

We can now calculate the overall noise figure for receiver and feeder. This is valid in this example because the antenna effective temperature is 290 K. The overall noise figure is

$$F = F_1 + (F_2 - 1)/A_{p1}$$

F_1 is the noise figure of the feeder $= L = 6$ dB, i.e. $F_1 = 3.98$. A_{p1} is the reciprocal of the feeder insertion loss and has a value of 0.251. F_2 is the receiver noise figure $= 6$ dB $= 3.98$.

Hence the overall noise figure is $F = 15.85$ (12 dB). The SNR at the receiver output will then be

$$\mathrm{SNR_{out}} = 52.5\,\mathrm{dB} - 12\,\mathrm{dB} = 40\,\mathrm{dB}$$

This figure is actually lower than the minimum acceptable SNR for reasonable picture quality, which is usually considered to be 47dB. The solution to this problem would be the use of a low-noise pre-amplifier between antenna and receiver. The pre-amplifier would have a typical power gain of 20 dB and a typical noise figure of 3 dB. It is possible to connect such an amplifier either directly to the antenna terminals (i.e. before the feeder) or directly to the receiver input (i.e. after the feeder). We suggest that the reader examines both cases and determines the output SNR when the pre-amplifier is included. It will be found that the output SNR will be 6 dB higher when the pre-amplifier is connected directly to the antenna terminals. This verifies the earlier conclusion that the first network in a cascaded system has the major effect on the overall noise figure.

In this example we represented the domestic television receiver as a single network with an overall noise figure. In fact the receiver will be divided into several distinct functional blocks including RF and IF amplifiers, vision detector, video frequency amplifier, etc. From our consideration of the superheterodyne receiver in Chapter 2 it is clear that the signal bandwidth at various points in a receiver (e.g. before and after the detector) is not necessarily constant. This means that possibilities exist for exchanging bandwidth for SNR at various points in a receiver, the theoretical relationship being given by Shannon's law. Thus although it is common practice to specify a single noise figure for a receiver it is more instructive to consider the individual sections of such a system.

A general telecommunications system can be represented in the form shown in Fig. 4.5. In this figure the bandwidth of the signal source is B_1 Hz. The signal enters the modulator and the bandwidth is changed to B_2 Hz. During the transmission noise is added to the signal. It is convenient to show this noise as a single input as in Fig. 4.5, but as we have shown, noise is actually added at every stage of a communications system. The SNR at the detector input is given the symbol SNR_i. At this point the detector transforms the bandwidth of the received signal back to its original value of B_1 Hz and in so doing produces an output signal-to-noise ratio SNR_0. Shannon's law states that whenever there is

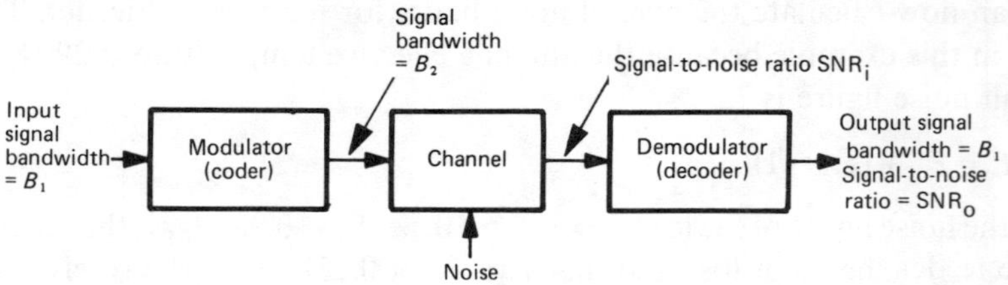

Fig. 4.5 General telecommunications system.

a change in signal bandwidth there should be an accompanying change in SNR. Shannon's law is, however, a theoretical law in which no allowance is made for any physical constraints that may exist in practical systems. We must therefore examine each transmission system on an individual basis and in order to do this it is necessary to develop an algebraic technique for specifying the effect of noise.

4.4 Algebraic Representation of Band-limited Noise

Equation (4.3) indicates that the noise power delivered by a matched source is $\eta\Delta f$ W. This means that the total noise delivered to the detector in a receiver of bandwidth B Hz is ηB Hz. The algebraic representation of this noise is derived by dividing the bandwidth B into small elements Δf and approximating the noise power within each element by a cosine wave. As the element of bandwidth $\Delta f \rightarrow 0$ this gives a very accurate representation of the noise signal. The technique is illustrated by Fig. 4.6. The noise voltage produced in the elemental bandwidth Δf is represented as

$$n(t) = A_n \cos[2\pi f_k t + \theta_k(t)] \tag{4.14}$$

where $\frac{1}{2}A_n{}^2 = \eta\Delta f$ and f_k is the centre frequency of the interval Δf. The phase angle $\theta_k(t)$ is an arbitrary random number. The total noise voltage produced over the entire bandwidth B is calculated by summing the contributions of each element Δf, i.e.

$$V_n(t) = \sum_k A_n \cos[2\pi f_k t + \theta_k(t)] \tag{4.15}$$

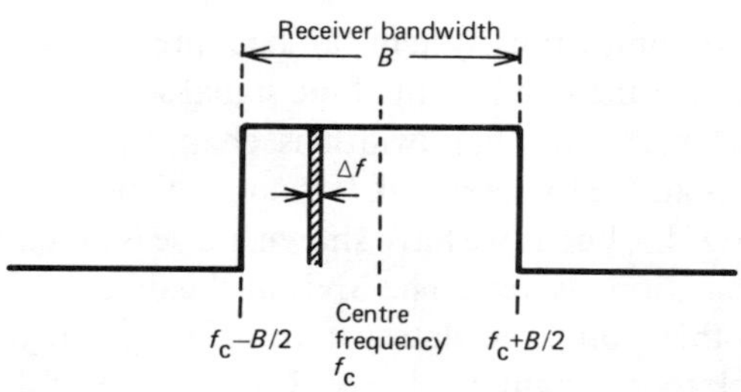

Fig. 4.6 Band-limited white noise.

It is convenient at this point to introduce the substitution $f_k = (f_k - f_c) + f_c$ where f_c is the centre frequency of the bandwidth B. Equation (4.14) then becomes

$$n(t) = A_n \cos[2\pi(f_k - f_c)t + \theta_k(t)]\cos(2\pi f_c t) - A_n \sin[2\pi(f_k - f_c)t + \theta_k(t)]\sin(2\pi f_c t)$$

which means that Eqn (4.15) can be re-written

$$V_n(t) = x(t)\cos(2\pi f_c t) + y(t)\sin(2\pi f_c t) \tag{4.16}$$

where

$$x(t) = \sum_k (2\eta\Delta f)^{1/2}\cos[2\pi(f_k - f_c)t + \theta_k(t)] \tag{4.16a}$$

and

$$y(t) = \sum_k (2\eta\Delta f)^{1/2}\sin[2\pi(f_k - f_c)t + \theta_k(t)] \tag{4.16b}$$

We can thus represent the noise voltage in terms of the sum of two amplitude modulated carriers in the phase quadrature. The carrier amplitudes are the random variables $x(t)$ and $y(t)$ that have mean square values $\overline{x^2(t)}$ and $\overline{y^2(t)}$, respectively. The total noise power in the bandwidth B is

$$P = \frac{\overline{x^2(t)}}{2} + \frac{\overline{y^2(t)}}{2}\ \text{W} \tag{4.17}$$

but

$$\overline{x^2(t)} = \overline{y^2(t)} = \sum_k \frac{2\eta\Delta f}{2}$$

hence

$$P = \overline{x^2(t)} \quad \text{or} \quad \overline{y^2(t)} \quad \text{i.e.} \quad P = \eta B\ \text{W} \tag{4.18}$$

Having derived this representation of band-limited noise we can now examine the effectiveness of various signal transmission systems in the presence of noise. This will be accomplished by comparing the SNR after detection (or decoding) with the value that exists before detection.

4.5 SNR Characteristics of Envelope Detected DSB-AM

The assumptions made in this section are

(i) the modulating signal is a single tone
(ii) the envelope detector has an ideal characteristic that infers that its output is directly proportional to instantaneous carrier amplitude.

Some care is required in defining SNR at the detector input; the DSB-AM signal consists of a carrier and two sidebands and we can specify the signal power in terms of combinations of these components. The DSB-AM signal plus noise is written as

$$V_{in}(t) = A_c[1 + m\cos(2\pi f_m t)]\cos 2\pi f_c t + x(t)\cos(2\pi f_c t) + y(t)\sin(2\pi f_c t) \tag{4.19}$$

If we normalize this voltage to be the voltage developed across a resistance of 1 Ω then the carrier power is $\frac{1}{2}A_c^2$ W, the sideband power is $\frac{1}{4}(mA_c)^2$ W and the total power is $\frac{1}{2}A_c^2\ (1 + \frac{1}{2}m^2)$ W. The noise power is $\frac{1}{2}\overline{x^2(t)} + \frac{1}{2}\overline{y^2(t)} = \overline{x^2(t)}$ W. There are three commonly used methods of specifying SNR at the detector input: these are

(1) carrier power-to-noise ratio,
(2) sideband power-to-noise ratio, and
(3) total power-to-noise ratio.

The carrier-to-noise ratio is $S_c/N = A_c^2/2\overline{x^2(t)}$, the sideband-to-noise ratio is $S_{sb}/N = (mA_c)^2/4\overline{x^2(t)}$, and the total SNR is

~~$S_t/N = A_c^2(1 + \frac{1}{2}m)^2/2\overline{x^2(t)}$~~ $S_t/N = A_c^2(1 + \frac{m^2}{2})/2\overline{x^2(t)}$

The resultant input to the detector is given by the vector sum of the amplitude modulated waveform and the noise components. The graphical addition is shown in Fig. 4.7, from which we can see that if the SNR is very large then the phase angle $\phi \to 0$. The input to the detector for large SNR is given approximately by

$$V_{in}(t) = A_c(1 + m\cos 2\pi f_m t) + x(t) \tag{4.20}$$

The detector, being ideal, will have an output

$$V_{out}(t) = aA_c(1 + m\cos 2\pi f_m t) + ax(t) \tag{4.21}$$

The output signal power is $\frac{1}{2}(amA_c)^2$ and the output noise power is $a^2\overline{x^2(t)}$. The SNR at the detector output is thus

$$\text{SNR}_{out} = (mA_c)^2/2\overline{x^2(t)} \tag{4.22}$$

If we compare this figure with the carrier-to-noise ratio at the detector input then

$$\text{SNR}_{out} = m^2 S_c/N$$

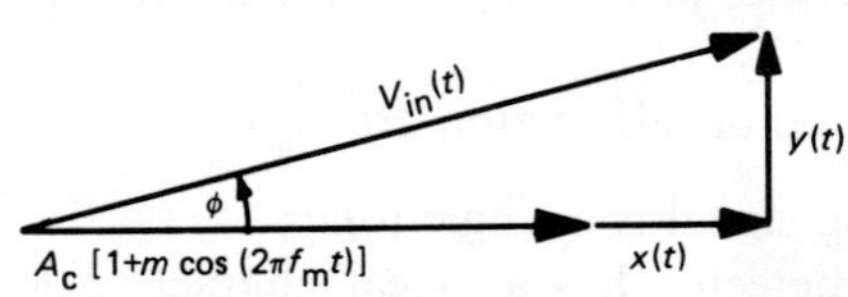

Fig. 4.7 Signal plus noise at the detector input.

This has its maximum when $m = 1$, i.e.

$$\mathrm{SNR}_{\mathrm{out(max)}} = S_c/N \tag{4.23}$$

This equation states that the SNR at the detector output has the same numerical value as the carrier-to-noise ratio at the detector input. It does not state that the SNR at the detector output is equal to the SNR at the detector input, which is the interpretation sometimes used in the literature.

If $\mathrm{SNR}_{\mathrm{out}}$ is compared with the sideband-to-noise ratio at the detector input, then

$$\mathrm{SNR}_{\mathrm{out}} = 2S_{sb}/N \tag{4.24}$$

The output SNR is 3 dB greater than the sideband power-to-noise ratio. It is not correct, however, to say that DSB-AM produces an SNR improvement of 3 dB.

To produce a realistic assessment of DSB-AM we must compare the output SNR with the total SNR at the detector input. When $m = 1$ then

$$\mathrm{SNR}_{\mathrm{out}} = \frac{2}{3}\frac{S_t}{N} \tag{4.25}$$

The SNR at the output of an ideal envelope detector is actually poorer than the SNR at the detector input. This is explained by the fact that even when $m = 1$, 66.6% of the total transmitted power is contained within the carrier component, which does not contribute at all to the signal power at the detector output. Thus DSB-AM does not obey Shannon's law in its strict theoretical statement. The situation is improved somewhat if the carrier is suppressed, as this shows an improvement of 3 dB in SNR as given by Eqn (4.24). If the carrier is suppressed, however, it is not possible to employ envelope detection.

4.6 SNR Characteristics of Coherently Detected DSB-AM

Equation (4.20) was derived on the assumption that the SNR at the detector input was very large. If this condition is not met the resulting values of $\mathrm{SNR}_{\mathrm{out}}$ given by Eqns (4.23), (4.24) and (4.25) are not valid. In fact at low values of SNR the performance of the envelope detector deteriorates rapidly. The envelope detector can therefore be employed only in good SNR conditions. This is usually taken to mean that the envelope detector has a signal-to-noise performance that is acceptable only if the SNR at the detector input is greater than 10 dB.

Coherent detection is an alternative demodulation technique for DSB-AM and an essential technique for suppressed carrier and SSB-AM. The performance of the coherent detector is maintained for all values of input SNR. The coherent detector multiplies the received signal by a locally produced reference signal $E\cos(2\pi f_c t)$. When the received signal is accompanied by noise, the detector output is

$$V_0(t) = EA_c[1 + m\cos(2\pi f_m t)]\cos^2(2\pi f_c t) + Ex(t)\cos^2(2\pi f_c t) + Ey(t)\cos(2\pi f_c t)\sin(2\pi f_c t) \tag{4.26}$$

The frequency terms produced by the multiplication, which are outside the bandwidth occupied by the modulating signal, are removed by the filter that follows the coherent detector, the resulting output being

$$V_{out}(t) = \tfrac{1}{2}EA_c + \tfrac{1}{2}EA_c m\cos 2\pi f_m t + \tfrac{1}{2}Ex(t) \tag{4.27}$$

The signal and noise powers in this expression are respectively $S_p = \frac{1}{8}(EA_c m)^2$ W and $N_p = \frac{1}{4}E^2\overline{x^2(t)}$ W. The SNR at the detector filter output is thus

$$\text{SNR}_{out} = (mA_c)^2/2\overline{x^2(t)} \tag{4.28}$$

This is the same result as for the envelope detector but there is no pre-condition that the SNR should be large. In other words, the coherent detector maintains its SNR performance for all values of input SNR and is therefore superior to the envelope detector in poor SNR conditions.

4.7 SNR Characteristics of DSB-SC-AM

The detection of DSB-SC-AM is discussed fully in Section 2.10. If we assume that such a detector has an ideal characteristic the output SNR will be 3 dB greater than input sideband power-to-noise ratio. In other words, DSB-SC-AM produces a 3 dB SNR improvement.

4.8 SNR Characteristics of SSB-AM

This form of modulation requires coherent detection. In this case the input to the detector will be one sideband only, plus noise. If we assume a single modulating tone of frequency f_m, the signal plus noise at the detector input will be

$$V_{in}(t) = A_c\cos[2\pi(f_c + f_m)t] + x(t)\cos(2\pi f_c t) + y(t)\sin(2\pi f_c t) \tag{4.30}$$

The input SNR is thus

$$\text{SNR}_{in} = A_c/2\overline{x^2(t)} \tag{4.31}$$

We should point out here that because the bandwidth of an SSB signal is approximately half that of a DSB signal, the noise power at the detector input will be half the equivalent noise power in a DSB system. The output of the SSB coherent detector after filtering is

$$V_{out}(t) = \tfrac{1}{2}EA_c\cos 2\pi f_m t + \tfrac{1}{2}Ex(t) \tag{4.32}$$

The SNR at the filter output is thus $SNR_{out} = A_c^2/2\overline{x^2(t)}$. Thus for SSB-AM

$$SNR_{out} = SNR_{in} \tag{4.33}$$

It is interesting to compare an SSB transmission with DSB-SC transmission when the transmitted power is the same in each case. We have shown that there is a 3 dB improvement in the DSB case, but because the noise power in the SSB system is only half that of the DSB system the two are actually equivalent in terms of output SNR when the transmitted power in each case is the same.

It would seem on a purely SNR basis that there is little to choose between DSB-SC-AM and SSB-AM. This is not entirely true: in certain situations (when ionospheric reflections are used, for example) severe distortion can be produced in DSB systems because components in the two sidebands can have differing phase velocities resulting in partial cancellation after detection. This problem is not encountered in SSB systems, when used for audio signal transmission, because the ear is insensitive to phase distortion. In these circumstances there is a considerable advantage in using SSB transmission.

4.9 SNR Characteristics of FM

We will establish the SNR properties of FM by again assuming an ideal detector, i.e. one that has an output voltage directly proportional to the 'instantaneous frequency' of the input signal. The algebraic representation of band-limited noise is identical to the AM case, but it should be borne in mind that the bandwidth of a frequency modulated signal is usually considerably greater than the bandwidth of an AM signal, and that noise power is directly proportional to bandwidth.

The voltage at the FM detector input will be the sum of signal and noise, i.e.

$$V_{in}(t) = A_c \cos[2\pi f_c t + \beta \sin(2\pi f_m t)] + x(t)\cos(2\pi f_c t) + y(t)\sin(2\pi f_c t) \tag{4.34}$$

The FM signal has constant amplitude and the signal power is also constant and independent of the amplitude of the modulating signal. This is a fundamental difference between FM and AM. The signal power at the detector input has a value of $A_c^2/2$ W. It is convenient to calculate the signal power at the detector output in the absence of noise, and the noise power in the presence of an unmodulated carrier (i.e. in the absence of signal).

If we assume that the modulating signal is a single tone of frequency f_m the instantaneous frequency of the FM waveform is $f_i = f_c + \Delta f \cos(2\pi f_m t)$, and this will produce a voltage at the detector output given by

$$V_s = b2\pi\Delta f \cos(2\pi f_m t) \tag{4.35}$$

The signal power at the detector output will thus be

$$S_0 = (b2\pi\Delta f)^2 \text{ watts} \tag{4.36}$$

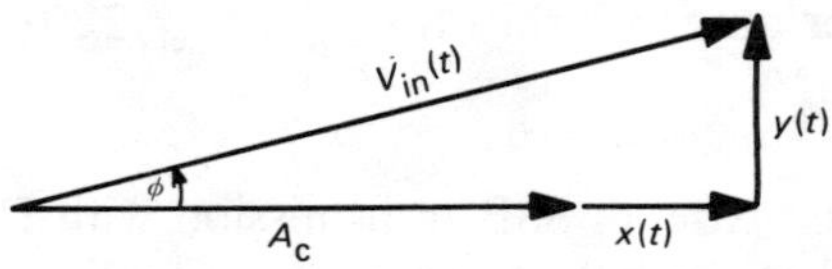

Fig. 4.8 Unmodulated FM carrier plus noise.

We calculate the noise voltage at the detector output using phasor methods. The phasor diagram for the unmodulated carrier and quadrature noise components is shown in Fig. 4.8. The FM detector produces an output proportional to frequency, which is the time differential of the phase angle ϕ. If we assume that the SNR at the detector input is large, then the phase angle is given by

$$\phi(t) \simeq \tan^{-1}\left[\frac{y(t)}{A_c}\right] \simeq \frac{y(t)}{A_c} \tag{4.37}$$

The "instantaneous frequency" produced by the noise is

$$\dot{\phi}(t) = \frac{\dot{y}(t)}{A_c} \tag{4.38}$$

which, if we assume $\dot{y}(t) \ll A_c$, is narrowband FM. The noise voltage at the detector output is

$$V_n = \frac{b}{A_c}\dot{y}(t) \tag{4.39}$$

where b is the detector constant of proportionality. The noise waveform $y(t)$ is itself the sum of many elemental noise components, i.e.

$$y(t) = \sum_k A_n \sin\left[2\pi(f_k - f_c)t + \theta_K(t)\right]$$

The noise component at a particular frequency f_k is

$$y_k(t) = A_n \sin\left[2\pi f t + \theta_k(t)\right] \tag{4.40}$$

where $f = (f_k - f_c)$ is the frequency difference between the noise component and the centre frequency, and can have both negative and positive values. The noise voltage at the detector output produced by an elemental noise component at the detector input is

$$\delta V_n = \frac{b}{A_c} 2\pi f A_n \cos\left[2\pi f t + \theta_k(t)\right] \tag{4.41}$$

where we have assumed that $\theta_k(t)$ varies very slowly and does not contribute to the output amplitude. The power at the detector output produced by this elemental component is

$$\Delta N_0 = \frac{(b 2\pi A_n f)^2}{2A_c^2} \text{ watts} \tag{4.42}$$

but

$$A_{\rm n} = (2\eta \Delta f)^{1/2}$$ cont.

i.e.

$$\Delta N_0 = \left(\frac{2\pi b}{A_{\rm c}}\right)^2 \eta f^2 \Delta f$$

Hence

$$\frac{\mathrm{d}N_0}{\mathrm{d}f} = K f^2 \quad \text{where} \quad K = \eta\left(\frac{2\pi b}{A_{\rm c}}\right)^2 \tag{4.43}$$

In Eqn (4.43) $\mathrm{d}N_0/\mathrm{d}f$ represents the power spectral density of the noise at the detector output and is no longer white but proportional to f^2. This means that the noise power at the output of an ideal FM detector increases with the square of the frequency difference between the centre frequency and the elemental noise frequency. The noise voltage spectral density $(\mathrm{d}N_0/\mathrm{d}f)^{1/2}$ is plotted as a function of f in Fig. 4.9.

The noise at the output of a frequency modulation detector will be produced by the sum of all components within the passband of the filter that follows the detector. If we assume that this filter has an ideal low-pass characteristic with a cut-off frequency of $\pm f_{\rm c}$ (the negative figure is required for negative values of $f = (f_k - f_{\rm c})$), then since the noise produces narrowband FM we can determine the total noise using super-position. The total noise power is

$$N_0 = \int_{-f_c}^{f_c} \mathrm{d}N_0$$

i.e.

$$N_0 = 2K \int_0^{f_0} f^2\,\mathrm{d}f$$

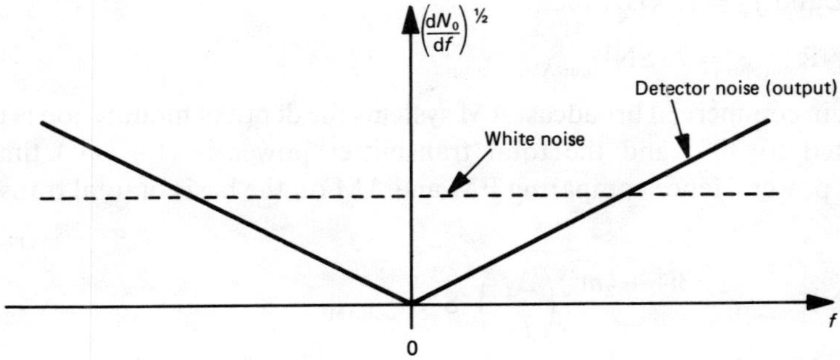

Fig. 4.9 FM detector noise spectral density.

which evaluates to

$$N_0 = \eta\left(\frac{2\pi b}{A_c}\right)^2 \tfrac{2}{3} f_0^3 \tag{4.44}$$

The signal power at the detector output is given by Eqn (4.36). Thus the SNR is

$$\mathrm{SNR}_{\mathrm{out}} = \frac{3S_c}{2\eta f_0}\left(\frac{\Delta f}{f_0}\right)^2 \tag{4.45}$$

It should be noted that $\Delta f/f_0 \neq \beta$ unless $f_0 = f_m$, which is not necessarily the case.

Equation (4.45) is interpreted as stating that the SNR at the output of a frequency modulation detector increases with the square of the carrier deviation, which is independent of the carrier power. It might be argued that the SNR could be increased indefinitely simply by increasing the carrier deviation. This overlooks the fact that increasing Δf increases the signal bandwidth with a consequent increase in the noise power, which is itself directly proportional to bandwidth. As Δf increases a threshold is reached at which the assumption $y(t) \ll A_c$ is no longer valid. Beyond this threshold a very rapid fall in SNR at the detector output is witnessed. This threshold effect is a characteristic of all wideband systems and is considered in more detail in section 4.11.

The relative performance of AM and FM systems in the presence of noise can be compared by reference to Eqn (4.45). In this equation, f_0 is the bandwidth of the filter following the detector and will be equal to the bandwidth of the modulating signal. The factor $2\eta f_0$ is thus the noise power that would occur at the output of an AM system with the same modulating signal bandwidth. If we compare FM and AM transmissions in which the AM carrier power equals the FM carrier power, the factor $S_c/2\eta f_0$ of Eqn (4.45) is equal to the output SNR for an AM system when the depth of modulation $m = 100\%$, i.e.

$$\mathrm{SNR}_{\mathrm{out(FM)}} = \mathrm{SNR}_{\mathrm{out(AM)}} 3(\Delta f/f_0)^2 \tag{4.46}$$

If we consider a typical FM commercial broadcast system in which $\Delta f =$ 75 kHz and $f_0 = 15$ kHz, then

$$\mathrm{SNR}_{\mathrm{out(FM)}} = 75\,\mathrm{SNR}_{\mathrm{out(AM)}}$$

In fact, in commercial broadcast AM systems the depth of modulation is usually restricted to 30% and the total transmitter power is $(1 + \frac{1}{2}m^2)$ times the carrier power. Hence comparing FM and AM on the basis of total transmitted power,

$$\mathrm{SNR}_{\mathrm{out(FM)}} = \frac{3(1 + \frac{1}{2}m^2)}{m^2}\left(\frac{\Delta f}{f_0}\right)^2 \mathrm{SNR}_{\mathrm{out(AM)}} \tag{4.47}$$

If

$$m = 0.3,\ \Delta f = 75\,\mathrm{kHz} \text{ and } f_0 = 15\,\mathrm{kHz},$$

then

$$\mathrm{SNR}_{\mathrm{out(FM)}} = 810.83\, \mathrm{SNR}_{\mathrm{out(AM)}}$$

Alternatively, to produce the same SNR the transmitted power in the FM case is 29 dB less than the required power in the AM case.

One conclusion we may draw from these figures is that for a given radiated power an FM transmitter will have a greater range than a DSB-AM transmitter, provided that the SNR at the FM detector input is sufficiently high for the noise to produce narrowband FM of the carrier. In other words, the FM system must be operating above the threshold level. This threshold level is related to both the frequency deviation and the SNR at the detector input. For large values of Δf, the threshold occurs at an SNR of about 13 dB. The conditions required for FM to exhibit its SNR improvement properties are that $\beta > 1/\sqrt{3}$ [assuming $f_0 = f_m$ in Eqn (4.46)] and the SNR at the detector input must exceed 13 dB. If the SNR at the detector input is below the threshold value, the output SNR decreases rapidly and ultimately becomes poorer than the equivalent AM value.

4.10 Pre-emphasis and De-emphasis

The analysis of the previous section was based on a single tone modulating signal and we assumed that irrespective of the frequency of this tone the carrier deviation had its maximum value Δf. In other words, we assumed that the FM signal occupied the maximum possible bandwidth for a given modulating signal. In reality the situation is somewhat different; the modulating signal is not a single tone but a complex signal with a particular power spectral density.

The power spectral density of natural speech is closely approximated by the graph shown in Fig. 4.10. We can see from this figure that above a certain frequency f_1 the power spectrum decreases at a rate approaching 6 dB/octave. If such a signal frequency modulates a carrier, the higher-frequency components will produce a lower carrier deviation than the low-frequency components. In

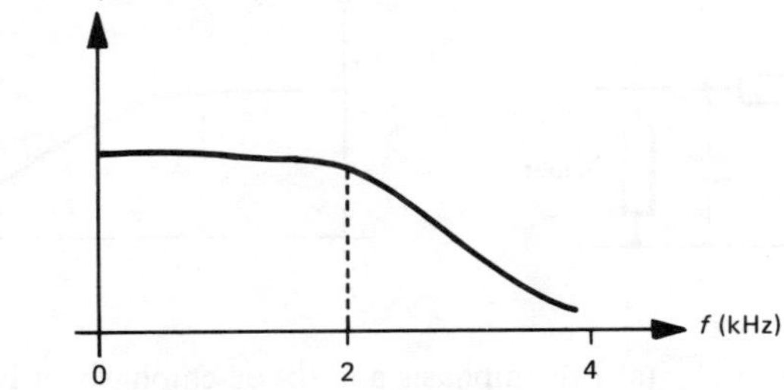

Fig. 4.10 Power spectrum of natural speech.

other words, the bandwidth occupied by a carrier frequency modulated by a signal of this type will be considerably less than if the power spectrum of the modulating signal were uniformly distributed. In practice, therefore, the FM system would not produce the maximum SNR improvement suggested by Eqn (4.46).

The theoretical SNR can be approached, however, if the power spectrum of the modulating signal is made uniform by emphasizing the higher-frequency components before modulation takes place. The spectrum of the modulating signal is restored after detection by applying a corresponding amount of the de-emphasis. The de-emphasis will, of course, operate on the noise produced by the detector as well as the signal, the overall effect being a reduction in the noise power at the detector output. The process of pre-emphasis and de-emphasis is illustrated graphically in Fig. 4.11.

The actual component values used in pre- and de-emphasis networks will vary from circuit to circuit. The amount of pre- and de-emphasis will depend upon the time constant of the filter used, and it is normal to express the pre- and de-emphasis in terms of this time constant. In the UK the value of the time constant used is 50 μs; in contrast the value used in the USA is 75 μs.

It is worth stating that the technique of pre- and de-emphasis does not produce SNR increases above the theoretical values specified by Eqn (4.46). The technique is one of conditioning the modulating signal in order that the maximum theoretical performance can be realized. It is shown by Schwartz[2] that the technique of pre- and de-emphasis is not restricted to FM but can be used in any system in which the power spectral density of the signal decreases

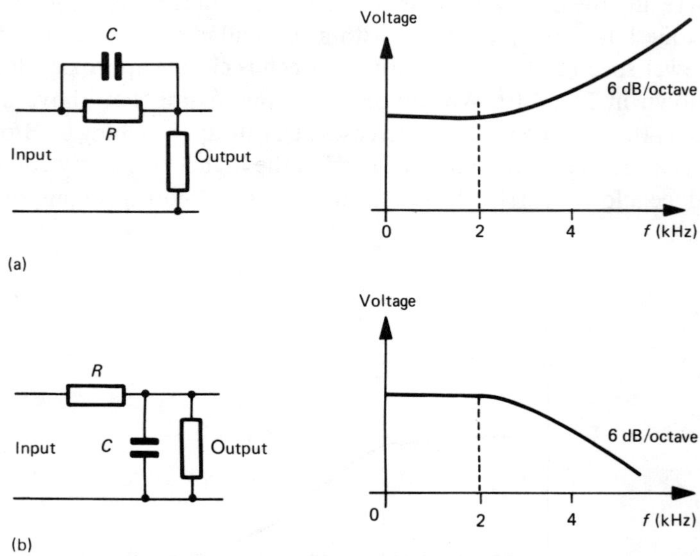

Fig. 4.11 (a) Pre-emphasis and (b) de-emphasis in FM.

more rapidly, as a function of frequency, than the noise power spectral density. In this context, pre- and de-emphasis could be used with AM systems to increase the overall SNR. It is not used in practice with AM systems because it is actually advantageous, from the point of view of adjacent channel separation, to have a signal power spectrum that decreases with frequency.

4.11 The FM Capture (Threshold) Effect

Consider the phasor diagram of Fig. 4.12a in which X represents the required FM carrier and Y represents an interfering FM carrier. (Y could be due to a second FM station at the receiver image frequency, or Y could represent the frequency modulated carrier produced by noise entering the detector.) It is assumed that the amplitudes X and Y are fixed and that the angle between the phasors ϕ_b is uniformly distributed in the range $-\pi \leqslant \phi_y \leqslant +\pi$. It is required to find the resultant phase angle ϕ_r which will be responsible for any interference generated at the detector output.

From Fig. 4.12a

$$\phi_r = \tan^{-1}\left[\frac{Y\sin(\phi_y)}{X + Y\cos(\phi_y)}\right] \tag{4.48}$$

ϕ_r will be a random quantity with mean square value

$$\overline{(\phi_r)^2} = \frac{1}{2\pi}\int_{-}^{+}\left\{\tan^{-1}\left[\frac{Y\sin(\phi_y)}{X + Y\cos(\phi_y)}\right]\right\}^2 \mathrm{d}\phi_y \tag{4.49}$$

Equation 4.49 can be solved numerically and the rms value of ϕ_r is plotted

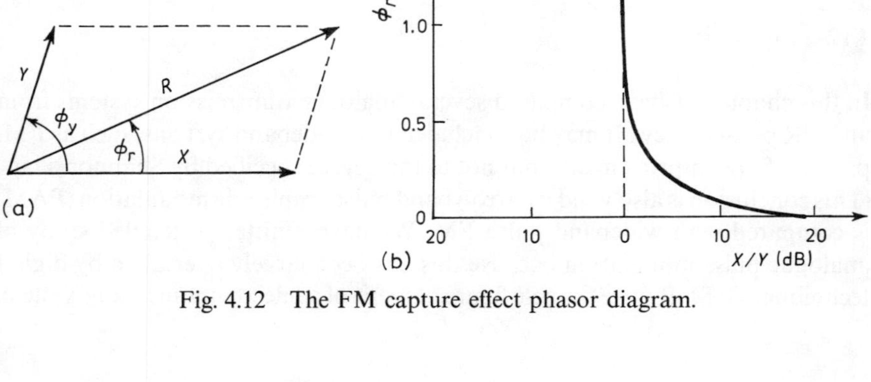

Fig. 4.12 The FM capture effect phasor diagram.

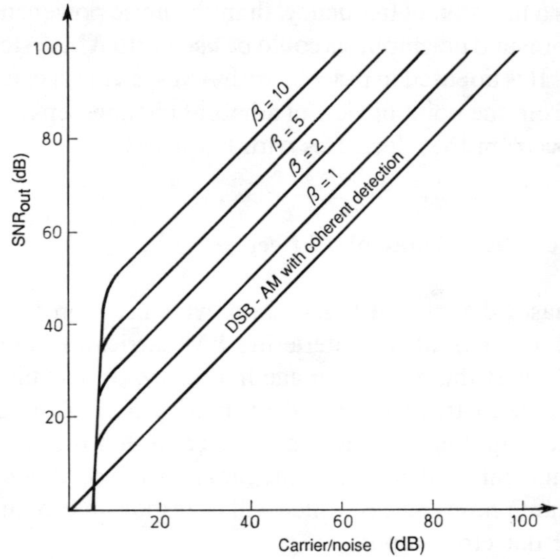

Fig. 4.13 Noise improvement obtained with FM.

against the ratio X/Y in Fig. 4.12b. It will be noted from this figure that a very rapid change in $\sqrt{(\overline{\phi_r})^2}$ occurs in the region $X = Y$. When $X/Y = 3$ the phase angle ϕ_r becomes very small. This means that the interference component produced by Y (which is proportional to $d\phi_r/dt$) also becomes very small. Hence when the ratio $X/Y > 3$ the carrier X takes over, or captures, the system. The FM receiver will thus discriminate in favour of the stronger signal. This is also true when $X/Y < 1$. In the latter case component Y captures the system. Hence if Y is due to noise, a very rapid deterioration in performance is observed. The FM threshold effect is clearly demonstrated in Fig. 4.13 which plots the SNR at the detector output as a function of the SNR at the detector input.

The FM capture effect is actually very useful and is employed in mobile cellular radio (see Fig. 13.21) to suppress interference from base stations in distant cells operating on the same frequency.

4.12 Conclusion

In this chapter we have compared several analogue transmission systems from an SNR point of view. It may be concluded that wideband systems, such as FM, produce SNR improvements, but not to the degree specified by Shannon's law. (This conclusion is also valid if narrowband pulse amplitude modulation (PAM) is compared with wideband pulse FM. We have omitted a detailed study of analogue pulse modulation because this has been largely overtaken by digital techniques.) If SNR were the sole figure of merit of a telecommunications system, it would be reasonable to conclude that wideband systems can offer superior performance to narrowband systems. There are, however, many other factors that influence the choice of modulation system. For example, the wide bandwidth of FM precludes its use in the medium-wave broadcast band because here the main requirement is to accommodate as many individual stations as possible in a relatively small bandwidth. In any case SNR problems are not usually significant in high-power broadcast transmissions.

There are many instances where the lower bandwidth of SSB is used in preference to DSB-AM. One such instance is the frequency multiplexing of telephone circuits for trunk transmission. In such a system it is possible to transmit a synchronizing signal from which the individual carriers required for coherent demultiplexing can be derived.

Traditionally medium-wave (i.e. local) broadcasts have used DSB-AM although bandwidth is at a premium. Historically DSB receivers were cheaper to produce and more reliable than SSB receivers. Modern technology has completely changed the situation where complexity and reliability are no longer closely related to cost. It is now perfectly possible both technically and economically to mass-produce reliable SSB receivers. It is unlikely that any moves will be made in this direction for some time because of the problems of compatibility, i.e. existing DSB receivers, of which there are many millions, would be unable to receive acceptable quality SSB transmission.

Thus, as we have stated, SNR is but one of many factors that influence the choice of a particular transmission system.

References

1. King, R., *Electrical Noise*, Chapman & Hall, London, 1966, Chapter 3.
2. Schwartz, M., *Information Transmission Modulation and Noise*, 3rd edn. McGraw-Hill, New York, 1980, p. 412.

Problems

4.1 Calculate the mean square output noise voltage when a signal generator with an output resistance of 600 Ω is connected to the input of the two-port network shown.

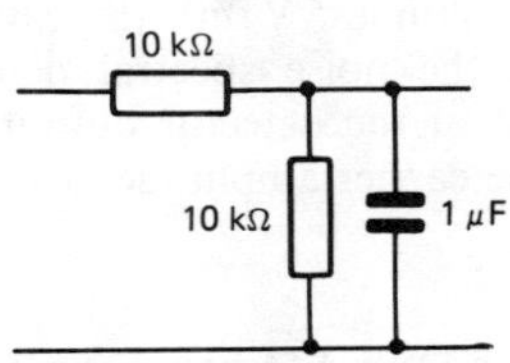

Answer: $4.002 \times 10^{-15}\ V^2$.

4.2 An amplifier is made up of three identical stages in cascade, each stage having equal input and output resistances. The power gain per stage is 8 dB and the noise figure per stage is 6 dB when the amplifiers are correctly matched. Calculate the overall power gain and noise figure for the cascaded amplifier.

Answer: 24 dB; 6.6 dB.

4.3 The noise figure of a receiver, relative to a matched source at a temperature of 290 K, is 0.9 dB. Calculate the effective noise temperature at the input of the receiver when an antenna of effective noise temperature 200 K is connected.

Answer: 266.6 K.

4.4 The following diagram represents a satellite receiving system coupled by a waveguide to an antenna of effective noise temperature 70 K.

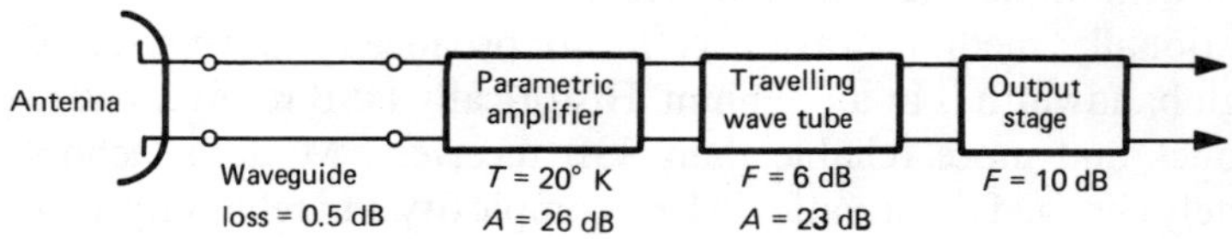

(a) Calculate the equivalent noise temperature of the waveguide and travelling wave tube (a device for amplifying frequencies in the GHz range).
(b) Calculate the SNR at the output of the receiver assuming the antenna radiation resistance to be 50 Ω and the available received power is 10 pW. The bandwidth of the system is 10 MHz. (Hint: find the equivalent SNR referred to the waveguide input.)

Answer: (a) 35 K, 870 K; (b) 27.5 dB.

4.5 A superheterodyne receiver is connected to an antenna with a noise temperature of 100 K by a co-axial feeder having a loss of 2 dB. The receiver characteristics are

RF bandwidth = 5 MHz
IF = 20 MHz
IF bandwidth = 1 MHz
noise figure = 4 dB

Calculate the total system noise temperature and the required signal power delivered by the antenna to give an SNR of 20 dB at the output of the IF stage. Assume the antenna and receiver are both matched to the co-axial line.

Answer: 964 K; 1.33 pW.

4.6 A single tone of amplitude 2 V rms and frequency 5.8 kHz is used to amplitude modulate a carrier of amplitude 5 V rms, the carrier and both sidebands being transmitted. Given that the noise spectral density at the detector input is 0.1 μW/Hz, find the SNR at the detector output. The audio bandwidth of the receiver is 10 kHz and the carrier amplitude at the detector input is 1 V rms.

Answer: 16 dB.

4.7 A vhf transmitter radiates a DSB-AM signal at a depth of modulation of 45% with an audio bandwidth of 15 kHz. This produces an SNR of 40 dB at the output of a receiver at a distance of 3 km from the transmitter. If the transmitter is switched to FM radiating the same total power, at a carrier deviation of 60 kHz, find the theoretical distance from the transmitter for the same SNR at the output of an FM receiver. Assume the noise spectral density at the receiver input is the same in each case and that the received power decreases as the square of the distance from the transmitter.

Answer: 48.5 km.

4.8 A radio station transmits a DSB-SC-AM signal with a mean power of 1 kW. If SSB-AM is used instead calculate the mean power for (a) the same signal strength and (b) the same SNR at the detector output.

A single tone of frequency 7.5 kHz forms the modulating signal for both a DSB-AM and a FM system, the power transmitted in each case being the same. When modulated, the peak deviation of the FM carrier is 60 kHz and the amplitude of the first pair of FM sidebands is equal to the sideband amplitude of the AM transmission. Assuming an audio bandwidth of 7.5 kHz for both AM and FM receivers, determine the SNR advantage of the FM receiver. If may be assumed that the noise spectral density has a constant value and is the same for each case.

Answer: (a) 2 kW, (b) 1 kW; 29.4 dB.

4.9 A frequency modulation receiver consists of a tuned amplifier of bandwidth 225 kHz that feeds a limiter and an ideal discriminator followed by a low-pass filter with a bandwidth of 10 kHz. The carrier-to-noise ratio at the discriminator input is 40 dB when the modulating signal is a 10 kHz tone, producing a carrier deviation of 50 kHz. Calculate the SNR at the output of the filter.

If the amplitude of the modulating signal is maintained at the same value when the frequency is changed to 1 kHz, find the new SNR at the filter output. What would be the SNR if the amplitude of the modulating signal is halved?

Answer: 69 dB; 69 dB; 63 dB.

4.10 The SNR at the output of a coherent detector is 25 dB when the input is an SSB-AM wave. If the input to the detector is transferred to DSB-AM with $m = 1$, find the increase in total power at the detector input to maintain an output SNR of 25 dB.

Answer: 4.8 dB.

5 / NOISE IN DIGITAL COMMUNICATIONS SYSTEMS

The primary interest in the study of analogue communications systems is the obscuring or masking of the transmitted signal by additive noise. This effect is most conveniently analysed by considering the spectral properties of the noise waveform. The situation in digital systems is quite different since only fixed signal levels are allowed. When these levels are obscured by additive noise the receiver is required to decide which of the allowed levels the noisy signal represents. If the receiver decides correctly the noise has no effect on the received signal whatever. If the receiver makes an incorrect decision the results can be catastrophic. Decision theory is based upon the statistical rather than the spectral properties of noise although, as one might expect, these properties are related. The most important statistical property of white noise is its amplitude distribution function.

5.1 The Amplitude Distribution Function of White Noise

White noise is a naturally occurring phenomenon produced by the superposition of many randomly occurring events. Consequently it is not possible to specify the instantaneous amplitude of the noise waveform at any given instant. The alternative is to determine the probability that the noise waveform amplitude will exceed a given value. This is possible and requires a knowledge of the amplitude distribution function of the noise waveform. Suppose that n independent samples are taken of a noise waveform. (In order for the samples to be statistically independent, these samples must be spaced in time by an interval not less than $1/B$ sec, where B is the noise bandwidth.) Suppose also that n_v of these samples have amplitudes between v_1 and v_2, i.e. the total number of samples in the range $\Delta v = (v_1 - v_2)$ is n_v. The probability that any noise sample will be in this range is simply n_v/n. If the interval $\Delta v \to 0$, a continuous distribution results. However, as this limit is approached, the probability that any sample will be in the range Δv also approaches zero. The problem is avoided if we represent probability as an area. The probability that a sample lies within the amplitude range Δv is represented as the area of a rectangle of base width Δv. The area of the rectangle thus has the value n_v/n and its height is $n_v/n\Delta v$ which approaches a finite value as $\Delta v \to 0$ and $n \to \infty$. If the height of the rectangle is plotted for all values of Δv the resulting curve is termed the probability density function. The probability that a sample will be within the range v_1 to v_2 is the area under the probability density function between these two limits. The probability density function $p_v(v)$ is formally defined by

$$p_v(v) = \lim_{\substack{\Delta v \to 0 \\ n \to \infty}} \left[\frac{n_v}{n \Delta v} \right]$$

Referring to Fig. 5.1, the probability that v has a value between v_1 and v_2 is

$$P = \int_{v_1}^{v_2} p_v(v)\, \mathrm{d}v \tag{5.1}$$

The central limit theorem is an important theorem in statistics which states that the probability density function (PDF) of the sum of many random variables approaches a Gaussian distribution regardless of the distributions of the individual variables. The PDF (also known as the amplitude distribution function) of white noise is therefore Gaussian and has the value

$$p_v(v) = \frac{\exp[-(v-a)^2/2\sigma^2]}{\sqrt{(2\pi\sigma^2)}} \tag{5.2}$$

In Eqn (5.2) a is known as the mean or expected value of v and σ is the standard deviation. The Gaussian PDF has a characteristic bell shape with a width proportional to σ. The mean value is given by

$$a = \int_{-\infty}^{\infty} v p_v(v)\, \mathrm{d}v \tag{5.3}$$

and the variance (σ^2) is given by

$$\sigma^2 = \int_{-\infty}^{\infty} (v-a)^2 p_v(v)\, \mathrm{d}v \tag{5.4}$$

The total area under the $p_v(v)$ curve must be unity, i.e. the probability that a noise sample has an amplitude between $-\infty$ and $+\infty$ is 1, hence

$$\int_{-\infty}^{\infty} p_v(v)\, \mathrm{d}v = 1 \tag{5.5}$$

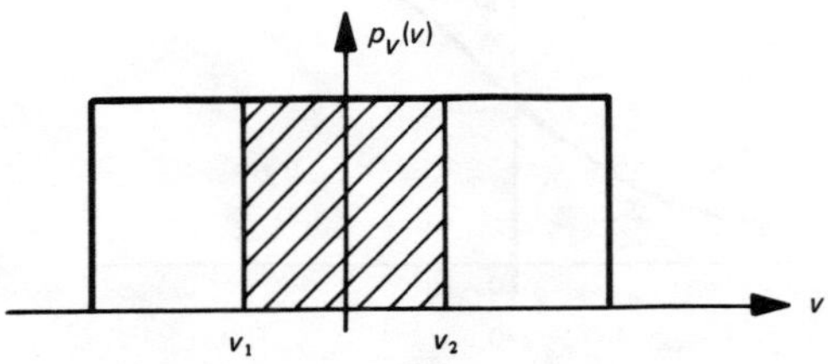

Fig. 5.1 Uniform PDF.

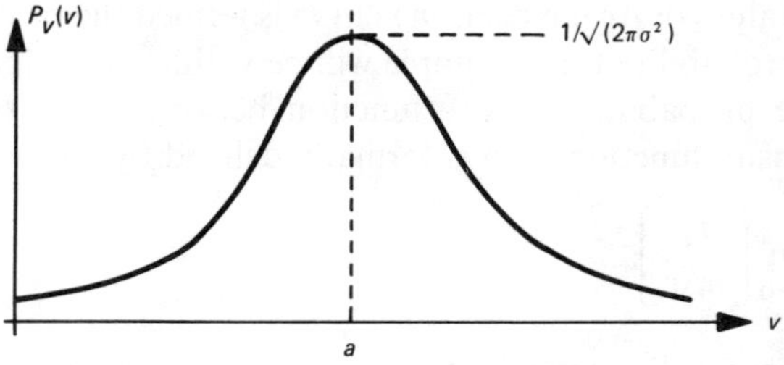

Fig. 5.2 Gaussian PDF.

This gives the Gaussian curve a peak value of $1/\sqrt{(2\pi\sigma^2)}$ at a value of $v = a$ (Fig. 5.2).

The mean value of white noise is zero; hence the variance may be interpreted as the mean square value. Alternatively σ = rms noise voltage.

A second distribution function, known as the cumulative distribution function, is defined by Eqn (5.6):

$$F_v(v) = \int_{-\infty}^{v} p_v(v)\, \mathrm{d}v \tag{5.6}$$

i.e.

$$F_v(v) = \int_{-\infty}^{v} \frac{\exp[-(v-a)^2/2\sigma^2]}{\sqrt{(2\pi\sigma^2)}}\, \mathrm{d}v \tag{5.7}$$

$F_v(v)$ is the probability that the noise will be less than some value v. Since (v) is symmetrical about $v = a$ it follows that $F_v(v) = 0.5$ when $v = a$. As shown in Fig. 5.3 the cumulative density function for white noise is symmetrical about $v = 0$.

The probability that the noise amplitude is less than some value $K\sigma$ is

$$\text{prob}(-K\sigma \leqslant v \leqslant K\sigma) = \int_{-K\sigma}^{K\sigma} \frac{\exp(-v^2/2\sigma^2)}{\sqrt{(2\pi\sigma^2)}}\, \mathrm{d}v \tag{5.8}$$

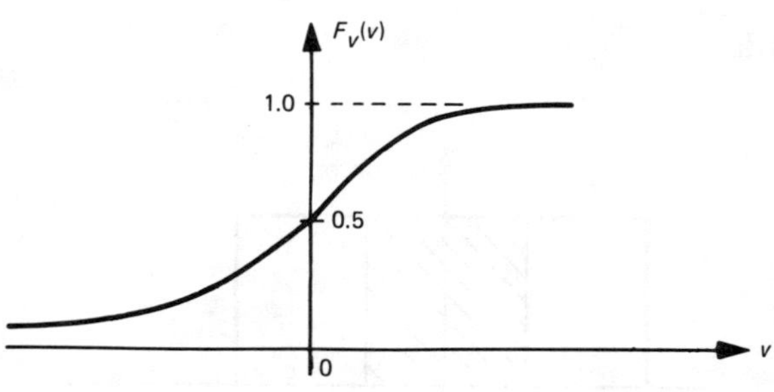

Fig. 5.3 Cumulative density function for white noise with zero mean.

Substituting $y = v/\sqrt{(2\sigma^2)}$ and noting that the Gaussian function is symmetrical about $v = 0$ gives

$$\text{prob}(-K\sigma \leqslant v \leqslant K\sigma) = \frac{2}{\sqrt{\pi}} \int_0^{K/\sqrt{2}} \exp(-y^2)\,dy \qquad (5.9)$$

This integral can be evaluated numerically for values of K and is known as the error function, i.e.

$$\text{erf}(P) = \frac{2}{\sqrt{\pi}} \int_0^{P} \exp(-y^2)\,dy \qquad (5.10)$$

The error function is tabulated in Appendix 3. The probability that v will exceed $K\sigma$ is

$$1 - \frac{2}{\sqrt{\pi}} \int_0^{K/\sqrt{2}} \exp(-y^2)\,dy$$

The complementary error function is defined as

$$\text{erfc}(P) = 1 - \frac{2}{\sqrt{\pi}} \int_0^{P} \exp(-y^2)\,dy \qquad (5.11)$$

Hence the probability that the noise voltage will be less than its rms voltage is, from Eqn (5.9),

$$\text{prob}(-\sigma \leqslant v \leqslant \sigma) = \text{erf}(1/\sqrt{2}) = 0.683$$

Alternatively, the probability that the noise voltage will exceed its rms value is $\text{erfc}(1/\sqrt{2}) = 0.317$. Equations (5.9) and (5.11) are of fundamental importance in determining the probability of decision errors in digital communications systems.

5.2 Statistical Decision Theory

This topic deals with the problem of developing statistical tests to determine which of M possible signals was transmitted over a communications link. In the case of binary communications $M = 2$, i.e. it is possible to transmit either a binary 0 or a binary 1. The receiver takes a single sample during each bit interval and then decides that 0 was transmitted or that 1 was transmitted. There are clearly two types of error that can occur, i.e. deciding 1 when 0 was transmitted or deciding 0 when 1 was transmitted. If the total probability of error is to be minimized, the minimization must be based on both types of error.

The decision rule is based on splitting all values of the received voltage v into two regions V_0 and V_1. The boundary between these regions is then chosen to minimize the total error probability. To illustrate this procedure we will assume that P_0 represents the probability of transmitting 0 and P_1 represents the probability of transmitting 1, clearly $P_0 + P_1 = 1$.

The probability that v will fall into region V_0 when a 1 is transmitted is

$$\int_{V_0} p_1(v)\,\mathrm{d}v$$

where $p_1(v)$ is the probability density function of the received voltage when a 1 is transmitted. The probability that v will fall into the region V_1 when a 0 is transmitted is

$$\int_{V_1} p_0(v)\,\mathrm{d}v$$

where $p_0(v)$ is the probability density function of the received voltage when a 0 is transmitted. The overall probability of error is then

$$P_e = P_0 \int_{V_1} p_0(v)\,\mathrm{d}v + P_1 \int_{V_0} p_1(v)\,\mathrm{d}v \tag{5.12}$$

The region $V_0 + V_1$ covers all values of v; hence

$$\int_{V_0+V_1} p_1(v)\,\mathrm{d}v = \int_{V_0} p_1(v)\,\mathrm{d}v + \int_{V_1} p_1(v)\,\mathrm{d}v = 1$$

Hence we can eliminate the integral over V_0 from Eqn (5.12) to give

$$P_e = P_1 + \int_{V_1} [P_0 p_0(v) - P_1 p_1(v)]\,\mathrm{d}v \tag{5.13}$$

If P_1 is known then P_e can be minimized by making the integral of Eqn (5.13) negative and as large as possible, i.e.

$$P_1 p_1(v) > P_0 p_0(v)$$

The required decision rule is

$$\frac{p_1(v)}{p_0(v)} > \frac{P_0}{P_1} \tag{5.14}$$

$\lambda = p_1(v)/p_2(v)$ is known as the likelihood ratio. In the binary case, if 0 is represented by zero volts and 1 is represented by A volts and the noise is Gaussian, the two density functions will be

$$p_0(v) = \frac{1}{(2\pi\sigma^2)^{1/2}} \exp[-(v)^2/2\sigma^2] \tag{5.15}$$

$$p_1(v) = \frac{1}{(2\pi\sigma^2)^{1/2}} \exp[-(v-A)^2/2\sigma^2]$$

The region V_1 is defined by all values of v for which $\lambda > P_0/P_1$, i.e.

$$\frac{\exp[-(v-A)^2/2\sigma^2]}{\exp[-v^2/2\sigma^2]} > \frac{P_0}{P_1} \tag{5.16}$$

Taking logarithms of Eqn (5.16)

$$v^2 - (v - A)^2 > 2\sigma^2 \ln P_0/P_1$$

The desired region V_1 is thus defined by all values of v corresponding to

$$v > \frac{A}{2} + \frac{\sigma^2}{A} \ln \frac{P_0}{P_1} \tag{5.17}$$

or the boundary between the regions V_0 and V_1 is

$$T = \frac{A}{2} + \frac{\sigma^2}{A} \ln \frac{P_0}{P_1} \tag{5.18}$$

If binary 1 and 0 are equi-probable the decision threshold becomes $T = A/2$ which is exactly half-way between the voltage levels representing 1 and 0. If 0 is transmitted more often than 1 the threshold moves towards V_1. This is illustrated in Fig. 5.4.

Once the decision threshold has been established the total probability of error can then be calculated. If binary 0 (0 volts) is sent, the probability that it will be received as a 1 is the probability that the noise will exceed $+A/2$ volts (assuming $P_0 = P_1$) i.e.

$$P_{e0} = \int_{A/2}^{\infty} \frac{\exp(-v^2/2\sigma^2)}{\sqrt{(2\pi\sigma^2)}} \, dv \tag{5.19}$$

If a 1 is sent (A volts) the probability that it will be received as a 0 is the

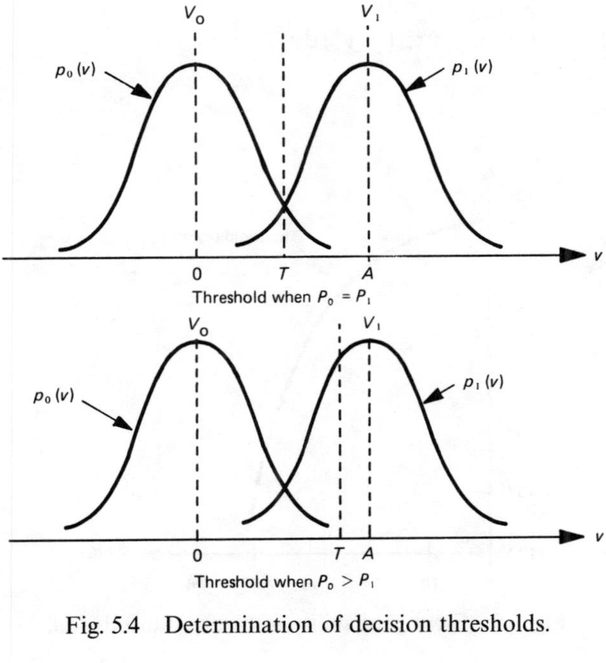

Fig. 5.4 Determination of decision thresholds.

probability that the noise voltage will be between $-A/2$ and $-\infty$, i.e.

$$P_{e1} = \int_{-\infty}^{-A/2} \frac{\exp(-v^2/2\sigma^2)}{\sqrt{(2\pi\sigma^2)}} \mathrm{d}v \tag{5.20}$$

These types of error are mutually exclusive, since sending a 0 precludes sending a 1. Because of the symmetry of the Gaussian function $P_{e0} = P_{e1}$, and the total probability of error, which is $P_0 P_{e0} + P_1 P_{e1}$, can then be written

$$P_e = P_{e1}(P_0 + P_1) = P_{e1}$$

i.e.

$$P_e = \int_{-\infty}^{-A/2} \frac{\exp(-v^2/2\sigma^2)}{\sqrt{(2\pi\sigma^2)}} \mathrm{d}v \tag{5.21}$$

This equation can be written in terms of two integrals

$$P_e = \int_{-\infty}^{0} \frac{\exp(-v^2/2\sigma^2)}{\sqrt{(2\pi\sigma^2)}} \mathrm{d}v - \int_{-A/2}^{0} \frac{\exp(-v^2/2\sigma^2)}{\sqrt{(2\pi\sigma^2)}} \mathrm{d}v \tag{5.22}$$

Using the fact that the Gaussian distribution is symmetrical about its mean value, Eqn (5.22) reduces to

$$P_e = \frac{1}{2} + \int_{0}^{-A/2} \frac{\exp(-v^2/2\sigma^2)}{\sqrt{(2\pi\sigma^2)}} \mathrm{d}v$$

If we substitute $y = v/\sqrt{(2\sigma^2)}$ this becomes

$$P_e = \frac{1}{2} + \frac{1}{\sqrt{\pi}} \int_{0}^{A/(2\sqrt{2}\sigma)} \exp(-y^2)\,\mathrm{d}y$$

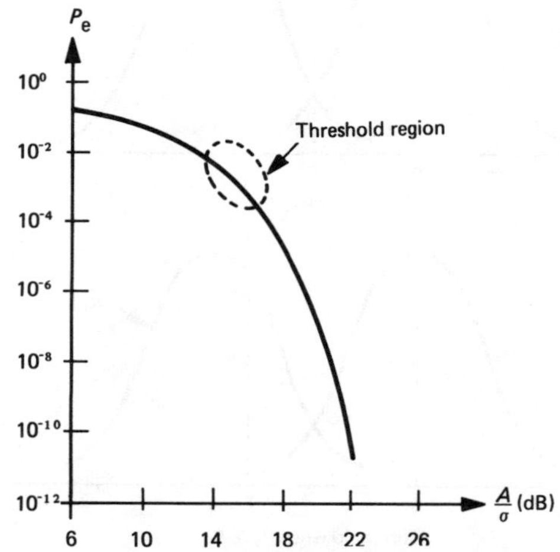

Fig. 5.5 Error probability in binary transmission.

i.e.

$$P_e = \tfrac{1}{2}\{1 - \mathrm{erf}[A/(2\sqrt{2}\sigma)]\} \tag{5.23}$$

The error probability therefore depends solely on the ratio of the peak pulse voltage A to the rms noise voltage σ. A graph of error probability against A/σ is plotted in Fig. 5.5.

This figures shows that when $A/\sigma = 17.4$ dB (a voltage ratio of 7.4:1), $P_e = 10^{-4}$, i.e. on average 1 digit in 10^4 will be in error. If A/σ is increased to 21 dB the error probability drops to $P_e = 10^{-8}$. Hence a very large decrease in error probability occurs for an increase in A/σ of only 3.6 dB. This decrease is much smaller for values of A/σ less than 14 dB. The characteristic thus exhibits a threshold effect for values of A/σ around 18 dB. An error probability of about 10^{-5} is usually considered acceptable for practical data communications systems. This corresponds to a pulse amplitude of about ten times the rms noise voltage.

5.3 Decision Errors in PCM

In Section 3.5 we made a somewhat unjustified assumption that the only source of signal distortion in PCM systems was due to quantization noise. This is only true if the probability of error during transmission is zero. In any PCM system decision errors will occur with a low but finite probability at each regenerative repeater. We will examine the effect of errors in a linearly quantized PCM system with a word length of 8 digits, i.e. 256 levels.

If an error occurs in the least significant digit, this will result in an error equal to the quantization step size δv after the 8-digit code word is converted to an analogue signal (see Fig. 3.11). If the next most significant digit is in error, the voltage error of the analogue signal will be $2\delta v$. An error in the rth digit will produce a voltage error of $\delta v 2^{r-1}$ volts. If we assume that only 1 digit in the 8-digit code is in error (a reasonable assumption if $P_e = 10^{-5}$) the mean square voltage error will be

$$\sum_{r=1}^{8} \frac{(\delta v 2^{r-1})^2}{8} = W \tag{5.24}$$

This is the mean "noise power" associated with each error. Since the error probability is P_e, the mean noise power at the output of a PCM link due to decision errors is $W \times P_e = \frac{1}{8}\sum_{r=1}^{8} P_e/(\delta v 2^{r-1})^2$ watts.

This noise will of course be added to the quantization noise already present on the PCM signal. In the case of PCM an error in the most significant digit is much more serious than an error in the least significant digit. In the transmission of high fidelity sound signals by PCM, the most significant digit is encoded in such a way that errors can be detected, and thus corrected before the digital signal is decoded. This scheme is used by the BBC in the distribution of stereophonic sound broadcast programmes in the UK.

5.4 Decision Errors in Carrier Modulated Data Signals

In this section we will examine the relative performance of ASK, PSK and FSK in noisy conditions. Unlike the comparison of the analogue modulation systems of Chapter 4, which was done on an SNR basis, we compare modulation methods for digital systems in terms of probability of error. All three modulation systems can in fact be detected using coherent detection, although it is more usual to use envelope detection for ASK and FSK. It is convenient, initially, to compare the performance of all three systems assuming coherent detection, because the coherent detector does not alter the PDF of the noise signal.

We consider first of all coherent detection of ASK. The input to the detector will be

$$V_{\text{in}}(t) = h(t)\cos(2\pi f_c t) + x(t)\cos(2\pi f_c t) + y(t)\sin(2\pi f_c t) \tag{5.25}$$

Where $h(t)$ is a binary function with value A or 0 and $[x(t)\cos(2\pi f_c t) + y(t)\sin(2\pi f_c t)]$ represents the band-limited white noise. It was noted in Section 4.4 that both $x(t)$ and $y(t)$ are random variables. We can apply the central limit theorem to Eqn (4.16a) which defines $x(t)$ as the sum of many sinusoidal components with random frequency and phase. The result is that the PDF of both $x(t)$ and $y(t)$ is Gaussian. If the ASK signal plus noise is coherently detected (i.e. multiplied by $\cos[2\pi f_c t]$) and filtered, the baseband output will be

$$V_{\text{out}}(t) = \tfrac{1}{2}h(t) + \tfrac{1}{2}x(t) \tag{5.26}$$

This is similar to Eqn (4.27) which describes the coherent detection of DSB-AM. Ignoring the factor of $\frac{1}{2}$ in Eqn (5.26) the output of the coherent detector is

$$V_{\text{out}} = (A \text{ or } 0) + x(t) \tag{5.27}$$

If binary 1 and 0 are equi-probable and σ^2 represents the variance of $x(t)$ the probability of error, assuming a decision threshold of $A/2$, is

$$P_{e(\text{ASK})} = \tfrac{1}{2}\{1 - \text{erf}[A/(2\sqrt{2}\sigma)]\} \tag{5.28}$$

which is identical to Eqn (5.23) for baseband signalling.

For PSK the value of $h(t)$ in Eqn (5.26) will be $\pm A$ and the decision threshold will be 0 volts. The error probability is then

$$P_{e(\text{PSK})} = \tfrac{1}{2}[1 - \text{erf}(A/\sqrt{2}\sigma)] \tag{5.29}$$

The peak signal-to-noise power ratio for PSK is 6 dB less than for ASK for a given error rate. If we bear in mind that ASK transmits no signal at all for binary 0 (i.e. there is zero signal for half the time) it is evident that to produce the same error rate PSK requires a mean signal-to-noise ratio 3 dB below the value required for ASK. In other words, PSK has a 3 dB advantage, in terms of signal-to-noise power ratio, over ASK and as is seen from Fig. 5.6 this can represent a significant improvement in terms of error probability.

FSK modems usually employ envelope detection and the block schematic of such a modem is given as Fig. 3.27. A similar schematic for an FSK modem employing coherent detection is presented in Fig. 5.7. The pre-detection filters,

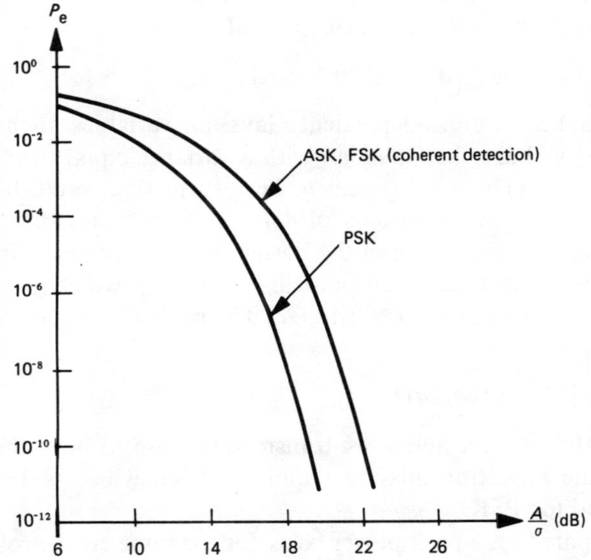

Fig. 5.6 Error probabilities for constant mean signal power.

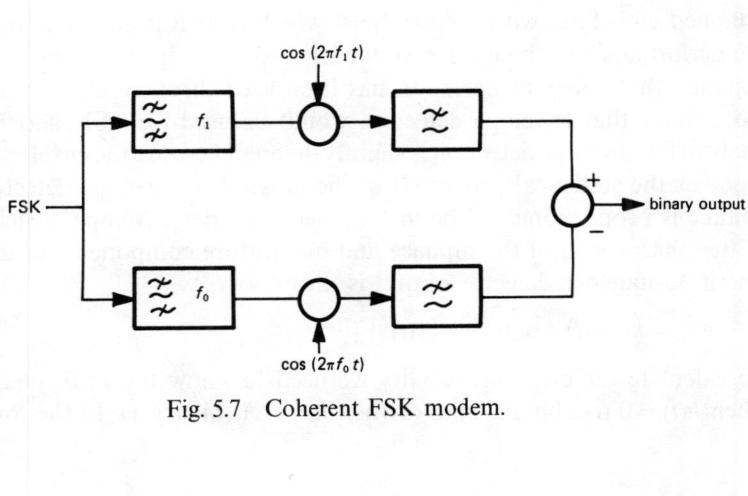

Fig. 5.7 Coherent FSK modem.

which are assumed to have an ideal band-pass characteristic, determine the bandwidth and hence the input noise power for each coherent detector.

The input voltage at each detector will be either

$$V_1(t) = h_1(t)\cos(2\pi f_1 t) + x_1(t)\cos(2\pi f_1 t) + y_1(t)\sin(2\pi f_1 t) \tag{5.30a}$$

and

$$V_0(t) = h_0(t)\cos(2\pi f_0 t) + x_0(t)\cos(2\pi f_0 t) + y_0(t)\sin(2\pi f_0 t) \tag{5.30b}$$

where $h_1(t)$ is the logical complement of $h_0(t)$ and $x_1(t)$ and $x_0(t)$ are independent random variables, provided that the passband of each pre-detection filter does not overlap that of the other.

The output of the composite detector will be

$$V_{out}(t) = (+A \text{ or } -A) + [x_1(t) - x_0(t)] \tag{5.31}$$

Since $x_1(t)$ and $x_0(t)$ are independent Gaussian variables, then the variable $[x_1(t) - x_0(t)]$ will also be Gaussian with a variance equal to the sum of the original variance. (This might seem to be a surprising result, but remember that if we subtract two sinusoids of *different* frequencies the power of the resultant waveform is the sum of the powers in the individual sinusoids.) The noise variance at the detector output will thus be $2\sigma^2$, which gives an effective peak signal-to-rms noise ratio of $2A/\sqrt{2}\sigma$. The probability of error in this case is then

$$P_{e(FSK)} = \tfrac{1}{2}[1 - \operatorname{erf}(A/2\sigma)] \tag{5.32}$$

This means that if FSK and PSK transmissions are to have the same error probability the FSK transmission requires a SNR which is 3 dB above the value required for PSK.

Thus, compared on a peak power basis, for the same error probability PSK requires an SNR that is 3 dB below that of FSK which in turn requires an SNR which is 3 dB below that of ASK. Compared on a mean power basis the performance of ASK and FSK are identical, but the same error rate can be obtained with PSK with a mean SNR which is 3 dB lower. A comparison of the performance of these three systems is given in Fig. 5.6, where it has been assumed that coherent detection has been used throughout. In practice it is more likely that envelope detection would be used for ASK and FSK. The analysis for envelope detection is slightly different because the envelope detector modifies the statistical properties of the noise. The envelope detector output voltage is proportional to the instantaneous carrier envelope, which is equal to the phasor sum of the inphase and quadrature components of Eqn (5.25). The instantaneous detector output is thus

$$V_{out} = k[\{h(t) + x(t)\}^2 + \{y(t)^2\}]^{1/2} \tag{5.33}$$

To calculate the error probability we need to know the PDF of Eqn (5.33) when $h(t) = 0$ (i.e. binary 0) and when $h(t) = A$ (binary 1). In the former case

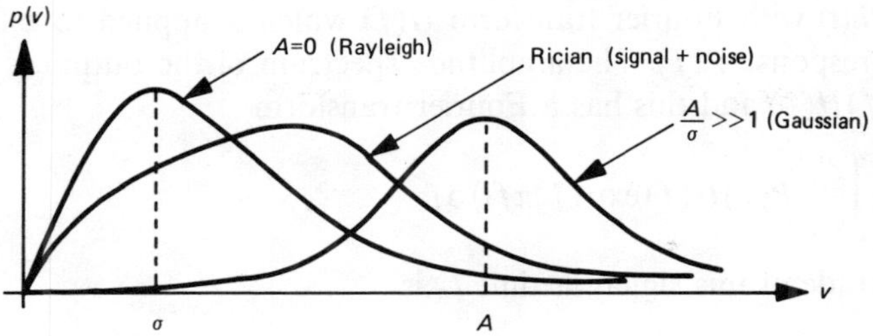

Fig. 5.8 Envelope distribution for signal + noise.

the density function has a Rayleigh distribution given by

$$p_0(v) = (v/\sigma^2)\exp(-v^2/2\sigma^2) \qquad \text{for} \qquad v \geqslant 0$$

and in the latter case the density function has a Rician distribution given by

$$p_1(v) = (v/\sigma^2)\exp[-(v^2 + A^2)/2\sigma^2]I_0(vA/\sigma^2) \qquad \text{for} \qquad v \leqslant 0$$

$I_0(x)$ is the modified Bessel function of zero order. The Rician distribution approaches the Gaussian distribution for $A/\sigma \gg 1$. The envelope probability density functions are illustrated in Fig. 5.8.

It has been shown[1] that the decision threshold for the envelope detector is

$$T = A^2/2\sigma^2 + \ln(P_0/P_1) \tag{5.34}$$

This reference also shows that for high SNRs envelope detection is marginally poorer for both ASK and FSK than coherent detection. For lower SNRs there is a significant deterioration in the performance of the envelope detected receiver. This, of course, is in agreement with the results for analogue transmissions considered in Chapter 4.

5.5 Matched Filtering and Correlation Detection

In all digital communication systems, baseband or carrier modulated, the probability of error ultimately depends upon the SNR. (This, of course, ignores the influence of intersymbol interference, which we considered in Section 3.7.) Thus it is reasonable to consider techniques that may be used to maximize the ratio A/σ. Matched filtering is one such technique, which emphasizes the signal voltage relative to the noise. In so doing, considerable distortion of the signal waveform usually results. Hence where waveform fidelity is important (as in analogue systems) matched-filtering techniques are not applicable. This is not the case with digital transmission, where the receiver is only required to decide between signal and noise.

The matched filter is designed to maximize the SNR at a precise interval t_0 after the signal is applied to its input. We will consider a general voltage waveform $h(t)$ with Fourier transform $H(f)$ which is applied to a filter with frequency response $P(f)$. The amplitude spectrum of the output is given by $G(f) = P(f)H(f)$ and this has a Fourier transform

$$g(t) = \int_{-\infty}^{\infty} P(f)H(f)\exp(j2\pi ft)\,\mathrm{d}f \tag{5.35}$$

The magnitude of this signal at time t_0 is

$$|g(t_0)| = \left|\int_{-\infty}^{\infty} P(f)H(f)\exp(j2\pi ft_0)\,\mathrm{d}f\right|$$

Hence the signal power (normalized) at time t_0 is

$$S_p = |g(t_0)|^2 = \left|\int_{-\infty}^{\infty} P(f)H(f)\exp(j2\pi ft_0)\,\mathrm{d}f\right|^2 \tag{5.36}$$

$P(f)$ is defined for both negative and positive frequencies and it is therefore necessary to use the double-sided power spectral density of $\eta/2$ W/Hz for the white noise.

The noise at the filter output will no longer be white but will have a power spectrum given by

$$N(f) = \frac{\eta}{2}|P(f)|^2$$

The noise power at the filter output is thus

$$N_p = (\eta/2)\int_{-\infty}^{\infty} |P(f)|^2\,\mathrm{d}f \tag{5.37}$$

The SNR at time t_0 is thus

$$\mathrm{SNR} = \frac{\left|\int_{-\infty}^{\infty} P(f)H(f)\exp(j2\pi ft_0)\,\mathrm{d}f\right|^2}{(\eta/2)\int_{-\infty}^{\infty} |P(f)|^2\,\mathrm{d}f} \tag{5.38}$$

The filter frequency response is then chosen to maximize the RHS of Eqn (5.38). This is accomplished by applying the Schwartz inequality,[2] which effectively states that the sum of two sides of a triangle must always be greater than or equal to the third side; in integral form this is usually written

$$\int_{-\infty}^{\infty} A^*(x)A(x)\,\mathrm{d}x \int_{-\infty}^{\infty} B^*(x)B(x)\,\mathrm{d}x \geqslant \left|\int_{-\infty}^{\infty} A^*(x)B(x)\,\mathrm{d}x\right|^2 \tag{5.39}$$

The inequality becomes an equality when $A(x)$ and $B(x)$ are co-linear, i.e. $A(x) = KB(x)$. If $A^*(x) = H(f)\exp(j2\pi ft_0)$ then its complex conjugate $A(x) = H^*(f)\exp(-j2\pi ft_0)$, and if $B(x) = P(f)$ then $B^*(x) = P^*(f)$.

Equation (5.39) can then be written

$$\int_{-\infty}^{\infty} |H(f)|^2\,\mathrm{d}f \int_{-\infty}^{\infty} |P(f)|^2\,\mathrm{d}f \geqslant \left|\int_{-\infty}^{\infty} P(f)H(f)\exp(j2\pi f t_0)\,\mathrm{d}f\right|^2 \tag{5.40}$$

i.e.

$$\int_{-\infty}^{\infty} |H(f)|^2\,\mathrm{d}f \geqslant \frac{\left|\int_{-\infty}^{\infty} P(f)H(f)\exp(j2\pi f t_0)\,\mathrm{d}f\right|^2}{\int_{-\infty}^{\infty} |P(f)|^2\,\mathrm{d}f}$$

Substituting in Eqn (5.38) we obtain

$$\frac{2}{\eta}\int_{-\infty}^{\infty} |H(f)|^2\,\mathrm{d}f \geqslant \mathrm{SNR}$$

But $\int_{-\infty}^{\infty}|H(f)|^2\,\mathrm{d}f$ is the energy of the signal, E. The maximum value of SNR is therefore

$$\mathrm{SNR}_{\max} = 2E/\eta \tag{5.41}$$

This value depends only on the ratio of signal energy to noise spectral density and is independent of the shape of the signal waveform. The SNR is a maximum when $A(x) = KB(x)$, i.e.

$$H^*(f)\exp(-j2\pi f t_0) = KP(f) \tag{5.42}$$

The impulse response of the matched filter is thus

$$p(t) = \frac{1}{K}\int_{-\infty}^{\infty} H^*(f)\exp(-j2\pi f t_0)\exp(j2\pi f t)\,\mathrm{d}t$$

i.e.

$$p(t) = \frac{1}{K}h(t_0 - t) \tag{5.43}$$

The impulse response is therefore the time reverse of the input signal waveform $h(t)$ with respect to t_0. Clearly the value of t_0 must be greater than the duration of the signal to which the filter is matched. This is the condition for physical realizability; in other words, the impulse response must be zero for negative values of t.

We will derive, as an example, the matched filter for a rectangular pulse of amplitude A and duration t_1. The Fourier transform of this pulse is

$$H(f) = \int_{-t_{1/2}}^{t_{1/2}} A\exp(-j2\pi f t)\,\mathrm{d}t$$

i.e.

$$H(f) = (At_1/2)\,\mathrm{sinc}\,(\pi f t_1)$$

Hence from Eqn (5.42) $P(f) = KH^*(f)\exp(-j2\pi f t_0)$ i.e.

$$P(f) = [\mathrm{sinc}\,(\pi f t_1)\exp(-j2\pi f t_0)]/K \tag{5.44}$$

This is a low-pass filter with a linear phase shift. Matched filters can be specified for any signal shape, e.g. when a FSK waveform is the input to a matched filter the SNR at the output is given by Eqn (5.41).

If we write Eqn (5.35) representing the matched filtering operation in terms of the convolution integral it becomes

$$g(t) = \int_{-\infty}^{\infty} h(t)h(t_0 - t)\,\mathrm{d}t$$

$h(t_0 - t)$ being the appropriate substitution for $p(t)$. We now compare this equation with the equation for the cross-correlation between two waveforms $h(t)$ and $y(t)$.

$$R_{hy}(\tau) = \int_{-\infty}^{\infty} h(t)y(t+\tau)\,\mathrm{d}t$$

If we make $y(t+\tau) = h(t_0 - t)$ we see that matched filtering is equivalent to cross-correlating the "noisy signal" $h(t)$ with a noise-free signal $h(t_0 - t)$, which is the impulse response of the matched filter. The correlation detector is based upon this principle.

The correlation coefficient between the waveforms $h(t)$ and $y(t)$ is defined as

$$R(\tau) = \frac{\displaystyle\int_{\infty}^{\infty} h(t)y(t+\tau)\,\mathrm{d}t}{\left|\displaystyle\int_{-\infty}^{\infty} h^2(t)\,\mathrm{d}t \int_{-\infty}^{\infty} y^2(t)\,\mathrm{d}t\right|^{1/2}} \tag{5.45}$$

The denominator of this equation is a normalizing factor that makes $R(\tau)$ independent of the actual mean square values of $h(t)$ and $y(t)$ and restricted to the range ± 1. Note when $h(t) = ky(t)$ then $R(\tau) = +1$, and when $h(t) = -ky(t)$ then $R(\tau) = -1$.

$R(\tau)$ is thus a measure of the similarity of the two waveforms $h(t)$ and $y(t)$. Correlation detection is frequently used in situations where the SNR < 1, e.g. space applications. The received waveform is correlated with several noise-free waveforms stored at the receiver. The one that produces the largest correlation coefficient is then assumed to have been the transmitted waveform.

It is found in most practical situations that the improvement obtained using matched filters, rather than conventional low- or band-pass filters, is marginal (usually less than 3 dB). Hence filters are usually designed to minimize intersymbol interference rather than maximize SNR. However, where SNR is of prime importance (e.g. radar signals) matched filtering and correlation techniques are employed.

5.6 Error Detection and Correction

The probability of error in a digital transmission system depends ultimately upon the SNR at the receiver input. This ratio is a maximum when matched filtering is used and this produces the lowest probability of error. A significant decrease in this figure can be produced by employing a type of code known as a block code for the digital signal. The purpose of coding the signal in this fashion is to introduce redundancy into the transmitted signal, which itself has no inherent redundancy. We observed in Section 1.13 that there are many sources of redundancy in the English language and we showed that the redundancy present reduced the information content but at the same time allowed the receiver to spot and correct errors. Redundancy has the same effect in digital signals. The simplest form of block code is the parity check 8-bit international standard code described in Section 3.11.

The international code represents each character by seven information digits and a single parity check digit. Either an even or an odd parity check can be used; in the former case the parity digit is chosen to make the number of 1s in the 8-bit word even, and in the latter case the parity digit is chosen to make the number of 1s in the 8-bit word odd. The receiver then checks each 8-bit word for odd or even parity depending upon the system in use. If the parity check fails the receiver notes that an error has occurred and requests the transmitter to repeat the 8-bit word. The parity check is carried out simply by exclusive-ORing the eight digits in each code word. If the number of 1s in the code word is odd the result of the exclusive-OR operation is binary 1. If the number of 1s is even, the result of the exclusive-OR operation is 0. A parity check circuit is given in Fig. 5.9.

The error-detecting properties of this code can be demonstrated by comparing the probability of error in the uncoded case with the probability of an undetected error in the coded case. We assume that the probability of error in the uncoded case is P_e. The probability of an undetected error in the coded case will be the probability of an even number of errors. The parity check at the receiver will fail, and therefore indicate an error, only if an odd number of errors occurs. We therefore have to compare the probability of error in the uncoded case with the probability of an even number of errors in the coded case.

If there are n digits in each code word the probability of one digit being in error is the joint probability of 1 incorrect digit and $(n-1)$ correct digits, i.e. the joint probability is

$$P_j = P_e(1 - P_e)^{n-1} \tag{5.45}$$

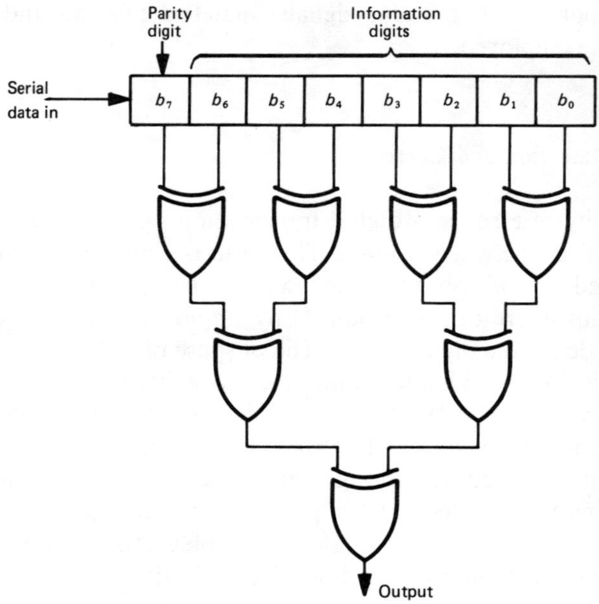

Fig. 5.9 Error detection using a single parity digit.

There are n digits in the group and Eqn (5.45) gives the probability that any one of these digits will be received incorrectly. The total probability in the n digits is thus

$$P = nP_e(1 - P_e)^{n-1}$$

As an example consider the uncoded case where $n = 7$. If $P_e = 10^{-4}$ the probability of one error in seven digits is

$$P_t = 7 \times 10^{-4} \times (0.9999)^6 \simeq 7 \times 10^{-4}$$

The probability of r errors in a group of n digits is the joint probability that r digits will be received incorrectly and $(n - r)$ will be received correctly, i.e.

$$P_j = P_e^r(1 - P_e)^{n-r}$$

There is a total of ${}^nC_r$ possible ways of receiving r digits incorrectly in a total of n digits where

$${}^nC_r = \frac{n!}{r!(n-r)!}$$

The total probability of error in this case is thus

$$P_t = {}^nC_r P_e^r(1 - P_e)^{n-r} \tag{5.46}$$

In the coded case, an extra digit is added to the original seven so that we require the probability of an even number of errors in eight digits.

$$\text{Probability of 2 errors} = {}^8C_2(10^{-4})^2(1-10^{-4})^6 = 2.8 \times 10^{-7}$$
$$\text{Probability of 2 errors} = {}^8C_4(10^{-4})^4(1-10^{-4})^4 = 7 \times 10^{-15}$$
$$\text{Probability of 2 errors} = {}^8C_6(10^{-4})^6(1-10^{-4})^2 = 2.8 \times 10^{-23}$$
$$\text{Probability of 2 errors} = {}^8C_8(10^{-4})^8 \qquad = 10^{-32}$$

Hence the total probability of an undetected error in the coded case is $2.8 \times 10^{-7} + 7.0 \times 10^{-15} + 2.8 \times 10^{-23} \simeq 2.8 \times 10^{-7}$. This is considerably less than the probability of error in the original uncoded group of seven digits. These numerical values show that a significant reduction in error probability is possible even with a very rudimentary error-checking system.

We have assumed in this example that the digit transmission rate is unchanged. This means that the rate of information transmission must necessarily be reduced to $\frac{7}{8}$ths of its value in the uncoded system. It would be more realistic to maintain a constant information rate which means that the transmission bandwidth should be increased to $\frac{8}{7}$ of its original value to accommodate the parity check digit. Such an increase would increase the noise power by the same ratio, i.e. the rms noise would be increased to 1.069 times its previous value. The effect of this increased noise on the error probability must be calculated. Assuming an uncoded probability of error $P_e = 10^{-4}$ for illustration then

$$\operatorname{erf}[(A/(2\sqrt{2}\sigma))] = 1 - 2 \times 10^{-4}$$

i.e.

$$A/(2\sqrt{2}\sigma) = 2.63$$

If the rms noise increases by a factor of 1.069, there is a corresponding decrease in this ratio, i.e. the ratio becomes 2.46, which produces an uncoded error probability of $P_e' = 2.51 \times 10^{-4}$. This value must now be a used in Eqn (5.46) to calculate the probability of an undetected error in the coded case, which is $P_t' = 1.75 \times 10^{-6}$. Thus even when the extra noise is taken into account the reduction in undetected errors is still considerable.

This is not always the case, however, if several parity digits are added to the information digits. The extra noise introduced by the increased bandwidth can cancel any advantage obtained by the error-detecting code. In such cases the error probabilty can sometimes be decreased by increasing the signal power. There are many cases where error-detecting codes do produce a significant advantage, and bearing in mind the threshold region shown by Fig. 5.5 we can only say that each case must be examined separately.

The error-detecting properties of a code can be enhanced to include error correction by increasing the redundancy. This becomes clear by considering an example that codes 16 information digits with 8 parity check digits, i.e. the code redundancy is 33%. In this example the 16 information digits are grouped into a 4 × 4 matrix. An even or odd parity check digit is transmitted for each row and column of the matrix

Column parity digits ↙						Column check fails ↙					
0	0	1	0			0	0	1	0		
0	1	1	1	1		0	0	1	1	1	← Row check fails
0	1	1	0	0		0	1	1	0	0	
1	0	0	0	1		1	0	0	0	1	
1	0	1	1	1		1	0	1	1	1	
				↑ Row parity digits							

If a single error occurs in the 16 information digits then a row and column parity check will both fail. The point of intersection of row and column then indicates which digit was in error. If a single parity digit is in error then only one column or one row will fail the parity check, and again the incorrect digit can be identified and therefore corrected.

The above example was chosen to illustrate the possibility of correcting errors, and was not a particularly efficient code. There are many more efficient error-correcting codes in use. The block codes are based upon the work of Hamming.[3] An example of Hamming's single-error-correcting code will now be considered. We assume that the original information digits are split up into blocks of m digits to which are added c parity check digits. For illustration assume that $m = 4$ and $c = 3$. Each coded word therefore contains seven digits

Table 5.1

Original message				Transmitted code						
M_1	M_2	M_3	M_4	M_1	M_2	M_3	M_4	C_1	C_2	C_3
0	0	0	0	0	0	0	0	0	0	0
0	0	0	1	0	0	0	1	0	1	1
0	0	1	0	0	0	1	0	1	0	1
0	0	1	1	0	0	1	1	1	1	0
0	1	0	0	0	1	0	0	1	1	0
0	1	0	1	0	1	0	1	1	0	1
0	1	1	0	0	1	1	0	0	1	1
0	1	1	1	0	1	1	1	0	0	0
1	0	0	0	1	0	0	0	1	1	1
1	0	0	1	1	0	0	1	1	0	0
1	0	1	0	1	0	1	0	0	1	0
1	0	1	1	1	0	1	1	0	0	1
1	1	0	0	1	1	0	0	0	0	1
1	1	0	1	1	1	0	1	0	1	0
1	1	1	0	1	1	1	0	1	0	0
1	1	1	1	1	1	1	1	1	1	1

and there are $2^7 = 128$ different words. Of these 128 words only $2^4 = 16$ words are required to transmit the original information. The 16 code words are chosen from the possible 128 to give a single-error-correcting capability. A possible choice of code words is shown in Table 5.1.

Each of the code words in Table 5.1 differs from any other in at least three positions. Such a code is said to have a Hamming distance of three. This means that at least three errors are required to convert any code word into one of the others. If a single error occurs in a received code word it will differ from the correct code word in one digit only and from all other code words in at least two digits. Hence the correct code word is chosen as the one with least difference from the received code word. This implies that the receiver requires a list of all allowed code words and then compares each received code word until a match is achieved. In practice this process can be reduced to one of parity checking.

The mathematical relationship between information digits and check digits is based on modulo 2 addition (exclusive-OR) and was devised by Hamming. The rules for a 7-digit block code containing four information digits and three check digits are

$$\begin{aligned} C_1 &= M_1 \oplus M_2 \oplus M_3 \\ C_2 &= M_1 \oplus M_2 \oplus M_4 \\ C_3 &= M_1 \oplus M_3 \oplus M_4 \end{aligned} \tag{5.47}$$

Remembering that $C_1 \oplus C_1 = 0$, etc., these equations can be written

$$\begin{aligned} C_1 \oplus M_1 \oplus M_2 \oplus M_3 &= 0 \\ C_2 \oplus M_1 \oplus M_2 \oplus M_4 &= 0 \\ C_3 \oplus M_1 \oplus M_3 \oplus M_4 &= 0 \end{aligned} \tag{5.48}$$

If there are no errors at the receiver the modulo 2 additions represented by Eqns (5.48) will all produce a binary 0 result. If there is a single error in any of the information or check digits then one or more of Eqns (5.48) will produce a binary 1. The parity checking circuit for the 7-digit block code is shown in Fig. 5.10.

If we represent a correct digit by Y and an incorrect digit by N the possible values of A, B, C for a single error are listed in Table 5.2.

The outputs A, B, C of the parity check circuits can be regarded as a 3×8 matrix. The number of columns in the matrix equals the number of check digits and each row, except the first, uniquely defines a single error position. This means that c check digits can define $2^c - 1$ error positions assuming only a single error occurs.

If there are m information digits and c check digits then for a single-error correcting code the relationship between m and c is

$$m + c \leqslant 2^c - 1 \tag{5.49}$$

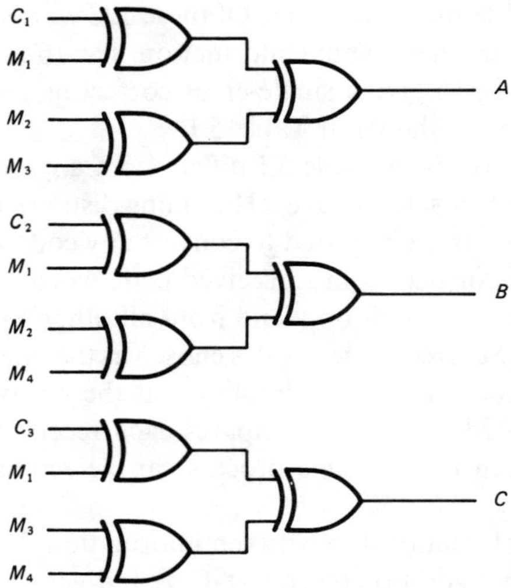

Fig. 5.10 Parity checking network.

Table 5.2

M_1	M_2	M_3	M_4	C_1	C_2	C_3	A	B	C
Y	Y	Y	Y	Y	Y	Y	0	0	0
Y	Y	Y	Y	Y	Y	N	1	0	0
Y	Y	Y	Y	Y	N	Y	0	1	0
Y	Y	Y	Y	N	Y	Y	0	0	1
Y	Y	Y	N	Y	Y	Y	1	1	1
Y	Y	N	Y	Y	Y	Y	1	1	0
Y	N	Y	Y	Y	Y	Y	1	0	1
N	Y	Y	Y	Y	Y	Y	0	1	1

Table 5.3

c	$n(=2^c-1)$	m	Code type (n, m)	Efficiency $m/(2^c-1)$
1	1	0	—	—
2	3	1	(3, 1)	0·33
3	7	4	(7, 4)	0·57
4	15	11	(15, 11)	0·73
5	31	26	(31, 26)	0·83
6	63	57	(63, 57)	0·90
7	127	120	(127, 120)	0·94
8	255	247	(255, 247)	0·97

Possible single-error-correcting codes are listed in Table 5.3, and we can see from this table that the larger the value of c the more efficient the code.

Returning to Table 5.2, in a practical system it is necessary to correct only errors in the information digits. This means that for error-correcting purposes only the last four rows of the parity check matrix need to be considered. In this context we note the following relationship:

$$X \oplus 1 = \bar{X} \qquad \text{and} \qquad X \oplus 0 = X$$

This means that if a received digit is in error, exclusive-ORing it with binary 1 produces the correct value. If a received digit is correct, exclusive-ORing it with binary 0 will preserve the received value. Using these conditions it is possible to devise a very simple error-correcting circuit consisting of exclusive-OR gates. This circuit is shown in Fig. 5.11. The binary inputs w, x, y, z are derived from the parity check matrix of Table 5.2.

$$w = \bar{A}BC$$
$$x = A\bar{B}C$$
$$y = AB\bar{C}$$
$$z = ABC$$

The (7, 3) code has a Hamming distance of 3 and can correct single errors. If we assume for illustrative purposes that an error occurs in both H_1 and H_2 the parity checking network will produce an output of $A=0$, $B=0$, $C=1$, which is equivalent to a single error in c_1. The fact that the parity checking network does not give $A=0$, $B=0$, $C=0$ means that a double error has been detected, but this will be incorrectly interpreted by the circuit of Fig. 5.11.

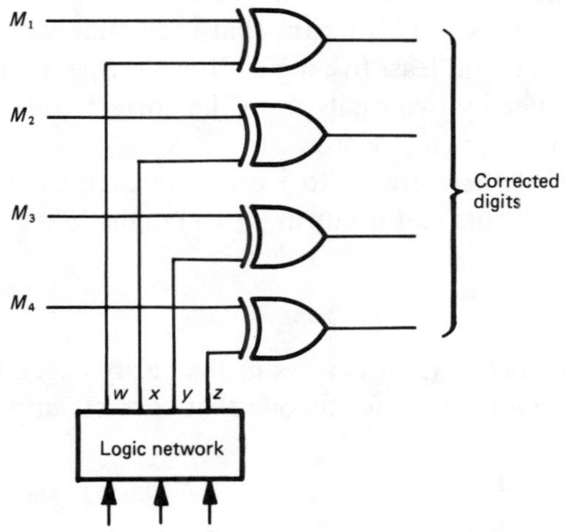

Fig. 5.11 Error-checking network.

Thus the (7, 4) code can detect a double error provided it is not required to correct a single error also. If we assume that the (7, 4) code is designed to correct a single error, i.e. using the hardware of Fig. 5.11 a double error will be ignored, then in order to evaluate the effectiveness of the (7, 4) code we must compare the probability of two errors in seven digits with the probability of one error in four digits. If we wish to maintain the same information rate the bandwidth must be increased in the ratio of 7:4, which increases the rms noise by a factor of 1.322.

We will consider two examples: in the first the uncoded error probability is assumed to be $P_e = 10^{-4}$ and in the second case $P_e = 10^{-5}$. The probability of a single error in four digits is thus either $P_t = 4 \times 10^{-4}$ or $P_t = 4 \times 10^{-5}$. If the rms noise is increased by a factor of 1.322 the corresponding error probability for an uncoded system is $P_e' = 2.4 \times 10^{-3}$ or $P_e' = 6{\cdot}31 \times 10^{-4}$. The probability of an undetected error in the coded case is the probability of two or more errors, i.e.

$$P_t' = {}^7C_2(P_e')^2(1-P_e')^5 + {}^7C_3(P_e')^3(1-P_e')^4 + {}^7C_4(P_e')^4(1-P_e')^3 + {}^7C_5(P_e')^5(1-P_e')^2 + {}^7C_6(P_e')^6(1-P_e') + {}^7C_7(P_e')^7$$

This evaluates to $P_t' = 1.29 * 10^{-4}$ or $P_t' = 8.36 * 10^{-6}$.

In the first case the error rate is reduced from 4×10^{-4} to 1.29×10^{-4} (a ratio of 3.1:1) and in the second case from 4×10^{-5} to $8{\cdot}36 \times 10^{-6}$ (a ratio of 4.7:1). These are fairly modest increases, the improvement being greater the smaller the initial error probability. If more efficient single-error-detecting codes are used (e.g. 127, 120 code) a very significant reduction in undetected error probability is obtained.

If the code is required to correct more than one error, then clearly the Hamming distance must be increased. The Hamming distance for a double-error-correcting code is 5. This means that each code word must differ from all other code words in at least five digits. If two errors are made, the received code word will differ by two digits from the correct code and at least three digits from all the other code words.

If we assume that there are up to y errors in each code word the number of ways that these errors can occur in $(m + c)$ digits is

$$\sum_{t=1}^{y} {}^{m+c}C_t$$

The c parity digits define $(2^c - 1)$ rows in the parity check matrix so that the relationship between m and c for a code that corrects up to y errors is

$$\sum_{t=1}^{y} {}^{m+c}C_t \leqslant 2^c - 1 \tag{5.50}$$

which is known as the Hamming bound. Clearly the greater the error-correcting capability of any code the greater the redundancy. The relationship between

Table 5.4

Hamming distance	Code properties
1	None
2	Single-error detecting
3	Single-error correcting or double-error detecting
4	Single-error correcting and double-error detecting
5	Double-error correcting or triple-error detecting
6	Double-error correcting and triple-error detecting
7	Triple-error correcting

the Hamming distance of a code and its error-combating capabilities is listed in Table 5.4 for Hamming distances up to 7. Block codes are just one example of several different types of error-detecting codes; convolution codes are another. These codes are discussed in detail by Virterbi.[4]

5.7 Multiple Decisions

We have assumed in this chapter that the receiver in a system is the only point at which a decision is made. This is not necessarily the case; in a PCM system, for example, a decision is made at each regenerative repeater. Hence when evaluating error probability in a multistage link, the probability of error at each stage must be taken into account. For simplicity we shall assume that the probability of error at each stage in such a link is the same, although this restriction need not apply in a real situation. A typical PCM system is illustrated in Fig. 5.12.

It is evident that errors that may occur at one repeater can be corrected by a second error in the same digit at a subsequent stage. If we consider a particular digit in its passage through a multistage link when it is subject to an even number of incorrect decisions it will be received correctly at the destination.

If there are L decision stages the probability of making U incorrect and $L - U$ correct decisions is

$$P = P_e{}^U \times (1 - P_e)^{L-U}$$

but there are ${}^LC_U$ possible ways of making U incorrect decisions in a total of L decisions. Hence for a single digit the probability of making U incorrect decisions in an L stage link is

$$P_t = {}^LC_U P_e \times (1 - P_e)^{L-U} \qquad (5.51)$$

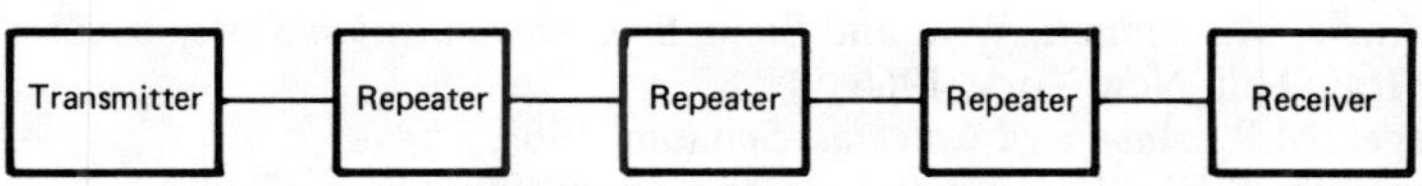

Fig. 5.12

The probability of a digit being incorrect at the receiver is found by summing Eqn (5.51) over all odd values of U, i.e.

$$P_m = \sum_{U(\text{odd})} {}^{L}C_U P_e{}^U (1 - P_e)^{L-U} \tag{5.52}$$

If $P_e \ll 1$ this approximates to $P_m \simeq LP_e$. Thus if P_e is small the probability of a single digit being received incorrectly at more than one stage is negligible. However, the use of L decision points makes the individual error L times as likely.

If error coding is employed in a multistage link, equations such as (5.46) should reflect the higher individual digit error probability, e.g.

$$P_t = {}^{n}C_r (LP_e)^r (1 - LP_e)^{n-r} \tag{5.53}$$

5.8 Conclusion

In this chapter we have been specifically interested in the effects of Gaussian noise on digital communications and have shown that the probability of error is related to the SNR at the receiver. This probability can be reduced either by employing SNR enhancement techniques such as matched filtering or by employing error-coding techniques. The major source of signal impairment on data networks is due to intersymbol interference, and the noise encountered tends to be impulsive rather than Gaussian. This leads to long bursts of errors that require long block codes with relatively high redundancy for error correction. It is not surprising, therefore, that error coding is usually restricted to error detection rather than error correction on data networks. SNRs are usually high, and tolerable error rates are achieved with simple parity checks. When an error is detected the transmitter is requested to repeat the data, which clearly requires the provision of a return path between receiver and transmitter. There are exceptions, however. In satellite transmissions, signal power is at a premium and the received noise as predominantly Gaussian. This is one example where extensive use of error correction is made.

A second example is covered in detail in Section 11.19, which deals with teletext transmission. In this case page and row address digits are error protected using a code with a Hamming distance of 4. Such a code can correct a single error and detect a double error.

References

1. Schwartz, M., Bennett, W.R. and Stein, S., *Communication Systems and Techniques*, McGraw-Hill, New York, 1966.
2. Spiegel, M.R., *Advanced Calculus*, Schaum, 1963.
3. Hamming, R.W., 'Error detecting and correcting codes', *Bell Systems Technical Journal*, **29**, 147 (1950).

4. Virterbi, A.J., 'Convolutional codes and their performance in communications systems', *IEEE Transactions on Communication Technology*, Com. **19**(5), (1971).

Problems

5.1 A binary waveform of amplitude $+2V$ or $-V$ is added to Gaussian noise with zero mean and variance σ^2. The *a priori* signal probabilities are $P(+2V)=\frac{1}{3}$ and $P(-V)=\frac{2}{3}$. If a single sample of signal-plus-noise is taken, determine the decision threshold that minimizes overall error probability.

Answer: $T = V/2 + 0.23\sigma^2/V$.

5.2 A teleprinter system represents each character by a 5-digit binary code that is transmitted as either $A/2$ V or $-A/2$ V. The binary signal is received in the presence of Gaussian noise with zero mean at an SNR of 11 dB. Determine the optimum decision level and the probability that the receiver will make an error.

Answer: $T = 0$ V; $P = 0.000193$.

5.3 In a binary transmission system, 1 is represented by a raised cosine pulse given by

$$v(t) = V[1 + \cos(2\pi t/T) \qquad \text{for} \qquad -T/2 < t < T/2$$

where T is the pulse duration. A binary 0 is represented by the same raised cosine waveform with amplitude $-V$. After transmission over a noisy channel the signal-to-noise power ratio at the detector is 6 dB. If the binary signal is reformed by sampling the received signal at the middle of each pulse interval, find the probability of error. Assume 1 and 0 are equi-probable and that the noise is white.

Answer: 0.000572.

5.4 Two binary communication links are connected in series, each link having a transmitter and a receiver. If the probability of error in each link is 0.00001, find the overall probability that:
(a) a 0 is received when a 1 is transmitted;
(b) a 0 is received when a 0 is transmitted;
(c) a 1 is received when a 0 is transmitted;
(d) a 1 is received when a 1 is transmitted.

Answer: (a) 0.00002; (b) 0.99998; (c) 0.99998; (d) 0.00002.

5.5 If a simple coding scheme is used in the previous question such that each individual digit is repeated three times, what is the probability of deciding incorrectly at the receiver if the following rule is used?
Decide 0 if the received code is 000, 001, 010, 100
Decide 1 if the received code is 111, 110, 101, 011

Answer: 0.000571.

5.6 The noise level on a channel produces an error probability of 0.001 during a binary transmission. In an effort to overcome the effect of noise each digit is repeated five times, and a majority decision is made at the receiver. Find the probability of error in this system if
(a) the digit rate remains constant;

(b) the information rate is half its original value.
Assume white Gaussian noise.

Answer: (a) 9.94×10^{-9}; (b) 1.445×10^{-4}.

5.7 Use Eqns (5.35) and (5.41) to determine the maximum signal and mean square noise voltage at the output of a matched filter.

Answer: $E, E\eta/2$.

5.8 The matched filter is a linear network. This means that when the input to such a filter is Gaussian noise the output noise will also be Gaussian but with a modified variance.
Using this fact, calculate the probability of error in a binary transmission system when the threshold detector is preceded by a matched filter. The transmission rate is 64 kbits/s with 0 and 1 having equal probability. The received waveform has a mean value (normalized) of 0 or 25 mV, and is accompanied by noise with a power spectral density of 1.68×10^{-9} W/Hz.

Answer: 0.0000205.

5.9 A signal which is a single pulse of amplitude A V and duration T sec is received masked by white Gaussian noise. Calculate the improvement in SNR produced by a matched filter as compared with the SNR at the output of a single-stage RC low-pass filter. Assume the 3 dB cut-off frequency of the RC network is $1/T$ Hz.

Answer: 0.9 dB.

5.10 A FSK receiver consists of two matched filters, one for each of the tone bursts f_0 and f_1. The FSK waveform is fed to both filters, each output being sampled at the instant of maximum signal. The samples are subtracted and then fed to a threshold detector.
Calculate the error probability in terms of the signal energy and noise power spectral density.

Answer: $1 - \text{erf}\sqrt{(E/\eta)}$.

6 / HIGH-FREQUENCY TRANSMISSION LINES

The study of transmission lines is the investigation of the properties of the system of conductors used to carry electromagnetic waves from one point to another. Here however, our attention will be limited to high-frequency applications, i.e. when the length of the transmission line is of at least the same order of magnitude as the wavelength of the signal. In this chapter an idealized model of the line will be used to represent the many different forms found in practice, ranging from twisted pairs to coaxial cables. The theory of transmission lines, which was developed in the early years of the study of electromagnetic propagation, is strictly applicable only to systems of conductors that have a "go" and "return" path, or that, in electromagnetic field terms, can support a TEM wave. Hollow-tube waveguides do not fall into this category, although, as we discuss in the chapter on microwaves, many of the concepts of transmission line theory can be applied to them. Transmission line theory is important to the communications engineer because it gives him the means for making the most efficient use of the power and equipment at his disposal. By applying his knowledge correctly he can ensure that a transmitting system is designed to transfer as much power as possible from the feeder line into the antenna, or he can take steps to ensure that a receiving antenna is correctly matched to the line that connects it to the receiver itself, so that no power is wasted. The range of systems over which transmission lines are used is as extensive as the subject of communications engineering itself, and the general theory we develop in this chapter may be used to solve a very wide variety of problems.

6.1 Voltage and Current Relationships on the Line

A transmission line consists of continuous conductors with a cross-sectional configuration that is constant throughout their length. A voltage is placed across the line at the sending end, and it is necessary to be able to determine how it changes with distance so that its value at the load, or at some other point of interest, can be found. Similarly, knowledge of the current on the line may also be required.

To calculate current and voltage values the line must be represented in some way that will allow circuit analysis to be used. However, in such analysis the parameters are discrete and considered to exist at one point, whereas in a transmission line they are evenly distributed throughout its length. This difficulty is overcome by considering a very short length of the line, as we shall see in the next few paragraphs.

The parameters used to describe the line are:

(i) Resistance (R). The conductors making up the line offer some resistance to the flow of current. R is usually made to include the total resistance of both conductors.

(ii) Inductance (L). The signal on the line is time varying and so there will be an inductive reactance associated with the line. The value of L depends on the cross-sectional geometry of the conductors.

(iii) Conductance (G). The conductors that form the line are held in position by a dielectric material that, because it cannot be a perfect insulator, will allow some leakage current to pass between them.

(iv) Capacitance (C). The conductors, and the dielectric between them, form a capacitor, and there will therefore be a capacity reactance to a time-varying signal.

These refer to only one unit length (usually a metre) of line. Consequently these per-unit-length values must be multiplied by the line length to find the total resistance, inductance, conductance and capacitance of the line.

The continuous distribution of these parameters is approximated by representing the line as a cascaded network of elements, each element having a very short length δz, as shown in Fig. 6.1.

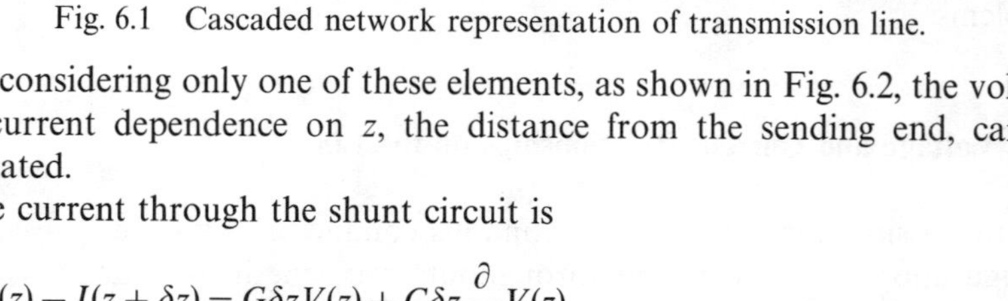

Fig. 6.1 Cascaded network representation of transmission line.

By considering only one of these elements, as shown in Fig. 6.2, the voltage and current dependence on z, the distance from the sending end, can be calculated.

The current through the shunt circuit is

$$I(z) - I(z + \delta z) = G\delta z V(z) + C\delta z \frac{\partial}{\partial t} V(z) \tag{6.1}$$

and the voltage drop in the series circuit is

$$V(z) - V(z + \delta z) = R\delta z I(z + \delta z) + L\delta z \frac{\partial}{\partial t} I(z + \delta z) \tag{6.2}$$

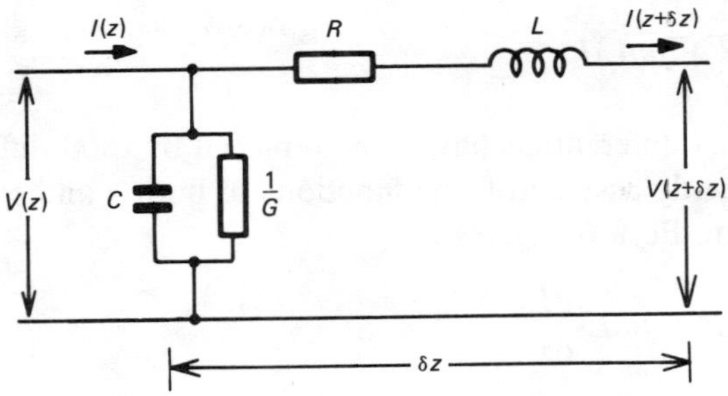

Fig. 6.2 Equivalent circuit for element of length δz.

Note that in these equations the impedance terms have been multiplied by the length of the element, δz. For small δz

$$I(z+\delta z) \simeq I(z) + \frac{\partial I}{\partial z}(z)\delta z$$

Then Eqn (6.1) becomes

$$\frac{\partial I}{\partial z}(z) = -\left(G + C\frac{\partial}{\partial t}\right)V(z) \tag{6.3}$$

Similarly, from Eqn (6.2),

$$\frac{\partial V(z)}{\partial z} = -\left(R + L\frac{\partial}{\partial t}\right)I(z+\delta z) \tag{6.4}$$

i.e.

$$\frac{\partial V(z)}{\partial z} = -\left(R + L\frac{\partial}{\partial t}\right)\left[I(z) + \frac{\partial I(z)}{\partial z}\delta z\right] \tag{6.5}$$

Assuming that the last term, $[\partial I(z)/\partial z]\delta z$, is very small compared with the other terms in Eqn (6.5), it can be ignored, and the equation written as

$$\frac{\partial V(z)}{\partial z} = -\left(R + L\frac{\partial}{\partial t}\right)I(z) \tag{6.6}$$

Equations (6.3) and (6.6) may now be combined to give solutions for voltage or current as functions of z. Further simplifications may be made to these equations if it is assumed that the signal on the line is sinusoidal, in which case the time dependence can be expressed by the term $\exp(j\omega t)$, and the time derivative, $\partial/\partial t$, replaced by $j\omega$. If this is done in Eqns (6.3) and (6.6), they become

$$\frac{dI}{dz} = -(G + j\omega C)V \tag{6.7}$$

$$\frac{dV}{dz} = -(R + j\omega L)I \tag{6.8}$$

where the partial differentials have been replaced by total differentials, and V and I are implicitly assumed to be functions of both z and t.

Differentiating Eqn (6.8) gives

$$\frac{d^2V}{dz^2} = -(R + j\omega L)\frac{dI}{dz} \tag{6.9}$$

and substituting dI/dz from Eqn (6.7)

$$\frac{d^2V}{dz^2} = (R + j\omega L)(G + j\omega C)V \tag{6.10}$$

More conveniently, Eqn (6.10) is expressed as

$$d^2V/dz^2 = \gamma^2 V \tag{6.11}$$

where

$$\gamma^2 = (R + j\omega L)(G + j\omega C) \tag{6.12}$$

Equation (6.11), which is a standard form of differential equation, has the solution

$$V = V_1 \exp(-\gamma z) + V_2 \exp(\gamma z) \tag{6.13}$$

V_1 and V_2 are determined by the initial conditions on the line. By a similar analysis, the current is related to z by

$$I = I_1 \exp(-\gamma z) + I_2 \exp(\gamma z) \tag{6.14}$$

It must be remembered that earlier we assumed a time dependence $\exp(j\omega t)$. It is conventional to omit that term from Eqns (6.13) and (6.14), and indeed from the equations in the rest of this chapter, but it is always implicitly assumed because of the time varying signal on the line. We shall return to this later, in Section 6.20.

Equation (6.13) can be expressed in a slightly different form if Eqn (6.12) is examined more closely. γ is a complex quantity and it can therefore be separated into real and imaginary parts, i.e.

$$\gamma = \sqrt{[(R + j\omega L)(G + j\omega C)]} = \alpha + j\beta \tag{6.15}$$

Then, if Eqn (6.15) is used in Eqn (6.13), the total voltage on the line is

$$V = V_1 \exp[-(\alpha + j\beta)z] + V_2 \exp(\alpha + j\beta)z \tag{6.16}$$

$$= V_1 \exp(-\alpha z)\exp(-j\beta z) + V_2 \exp(\alpha z)\exp(j\beta z) \tag{6.17}$$

Thus, the voltage at any point z from the sending end is the sum of two components:

(i) $V_1 \exp(-\alpha z) \exp(-j\beta z)$. The initial voltage V_1 at $z=0$, is attenuated as it travels down the line, i.e. the amplitude decreases with z as $\exp(-\alpha z)$. The term $\exp(-j\beta z)$ is a phase term and does not influence the amplitude of the voltage. This component is called the forward, or incident, wave.

(ii) $V_2 \exp(\alpha z) \exp(j\beta z)$. Here the amplitude term is of the form $V_2 \exp(\alpha z)$. The component $\exp(\alpha z)$ increases with increasing z, so z must decrease because the voltage must be attenuated as it travels along the line. This component is called the backward, or reflected, wave. It is, as we shall see, produced by a mismatch between the transmission line and the load.

We can therefore conclude that the voltage at a point on the line a distance z from the sending end is the sum of the voltages of the forward and reflected waves at that point.

6.2 Line Parameters

Referring again to Eqn (6.16), α and β are parameters determined by the characteristics of the line itself, and they have the following interpretation:

α *Attenuation coefficient.* Both the foward and reflected waves are attenuated exponentially at a rate α with distance travelled. α, the real part of Eqn (6.15), is a function of all the line parameters. If R, L, G and C are in their normal units, α is in nepers/m, although it is more usually expressed in dB/m.

β *Phase constant.* β, the imaginary part of Eqn (6.15), shows the phase dependance with z of both the forward and backward waves. If z changes from z_1, by a wavelength λ, to $z_1 + \lambda$, the phase of the wave must change by 2π. Therefore

$$\beta(z_1 + \lambda) - \beta z_1 = 2\pi$$

and

$$\beta = 2\pi/\lambda \tag{6.18}$$

γ *Propagation constant.* The complex sum of the attenuation and phase coefficients, as given by Eqn (6.15), is called the propagation constant because it determines, from Eqn (6.13), how the voltage on the line changes with z.

6.3 Characteristic Impedance

In Eqn (6.14) the current on the line is given as the sum of two component current waves, but by using Eqns (6.8), (6.13) and (6.15) it can be expressed in terms of voltage and the line parameters. From Eqn (6.8)

$$I = -\frac{1}{R + j\omega L}\frac{\mathrm{d}V}{\mathrm{d}z}$$

From Eqn (6.13)

$$\frac{\mathrm{d}V}{\mathrm{d}z} = \gamma[V_2 \exp(\gamma z) - V_1 \exp(-\gamma z)]$$

then

$$I = \frac{\gamma}{R + j\omega L}[V_1 \exp(-\gamma z) - V_2 \exp(\gamma z)] \tag{6.19}$$

Substituting for γ from Eqn (6.15) gives

$$I = \sqrt{\left(\frac{G + j\omega C}{R + j\omega L}\right)}[V_1 \exp(-\gamma z) - V_2 \exp(\gamma z)] \tag{6.20}$$

This equation is of the form: current $= K \times$ voltage and, by analogy with Ohm's Law, $[(R + j\omega L)/(G + j\omega L)]^{1/2}$ is an impedance. Its value at any particular frequency is determined entirely by the line parameters R, L, G and C, so it is called the characteristic impedance of the line, Z_0, where

$$Z_0 = \sqrt{[(R + j\omega L)/(G + j\omega C)]} \tag{6.21}$$

It is sometimes helpful to think of Z_0 in one of the following ways:

(i) it is the value which the load impedance must have to match the load to the line;
(ii) it is the impedance seen from the sending end of an infinitely long line;
(iii) it is the impedance seen looking towards the load at any point on a matched line – moving along the line produces no change in the impedance towards the load.

The concept of matching is explained further in the following section.

Having produced the fundamental analysis, we need to look at real problems on transmission lines, and the remainder of this chapter will make considerable use of the equations developed. It is therefore worthwhile to note the assumptions that have been made as we have built up our model of a

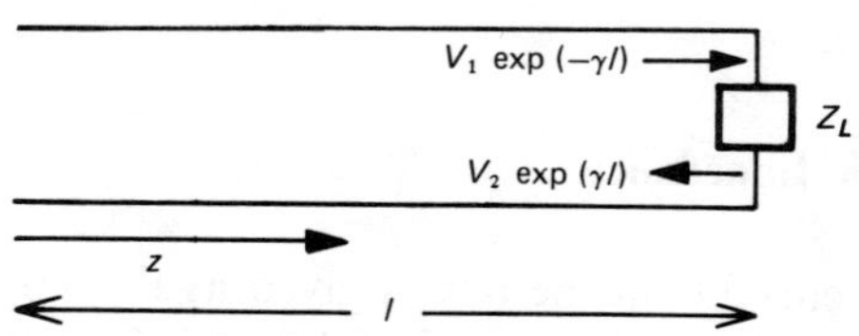

Fig. 6.3 Reflection at the load.

transmission line:

(i) the line is uniform, homogeneous and straight;
(ii) the line parameters R, L, G and C do not vary with change in ambient conditions, such as temperature and humidity;
(iii) the line parameters are not frequency dependent;
(iv) the analysis applies only between the junctions on the line because the circuit model used in Fig. 6.3 is not valid across a junction.

It is important to be aware that these assumptions have been made and to appreciate that there may be occasions when they should be taken into account.

6.4 Reflection From the Load

In the last section we noted that a line is matched if it is terminated in a load equal to its characteristic impedance. When we say the line is matched we mean that the forward wave is totally absorbed by the load; consequently there is no reflected wave, and V_2 is zero.

Usually, however, the load will be different from Z_0, say Z_L, and some of the incident wave will be reflected back down the line. The size of the reflected wave will depend on the difference between Z_0 and Z_L as will be shown in the next section.

6.5 Reflection Coefficient ρ

The amount of reflection caused by the load is expressed in terms of the voltage reflection coefficient ρ. It is defined as the ratio of the reflected voltage to the incident voltage, at the load terminals.

The load is at the position $z = l$, as shown in Fig. 6.3, and from Eqn (6.13) the voltage there is

$$V_L = V_1 \exp(-\gamma l) + V_2 \exp(\gamma l) \tag{6.22}$$

The terms $V_1 \exp(-\gamma l)$ and $V_2 \exp(\gamma l)$ are the incident and reflected voltages respectively, at the load.

So, from the definition of ρ,

$$\begin{aligned}\rho &= V_2 \exp(\gamma l)/V_1 \exp(-\gamma l)\\ &= (V_2/V_1)\exp(2\gamma l)\end{aligned} \tag{6.23}$$

ρ will usually be a complex quantity and it is often convenient to express it in a polar form as

$$\rho = |\rho| \exp(j\psi) \tag{6.24}$$

ψ is referred to as the angle of the reflection coefficient. Now we can find an expression for ρ in terms of the load and the characteristic impedance of the line.

The current at the load, from Eqns (6.20) and (6.21), is

$$I_L = \frac{V_1}{Z_0}\exp(-\gamma l) - \frac{V_2}{Z_0}\exp(\gamma l) \tag{6.25}$$

and the load impedance is

$$Z_L = \frac{V_L}{I_L}$$

Therefore, from Eqns (6.25) and (6.22),

$$Z_L = \frac{V_1\exp(-\gamma l) + V_2\exp(\gamma l)}{(V_1/Z_0)\exp(-\gamma l) - (V_2/Z_0)\exp(\gamma l)}$$

Dividing by $V_1\exp(-\gamma l)$

$$Z_L = Z_0\left[\frac{1 + (V_2/V_1)\exp(2\gamma l)}{1 - (V_2/V_1)\exp(2\gamma l)}\right]$$

and from Eqn (6.23)

$$Z_L = Z_0\left(\frac{1+\rho}{1-\rho}\right) \tag{6.26}$$

or

$$\rho = (Z_L - Z_0)/(Z_L + Z_0) \tag{6.27}$$

Consider the following three cases:

(i) *Short-circuit load*, $Z_L = 0$
From Eqn (6.27), $\rho = -1$. Hence, from Eqn (6.24), $|\rho| = 1$ and $\psi = \pi$.

(ii) *Open-circuit load*, $Z_L = \infty$
From Eqn (6.27), $\rho = 1$, so $|\rho| = 1$ and $\psi = 0$.

(iii) $Z_L = 150 + j100\,\Omega$, $Z_0 = 50\,\Omega$
Then

$$\rho = (100 + j100)/(200 + j100)$$

from which $\rho = 0.6 + j0.2$
which gives $|\rho| = 0.63$ and $\psi = 18.43°$.

Each of these cases is shown diagrammatically in Fig. 6.4 in terms of incident and reflected voltages.

So far we have been interested in the voltage on the line and the voltage reflection coefficient. However, we would expect the current wave to be reflected also, and by an analysis similar to that used above the amount of reflection can be seen to depend on the load impedance, as in the voltage case. This

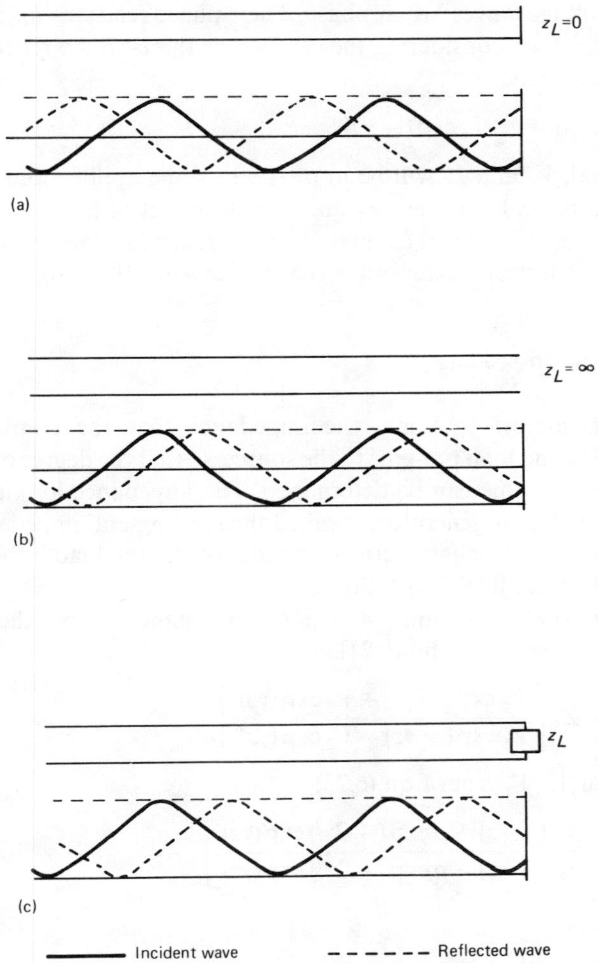

Fig. 6.4 (a) Voltage reflections at a short circuit, $|\rho|=1, \psi=\pi$; (b) voltage reflections at an open circuit, $|\rho|=1$, $\psi=0$; (c) voltage reflections at a typical load z_L, $\rho=|\rho|\exp(j\psi)$.

time, however, the current reflection coefficient, $\rho_I=|\rho_I|\exp(j\psi_I)$ is given by

$$\rho_I=(Z_0-Z_L)/(Z_0+Z_L)$$

For a short-circuit load $\rho_I=1$ ($|\rho_I|=1, \psi=0$), which means that there is no change of phase between the incident and reflected waves – as we would expect.

From the discussion about reflection coefficients it is evident that the phase of the reflected wave relative to the incident wave is determined by Z_L and Z_0. However, we have not justified our implicit assumption that the incident current and voltage waves are in phase. Their phase relationship depends on Z_0, as can be seen by considering the case where there are no reflected waves, i.e. $V_2 = I_2 = 0$. Then

$$Z_0 = V_1 \exp(-\gamma z)/I_1 \exp(-\gamma z)$$

and if Z_0 is real, V_1 and I_1 will be in phase. In some applications Z_0 cannot be assumed to be real and then V_1 and I_1 will be out of phase.

In the following part of this chapter, the term reflection coefficient will apply to the voltage reflection coefficient given by Eqns (6.23), (6.24) and (6.27).

6.6 Sending-end Impedance

It is often of interest to know the impedance that the combination of transmission line and load presents to the source, so that the degree of mismatch between source and line can be determined. The impedance looking into the line from the source, or generator, is called the sending-end impedance. It can be found in terms of the characteristic impedance Z_0, the load impedance Z_L, and the length of the line, l, as follows.

The impedance at some point A (Fig. 6.5) a distance z from the generator is, from Eqns (6.13), (6.20) and (6.21),

$$Z_A = \frac{V_A}{I_A} = Z_0\left[\frac{V_1 \exp(-\gamma z) + V_2 \exp(\gamma z)}{V_1 \exp(-\gamma z) - V_2 \exp(\gamma z)}\right]$$

Substituting for V_2/V_1 from Eqn (6.23),

$$Z_A = Z_0\left[\frac{\exp(-\gamma z) + \rho \exp(-2\gamma l)\exp(\gamma z)}{\exp(-\gamma z) - \rho \exp(-2\gamma l)\exp(\gamma z)}\right]$$

or

$$Z_A = Z_0\left[\frac{\exp \gamma(l-z) + \rho \exp -\gamma(l-z)}{\exp \gamma(l-z) - \rho \exp -\gamma(l-z)}\right]$$

From Fig. 6.5, $x = l - z$; then

$$Z_A = Z_0\left[\frac{\exp(\gamma x) + \rho \exp(-\gamma x)}{\exp(\gamma x) - \rho \exp(-\gamma x)}\right]$$

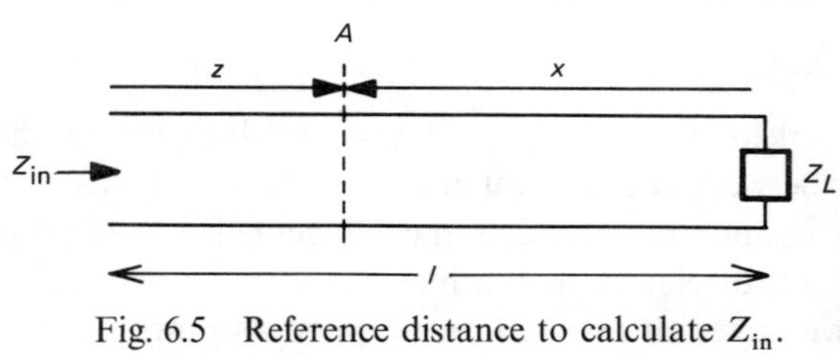

Fig. 6.5 Reference distance to calculate Z_{in}.

Replacing ρ from Eqn (6.27)

$$Z_A = Z_0\left[\frac{(Z_L + Z_0)\exp(\gamma x) + (Z_L - Z_0)\exp(-\gamma x)}{(Z_L + Z_0)\exp(\gamma x) - (Z_L - Z_0)\exp(-\gamma x)}\right]$$

Rearranging terms

$$Z_A = Z_0\left[\frac{Z_L[\exp(\gamma x) + \exp(-\gamma x)] + Z_0[\exp(\gamma x) - \exp(-\gamma x)]}{Z_L[\exp(\gamma x) - \exp(-\gamma x)] + Z_0[\exp(\gamma x) + \exp(-\gamma x)]}\right]$$

The exponentials can be replaced by hyperbolic functions to give

$$Z_A = Z_0\left[\frac{Z_L\cosh\gamma x + Z_0\sinh\gamma x}{Z_L\sinh\gamma x + Z_0\cosh\gamma x}\right]$$

or

$$Z_A = Z_0\left[\frac{Z_L + Z_0\tanh\gamma x}{Z_L\tanh\gamma x + Z_0}\right] \tag{6.28}$$

The sending-end impedance, Z_{in}, is obtained from Eqn (6.28) by putting $x = l$, i.e.

$$Z_{in} = Z_0\left[\frac{Z_L + Z_0\tanh\gamma l}{Z_L\tanh\gamma l + Z_0}\right] \tag{6.29}$$

For general application of this equation it is often more useful if it is normalized to the characteristic impedance of the line. Normalized impedances will be shown as lower-case letters, e.g.

$$z_{in} = Z_{in}/Z_0 \qquad \text{and} \qquad z_L = Z_L/Z_0$$

From Eqn (6.29),

$$\frac{Z_{in}}{Z_0} = \left[\frac{(Z_L/Z_0) + \tanh\gamma l}{1 + (Z_L/Z_0)\tanh\gamma l}\right]$$

or

$$z_{in} = \frac{z_L + \tanh\gamma l}{1 + z_L\tanh\gamma l} \tag{6.30}$$

6.7 Lines of Low Loss

In Eqn (6.12) we noted that the propagation constant is related to the line parameters by

$$\gamma = (R + j\omega L)^{1/2}(G + j\omega C)^{1/2}$$
$$= j\omega\sqrt{(LC)}\left(1 + \frac{R}{j\omega L}\right)^{1/2}\left(1 + \frac{G}{j\omega C}\right)^{1/2} \tag{6.31}$$

Using the binomial series, this equation can be expanded to

$$\gamma = j\omega\sqrt{(LC)}\left(1+\frac{R}{2j\omega L}-\frac{1}{4}\frac{R^2}{(j\omega L)^2}+\cdots\right)\left(1+\frac{G}{2j\omega C}-\frac{G^2}{4(j\omega C)^2}+\cdots\right) \tag{6.32}$$

Now, in low-loss lines the impedance parameters R and G will be very small, allowing all terms in R^2 and G^2, and higher powers, to be ignored. Then

$$\gamma \simeq j\omega\sqrt{(LC)}\left[1-\frac{RG}{4\omega^2 LC}-j\frac{R}{2\omega L}-j\frac{G}{2\omega C}\right] \tag{6.33}$$

and since $\gamma = \alpha + j\beta$ we find that

$$\alpha \simeq \frac{R}{2}\sqrt{\frac{C}{L}}+\frac{G}{2}\sqrt{\frac{L}{C}} \tag{6.34}$$

and

$$\beta \simeq j\omega\sqrt{(LC)}\left[1-\frac{RG}{4\omega^2 LC}\right] \tag{6.35}$$

Since R and G are small, we can reasonably assume that at high frequencies the second term in the bracket is negligible. Then

$$\beta \simeq j\omega\sqrt{(LC)} \tag{6.36}$$

By a similar argument

$$Z_0 = (R+j\omega L)^{1/2}(G+j\omega C)^{-1/2}$$

$$= \sqrt{\left(\frac{j\omega L}{j\omega C}\right)}\left(1+\frac{R}{j\omega L}\right)^{1/2}\left(1+\frac{G}{j\omega C}\right)^{-1/2}$$

$$\simeq \sqrt{\left(\frac{L}{C}\right)}\left[\left(1+\frac{R}{2j\omega L}\right)\left(1-\frac{G}{2j\omega C}\right)\right]$$

if, as before, all terms in R^2, G^2 and above are ignored.

Expanding, and again assuming that the term in RG/ω^2 is negligible,

$$Z_0 \simeq \sqrt{\left(\frac{L}{C}\right)}\left[1-\frac{jR}{2\omega L}-\frac{jG}{2\omega C}\right]$$

For many applications $R/\omega L$ and $G/\omega C$ will be very small, and a reasonable approximation for Z_0 is

$$Z_0 \simeq \sqrt{(L/C)} \tag{6.37}$$

Note that when this approximation is valid, Z_0 is real. Using Eqn (6.37) in Eqn (6.34) gives

$$\alpha \simeq \frac{R}{2Z_0}+\frac{GZ_0}{2} \tag{6.38}$$

6.8 Loss-less Lines

In many communications problems concerning transmission lines it is often a reasonable approximation to assume that the line is loss-less, i.e. the attenuation is zero. The advantage of making this assumption is that it greatly simplifies the calculations involved, and the disadvantage of a loss in accuracy is not serious when the line is relatively short and operating at very high frequencies.

The effect of zero attenuation is to make the characteristic impedance real, as in Eqn (6.37), and to make the propagation constant imaginary, i.e.

$$\gamma = j\beta$$

We can see immediately the simplification that can result if we replace γ in Eqn (6.29) by $j\beta$. The sending-end impedance is then given by

$$Z_{in} = Z_0\left[\frac{Z_L + jZ_0 \tan \beta l}{Z_0 + jZ_L \tan \beta l}\right] \tag{6.39}$$

using

$$\tanh j\beta l = j \tan \beta l.$$

Similarly, Eqn (6.30) simplifies to

$$z_{in} = \frac{z_L + j \tan \beta l}{1 + jz_L \tan \beta l} \tag{6.40}$$

6.9 Quarter-wave Transformer

The relationship between the input and load impedances of a quarter-wave loss-less line can be found easily from Eqn (6.39), i.e.

$$Z_{in} = Z_0\left[\frac{Z_L + jZ_0 \tan \beta l}{Z_0 + jZ_L \tan \beta l}\right] \tag{6.41}$$

When l is a quarter-wavelength long, $\tan \beta l = \tan \pi/2 = \infty$. Then

$$Z_{in} = Z_0{}^2/Z_L \tag{6.42}$$

or

$$Z_{in}Z_L = Z_0{}^2 \tag{6.43}$$

We therefore have a method for matching a known load to a known input (line) impedance by placing between them a quarter-wave section of loss-less line of characteristic impedance calculated from Eqn (6.43).

This type of transformer is very frequency sensitive because $\tan \theta$ varies rapidly about $\theta = \pi/2$. Problem 6.10 looks further into this sensitivity.

6.10 Stubs

From Eqns (6.30) and (6.40) we can see that the input impedance of a transmission line varies with its length. This property can be used in short lengths of line, known as stubs, to provide an adjustable impedance (or admittance) for use in matching applications, as we shall discuss later. Stubs are terminated in either a short-circuit or open-circuit load. If the load is a short circuit, $z_L = 0$, and from Eqn (6.40), if we assume that the stub is formed from a loss-less length of line, the input impedance of the line is

$$z_{in} = j \tan \beta l \tag{6.44}$$

If the load is an open circuit, $z_L = \infty$, and

$$z_{in} = -j \cot \beta l \tag{6.45}$$

The variation of the input impedance with the length of the stub is shown in Fig. 6.6. It can be seen that the impedance is always reactive and that its value is repeated at half-wavelength intervals.

The use of stubs as matching devices will be discussed in a later section.

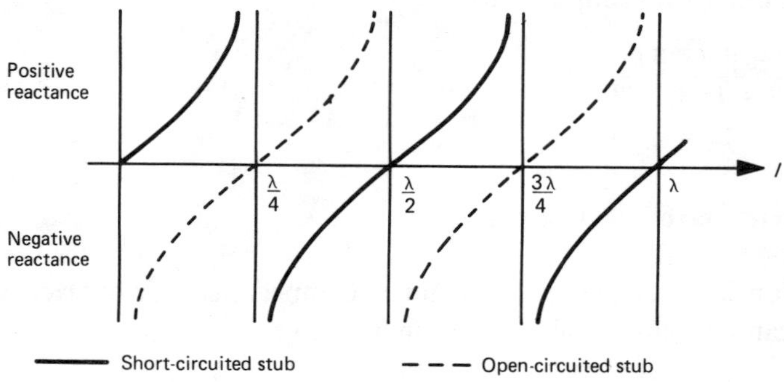

Fig. 6.6 Stub reactance.

6.11 Standing Waves

The total voltage on a loss-less line at some point z from the sending end is, from Eqn (6.17),

$$V = V_1 \exp(-j\beta z) + V_2 \exp(j\beta z)$$

and, remembering the time dependence,

$$V = [V_1 \exp(-j\beta z) + V_2 \exp(j\beta z)] \exp(j\omega t)$$

Rearranging the right-hand side, and substituting for ρ from Eqn (6.23),

$$V = V_1 \exp(j\omega t)\exp(-j\beta l)[\exp(j\beta x) + \rho \exp(-j\beta x)] \tag{6.46}$$

where $x = l - z$ as shown in Fig. 6.5.

This equation represents a voltage standing wave, i.e. a stationary wave composed of two component travelling waves, one in the forward direction and the other in the backward direction, reflected from the load.

The precise shape of the voltage standing wave defined by Eqn (6.46) will depend on the value of ρ, which may be complex. In practice, a simple analytical solution of that equation may not be easy to obtain. However, as we noted earlier, when the load is a short or open circuit, ρ has the values -1 or $+1$, respectively, and in each of these cases simple trigonometric solutions of Eqn (6.46) can be found. Without the time dependence, Eqn (6.46) is

$$V = V_1 \exp(-j\beta l)[\exp(j\beta x) + \rho \exp(-j\beta x)] \tag{6.47}$$

which, for a short-circuit load, becomes

$$\begin{aligned} V &= V_1 \exp(-j\beta l)[\exp(j\beta x) + \beta \exp(-j\beta x)] \\ &= j2V_1 \exp(-j\beta l)\sin\beta x \end{aligned} \tag{6.48}$$

This standing wave could be measured by using a standing-wave detector which displays the real part of the modulus of Eqn (6.48), i.e.

$$|V| = \mathrm{Re}[2V_1 \exp(-j\beta l)|\sin\beta x|] \tag{6.49}$$

as shown in Fig. 6.7(a).

From Eqn (6.49) we can see that minima will occur on the line when $\beta x = (n-1)\pi$, where n is any positive integer. From this expression we can see that adjacent minima will be separated by π/β; that is, by $\lambda/2$, because $\beta = 2\pi/\lambda$. The

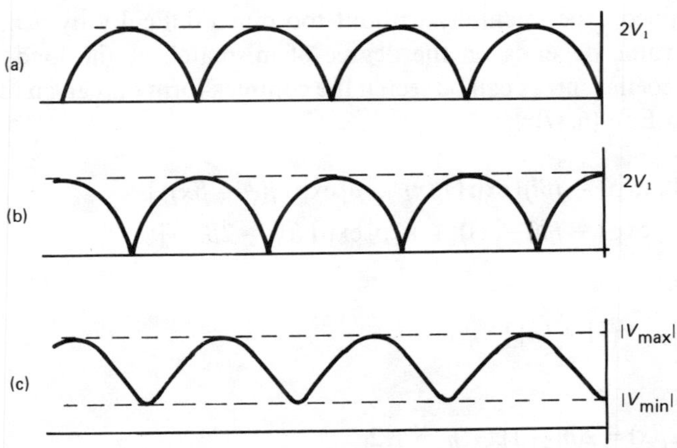

Fig. 6.7 Standing waves: (a) produced by a short circuit; (b) produced by an open circuit; (c) produced by a load z, with reflection coefficient $0.6 + j0.3$.

first voltage minimum will be at the load terminals, $x = 0$, as we have concluded earlier.

In a similar way we can use Eqn (6.47) to show that if there is an open circuit at the load the detected voltage will be

$$|V| = \mathrm{Re}\,[2V_1 \exp(-j\beta l)|\cos \beta x|]$$

which again has minima separated by $\lambda/2$, but with a maximum at the load. The standing wave due to an open circuit is shown in Fig. 6.7(b).

For any other load, ρ may be complex and, as an example, the standing wave for $\rho = 0.6 + j0.3$ is shown in Fig. 6.7(c). It can be seen that the minima do not fall to zero and the maxima do not rise to $2V_1$. However, adjacent minima are still half a wavelength apart, and indeed they will be for all values of ρ.

Current standing waves also exist and they can be discussed in exactly the same way but they will not be considered any further here.

6.12 Voltage Standing Wave Ratio (VSWR)

The general shape of the voltage standing wave pattern on a loss-less line is shown in Fig. 6.7(c). By using a detector, and moving it along the line, values for V_{max} and V_{min} can be obtained. The ratio of these two is called the voltage standing wave ratio, or VSWR. We shall use the symbol S to denote VSWR and define it as

$$S = \frac{|V_{max}|}{|V_{min}|} \tag{6.50}$$

which allows S to take values in the range $1 \leqslant S \leqslant \infty$. The value of S can often be determined experimentally without too much difficulty by using a slotted line.[1] Its value depends on the degree of mismatch at the load, i.e. on the reflection coefficient, as can be seen if the complex form of ρ given in Eqn (6.24) is put into Eqn (6.47):

$$\begin{aligned} V &= V_1 \exp(-j\beta l)[\exp(j\beta x) + |\rho| \exp(j(\psi - \beta x))] \\ &= V_1 \exp(-j\beta(l - x))[1 + |\rho| \exp(j(\psi - 2\beta x))] \end{aligned} \tag{6.51}$$

so that

$$|V_{max}| = V_1[1 + |\rho|] \tag{6.52}$$

when

$$(\psi - 2\beta x) = 2(m-1)\pi \quad m = 1, 2, 3 \ldots$$

and

$$|V_{min}| = V_1[1 - |\rho|] \tag{6.53}$$

when

$$(\psi - 2\beta x) = (2m - 1)\pi \qquad m = 1, 2, 3\ldots$$

Therefore, from Eqns (6.52) and (6.53), the standing wave ratio is

$$S = \frac{|V_{max}|}{|V_{min}|} = \frac{1 + |\rho|}{1 - |\rho|} \tag{6.54}$$

We will find it convenient later to use this expression in the form

$$|\rho| = \frac{S - 1}{S + 1} \tag{6.55}$$

6.13 Impedance at a Voltage Minimum and at a Voltage Maximum

The normalized impedance at a voltage minimum can be derived from Eqn (6.51):

$$V = V_1 \exp[-j\beta(l - x)]\{1 + |\rho| \exp[j(\psi - 2\beta x)]\}$$

A voltage minimum occurs when

$$\exp[j(\psi - 2\beta x)] = \exp(j\pi)$$

and this coincides with an impedance minimum.

At the first voltage minimum from the load, when $x = x_{min}$,

$$\psi = \pi + 2\beta x_{min}$$

and at x_{min}

$$Z = Z_{min} = \left(\frac{V}{I}\right)_{x_{min}}$$

Putting $x = x_{min}$ in Eqn (6.51) gives $V_{x_{min}}$ and $I_{x_{min}}$ is found by putting $z = l - x_{min}$ in Eqn (6.20).

Then

$$Z_{in} = \frac{\{V_1 \exp[-j\beta(l - x_{min})]\}\{1 + |\rho| \exp[j(\psi - 2\beta x_{min})]\}}{\{V_1 \exp[-j\beta(l - x_{min})]\}\{1 - |\rho| \exp[j(\psi - 2\beta x_{min})]\}}$$

$$= Z_0 \frac{[1 + |\rho| \exp(j\pi)]}{[1 - |\rho| \exp(j\pi)]} = Z_0 \left[\frac{1 - |\rho|}{1 + |\rho|}\right] = Z_0/S \tag{6.56}$$

or

$$z_{min} = 1/S \tag{6.57}$$

By a similar argument, we can show that

$$Z_{max} = Z_0 S \tag{6.58}$$

and

$$z_{max} = S \tag{6.59}$$

6.14 Load Impedance on a Loss-less Line

There are often occasions when the load impedance is not known, but its value is required. It can be found, if the line is assumed to be loss-less, by measuring the VSWR, the wavelength and the distance from the load to the nearest voltage minimum.

Rewriting Eqn (6.51)

$$V = V_1 \exp[-j\beta(l-x)]\{1 + |\rho| \exp[j(\psi - 2\beta x)]\}$$

and, as we noted in the last section, this is a minimum when

$$\exp[j(\psi - 2\beta x)] = -1$$

i.e. when

$$\psi - 2\beta x = (2m-1)\pi \qquad m = 1, 2, 3 \ldots$$

The solution of interest is that for the smallest value of x, which we have called x_{min}. It occurs when $m = 0$; then

$$\psi - 2\beta x_{min} = \pi$$

or

$$\psi = \pi + 2\beta x_{min} \tag{6.60}$$

The load impedance and the reflection coefficient are related by

$$Z_L = Z_0\left(\frac{1+\rho}{1-\rho}\right)$$

i.e. using Eqn (6.24),

$$Z_L = Z_0\left(\frac{1+|\rho|\exp(j\psi)}{1-|\rho|\exp(j\psi)}\right)$$

and substituting for $|\rho|$ from Eqn (6.55),

$$Z_L = Z_0\left(\frac{1+[(S-1)/(S+1)]\exp(j\psi)}{1-[(S-1)/(S+1)]\exp(j\psi)}\right)$$

ψ can be replaced by Eqn (6.60):

$$Z_L = Z_0\left[\frac{1+[(S-1)/(S+1)]\exp[j(\pi+2\beta x_{min})]}{1-[(S-1)/(S+1)]\exp[j(\pi+2\beta x_{min})]}\right] \tag{6.61}$$

Rearranging Eqn (6.61), and remembering that $e^{j\pi} = -1$,

$$Z_L = Z_0\left[\frac{S[1-\exp(j2\beta x_{min})]+[1+\exp(j2\beta x_{min})]}{S[1+\exp(j2\beta x_{min})]+[1-\exp(j2\beta x_{min})]}\right]$$

Dividing throughout by $\exp(j\beta x_{\min})$ gives

$$Z_L = Z_0\left[\frac{S[\exp(-j\beta x_{\min}) - \exp(j\beta x_{\min})] + [\exp(-j\beta x_{\min}) + \exp(j\beta x_{\min})]}{S[\exp(-j\beta x_{\min}) + \exp(j\beta x_{\min})] + [\exp(-j\beta x_{\min}) - \exp(j\beta x_{\min})]}\right]$$

or

$$Z_L = Z_0\left[\frac{-jS\sin\beta x_{\min} + \cos\beta x_{\min}}{S\cos\beta x_{\min} - j\sin\beta x_{\min}}\right]$$

$$= Z_0\left[\frac{1 - jS\tan\beta x_{\min}}{S - j\tan\beta x_{\min}}\right] \tag{6.62}$$

The normalized load impedance is

$$z_L = \frac{1 - jS\tan\beta x_{\min}}{S - j\tan\beta x_{\min}} \tag{6.63}$$

6.15 Smith Transmission Line Chart

Transmission line problems can be solved using the equations we have derived in the earlier part of this chapter, but often there is a considerable amount of tedious algebraic manipulation involved that can make calculations rather lengthy. To overcome this problem, several graphical techniques have been developed. We will study only one—the Smith chart, named after P.H. Smith who introduced it in 1944.[2] It is the best known and most comprehensive of all graphical methods, and compared with analysis it has the considerable additional benefit of providing a picture of the conditions on the line. This allows us to assess what sort of adjustments might be necessary to produce the conditions we require.

The derivation of the chart is not difficult to follow, but it is easier to appreciate after some familiarity with its use has been developed, so a discussion on why it has its particular shape will be deferred until a later section. The most commonly used version of the Smith chart is shown in Fig. 6.8. Some time should be spent in reading the various legends inside the chart and on the circumferential scales. You will notice that there are three sets of circular arcs in the chart, and one straight line. The chart is designed to allow any value of impedance or admittance to be represented, and in the examples that we study later it will be used in both modes. The horizontal line dividing the chart into an upper and a lower half is the locus of impedances and admittances having no reactive or susceptive component. Along that line the following equations are satisfied

$$z = r + j0$$
$$y = g + j0$$

Figure 6.9(a) shows one set of circles. They all have their centres on the

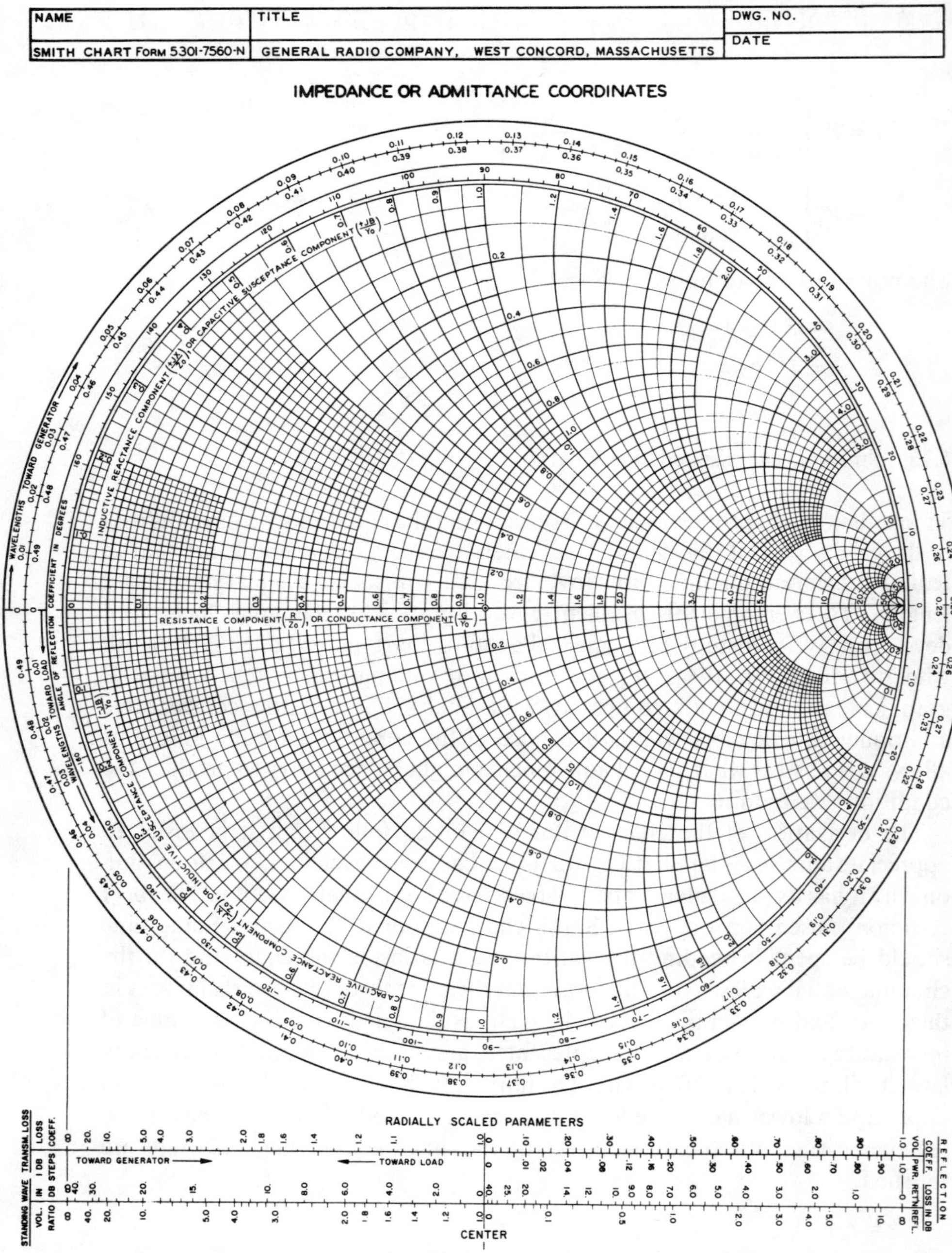

Fig. 6.8 Smith chart. Reproduced by permission of Phillip H. Smith, Analog Instruments Company, New Providence, N.J.

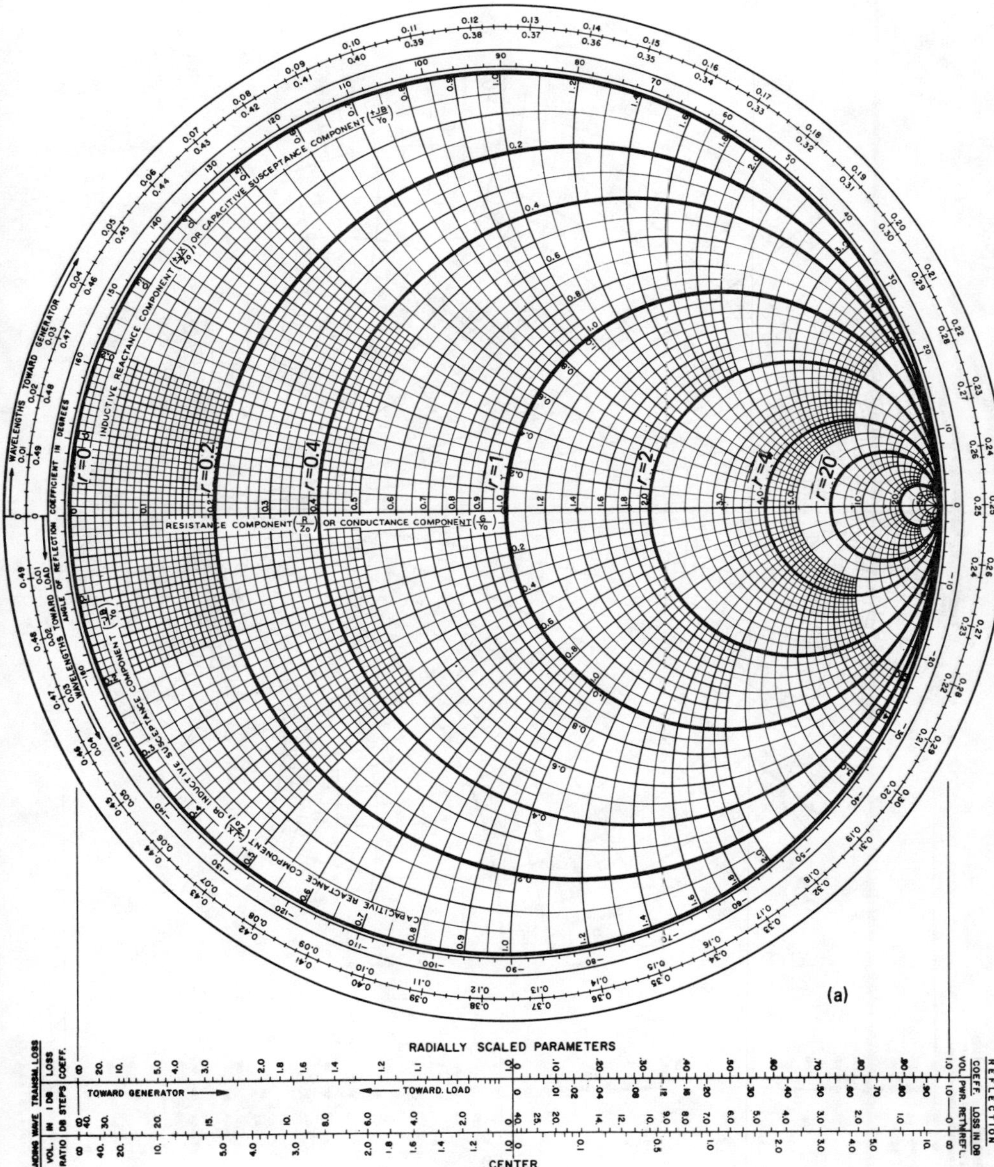

Fig. 6.9 (a) Circles of constant resistance (or conductance); (b) circles of constant positive reactance (or susceptance), and circles of constant negative reactance (or susceptance).

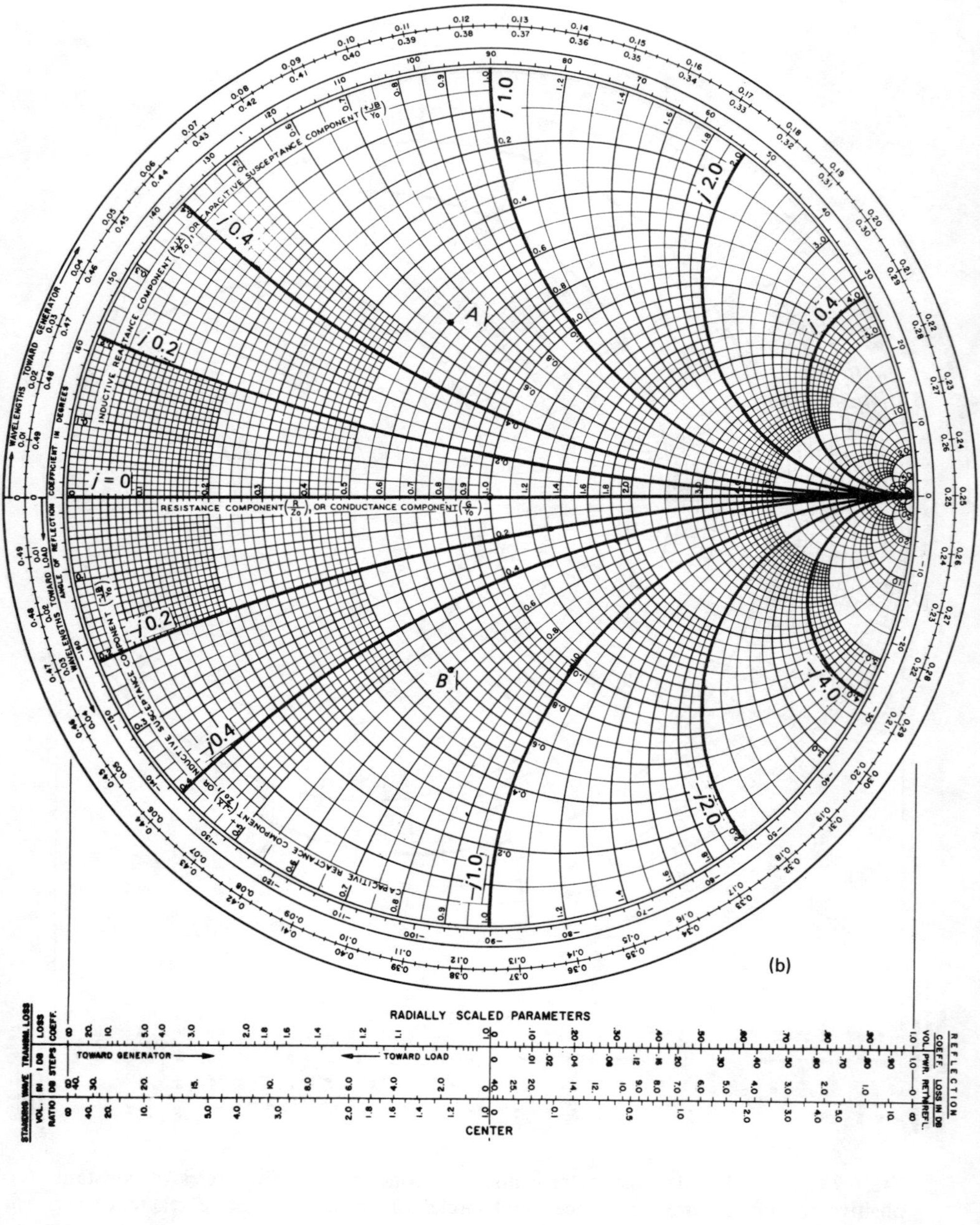

WAVELENGTHS TOWARD GENERATOR
WAVELENGTHS TOWARD LOAD
ANGLE OF REFLECTION COEFFICIENT IN DEGREES
INDUCTIVE REACTANCE COMPONENT (+jX/Zo), OR CAPACITIVE SUSCEPTANCE COMPONENT (+jB/Yo)
CAPACITIVE REACTANCE COMPONENT (−jX/Zo), OR INDUCTIVE SUSCEPTANCE COMPONENT (−jB/Yo)
RESISTANCE COMPONENT (R/Zo), OR CONDUCTANCE COMPONENT (G/Yo)
j = 0
j 0.2
j 0.4
j 1.0
j 2.0
j 4.0
−j 0.2
−j 0.4
−j 1.0
−j 2.0
−j 4.0
A
B
RADIALLY SCALED PARAMETERS
TOWARD GENERATOR
TOWARD LOAD
CENTER
STANDING WAVE
VOL. RATIO
IN DB
1 DB STEPS
TRANSM. LOSS
LOSS COEFF.
REFLECTION
COEFF. VOL. PWR.
LOSS IN DB RETN. REFL.

(b)

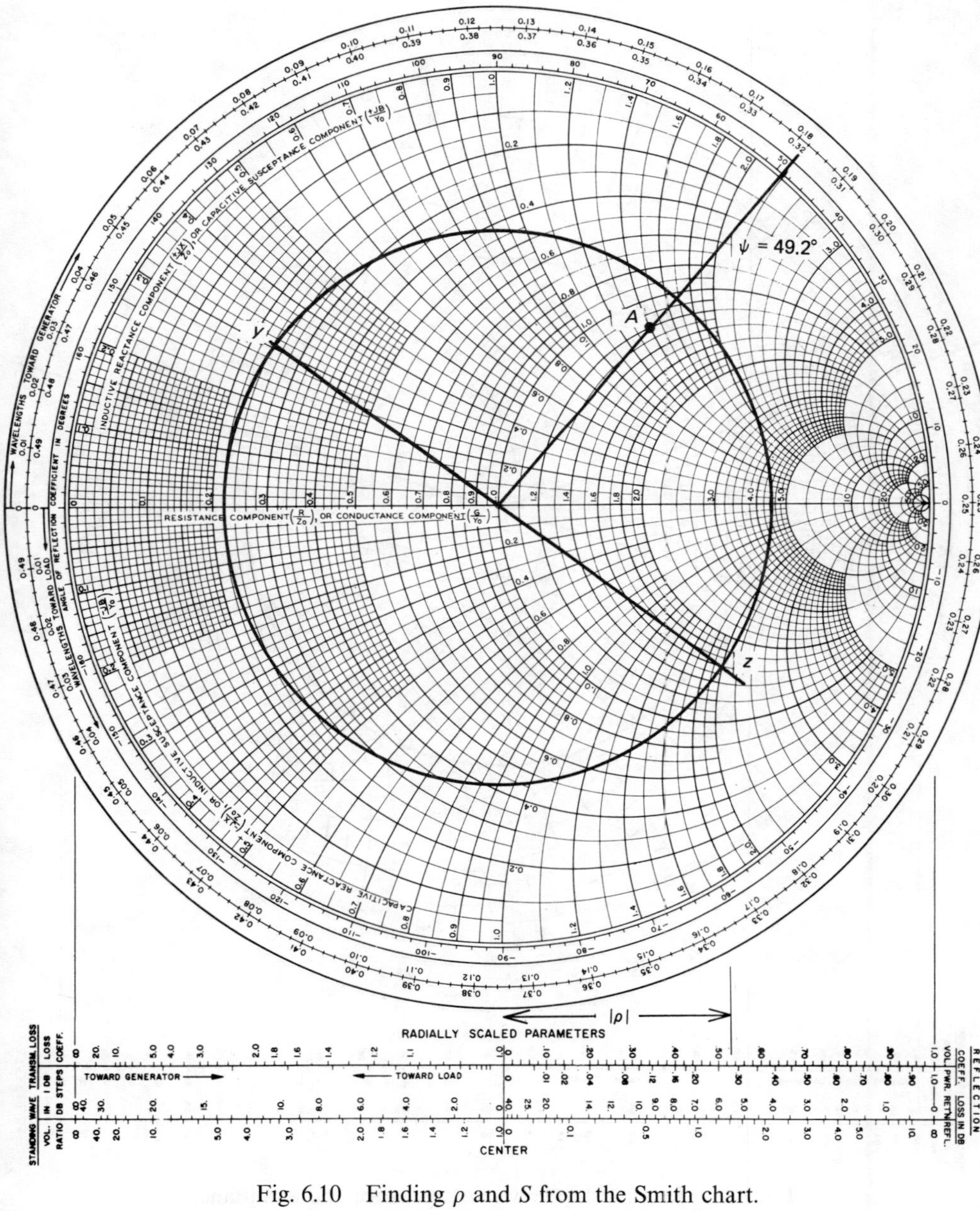

Fig. 6.10 Finding ρ and S from the Smith chart.

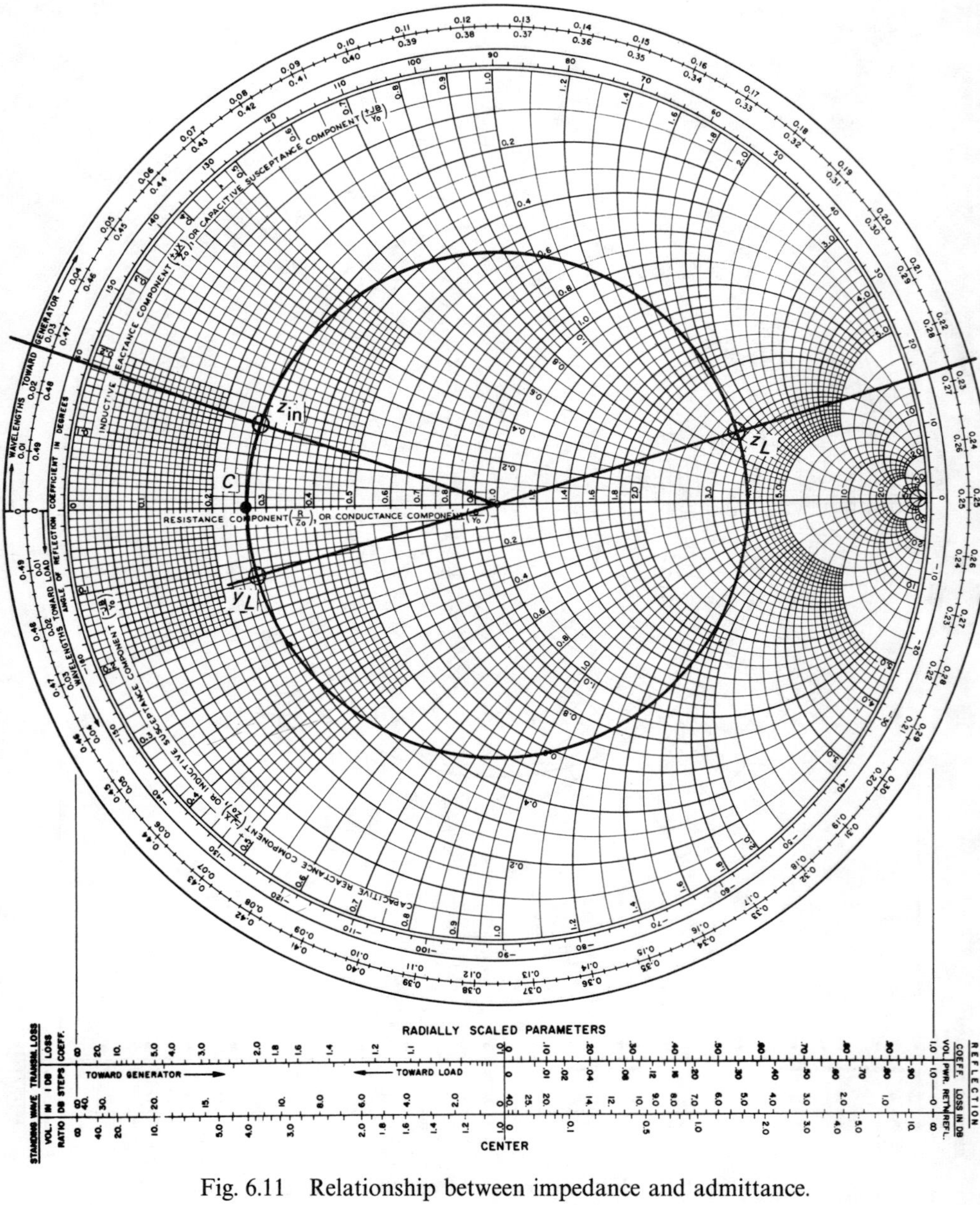

Fig. 6.11 Relationship between impedance and admittance.

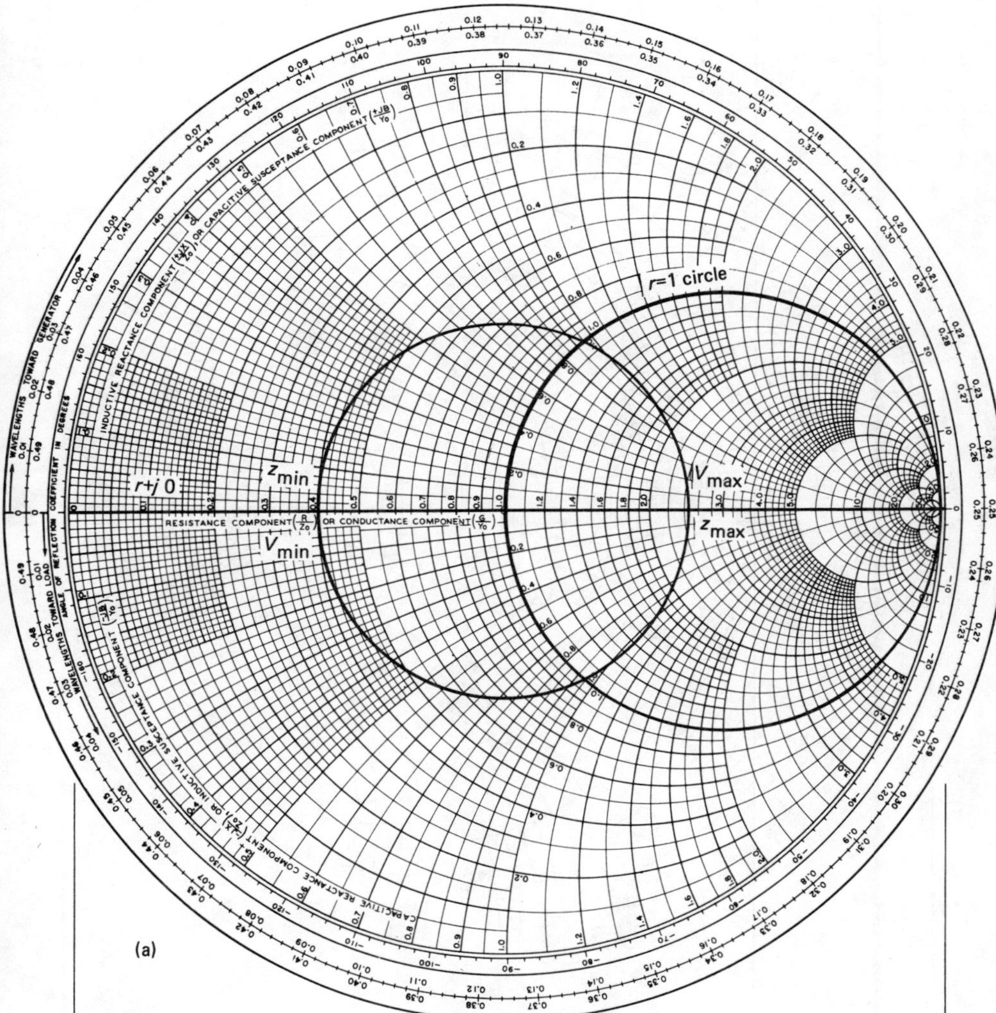

Fig. 6.12 (a) Position of voltage minimum on impedance chart; (b) position of voltage minimum on admittance chart.

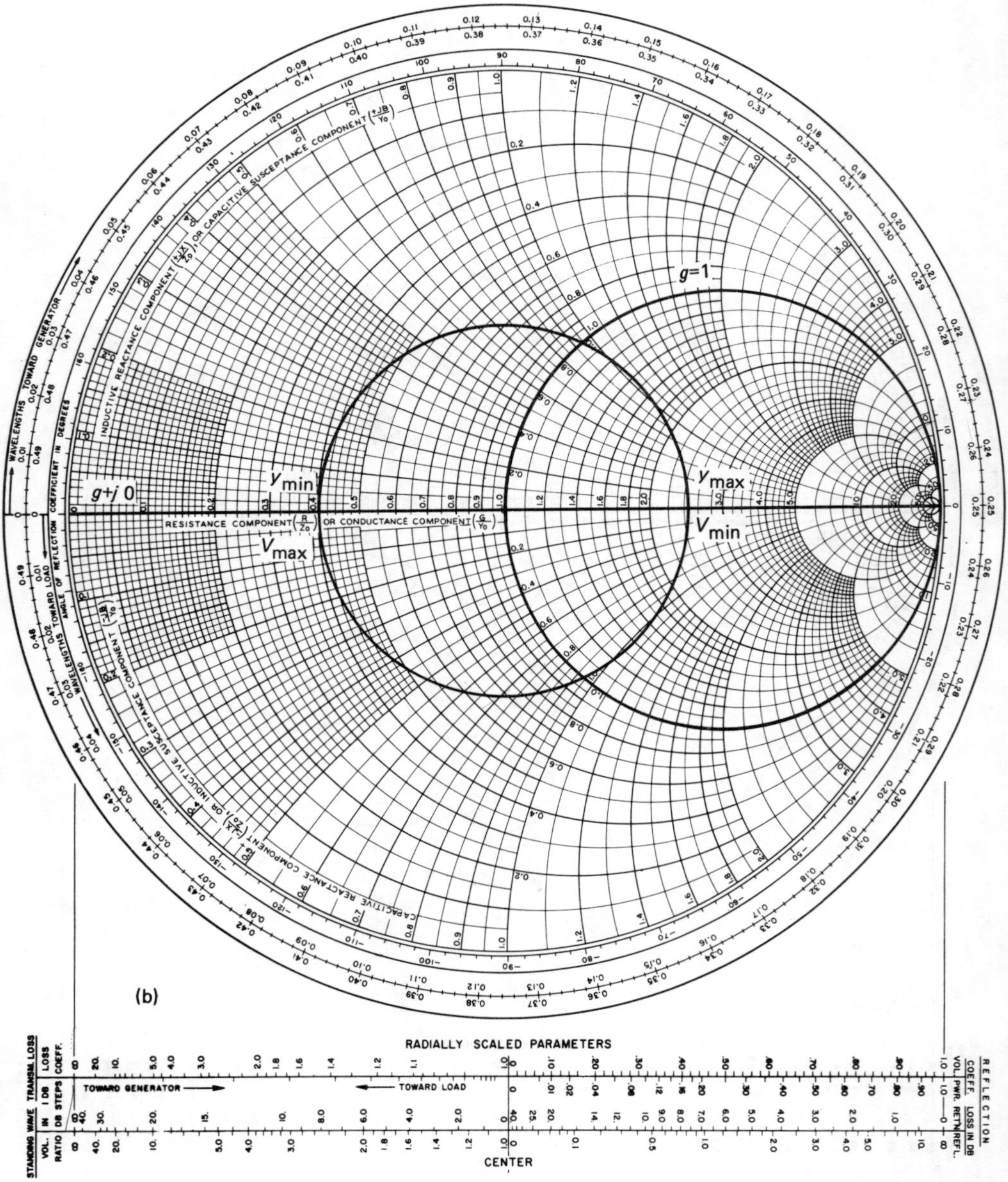

g=1
y_min
y_max
V_max
V_min
g+j 0
RADIALLY SCALED PARAMETERS
TOWARD GENERATOR
TOWARD LOAD
CENTER

(b)

horizontal line, and they are contiguous at the right-hand end of the chart. These circles give the normalized resistive or conductive component of impedance or admittance, i.e. the r or g in the above equations. The centre of the chart, where the circle representing unity normalized resistance or conductance crosses the horizontal line, has the coordinates $(1, 0)$. It therefore represents the impedance of the line under consideration. When we discuss methods of matching, the object will be to change the impedance at the matching point to the value represented by $(1, 0)$.

The resistance, or conductance, circles range from zero at the outer periphery to infinity at the right-hand side, although the highest value of normalized resistance or conductance shown is 50. Another set of circular arcs is shown in Fig. 6.9(b). These are lines that represent the positive imaginary components of impedance or admittance. For example, the point A has an impedance or admittance of $0.6 + j0.6$.

The third set of arcs, shown in Fig. 6.9(b), is in the lower half of the chart and represents negative imaginary components of impedance or admittance. Point B has an impedance or admittance of $0.6 - j0.6$.

The arcs for the imaginary components are parts of circles that have their centres on a vertical line through the right-hand end of the horizontal line and they all touch at that point (see Section 6.19). Only those parts of the arcs that occur in the range of positive real components are shown.

Another feature that will be used regularly can be seen from Fig. 6.10. Plot $z = 1.6 - j2.0$ on the chart. Then draw a circle, with centre at $(1, 0)$ through that point. From z draw a diameter to cut the circle at the opposite side. The point of intersection has a value $y = 0.25 + j0.31$ and comparing y and z it is found that $z = 1/y$. This relationship applies in general, i.e. the relative impedance and admittance are at opposite ends of a diameter of the circle, with centre $(1, 0)$ passing through them. The inner scale round the chart is marked "angle of reflection coefficient in degrees". It relates the characteristic impedance of the line, at $(1, 0)$, to any load impedance, e.g. in Fig. 6.10, if point A represents a load of $1.2 + j1.4$, the angle of the reflection coefficient can be read from the intersection of a line from $(1, 0)$ through A, and the inner scale at B, to be $\psi = 49.2°$.

The magnitude of the reflection coefficient, $|\rho|$, is related to the distance between $(1, 0)$ and A. Below the chart there are eight radial scales, and the uppermost one on the right is marked "Voltage Reflection Coefficient". $|\rho|$ is found by measuring the distance from $(1, 0)$ to A along that scale from the centre. In the example given $|\rho| = 0.541$.

The other important radial scale, which is related through Eqn (6.54) to that of $|\rho|$, is the lowest one on the left, marked "Standing Wave Voltage Ratio". We know that if $|\rho|$ is constant, S is also represented by a circle. As an example, Fig. 6.10 shows the circle for $S = 4.5$. Its radius is found from the bottom-left radial scale, but we can see from the figure that the circle passes through the point 4.5 on the horizontal line across the chart. The resistance scale to the right of the centre of the chart is identical with the VSWR scale, and either may be used.

It will help if we consider an example of how these various aspects of the Smith chart are used.

EXAMPLE: A loss-less transmission line of characteristic impedance $50\,\Omega$ is terminated in a load of $150 + j75\,\Omega$. Find the reflection coefficient, the load admittance, the VSWR, the distance between the load and the nearest voltage minimum to it, and the input impedance, if the line is 92 cm long and the wavelength of the signal on the line is 40 cm.

The normalized load impedance is z_L where

$$z_L = \frac{150 + j75}{50} = 3 + j1.5$$

and it is plotted on the chart as shown in Fig. 6.11. Then the S circle is drawn through z_L. We can now read the value of VSWR to be $S = 3.8$. By drawing a line from (1, 0) through z_L to the Angle of Reflection Coefficient scale we can read that $\psi = 16°$; $|\rho|$ is measured on the appropriate radial scale to be 0.585. Thus we have found that $\rho = 0.585 \exp(j0.28)$.

The load admittance y_L is on the same S circle as z_L, and at the opposite end of a diameter. Thus, from the chart, $y_L = 0.27 - 0.13$.

The line is 92 cm long which is equivalent to $(92/40)\lambda = 2.3\lambda$. Therefore, to find the input impedance z_{in} we must go from the load a distance 2.3λ towards the generator, on a constant S circle. This we can do by noting where the line from the centre through z cuts the Wavelengths Towards Generator scale. Reading from the chart, this occurs at 0.227λ. The input impedance is therefore at $(2.3 + 0.227)\lambda$ on the Wavelengths Towards Generator scale, i.e. at 2.527λ. One complete revolution on an S circle is 0.5λ; therefore in 2.527λ there are 5 revolutions $+ 0.027\lambda$. Hence the input impedance is at 0.027λ on the Wavelengths Towards Generator scale and reading from the chart, $z_{in} = 0.27 + j0.16$.

Finally, we want to know the distance of the nearest voltage minimum to the load from the load itself, x_{min}. The normalized impedance at a voltage minimum, which has been shown to be $1/S$ [Eqn (6.57)], must lie on the horizontal line to the left of (1, 0). Then x_{min} is found from the chart by reading the distance from z_L to V_{min} on the Wavelengths Towards Generator scale, and V_{min} will be at point C on the chart which reads 0.0 on the scale. Therefore V_{min} is 0.273λ from the load.

All of these results can be obtained by using the equations which were developed earlier, and to do so would be a very useful exercise.

The decision on whether to use the chart in the admittance or impedance mode depends on the application, If lines are joined in series then, because the impedances at the junction can be added, the impedance chart should be used. Alternatively, if there are parallel junctions on the line, then admittances would be used. The differences in the use of the chart can be summarized as follows (see Fig. 6.12).

1. When the chart is used for impedance
   (a) The intersection of the circle S with the line $r + j0$ gives the maximum impedance on the line at the intersection to the right of (1, 0), and the minimum impedance at the intersection to the left.
   (b) Voltage minima are at impedance minima.
2. When the chart is used for admittance
   (a) The intersection of the circle S with the line $g + j0$ gives the maximum admittance at the intersection to the right of (1, 0), and the minimum admittance at the intersection to the left.
   (b) Voltage minima are at admittance maxima.

6.16 Stub Matching

The presence of a reflected wave and a high VSWR, due to a mismatched load, can have undesirable effects on both the generator and the line. The reflected wave may either interfere with the performance of the generator or cause it some damage. The voltage maximum of the standing wave may, in high-voltage applications, produce breakdown stresses in the dielectric. Therefore there is some value in arranging for as much of the line as possible to have a VSWR of unity so that there is no reflected wave at the generator, and this can be done by using stubs to match the line to the load at some point near the load.

6.17 Single-stub Matching

Moving from the load towards the generator the immittance (impedance or admittance) of the line will, at some point, have a normalized real part of unity. If a stub is inserted there and its length adjusted so that it presents an imaginary imittance equal to, and of opposite sign from, that of the line at that point the line will be matched on the generator side of the stub. In this method there are two distances to be determined: first, the position of the stub relative to the load and, second, the length of the stub itself. The stub, as described in Section 6.8, may be either open or short circuited, and may be joined to the line in either series or parallel.

EXAMPLE: Consider a line terminated in an unknown load. Measurements show that the VSWR is 3.5, adjacent minima are 15 cm apart, and $x_{\min} = 1.2$ cm. A short-circuit shunt stub is to be used to match the line, as shown in Fig. 6.13(a).

Because a shunt stub is being used, from the information given, calculations should be made in admittances. The VSWR from the load to y_1 is given as 3.5, and this is drawn on the Smith chart as shown in Fig. 6.13(b). y_{in} provides a

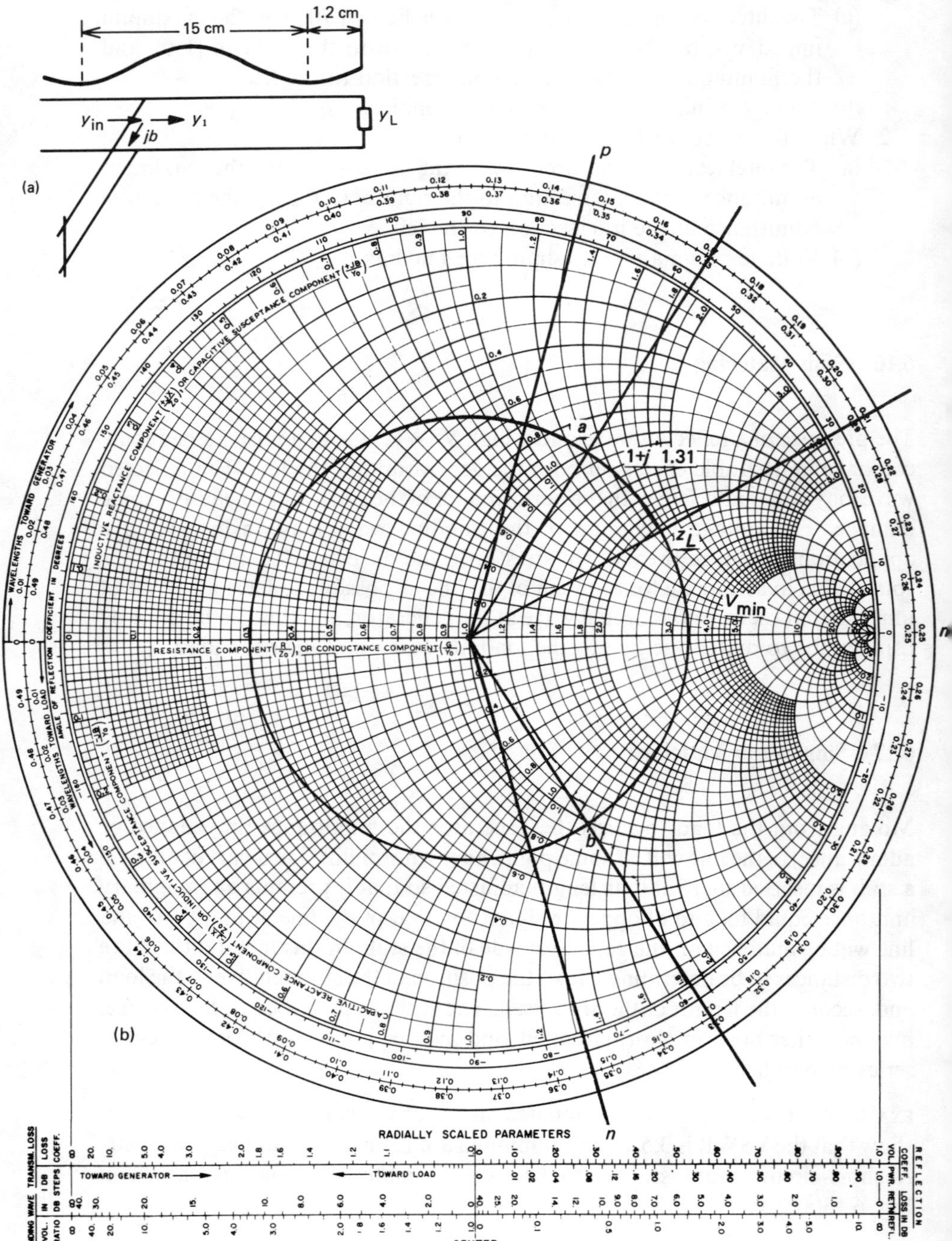

Fig. 6.13 Single-stub matching. (a) Example of shunt stub; (b) Smith chart representation.

match to the line, i.e.

$$y_{in} = 1 + j0$$

The stub can only add susceptance, shown as jb, and therefore

$$y_{in} = y_1 + jb$$

or

$$1 + j0 = y_1 + jb$$

and

$$y_1 = 1 - jb$$

Hence y_1 must be on the locus of unity conductance, i.e. the $g = 1$ circle, and as it must also lie on the $S = 3.5$ circle it is situated at the points at which these two intersect, shown in Fig. 6.13(b) as a and b.

Consider point a

The admittance of a is, from the chart, $1 + j1.31$. Therefore the stub must supply a susceptance of $-j1.31$. The load is not given, so its position on the chart is unknown, but the position of V_{min} is known – it is at the point shown on the chart, which is $0.421\lambda(= 12.63\,\text{cm})$ towards the load from a. The voltage minimum is given as 1.2 cm from the load, making point a 13.83 cm from the load.

The length of the stub can be found from the chart with reference to Fig. 6.13(b). The short-circuit load produces a VSWR of ∞ in the stub, which is represented on the chart by the outermost circle of the chart proper. The load admittance of the stub is $y_L = \infty$ which is at the right-hand end of the horizontal line, point m in Fig. 6.13(b). The stub length is therefore determined by moving from m towards the generator on the $S = \infty$ circle to the point at which susceptance is $-j1.31$, shown as point n in the diagram. Thus mn is the stub length required.

From the chart, $mn = 0.104\lambda$, and since $\lambda = 30\,\text{cm}$, $mn = 3.12\,\text{cm}$

Hence one solution to the problem is to make the short-circuited stub 3.12 cm long and place it 13.83 cm from the load.

Alternatively, consider point b

The susceptance at b is $-j1.31$ so the stub susceptance must be $j1.31$.

Then, following the same proceudre as before, the stub length is given by the distance from m to the point on the $S = \infty$ circle where susceptance is $j1.31$, i.e. point p.

From the chart, $mp = 0.396\lambda = 11.88\,\text{cm}$

The stub position, from V_{min} to b, is the same distance as a from V_{min}, but towards the generator. V_{min} to b is 0.079λ, making the total distance from the load to the stub $(0.079 + 0.04)\lambda = 2.49$ cm.

The other solution, therefore, is to make the stub 11.88 cm long, and place it 2.49 cm from the load.

6.18 Double-stub Matching

The use of a single stub may not be convenient for some applications because of the need to be able to adjust the position of the stub relative to the load if the load impedance is changed. By using two stubs the two variables necessary to produce a match are the lengths of the stubs, and the stub positions can be fixed. Their position is not entirely arbitrary, as will be seen later. This method of matching is referred to as double stub, and the basic idea behind it can be understood with the help of Fig. 6.14. Admittances are used again, because short-circuited shunt stubs are shown. If series stubs were being used, it would be necessary to work in impedances.

The input admittance of the stub nearest to the generator, y_{in}, must match the line, so

$$y_{in} = 1 + j0$$

which is at the centre of the chart. The admittance on the load side of the first stub, y_1, must be given by

$$y_{in} = y_1 + jb_1$$

or

$$y_1 = 1 - jb_1$$

which lies somewhere on the unity conductance $g = 1$ circle. y_d, the admittance on the generator side of stub 2, must lie on the same VSWR circle as y_1, since it

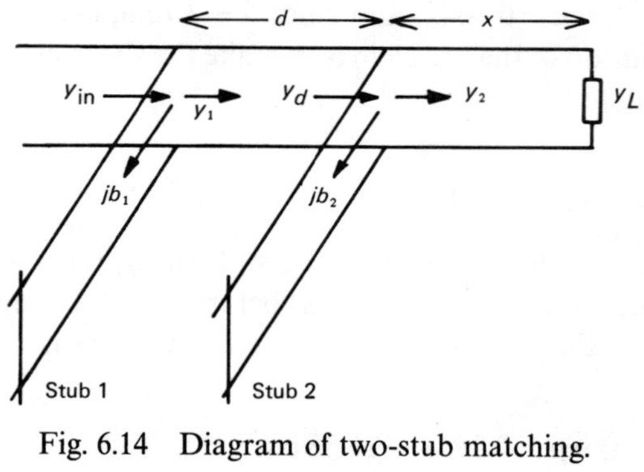

Fig. 6.14 Diagram of two-stub matching.

belongs to the same section of line, and it can be found by moving from y_1 on a constant S circle a distance d towards the load. However, the precise position of y_1 is not known. All that we know is that it lies on the $g = 1$ circle, so y_d must lie on the $g = 1$ circle translated through distance d towards the load. This circle we will call, after Everitt and Anner,[3] the L circle.

If y_d is on the L circle, y_1 lies on the $g = 1$ circle, as required, because each point on the L circle has a corresponding point, distance d towards the generator, on the $g = 1$ circle. The position of the L circle is determined by the stub separation d, and it can be drawn on the chart, once a value for d has been fixed, by drawing a circle of the same radius as the $g = 1$ circle on a centre equal to the $g = 1$ circle moved a distance d, on a constant S circle, towards the load.

The function of the first stub is to add the susceptance or reactance required to cancel that of y_1, in much the same way as was done with the single stub in the previous method.

EXAMPLE: A loss-less transmission line, of characteristic impedance 50 Ω, is terminated in a load of $100 + j25$ Ω. Two short-circuited series stubs are to be used to match the line. One stub is 8 cm and the other 32 cm from the load. The signal on the line is at 750 MHz. Find the lengths of the stubs.

Figure 6.15(a) shows the arrangement. Because series stubs are used, the problem will be solved by using impedances.

First, the stub positions must be fixed in terms of wavelength. 750 MHz is equivalent to a wavelength of 40 cm. So the first stub is 0.2λ from the load, and the distance between the stubs is 0.6λ.

The L circle is drawn with its centre $0.6\lambda(\equiv 0.1\lambda)$ towards the load from the $x = 1$ circle, as shown in Fig. 6.15(b). The normalized load, $2 + j0.5$, is fixed on the chart, and the VSWR circle is drawn.

The value of z_2 can be found by moving on the constant S circle a distance of 0.2λ towards the generator from z_L, as shown in Fig. 6.15(b). Thus the impedance z_2, at the load side of the terminals of stub 2 is

$$z_2 = 0.55 - j0.36$$

The stub must add a reactance to z_2 which is sufficient to put z_d on the L circle. Moving on a constant resistance circle from z_2, z_d is on the L circle at the point shown. From the chart

$$z_d = 0.55 + j0.25$$

The difference in reactance between z_d and z_2 is that provided by stub 2. From the chart, then, stub 2 adds reactance of $j0.61$ to z_2 to give z_d on the L circle.

z_1 is found by moving towards the generator a distance d on a constant S circle from z_d, i.e. by moving on a constant S circle from z_d, on the L circle, to the corresponding point, d wavelengths towards the generator, on the $x = 1$ circle. Note that once z_d is fixed there is only one point on the $x = 1$ circle that corresponds to z_1. From Fig. 6.15, $z_1 = 1 + j0.7$ and stub 1 must add the reactance required to reduce the reactance of z_1 to zero, i.e. $-j0.7$.

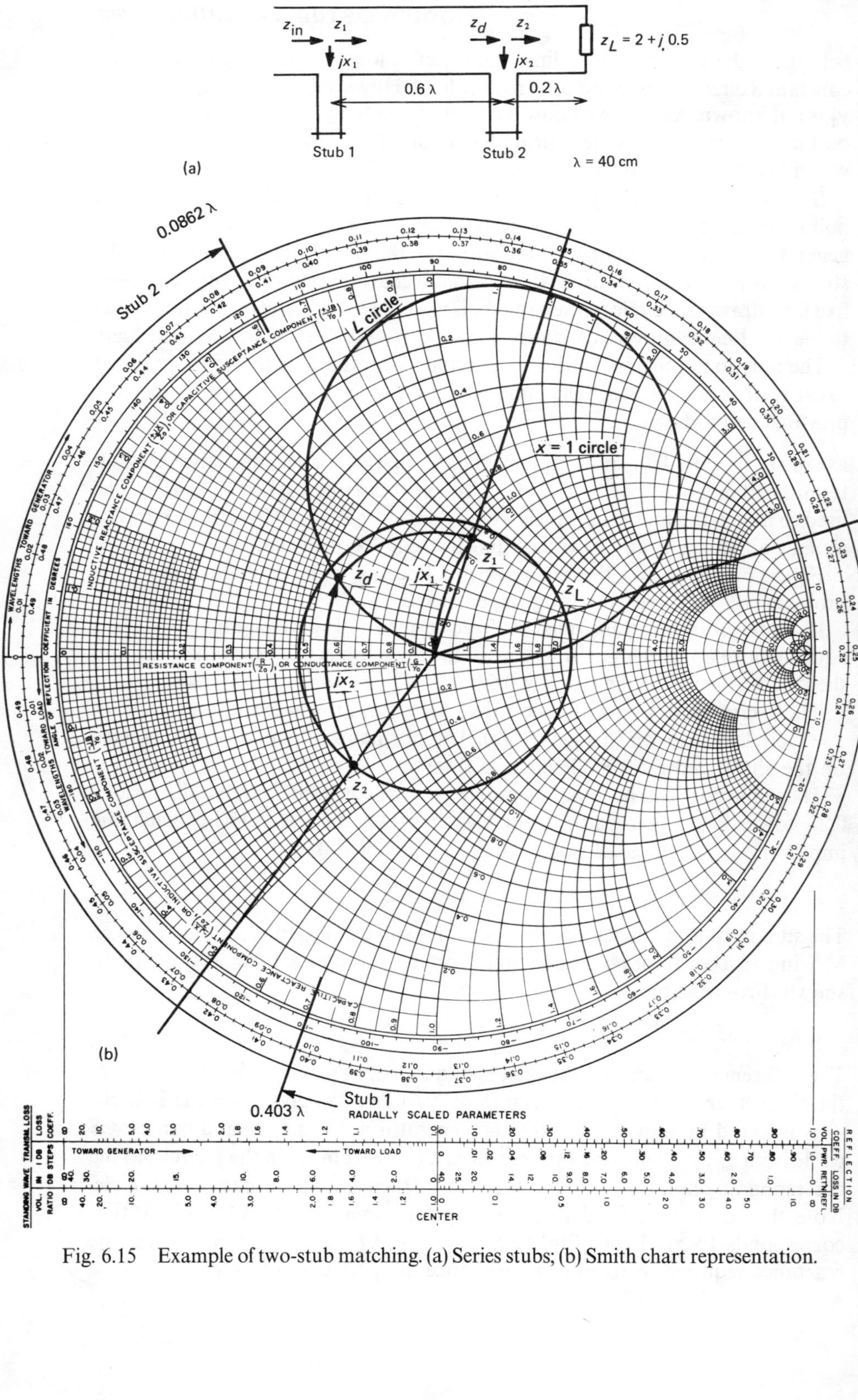

Fig. 6.15 Example of two-stub matching. (a) Series stubs; (b) Smith chart representation.

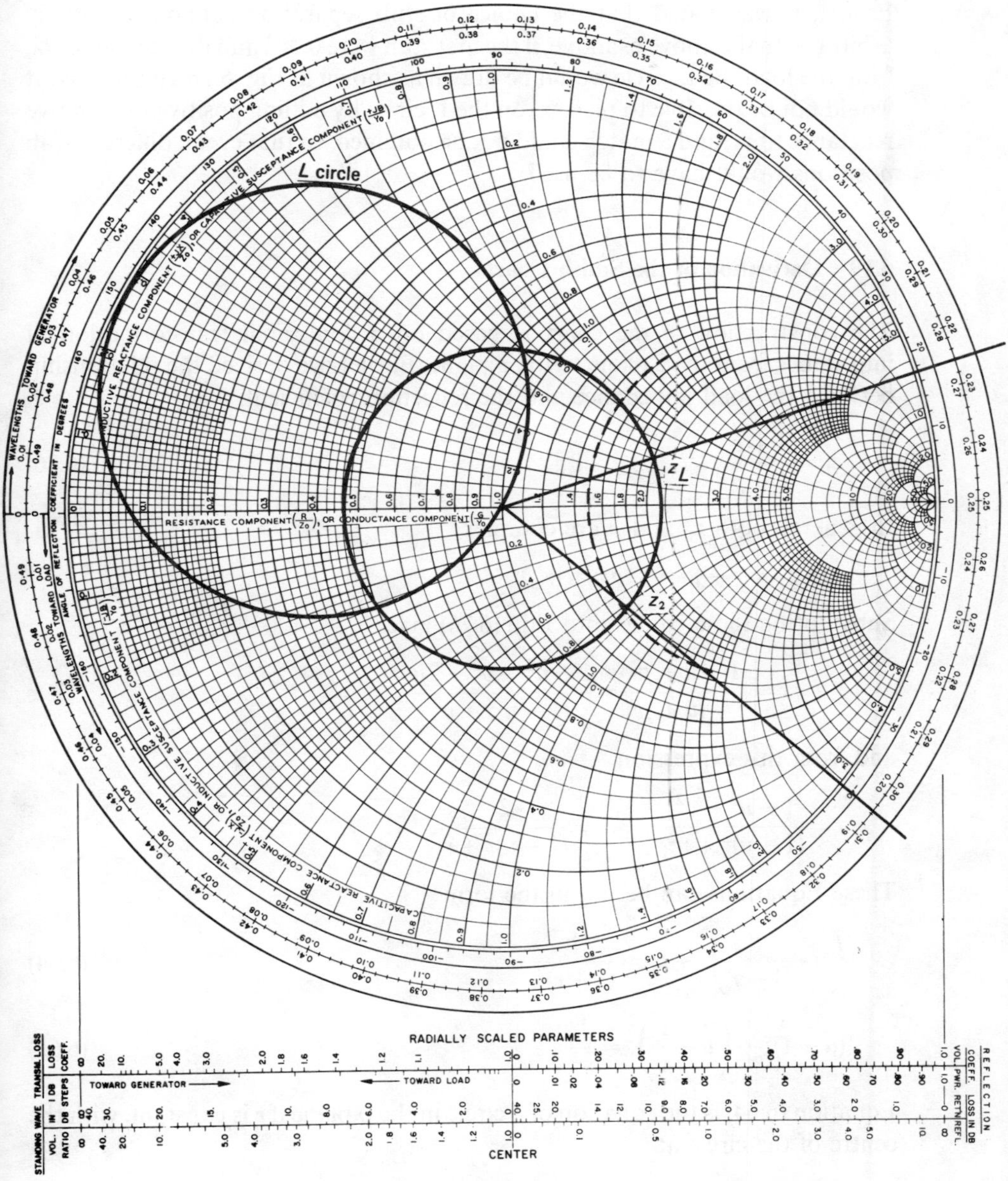

Fig. 6.16 Limitation on two-stub matching arrangement.

The stubs are short-circuited, so to find their length it is necessary to start at point $z = 0$, and go round the $S = \infty$ circle towards the generator to the required reactance.

Stub 1 is 0.6λ from the load, and 16.12 cm long.
Stub 2 is 0.2λ from the load, and 3.45 cm long.

Earlier, it was stated that the choice of stub separation cannot be entirely arbitrary. In the above example, if the first stub were 0.08λ and the second 0.29λ, from the load the L circle would be drawn as shown in Fig. 6.16. In that case it would not be possible to place z_d on the L circle by adding positive or negative reactance to z_2, and a match could not be obtained. In that event, different stub positions would have to be used.

6.19 Derivation of the Smith Chart

The construction of the chart is based on a conformal transformation of impedance from the z plane to the ρ plane. Earlier, we derived the relationship between the load impedance and the reflection coefficient:

$$z_L = \frac{1+\rho}{1-\rho} \tag{6.26}$$

In general, both z_L and ρ are complex.

Let $z_L = r + jx \qquad \rho = u + jv$

Substituting these values into Eqn (6.26) gives

$$r + jx = \frac{1 + u + jv}{1 - (u + jv)} = \frac{1 - u^2 - v^2 + 2jv}{(1-u)^2 + v^2}$$

and from this expression

$$r = \frac{1 - u^2 - v^2}{(1-u)^2 + v^2} \qquad x = \frac{2v}{(1-u)^2 + v^2}$$

These Equations can be put in the form

$$\left(u - \frac{r}{1+r}\right)^2 + v^2 = \frac{1}{(1+r)^2} \tag{6.64}$$

$$(u-1)^2 + \left(v - \frac{1}{x}\right)^2 = \frac{1}{x^2} \tag{6.65}$$

Equation (6.64) is the equation of a circle in the ρ plane if r is constant, with the centre of the circle at

$$u = \frac{r}{1+r} \qquad v = 0$$

r cannot be negative, so the centre of the circle must lie on the horizontal axis between $u = 0$ and $+1$. The radius of the circle is $1/(1+r)$.

Similarly, Eqn (6.65) is also the equation of a circle in the ρ plane, when x is constant. Now the centre of the circle is at

$$u = 1, \qquad v = 1/x$$

which lies on a vertical line, one unit to the right of the origin. If x is positive, the centre points will be above $v = 0$, and if x is negative they will lie below that point.

The Smith chart therefore consists of plots of Eqns (6.64) and (6.65) for various values of r and x. The result is a series of orthogonal circles, with resistance circles having their centres on the horizontal line, and all passing through the point $u = 1, v = 0$, at the right-hand edge of the chart; and reactance circles having their centres on the vertical line $u = 1$, and their perimeters also tangential at the point $u = 1, v = 0$. As we noted earlier, all $r =$ constant circles are shown, but only those segments of the $x =$ constant circles situated in the region $r \geqslant 0$. On the chart, apart from $v = 0$, which corresponds to the line $x = 0$, there are no u or v values shown. The numbers given on the circles are for r and x, but if it is remembered that the radius of the circle $r = 0$ is one unit of u, then Eqns (6.64) and (6.65) are consistent with the normalized values on the chart.

The assumption that $\rho = u + jv$ means that $\tan \psi = v/u$. Therefore, any line from the centre of the chart to the outer circumference will be of constant ψ. Angular movement round the chart therefore represents a movement with ψ, and so the inner circumferential scale is marked "Angle of Reflection Coefficient in Degrees".

Any circle with centre at the origin must satisfy the equation

$$u^2 + v^2 = |\rho|^2$$

and therefore it has radius $|\rho|$. From Eqn (6.54) the VSWR and $|\rho|$ are directly related, so constant $|\rho|$ means constant S.

6.20 Travelling Waves on a Loss-less Line

In Section 6.1 we assumed that forward and backward travelling waves would be produced by the time-varying input signal, and in deriving the voltage equations we ignored the effect of time. Now we will consider again our earlier ideas, and show that we do indeed have waves travelling along the line.

Referring to Eqns (6.3) and (6.6), i.e.

$$\frac{\partial I}{\partial z} = -\left(G + C\frac{\partial}{\partial t}\right)V \tag{6.3}$$

and

$$\frac{\partial V}{\partial z} = -\left(R + L\frac{\partial}{\partial t}\right)I \tag{6.6}$$

Differentiating Eqn (6.3) with respect to t, and Eqn (6.6) with respect to z,

$$\frac{\partial^2 I}{\partial t \partial z} = -\left(G + C\frac{\partial}{\partial t}\right)\frac{\partial V}{\partial t} \tag{6.66}$$

and

$$\frac{\partial^2 V}{\partial z^2} = -\left(R + L\frac{\partial}{\partial t}\right)\frac{\partial I}{\partial z} \tag{6.67}$$

6.21 Wave Equation for a Loss-less Transmission Line

In a loss-less line it is assumed that R and G can be ignored. Then Eqns (6.66) and (6.67) become

$$\frac{\partial^2 I}{\partial t \partial z} = -C\frac{\partial^2 V}{\partial t^2} \tag{6.68}$$

and

$$\frac{\partial^2 V}{\partial z^2} = -L\frac{\partial^2 I}{\partial t \partial z} \tag{6.69}$$

Substituting in Eqn (6.69) for $\partial^2 I/(\partial t \partial z)$ from Eqn (6.68),

$$\frac{\partial^2 V}{\partial z^2} = LC\frac{\partial^2 V}{\partial t^2} \tag{6.70}$$

This is the usual form of what is known as the wave equation. It is satisfied by any function of the form

$$f\left(t - \frac{z}{v}\right)$$

so that, if

$$V = V_1 f\left(t - \frac{z}{v}\right)$$

then

$$\frac{\partial^2 V}{\partial z^2} = \frac{V_1}{v^2} \quad \text{and} \quad \frac{\partial^2 V}{\partial t^2} = V_1$$

and Eqn (6.70) is satisfied if

$$1/v^2 = LC$$

A similar argument shows that with the same condition on V_2, the function $f[t + (z/v)]$ is also a solution of Eqn (6.70), so that the total voltage on the line is

$$V = V_1 f\left(t - \frac{z}{v}\right) + V_2 f\left(t + \frac{z}{v}\right) \tag{6.71}$$

The function $f[t-(z/v)]$ can be seen to represent a travelling wave if it is considered that in order to retain a position of constant phase on the line it is necessary that $[t-(z/v)]$ is constant, i.e. that a movement in the z direction of δz must be related to the time it takes to make the movement δt, in such a way that

$$t-\frac{z}{v}=t+\delta t-\frac{z+\delta z}{v}$$

or

$$\delta t=\frac{\delta z}{v}$$

Hence the velocity of propagation, defined as the change in z in unit time, is

$$v=\frac{\delta z}{\delta t}$$

Thus v in Eqn (6.71) is the velocity of the wave and is given by

$$v=1/\sqrt{(LC)} \tag{6.72}$$

v is referred to as the *phase velocity* of the wave.

If a sinusoidal signal is placed on the line, the functions $f[t\pm(z/v)]$ are written

$$\exp[j\omega(t\pm\beta z)] \tag{6.73}$$

for a loss-less line, as noted earlier. Then,

$$V=V_1\exp[j\omega(t-\beta z)]+V_2\exp[j\omega(t+\beta z)] \tag{6.74}$$

Comparing Eqns (6.71) and (6.74),

$$\frac{\beta}{\omega}=\frac{1}{v} \quad \text{i.e.} \quad v=\frac{\omega}{\beta} \tag{6.75}$$

so, from Eqn (6.72),

$$\beta=\omega\sqrt{(LC)} \tag{6.76}$$

6.22 Conclusion

In this chapter we have introduced and examined several concepts such as reflection, standing waves, line impedance, characteristic impedance, matching and travelling waves. We shall see in some of the following chapters that these same ideas occur in different contexts, but with similar meaning, and it is this wide applicability of the main features of transmission line theory that makes it so important to the study and design of communications transmission systems in general.

References

1. Chipman, R.A., *Transmission Lines*, Schaum–McGraw-Hill, New York, 1968.
2. Smith, P.H., 'An improved transmission line calculator', *Electronics*, Jan 1944, 130.
3. Everitt, W.E. and Anner, G.E., *Communication Engineering*, McGraw-Hill, New York, 1956.
4. Davidson, C.W., *Transmission Lines for Communications*, Macmillan, London, 1978.

Problems

6.1 A loss-less transmission line of $Z_0 = 100\,\Omega$ is terminated by an unknown impedance. The termination is found to be at a maximum of the voltage standing wave, and the VSWR is 5. What is the value of the terminating impedance?

Answer: $500\,\Omega$.

6.2 A load of $90 - j120\,\Omega$ terminates a $50\,\Omega$ loss-less transmission line. Find, both analytically and by using the Smith chart,
(a) the reflection coefficient;
(b) the VSWR;
(c) the distance from the load to the first voltage minimum
if the operating frequency is 750 MHz.

Answer: (a) $0.69\,e^{-0.54}$; (b) 5.4; (c) 8.28 cm.

6.3 Calculate the impedance at terminals AA of the loss-less transmission line system shown. All three lines have a characteristic impedance of $50\,\Omega$. The reflection coefficients, at an operating frequency of 750 MHz, are $\rho_1 = 0.6 - j0.5$ and $\rho_2 = 0.8 + j0.6$.

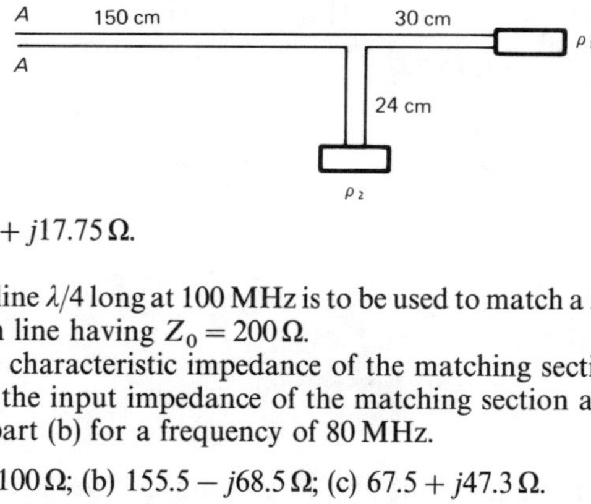

Answer: $0.9 + j17.75\,\Omega$.

6.4 A section of line $\lambda/4$ long at 100 MHz is to be used to match a $50\,\Omega$ resistive load to a transmission line having $Z_0 = 200\,\Omega$.
(a) Find the characteristic impedance of the matching section.
(b) What is the input impedance of the matching section at 120 MHz?
(c) Repeat part (b) for a frequency of 80 MHz.

Answer: (a) $100\,\Omega$; (b) $155.5 - j68.5\,\Omega$; (c) $67.5 + j47.3\,\Omega$.

6.5 A load Z produces a VSWR of 5 on a $50\,\Omega$ loss-less transmission line. The first minimum is 0.2λ from the load and the frequency of operation is 750 MHz. Without using a Smith chart, find the position and length of a short-circuit shunt stub that will match the load to the line.

Answer: Either 3.252 cm long, 5.608 cm from load, or 16.478 cm long, 10.948 cm from load.

6.6 A 50 Ω loss-less transmission line is terminated by a load $60 + j90\,\Omega$. Find the position and length of an open-circuit series stub that will provide a match at 800 MHz.

Answer: Either 5.25 cm from load, 2.7 cm long or 18.375 cm from load, 10.2 cm long.

6.7 A loss-less transmission line is carrying a signal of 47 cm wavelength, $S = 4.5$ and x_{min} is 5 cm from the load. Find the lengths of the two open-circuited series stubs that will match the line if one is placed 0.43λ and the other 0.22λ from the load.

Answer: 11.49 cm at 0.22λ, 19.27 cm at 0.43λ.

6.8 A loss-less transmission line with $Z_0 = 200\,\Omega$ is terminated by a load Z_L. If the VSWR is 6.5, and x_{min} is 0.168λ, find Z_L and the length of the two short-circuited shunt stubs required to match the line to the load if one stub is placed at the load and the other one-quarter of a wavelength away.

Answer: $Z_L = 120 - j320\,\Omega$, $l_1 = 0.336\lambda$, $l_2 = 0.423\lambda$.

6.9 A double-stub matching system is to be used on a loss-less transmission line of load impedance z_L. If the distance from the load to the stubs is s and $d + s$, show, using a Smith chart if necessary, that there are values of s for which the system cannot be used.

6.10 A transmission line is matched by two stubs, one open circuited and placed in parallel with the line 0.15λ from the load, and the other short circuited, 0.3λ long, and placed in series with the line a distance d towards the generator from the first stub. Find the length of the parallel stub, and the distance d, if the line is assumed to be loss-less and the wavelength of the signal is 40 cm. The first voltage minimum is 0.7λ from the load, and the VSWR at the load is 3.

Answer: Stub length = 12.78 cm, $d = 6.47$ cm.

7 ANTENNAS

In the last chapter we discussed transmission lines, assuming that each line was terminated in a load. That load was drawn as a lumped circuit, and we did not enquire about its purpose, or its form. In practical systems, a transmission line is often used to connect either a power source to a transmitting antenna, or a receiving antenna to a detecting circuit. Antennas are reciprocal devices, so the ideas we will develop in this chapter apply to both transmitting and receiving systems. However, to avoid unnecessary repetition we will only discuss antennas in the transmitting mode.

When a transmission line feeds an antenna, it 'sees' the impedance of the antenna, which will be frequency dependent. If the antenna is matched to the transmission line, all the energy transmitted along the line is absorbed by the antenna and, apart from losses within the antenna itself, is radiated.

From an antenna designer's point of view, there are several parameters that contribute to a design specification. Those of particular interest are polar diagram, beamwidth, bandwidth and impedance. Other quantities relating to the mechanical specification of the antenna will also be important, such as its size, weight, shape and material.

Mechanisms involved in the propagation of radio waves have an important effect on the thinking of the antenna designer. This chapter concludes with some of the fundamental features of radio propagation and indicates the extent to which it is dependent on the frequency of operation.

7.1 Radiation Pattern (Polar Diagram)

The radiation from an antenna is an electromagnetic wave, and for most applications our interest is in the strength of this wave at a distant point. In terms of metres, what constitutes a distant point depends on the size of the antenna and the wavelength of the radiation, but in principle it refers to the region in space in which the wave can be considered to be plane, and normal to the direction of the antenna. There will be just two field components, an E field vector and an H field vector, as shown in Fig. 7.1.

The radiation pattern is the variation of this distant electric field as a function of angle; an example is shown in Fig. 7.2. A detector is placed at a fixed distance

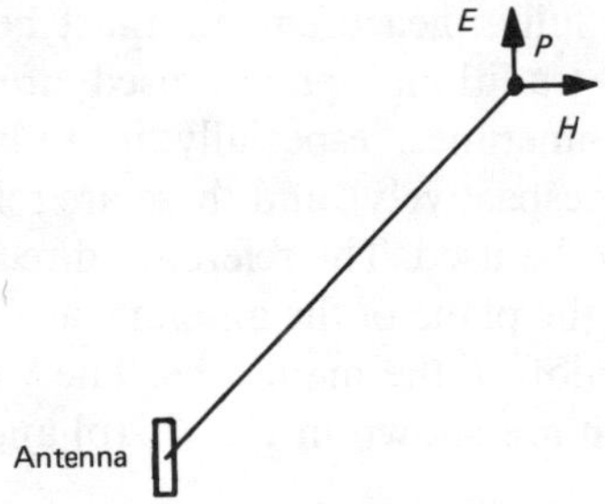

Fig. 7.1 Field components due to an antenna at a distant point.

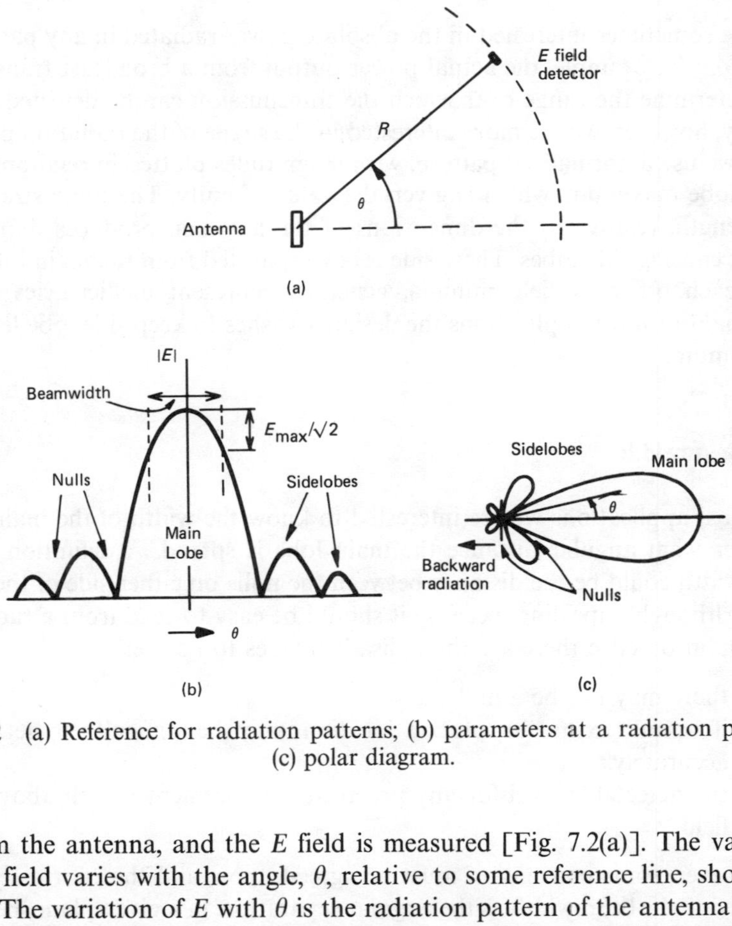

Fig. 7.2 (a) Reference for radiation patterns; (b) parameters at a radiation pattern; (c) polar diagram.

R from the antenna, and the E field is measured [Fig. 7.2(a)]. The value of this E field varies with the angle, θ, relative to some reference line, shown as $\theta = 0$. The variation of E with θ is the radiation pattern of the antenna in the particular plane in which the measurement is taken. The radiation pattern can be represented in either cartesian coordinates [Fig. 7.2(b)], or in polar coordinates, when it is usually referred to as a polar diagram [Fig. 7.2(c)]. Clearly, the radiation pattern of any real antenna is three-dimensional and therefore, to represent it fully, measurements must be made in at least two orthogonal planes. The traditional planes used are the vertical and the horizontal (which are sometimes, especially in radar applications, called elevation and azimuthal, respectively), and these are related to the orientation in which the antenna is to be used. The reference direction, $\theta = 0$, is generally assumed to be normal to the plane of the antenna, as shown. For symmetrical antennas this is the direction of the main lobe. The various terms associated with the radiation pattern are shown in Fig. 7.2(b) and (c).

7.2 Lobes

We are sometimes interested in the absolute power radiated in any particular direction; for example, the actual power output from a broadcast transmitter will determine the range over which the transmission can be detected. More usually, however, we are more interested in the shape of the radiation pattern. Then we use a normalized pattern, with magnitudes plotted in relation to the main lobe maximum, which is given the value of unity. The finite size of the wavelength, relative to the dimensions of the antenna, produces diffraction effects, causing side lobes. These side lobes, separated from the main lobe and from each other by field minima, generally represent inefficiencies in the antenna. For most applications the designer wishes to keep side lobe levels to a minimum.

7.3 Beamwidth

For most applications we are interested to know the width of the main lobe, i.e. over what angular distance the main lobe is spread. A definition of the beamwidth could be the distance between the nulls on either side of the main lobe. Although appealing, because it should be easy to read from a radiation pattern, in practice there are three disadvantages to its use:

(i) there may not be a null;
(ii) if a null exists, its precise location may be difficult to determine accurately;
(iii) the detectable level for any practical measurement is well above zero field.

However, some convention must be agreed upon, and that normally used is to define the beamwidth as the angular separation, in a given plane, between the points on either side of the main lobe that are 3 dB in power below that of the maximum. On an E field radiation plot, which has a normalized maximum of unity, the beamwidth is between those points on each side of the main lobe which have field values of $1/\sqrt{2}$.

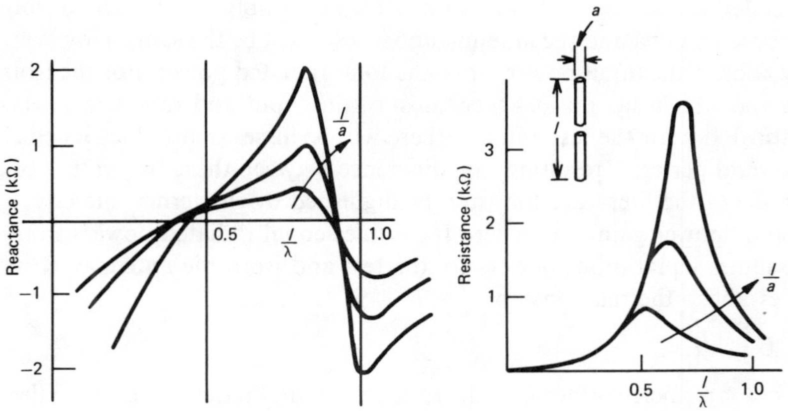

Fig. 7.3 Dipole impedance variation with dipole length.

7.4 Antenna Impedance and Bandwidth

Antennas can be considered to belong to one of two categories, resonant or non-resonant. Resonant antennas have a dimension that is near to a half-wavelength, or one of its multiples, whereas non-resonant systems do not. The difference between the two is particularly noticeable in respect of antenna impedance and bandwidth. In resonant systems, the impedance and bandwidth vary significantly with small changes in operating frequency; the impedance changes from negative reactance, through pure resistance, to positive reactance, as the frequency changes from below to above resonance. Figure 7.3 shows the impedance as a function of wavelength for a dipole. From it we can see that at about resonance, $l \simeq \lambda/2$, there is a rapid change of impedance with frequency resulting in a small bandwidth, which means that the characteristics of the antenna are strongly frequency dependent. In non-resonant antennas the radiation pattern and the input impedance are not so variable with frequency, and these antennas are sometimes referred to as broadband.

7.5 Gain

The final parameter of interest to the systems designer is antenna gain. In any physically realizable antenna there will be one direction in which the power density is highest. To give a quantitative value to this maximum, it is compared with some reference, the most common being the power density, in the same direction, from an isotropic antenna. An isotropic antenna has a spherical radiation pattern. There is no practical device that can produce such a pattern, but it is a useful notional tool for gain comparisons. There is a distinction to be made between two types of gain. Understandably, the power to both the isotropic antenna and the antenna under test must be the same. However, that may refer to the input power, or to the total radiated power. For the isotropic antenna, which has no loss mechanisms, the input and radiated powers are identical, but for the test antenna there will be losses that reduce its efficiency by several per cent, resulting in a difference between these two values of gain. In some of the literature they are distinguished by the terms "directive gain" (D) and "power gain" (G), where D assumes equal radiated powers from, and G assumes equal input powers to, the test and isotropic antennas. Thus, we can establish the relationship

$$G = kD \tag{7.1}$$

where k, an efficiency factor, is the ratio of the total radiated power to the total input power for the antenna under test.

For applications in which the power transmitted in a specified direction is of particular interest (e.g. in a microwave link, or a satellite earth station), the gain is an indicator of the antenna's effectiveness.

7.6 Field Due to a Filament of Current

By considering the simplest possible antenna, we can obtain some insight into radiated fields, and derive some general ideas that are applicable to all antennas. In Fig. 7.4 we see a short, very thin element of length l, carrying a sinusoidal current $I \sin \omega t$. The current flowing along the filament induces a magnetic field in the surrounding air. This magnetic field has an associated electric field and the result is a movement of electromagnetic energy away from the filament. We will examine the field at some point P, a distance r from the centre of the filament at O. Imagine that P is situated on a sphere centre O. Then P is related to O by r, θ and ϕ. In spherical coordinates, the field components at P are E_r, E_θ, E_ϕ and H_r, H_θ, H_ϕ. Allowing the current through the filament

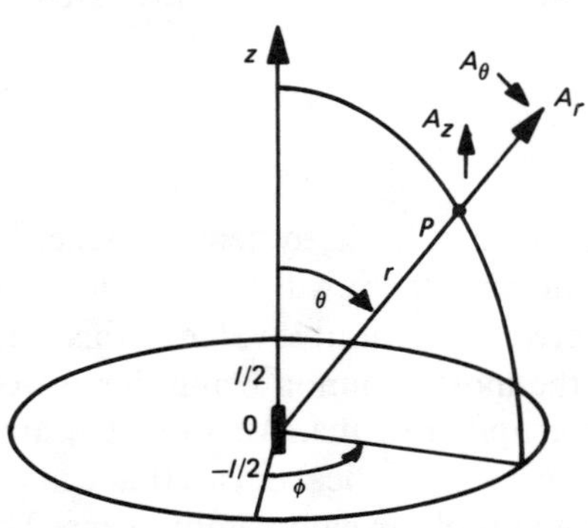

Fig. 7.4 Coordinate system-relating fields at distant point P to antenna current.

to be $I \sin \omega t$, the field at P will be produced by the current flowing through the filament at time $t - r/c$, where r/c is the time taken for the field to travel from the filament to P at the velocity of light, c.

To find the field at P we use the concept of a retarded vector potential, $\bar{A}$, which is related to the current in the filament by[1]

$$\bar{A} = \frac{1}{4\pi} \int_V \frac{i[t-(r/c)]}{r} \mathrm{d}V \tag{7.2}$$

In this example, the current has only one component, in the z direction (see Fig. 7.4), which is constant throughout the length of the filament. We can therefore replace $\bar{A}$ by A_z, and the volume integral by the linear integral over the range $-l/2 \leqslant z \leqslant l/2$, giving

$$A_z = \frac{1}{4\pi} \int_{-l/2}^{l/2} \frac{I \sin \omega[t-(r/c)]}{r} \mathrm{d}z \tag{7.3}$$

$$= \frac{Il}{4\pi r} \sin \omega[t-(r/c)] \tag{7.4}$$

The vector potential is a mathematical device. It allows us, in some cases, to find the electromagnetic field components more easily than would be possible if we were to try to relate the fields directly to the electric and magnetic charge on the antenna. The vector potential we are using here is related to the current in the antenna (the thin filament) through Eqn (7.4), and also to the magnetic field $\bar{H}$. The quantities $\bar{A}$ and $\bar{H}$ are, by definition, related by

$$\bar{\nabla} \times \bar{A} = \bar{H} \tag{7.5}$$

Hence, knowing the components of $\bar{A}$, those of $\bar{H}$ can be found. From Eqn (7.4) and Fig. 7.4, the spherical coordinate components of $\bar{A}$ are

$$\left.\begin{aligned} A_r &= A_z \cos\theta = \frac{Il}{4\pi r} \sin\omega\left(t - \frac{r}{c}\right)\cos\theta \\ A_\theta &= -A_z \sin\theta = \frac{-Il}{4\pi r} \sin\omega\left(t - \frac{r}{c}\right)\sin\theta \\ A_\phi &= 0 \end{aligned}\right\} \tag{7.6}$$

By expanding Eqn (7.5), we can see that the only component of $\bar{H}$ is H_ϕ. Both H_θ and H_r are zero. Appendix 7.1 gives a fuller derivation of the electric and magnetic field components at P. We can see from there that

$$H_\phi = \frac{Il \sin\theta}{4\pi}\left[\frac{\omega}{cr}\cos\omega\left(t - \frac{r}{c}\right) + \frac{1}{r^2}\sin\omega\left(t - \frac{r}{c}\right)\right] \tag{7.7}$$

$$E_\theta = \frac{Il \sin\theta}{4\pi\varepsilon}\left[\frac{\omega}{c^2 r}\cos\omega\left(t - \frac{r}{c}\right) + \frac{1}{cr^2}\sin\omega\left(t - \frac{r}{c}\right) - \frac{1}{\omega r^3}\cos\omega\left(t - \frac{r}{c}\right)\right] \tag{7.8}$$

$$E_r = \frac{Il\cos\theta}{2\pi\varepsilon}\left[\frac{1}{cr^2}\sin\omega\left(t-\frac{r}{c}\right) - \frac{1}{\omega r^3}\cos\omega\left(t-\frac{r}{c}\right)\right] \tag{7.9}$$

We are not really interested in the magnitudes of these expressions, but in their form. Within each of the brackets there are terms related to powers of $1/r$, and these allow us to divide the space surrounding the current-carrying element into two parts. In the region near the antenna, terms in $1/r^3$ and $1/r^2$ will predominate, but at large distances these terms will be negligible, leaving only the terms in $1/r$. The fields in these two regions are referred to as the induction and radiation fields, respectively.

7.7 Induction Field

The near-field region, or Fresnel zone, is occupied chiefly by the induction field, which does not radiate. Energy is stored in the field terms having $1/r^2$ coefficients for one half-cycle, and passes back to the antenna for the other half-cycle. The $1/r^3$ terms are confined to a region very close to the antenna before becoming negligible as r increases. We will not discuss them further here.

7.8 Radiation Field

In the far-field, or Fraunhofer region, only the $1/r$ terms are significant, and, as we shall see below, it is those that provide the energy outflow from the antenna.

We noted earlier that in most applications it is the far field that is of interest, and radiation patterns refer to that area. The distance from the antenna at which the radiation field predominates can be found from the above equations by comparing the coefficients of the $1/r$ and $1/r^2$ terms for various values of r at the wavelength of interest.

7.9 Power Radiated From a Current Element

We have called the fields in the Fraunhofer region the radiation fields, thus suggesting that it is the $1/r$ terms in Eqns (7.7) and (7.9) that are the components of power flow. We can justify this claim by considering the Poynting vector which, by taking the vector cross-product of the electric and magnetic fields, gives the value and direction of the net power flow from the antenna.

In vector notation, the Poynting vector $\bar{P}$ gives the power density in terms of the electric and magnetic field vectors:

$$\bar{P} = \bar{E} \times \bar{H} \tag{7.10}$$

For the short current element, the fields that exist at a point outside the antenna are E_θ, E_r, and H_ϕ; therefore

$$P_\phi = 0$$

and

$$P_\theta = -E_r H_\phi$$

i.e., from Eqns (7.7) and (7.9),

$$P_\theta = -\frac{Il\cos\theta}{2\pi\varepsilon}\left[\frac{\sin\omega[t-(r/c)]}{r^2c} - \frac{\cos\omega[t-(r/c)]}{\omega r^3}\right]\frac{Il\sin\theta}{4\pi}$$
$$\times\left[\frac{\omega\cos\omega[t-(r/c)]}{rc} + \frac{\sin\omega[t-(r/c)]}{r^2}\right]$$

Since the time average of terms in $\cos 2\omega[t-(r/c)]$ and $\sin 2\omega[t-(r/c)]$ is zero, P_θ is zero. This leaves the radial power density component

$$P_r = E_\theta H_\phi \tag{7.11}$$

i.e. from Eqns (7.7) and (7.8),

$$P_r = \frac{I^2l^2\sin^2\theta}{16\pi^2\varepsilon}\left\{\left[\frac{\omega}{rc^2}\cos\omega[t-(r/c)] + \frac{1}{r^2c}\sin\omega[t-(r/c)\right]\right.$$
$$\left. - \frac{1}{\omega r^2}\cos[t-(r/c)]\left[\frac{\omega}{rc}\cos\omega[t-(r/c)] - \frac{1}{r^2}\sin\omega[t-(r/c)]\right]\right\} \tag{7.12}$$

Again, the terms in $\sin 2\omega[t-(r/c)]$ and $\cos 2\omega[t-(r/c)]$ have zero time average, leaving

$$P_r = \frac{\omega^2 I^2 l^2 \sin^2\theta}{16\pi^2\varepsilon r^2 c^3}\frac{1}{2}$$

giving

$$P_r = \frac{\eta I^2 l^2 \sin^2\theta}{8\lambda^2 r^2} \tag{7.13}$$

where $\eta =$ the space impedance, $\sqrt{(\mu_0/\varepsilon_0)}$, $f =$ the frequency of the radiation $= \omega/2\pi = c/\lambda$ and $c =$ the velocity of the radiated wave $= 1/\sqrt{(\mu_0\varepsilon_0)}$.

This remaining component of the Poynting vector gives the power density flowing from the antenna. Before finding the actual power from the filament, it is worth noting that P_r in Eqn (7.13) is formed from the $1/r$ terms in E_θ and H_ϕ, which is why they are called the radiation terms. The positive sign of P_r indicates a flow of power outwards from the antenna.

We can find the total radiated power from the current element by integrating the power density at a distant point P over the surface of a sphere through P.

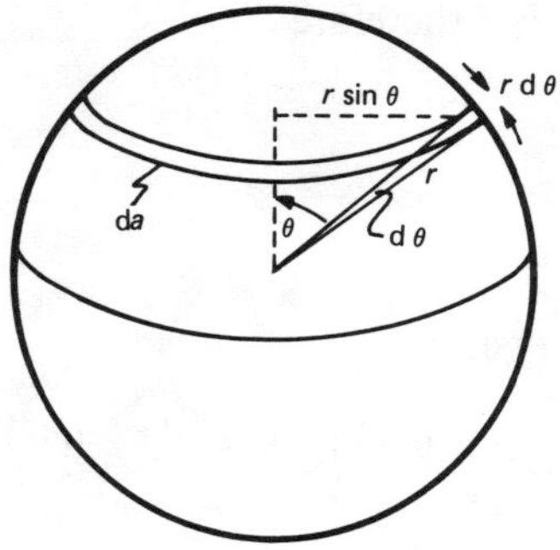

Fig. 7.5 Power radiated from an antenna at the centre of a sphere.

Referring to Fig. 7.5 the total power P_T is given by

$$P_T = \int P_r \, da$$

integrated over the area of the sphere.

From the diagram, $da = 2\pi r^2 \sin\theta \, d\theta$. Hence, from eqn (7.13),

$$P_T = \int_0^\pi \frac{\eta I^2 l^2 \sin^2\theta}{8\lambda^2 r^2} 2\pi r^2 \sin\theta \, d\theta$$

$$= \frac{\eta I^2 l^2 \pi}{4\lambda^2} \int_0^\pi \sin^3\theta \, d\theta$$

$$= \frac{\eta I^2 l^2 \pi}{8\lambda^2} \left[-\frac{\cos\theta}{3} (\sin^2\theta + 2) \right]_0^\pi$$

$$= \frac{\eta I^2 l^2 \pi}{3\lambda^2} \tag{7.14}$$

I is the peak value of current in the filament. If, instead, the rms value is used, the total power from the element is

$$P = \frac{2\eta I_{rms}^2 l^2 \pi}{3\lambda^2} \quad \text{W} \tag{7.15}$$

In free space, the impedance η has the value $120\pi = 377\ \Omega$, and using this value

$$P = I_{rms}^2 80\pi^2 (l/\lambda)^2 \quad \text{W} \tag{7.16}$$

We can see immediately that the quantity $80\pi^2(l/\lambda)^2$ has the dimensions of resistance. It is known usually as the radiation resistance of the antenna, and for this particular antenna it is related to the electrical length of the element, l/λ.

The idea of radiation resistance is applicable to any antenna. Its value for a particular antenna can be found if the current distribution along its length is known.

In the short element considered above, we assumed that the current was constant over the length of the antenna. Usually, however, this assumption is not valid, and in order to determine the amount of power radiated, the current distribution must be known. There are, then, two difficulties associated with finding the power radiated from a practical antenna.

(i) The current distribution cannot be determined very readily. Experimental measurements are not easy to obtain, and theoretical analysis, in which Maxwell's equations are solved for the antenna surface, is very complex. Current distributions have been calculated for simple, idealized shapes, but even then the solutions are complicated and difficult to use. In any practical analysis, unless lengthy numerical methods are employed, the usual approach is to build up an approximate model from simpler shapes, but even then the analysis is not easy.

(ii) Once the current distribution is known, the radiation fields must be calculated. That, too, may be a very difficult analytical process which must sometimes be approached by making realistic simplifications.

We will not concern ourselves with the solution of such problems, but we will accept that any antenna has

(i) a radiation field at some distance from the antenna, appearing as a plane wave moving radially outwards,

(ii) an induction field near to the antenna,

(iii) a radiation resistance that gives a measure of the amount of power radiated by the antenna,

and that the distance from the antenna at which the near field merges into the far field depends on the electrical length of the antenna.

7.10 Radiation Pattern of a Short Dipole

The short current element we have just considered is like a short dipole. Equation (7.41) gives E_θ, the electric field component at a distant point. If the range from the antenna to the point of interest is in the far field, only the radiation term will be significant, i.e.

$$E_\theta = \frac{\eta Il}{2\lambda r} \sin\theta \cos\omega[t-(r/c)]$$

which has the magnitude

$$\frac{60\pi I}{r}\frac{l}{\lambda}\sin\theta \qquad (7.17)$$

where, as before, η has been replaced by 120π.

Referring again to Fig. 7.4 we can see that for fixed θ the magnitude will

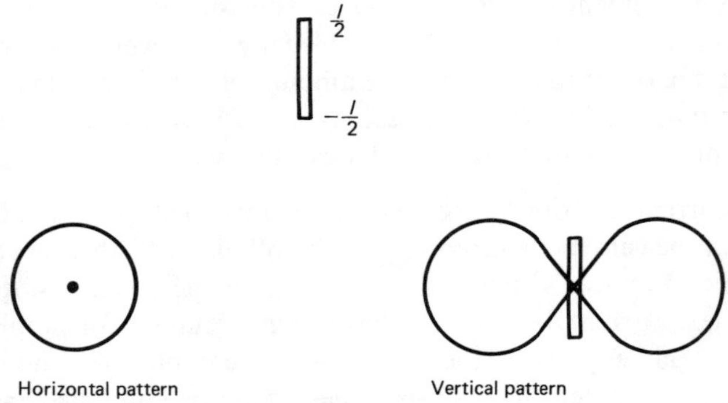

Fig. 7.6 Polar diagram of short dipole.

be constant for all values of ϕ. As θ is varied, the magnitude will vary from zero at $\theta = 0$ and π, to a maximum at $\theta = \pi/2$ from Eqn (7.17). If the dipole is placed vertically, the horizontal radiation pattern, with θ constant, is a circle, and the vertical radiation pattern, with ϕ constant, is a figure of eight. We can see that the three-dimensional radiation pattern is a solid of revolution around the dipole axis (Fig. 7.6). From a practical point of view this type of antenna could be used where no preference was to be given to any direction, or where nulls were required, as in some direction-finding methods. The basic shape of the radiation pattern in Fig. 7.6 is maintained as the value of l increases into the resonant region of $\lambda/2$, the half-wave dipole. The impedance of the antenna changes quite dramatically as l is increased, as we saw in Fig. 7.3.

7.11 Antenna Arrays

The circular horizontal radiation pattern of a dipole antenna may not be satisfactory for some applications. By placing several dipoles in line, the radiation pattern can be modified to give a large variety of radiation patterns depending on the parameters of the array. Here we shall start with the simplest array of two elements, and then consider arrays of several elements, equally spaced and in line. Other, more complex, arrays have been studied, and details can be found in, for example, Kraus.[2]

7.12 Two-element Array

In the arrangement of Fig. 7.7, the elements are fed with currents I_0 and I_1, which we will assume to be of equal magnitude, but out of phase, i.e.

$$I_1 = I_0 \underline{|\alpha}$$

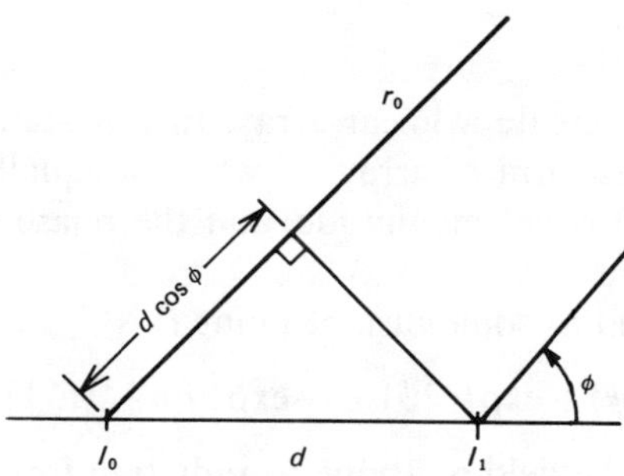

Fig. 7.7 Two-element array.

If P, the point of observation, is well into the far field, the path-length difference to P from the elements will be approximately equal to $d \cos \phi$, where d is the element spacing.

In the position shown, the phase of the radiation at P from element 1 will lead that from element 0 by ψ, where

$$\psi = \beta d \cos \phi + \alpha$$

β is the phase constant of the transmitted wave.

Given that the field at P due to element 0 is E_0, then the total field E_P is given by

$$\begin{aligned} E_P &= E_0[1 + \exp(j\psi)] \\ &= E_0[\exp(-\tfrac{1}{2}j\psi) + \exp(\tfrac{1}{2}j\psi)]\exp(j\psi/2) \end{aligned}$$

which has a magnitude

$$E = 2E_0 \cos \tfrac{1}{2}\psi \tag{7.18}$$

i.e.

$$\begin{aligned} E &= 2E_0 \cos \tfrac{1}{2}(\beta d \cos \phi + \alpha) \\ &= 2E_0 \cos\left(\frac{\pi d}{\lambda}\cos \phi + \frac{\alpha}{2}\right) \end{aligned} \tag{7.19}$$

Thus, for any given separation and phase difference, the radiation pattern, which is the variation of E with ϕ, can be found. Major changes in the polar diagram can be produced by varying d/λ and α. Problem 7.1 clearly demonstrates these changes, and it is a useful exercise to plot the radiation pattern for each value, and then to compare the results. Jordan,[1] Kraus[2] and Jasik[3] show many of these patterns.

7.13 Linear Arrays

Several elements in line create a linear array. In this section we restrict our attention to the simplest form of array, in which n equally spaced elements are fed with currents of equal magnitude, and the phase difference between adjacent elements is constant.

From Fig. 7.8 the field at some distant point P is

$$E_p = E_0[1 + \exp(j\psi) + \exp(j2\psi) + \cdots \exp[j(n-1)\psi]] \tag{7.20}$$

where, as before, E_0 is the field at P due to radiation from element 0, and ψ is the phase difference at P due to radiation from element 1 compared with that from element 0, i.e.

$$\psi = \beta d \cos\phi + \alpha \tag{7.21}$$

To find the magnitude of E_P we can use the relationship

$$1 + \exp(j\psi) + \cdots \exp[j(n-1)\psi] = \frac{1 - \exp(jn\psi)}{1 - \exp(j\psi)} \tag{7.22}$$

Then, the magnitude of Eqn (7.22) is

$$E = E_0\left|\frac{1 - \exp(jn\psi)}{1 - \exp(j\psi)}\right|$$

$$= E_0\left|\frac{\sin n\psi/2}{\sin\psi/2}\right| \tag{7.23}$$

The quantity $|(\sin n\psi/2)/(\sin\psi/2)|$ is known as the array factor because it determines the shape of the radiation pattern. It has a maximum, as we could verify by using L'Hospital's Rule, when $\psi = 0$, i.e., when

$$\beta d \cos\phi = -\alpha$$

By choosing α correctly we can place the maximum where required. Consider two examples.

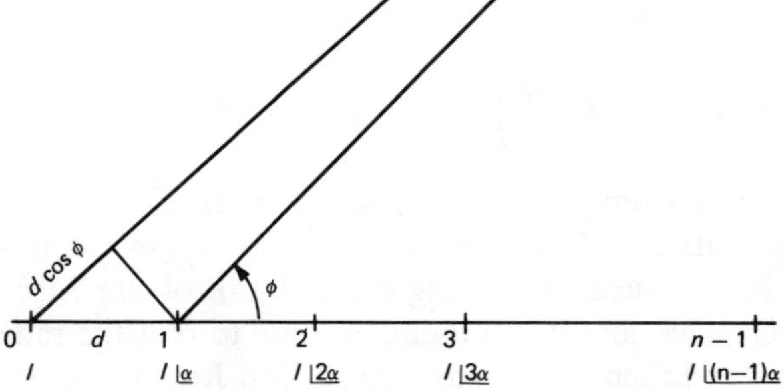

Fig. 7.8 Uniform linear array of n elements.

Broadside array. The maximum is normal to the line of the elements, in the direction $\phi = \pi/2$. The radiation from the elements must be in phase in this direction, and from Eqn (7.21) this requires, as we would expect, that the elements are fed in phase, i.e. $\alpha = 0$.
End-fire array. In this case the maximum is placed along the line of the elements, i.e. in the direction $\phi = 0$. Here the radiation from adjacent elements will add in one direction, and subtract in the reverse, backward, direction. Again, we are not surprised to find from Eqn (7.21) that the phase relationship between adjacent elements is

$$\alpha = -\beta d$$

so that the phase difference in the currents between, for example, elements 0 and 1 is cancelled by their separation.

In both these cases, and for any linear array, the width of the main lobe is reduced by increasing the number of elements, which is similar to making the array physically longer.

To find the shape of the radiation pattern of a particular array it is often sufficient to establish the direction of the maximum, from Eqn (7.23), and to find the position of the nulls. These null directions, where the field at P is zero, can be determined much more easily than the directions of the maxima of the secondary lobes. From Eqn (7.23) we can see that a zero field occurs when the array factor is zero, i.e. when

$$\frac{n\psi}{2} = \pm k\pi$$

or

$$\tfrac{1}{2}n(\beta d \cos\phi + \alpha) = \pm k\pi \tag{7.24}$$

where $k = 1, 2, 3 \cdots$. Further insight into the behaviour of these arrays can be found in Problems 7.3 and 7.4 at the end of the chapter.

A detailed evaluation of the array factor is necessary if a complete radiation pattern is required. Often, however, it is sufficient to estimate the minor lobes by calculating one or two values for sample angles between adjacent nulls.

7.14 Pattern Multiplication

There are some more complex linear arrangements of elements that can be analysed by decomposing the array into components that have a known, or easily found, radiation pattern, and then recombining these results to find the overall pattern. This method of pattern multiplication can be described most easily by using an example. Consider the array of elements shown in Fig. 7.9(a). Assuming that the elements are isotropic, we can decompose the array into

(i) a three-element linear array with $d = 3\lambda/4$, and $\alpha = 30°$ [Fig. 7.9(b)];
(ii) each of these three elements is in fact a two-element array with

$$d = 0{\cdot}3\lambda \qquad \text{and} \qquad \alpha = 0 \qquad \text{[Fig. 7.9(c)].}$$

In this example the six elements of the initial array, $ABCDEF$ have been replaced by three elements, $R\,S\,T$, where R has replaced A and B, S has replaced C and D, and T has replaced E and F.

The radiation pattern of RST, assuming each element to be isotropic, is derived from Section 7.6. The result is shown in Fig. 7.9(b). The radiation pattern of R (or S or T) is found from Eqn (7.19) and it has the shape given in Fig. 7.9(c).

The resultant radiation pattern for the six-element array is therefore the product of the array pattern [Fig. 7.9(b)] and the group pattern [Fig. 7.9(c)]. We can see that the number of nulls in the resultant pattern is the sum of the

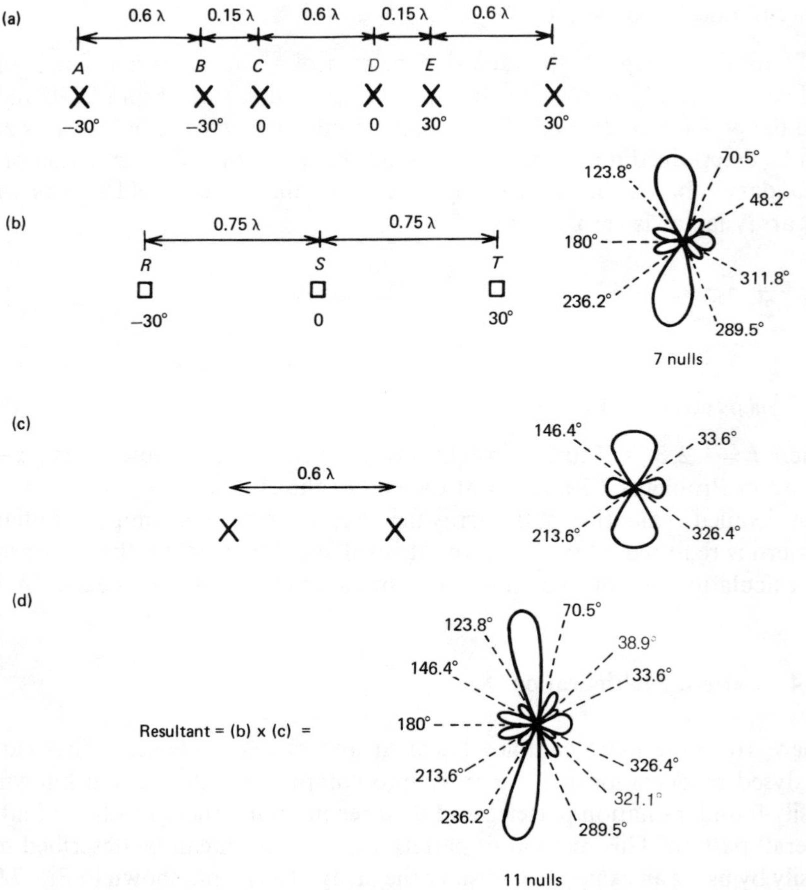

Fig. 7.9 Application of pattern multiplication.

nulls in the component patterns, i.e. four nulls from the group pattern and seven nulls from the array pattern, giving eleven in all [Fig. 7.9(d)].

7.15 Antenna Matching

Efficient connection between the feeder transmission line and the antenna depends on matching the impedance of one to the other. For narrowband applications, the antenna impedance can be calculated or measured, and an appropriate transmission line, with matching unit if necessary, can be used. In wideband applications, however, such matching may not be possible. Non-resonant antennas have a wide bandwidth capability because their impedance variation with frequency is less marked than that of resonant structures. We can see the effect of resonance on antenna terminal impedance by considering a simple half-wave dipole. Figure 7.3 shows the typical form of the variation for several thicknesses of antenna rod.

The resonant length is that which produces a purely resistive impedance, and we can see from the diagram that the reactance is zero at length slightly less than $\lambda/2$, depending on the thickness of the dipole rod. As the rod diameter falls to zero the resonant length approaches $\lambda/2$. This is due in part to the capacitive effect of the rod ends.

The half-wave dipole impedance at resonance is about 73 Ω, and therefore a cable having a nominal characteristic impedance of 75 Ω will provide a good match. For some applications, however, a 300 Ω balanced cable is more appropriate, and this does not match with the 73 Ω dipole. If the dipole is folded, as in Fig. 7.10, the impedance is approximately 300 Ω, which will match the characteristic impedance of the line. There are other advantages of using a folded dipole; it is mechanically more robust, it can be attached more easily to a support, and the feed area is free from supporting brackets.

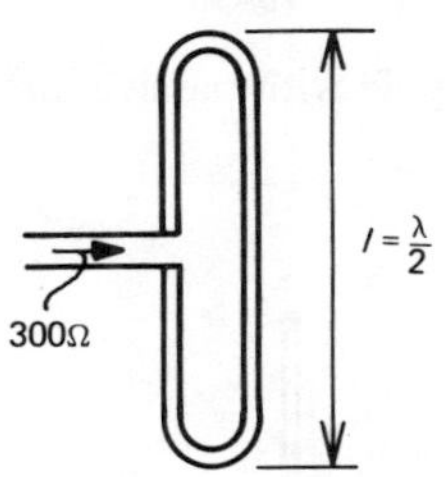

Fig. 7.10 Folded dipole.

7.16 Parasitic Elements

In discussing the dipole, we established that its horizontal polar diagram is circular, and therefore has no directive properties. By using several dipoles in line a more directive antenna can be produced, but that is not always convenient. The feeding arrangements, requiring specified magnitude and phase of the current at each element, add to the complexity and hence the expense of the antenna.

For domestic broadcasting, in the frequency ranges where a half-wave dipole is a manageable size, parasitic elements are used to distort the dipole field to produce a directive radiation pattern. A parasitic element does not have any feeder cable. It is placed near to the driven element, and picks up current from it by mutual inductance. The parasitic then acts as a separate, weaker, radiator and the resulting radiation pattern is a combination of the radiation from both the driven and parasitic dipoles. The position of the parasitic element, relative to the direction of the main beam, determines its function. In Fig. 7.11 we see the parasitic element acting as a director, being placed at such a distance in front of the driven element that the fields from the two add along the line joining them, in the direction of the required maximum. Alternatively, a parasitic element placed behind the driven element acts as a reflector, with the reradiated energy adding to that of the driven element, again in the forward direction (Fig. 7.12).

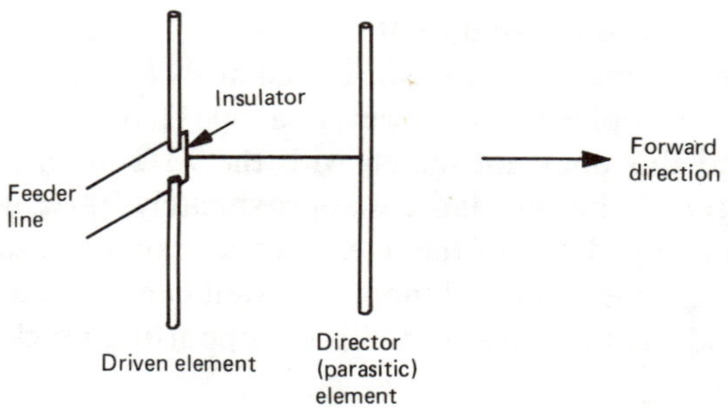

Fig. 7.11 Dipole with passive director element.

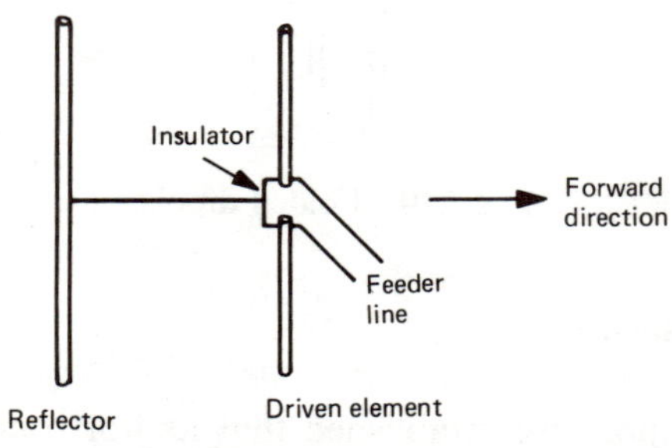

Fig. 7.12 Dipole with passive reflector element.

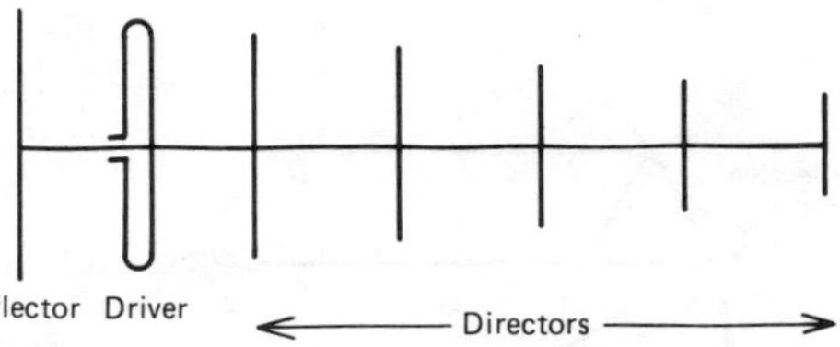

Fig. 7.13 Yagi array.

By using many director elements, as is common in a VHF antenna (Fig. 7.13), a highly directional beam results. When used as a receiver, this type of antenna is most sensitive in the forward direction, and therefore discriminates against signals from other directions. The beamwidth decreases as the number of elements increases.

We are not attempting to analyse these antennas. The mutual coupling is complex, depending on the spacing, length, diameter, and material of each element, and it exists not only between the driven and parasitic elements, but between the parasitic elements themselves. The design of multi-element antennas is based on empirical experience, rather than analysis, because of the difficulty in modelling the interaction between the elements.

7.17 Microwave Antennas

At microwave frequencies the antennas are generally significantly larger than a wavelength, and this allows techniques akin to those of optical devices, such as reflectors and lenses, to be used, thus making some microwave antennas different in kind from those possible at lower frequencies. The range of applications of microwaves in radar and communications is very large, and growing, and hence the demand for efficient, robust and inexpensive antennas, particularly in the millimetre-wave range, will increase.

Section 14.5 discusses some of the design problems related to satellite antennas.

7.18 Parabolic Reflector

Probably the most commonly used microwave antenna is the parabolic dish. It reflects energy from a primary source placed at its focus. The principle is the same as that used in an optical mirror; when the source is at the focus, on the axis of the parabola, the reflected beam is parallel to the axis. Figure 7.14 shows the main features. Assuming that ray theory applies, in order to produce a parallel beam, all rays leaving the focus F and striking the reflecting surface must be in phase in some plane normal to the axis of the

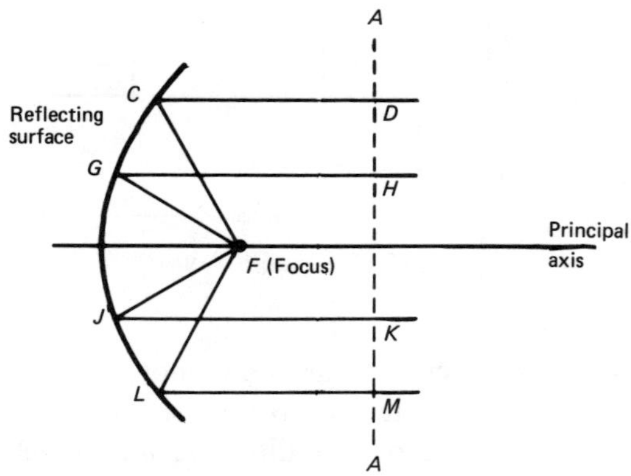

Fig. 7.14 Geometry of a parabolic dish antenna.

reflector, such as AA in the diagram. Typical rays would be FCD, FGH, FJK and FLM as shown. The requirement that DHK and M are in phase means that

$$FC + CD = FG + GH = FJ + JK = FL + LM$$

and this relationship is satisfied by a parabolic reflecting surface. The usual shape of antenna is the paraboloid, formed by revolving a parabola on its main axis, and designed to give a symmetrical radiation pattern. Two main factors prevent the reflected beam from being parallel with the axis: firstly, the feed cannot be a point source; and second, the wavelength of the radiation is not so small that diffraction effects can be ignored.

The feed introduces some problems. It has a comparatively large size and therefore blocks energy reflected from the paraboloid, and its own radiation pattern may cause destructive interference with energy from the reflector.

The comparatively large wavelength of microwave radiation produces diffraction effects at the edge of the reflector, allowing energy to spill round the back of the dish, and be wasted. One way of limiting such loss is to ensure, by design of the primary feed, that the reflector is not illuminated uniformly. A typical design figure is that the energy from the source, incident at the edge of the reflector, is only 10% of that along the axis.

The primary feeds normally used are shown in Fig. 7.15. The dipole, with parasitic reflector to prevent radiation in the forward direction, has a less directive radiation pattern than the horn antenna, and therefore illuminates the edge of the reflector too strongly, but it has a much smaller supporting structure than that required for the horn, and so does not block as much of the reflected wave. To reduce this blocking by the support, a Cassegrain antenna is sometimes used, particularly in receiving applications. The Cassegrain antenna (Fig. 7.16) consists of a primary horn feed, set in the surface of the reflector, and a sub-reflector of hyperbolic shape placed just in front of the focus of the parabola. Energy from the horn strikes the sub-reflector and is

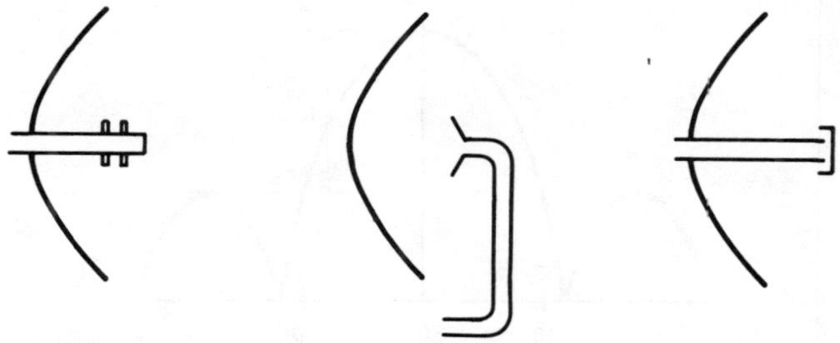

Fig. 7.15 Parabolic dish feeds.

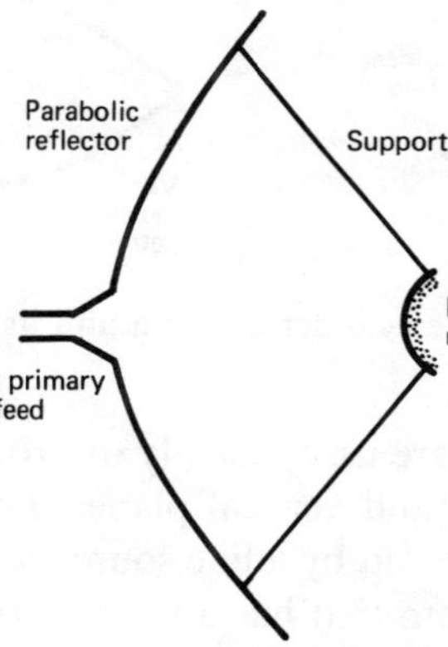

Fig. 7.16 Cassegrain antenna.

then redirected to the surface of the parabolic dish. The supports used for the sub-reflector are much less robust than the waveguide feed necessary for a horn-type primary source.

The beamwidth of the parabolic reflector is related to the wavelength of the signal, and the diameter of the dish at its rim, as well as to the illumination variation across the reflector surface. A typical radiation pattern, assuming uniform illumination, is shown in Fig. 7.17. The first side lobes rise to only 13% of the main lobe intensity, and the beamwidth is a few degrees. As the diameter is increased, the beamwidth decreases, according to the relationship

$$\theta = k\lambda/D \tag{7.25}$$

where θ is the beamwidth, D is the diameter of the reflector at the rim and λ is the radiation wavelength. k is a constant depending on the illumination from the primary feed, with a value of about 60. Thus, for $\theta = 10°$, D will need to be about 6λ.

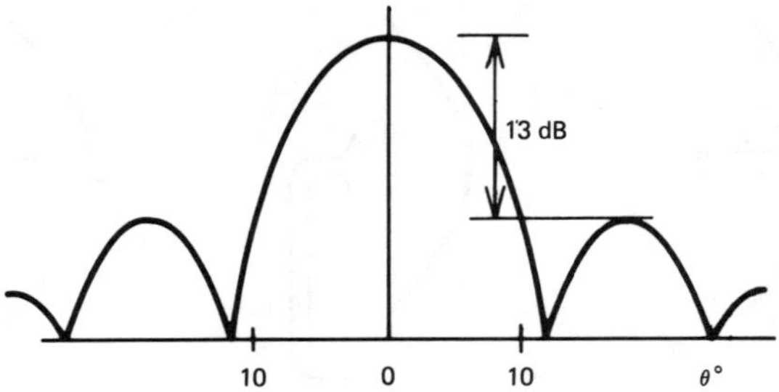

Fig. 7.17 Radiation pattern of parabolic dish with uniform illumination.

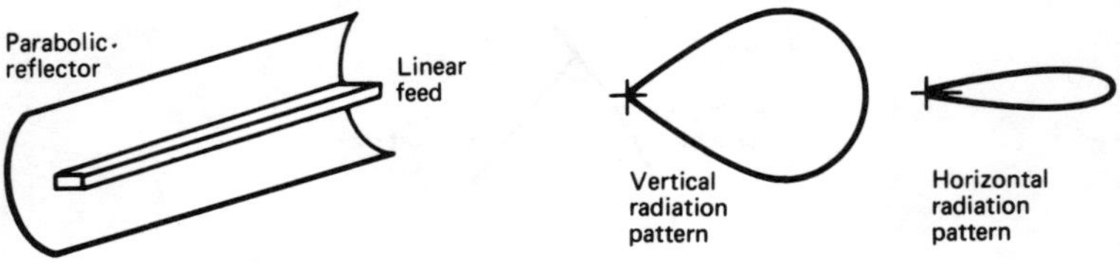

Fig. 7.18 Parabolic cylinder antenna and its polar diagrams.

Other shapes of reflector are used, usually to produce beamshapes that are not equal in the horizontal and vertical planes. For example, the parabolic cylinder shown in Fig. 7.18 is fed by a line source, such as an array of dipoles, and produces a polar diagram that has a very narrow horizontal beamwidth and a broad vertical beamwidth. Used as a radar antenna it can therefore identify the horizontal direction of a target with reasonable accuracy, but it will have no discrimination in elevation.

7.19 Horn Antennas

If a waveguide is unterminated it will radiate energy, producing a broad radiation pattern, as in Fig. 7.19. By flaring the end of the guide a more directional pattern is produced.

The flare can be in the E plane [Fig. 7.20(a)], the H plane [Fig. 7.20(b)] or both [Fig. 7.20(c)], producing a beam which is narrow in the direction of the

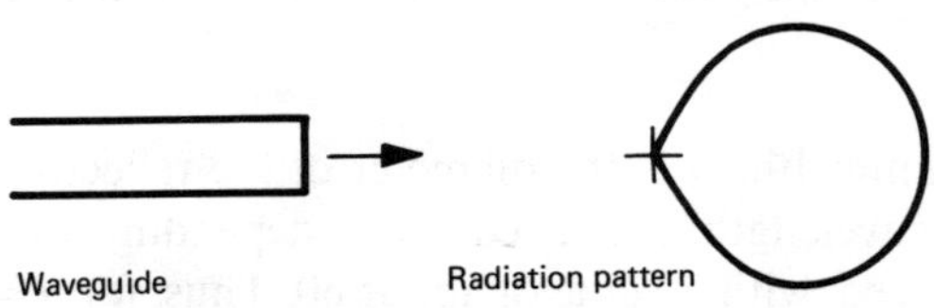

Fig. 7.19 Radiation pattern from waveguide end.

flare. It is worth noting that radiation from a horn antenna is polarized, that is the direction of the E field is related to the orientation of the wave guide, hence the examples shown in Fig. 7.20 are all vertically polarized.

The shape of the polar diagram from a waveguide horn will depend on two factors – the flare angle θ and the length of the horn L (Fig. 7.21). By using a large flare angle, the horn aperture can be made large in a short distance from the throat. However, a large θ will result in a large phase difference across

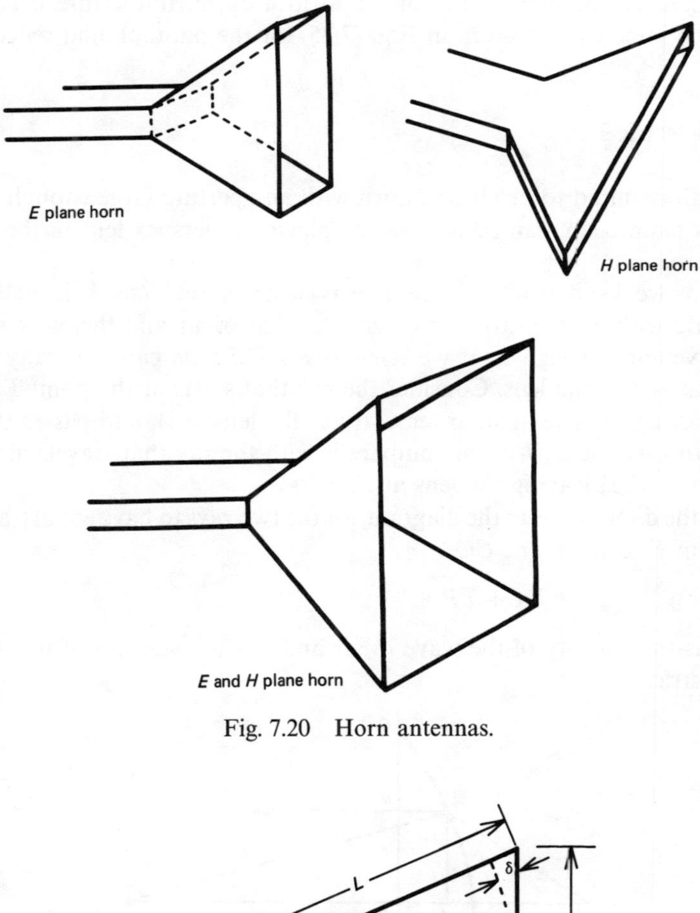

Fig. 7.20 Horn antennas.

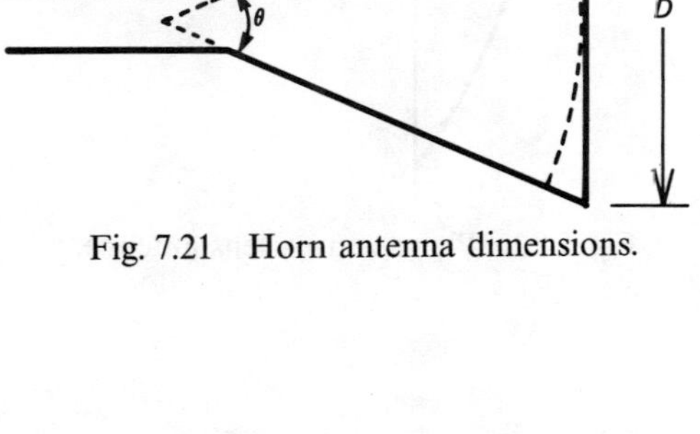

Fig. 7.21 Horn antenna dimensions.

the wavefront at the horn aperture, shown as δ in Fig. 7.21. A reasonable limit for δ is $\pi/4$, in which case from the geometry of the horn the relationship between L, D and the wavelength λ is

$$L = D^2/2\lambda \tag{7.26}$$

At small wavelengths, in the millimetre range for example, Eqn (7.26) does not impose difficult conditions on the size of the horn, but at longer wavelengths the value of L is so large that any practical horn would be too long, too bulky and too heavy. The dependence of beamwidth on horn aperture is approximately the same as that given in Eqn (7.25) for the paraboloidal reflector.

7.20 Dielectric Lens

The length required to produce a horn with an aperture large enough to give a narrow main beam can be reduced by placing a perspex lens in the mouth of the horn.

The lens works in exactly the same way as an optical lens. It is made from a dielectric with a permittivity greater than that of air and therefore slows a wave travelling through it. If we refer to Fig. 7.22 we can see a ray theory approximation of the lens. Consider the ray that starts at the point F on the convex side of a convex-planar lens, strikes the lens at A, and passes through the lens to leave at B. We can compare it with the ray that travels along the axis from F to T, leaving the lens at P.

Given the dimensions in the diagram, for the two rays to have equal travelling times from F to the plane CC,

$$FA/c + AB/v = FT/c + TP/v \tag{7.27}$$

where c is the velocity of the wave in air and v is the velocity of the wave in the dielectric.

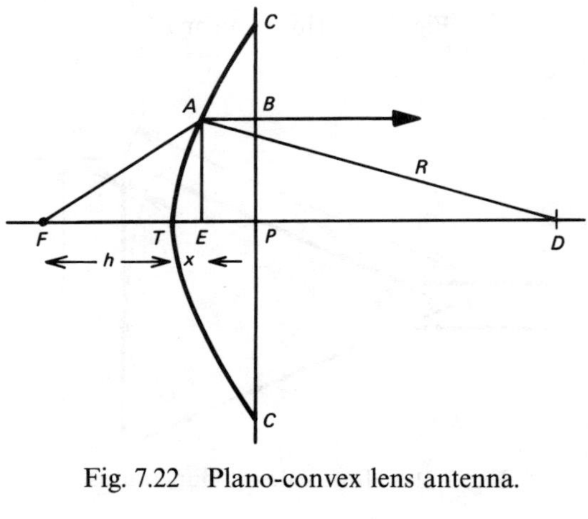

Fig. 7.22 Plano-convex lens antenna.

Equation (7.27) can be written

$$FA/c = FT/c + (TP - AB)/v$$

From Fig. 7.22, FA, FT, and $TP - AB$ can be replaced by r, h and x, giving

$$r/c = h/c + x/v \tag{7.28}$$

The velocity v is related to c by

$$v = c/\sqrt{(\varepsilon_r)}$$

where ε_r is the relative permittivity of the lens material. The quantity $\sqrt{(\varepsilon_r)}$ is called the refractive index, n, so $v = c/n$. Using this expression in Eqn (7.28) allows the velocity of the wave in air, c, to be cancelled, leaving

$$r = h + nx \tag{7.29}$$

From the triangle TAF, x is related to r, θ and h by

$$x = r\cos\theta - h$$

Hence Eqn (7.29) becomes

$$r = h + n(r\cos\theta - h)$$

Rearranging terms

$$r(n\cos\theta - 1) = h(n - 1)$$

or

$$r = h(n-1)/(n\cos\theta - 1)$$

This equation is the polar coordinate expression for a hyperbola, and it defines the curve of the convex surface of the lens.

If the focal length h is short, the thickness of the lens at the axis (TP) will be substantial, resulting in a device which is both bulky and heavy. Provided that the application is at a fixed frequency, the weight and structure of the lens can be improved by taking away zones that are a wavelength deep. The zones can be removed from either the convex or plane side of the lens, as shown in Fig. 7.23.

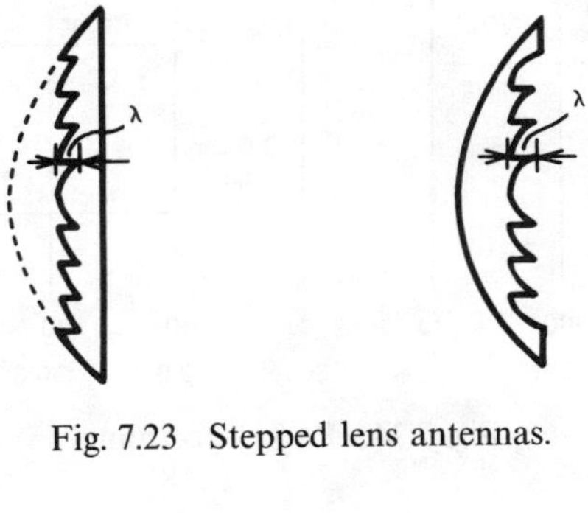

Fig. 7.23 Stepped lens antennas.

7.21 Radio Wave Propagation

The mechanism whereby a radio signal transmits through the air between transmitters depends chiefly on the frequency of the wave. In the electromagnetic spectrum shown in Fig. 7.24 the bands of frequencies used for various applications are indicated. For domestic radio broadcasts the medium and long wave bands are used, and broadcast television frequencies are normally in the VHF and UHF bands. Between these two ranges there is the commonly used shortwave radio or HF band between 1 and 30 MHz. Other bands of common interest are also indicated in the diagram. It is noteworthy that there is comparatively little interest at present in the enormous range between millimetre radar (100 GHz) and optical fibre transmission frequencies. This is due mainly to a lack of suitable sources and detectors, and the problems of transmission.

The propagation mechanisms involved in radio transmission are (a) groundwave (b) skywave (c) troposphere scatter and (d) line of sight. At low frequencies the ground wave is the most important; at HF frequencies it is the skywave that is predominant and at higher frequency still, line of sight becomes the most important mechanisms.

At very low frequencies the atmosphere between the earth's surface and the ionosphere acts as a waveguide which allows the low frequency energy to be guided around the surface of the earth. This mechanism allows for propagation distances of thousands of kilometres and is therefore greatly used for long distance navigation.

The groundwave becomes increasingly attenuated as the frequency rises. It is still predominant at the radio broadcast frequencies but at the upper end of the MF band the groundwave has become a surface wave which attenuates rapidly if the surface has high conductivity. In practice the conductivity of the surface of the earth is complicated by firstly the large range of materials of which the earth's surface is constituted, and secondly the presence of water

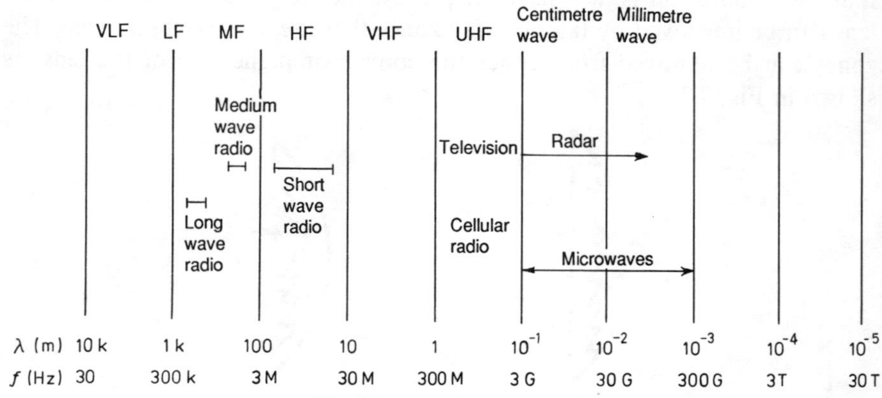

Fig. 7.24 Radio spectrum.

contaminated by salt solutions. Both the dielectric constant and the conductivity of the earth's surface are much dependent on the moisture content and the nature of the moisture. This has the effect that propagation over the sea is better than over the land and the depth of penetration of the radiowave over the earth is highly related to the dryness of the ground, its constituent materials, and the presence of water below its surface. A particulary useful discussion of groundwaves is given in reference 1 to which the reader is referred for further information.

Skywaves

In the HF frequency band the main mode of propagation is by skywave. Radiation from the earth is propagated outwards into the sky and reflected or refracted back to earth where it can be detected. The mechanism which causes the refraction or reflection is the ionosphere. The ionosphere consists of a number of layers of ionized gas which are produced at various heights due to the absorption of radiation from the sun. The predominant layers are known as D, E, F1 and F2, with the D layer at the lowest altitude, of round about 50 km, extending to approximately 300 km for the F2 layer. Because these layers are generated by the sun's activities the lower layers are in place only during the day; at night the ionization disperses and there are no free electrons. At the upper reaches of the atmosphere the sun is effective throughout the 24 hours. The ionized gases in the various layers act like a refractive medium which, depending on the frequency and angle of presentation of the radiowave, will cause reflection or refraction to a degree dependent on both the density of the free electrons and the frequency of the wave. We can see from Fig. 7.25 that there is a distance measured along the ground known as the skip distance; this is the distance between the transmitting antenna and the first position at which the wave can be received. To some extent the area around the transmitter will be covered by the groundwave but there will be a significant region which does not receive any signal.

A simple argument for the refraction mechanism, based on geometric optics and a very simplified model of the ionospheric layers is as follows.

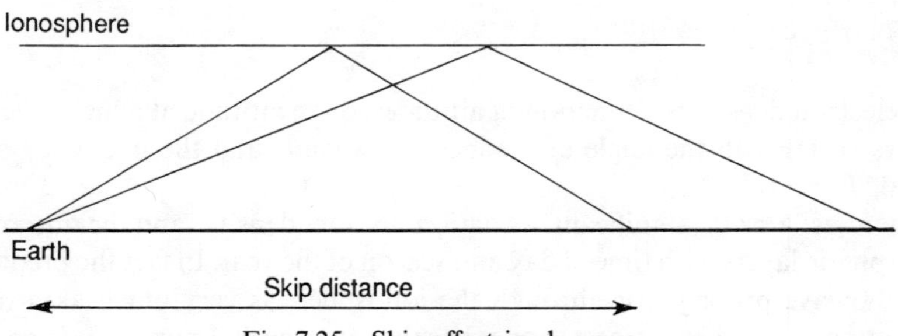

Fig. 7.25 Skip effect in skywave.

To a first approximation the refractive index of an ionized medium, relative to that of free space is related to the electron density, the electron mass and charge, and the frequency of the incident wave by

$$\varepsilon_r \simeq 1 - \frac{Ne^2}{\varepsilon_0 m\omega^2}$$

$$\simeq 1 - \frac{81N}{f^2}$$

using $e = 1.59.1.0^{-19}$ coulomb, $m = 9 \times 10^{-31}$ kg, $\varepsilon_0 = 10^{-9}/36\pi$ farads/m and $\omega = 2\pi f$. N is measured in electron/m^3.

We can model the ionosphere as a series of layers, each with an increasing electron density, as shown in Fig. 7.26. The thickness of each layer can be made arbitrarily small. The angle of incidence θ_i at the ionosphere is related to the refractive index of the first layer by Snell's Law of Refraction:

$$n_0 \sin\theta_i = n_1 \sin\theta_1$$

Assuming $n_0 = 1$

$$\sin\theta_i = n_1 \sin\theta_1$$

Repeating this equation for each boundary interface,

$$\sin\theta_i = n_n \sin\theta_n$$

The refractive index $= \sqrt{\varepsilon_r}$.

Hence, if N is the electron density in the nth layer,

$$\sin\theta_i = \left[1 - \frac{81N}{f^2}\right]^{1/2} \sin\theta_n$$

As the electron density increases there will be some layer for which the above relationship cannot be satisfied, i.e. when $\sin\theta_n \geqslant 1$ and the wave no longer passes through the ionosphere but is reflected back to earth. For a given value of incident angle θ_i, this occurs, from the last equation, when N satisfies the relationship

$$N = \frac{f^2}{81}\cos^2\theta_i$$

The electron density is a function of altitude, so the altitude at which reflection occurs varies with the angle of the incident wave θ_i and the frequency of the signal, f.

There is a very significant variation in the density and height of the ionospheric layers with time of day and season of the year. In fact the prediction of radiowave propagation through the ionosphere is very much akin to the prediction of weather forecasts and it has the same level of success. It is possible to forecast reasonably well how the ionosphere is likely to behave and therefore to choose those frequencies which will produce the most effective reception. This generally means that through the 24-hour period the transmission frequency has to vary significantly; it would not be unusual for the frequency to have to change by a factor of 2 over a very small period of time to respond to, say, the end of darkness. Nowadays these considerations of ionospheric condition are studied by computer and graphs of recommended transmission frequency against time of day are plotted.

Apart from the skip distance problem, which can be overcome to some extent by having a broad elevation in the radiated pattern, there is another problem with multiple reflections. Under certain circumstances a wave can be reflected by the ionosphere and then reflected once at the ground and again at one of the ionospheric layers so that a multipath signal is eventually received. In that case interference will result between the various signals. A useful discussion on the general approach to coping with ionospheric problems is given in reference 7.

Line of sight

At frequencies above the VHF range the most important propagation mechanism is line of sight. At these frequencies the ionosphere ceases to refract the wave which goes straight through the various layers; and the groundwave is attenuated very rapidly. This means that from a practical point of view transmitter and receiver must be in visible contact, although there will be a small degree of defraction around objects at the lower frequencies. This results in relatively poor over-the-horizon communication and it raises a problem for coverage in the higher television broadcast bands where transmitter and customers are in hilly regions.

Tropospheric Scatter

At frequencies above about 400 MHz there is significant scatter in the

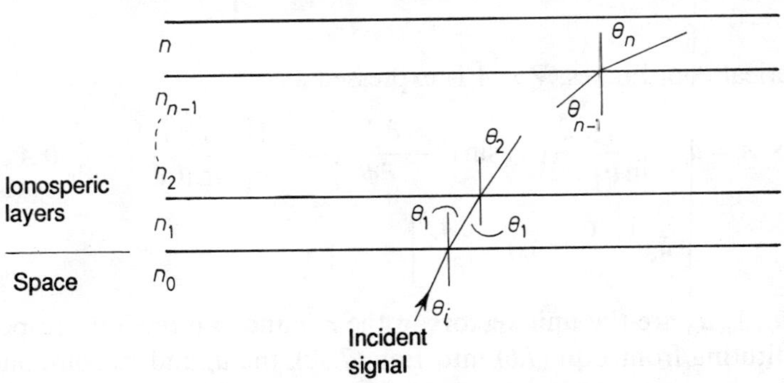

Fig. 7.26 Simple ionospheric model.

troposphere (below the ionosphere). This provides the means for extending the range of a communication link, as shown in Fig. 7.26. The signal level may be low, and the link is subject to fading due to variations in the tropospheric height, but for some applications it is the most practical method of providing a high capacity communications channel.

References

1. Jordan, E.C., *Electromagnetic Waves and Radiating Systems*, Prentice-Hall, London, 1968.
2. Kraus, J.D., *Antennas*, McGraw-Hill, Maidenhead, 1950.
3. Jasik, H., *Antenna Engineering Handbook*, McGraw-Hill, Maidenhead, 1961.
4. Fradin, A.Z., *Microwave Antennas*, Pergamon, Oxford, 1961.
5. Weeks, W.L., *Antennas Engineering*, McGraw-Hill, Maidenhead, 1968.
6. Rudge, A.W., Olver, A.D., Milne, K. and Knight, P., *The Handbook of Antenna Design* (Vol. I), Peter Peregrinus, London, 1982.
7. Rudge, A.W., Olver, A.D., Milne, K. and Knight, P., *The Handbook of Antenna Design* (Vol. II), Peter Peregrinus, London, 1983.

Appendix 7.1 Field Components Produced by a Short Current Element

In Section 7.3 we used Eqn (7.5) to represent the relationship between the retarded vector potential and the magnetic field $\bar{H}$, viz.

$$\nabla \times \bar{A} = \bar{H} \tag{7.5}$$

and the components of $\bar{A}$ for the current filament shown in Fig. 7.4 are

$$\left.\begin{aligned} A_r &= A_z \cos\theta = \frac{Il}{4\pi r}\sin\omega[t-(r/c)]\cos\theta \\ A_\theta &= -A_z \sin\theta = -\frac{Il}{4\pi r}\sin\omega[t-(r/c)]\sin\theta \\ A_\phi &= 0 \end{aligned}\right\} \tag{7.6}$$

In spherical coordinates, $\nabla \times \bar{A}$ is expressed as

$$\nabla \times \bar{A} = \bar{a}_r \frac{1}{r\sin\theta}\left[\frac{\partial}{\partial\theta}(A_\phi \sin\theta) - \frac{\partial A_\theta}{\partial\phi}\right] + \bar{a}_\theta \frac{1}{r}\left[\frac{1}{\sin\theta}\frac{\partial A_r}{\partial\phi} - \frac{\partial}{\partial r}(rA_\phi)\right] + \bar{a}_\phi \frac{1}{r}\left[\frac{\partial}{\partial r}(rA_\theta) - \frac{\partial A_r}{\partial\theta}\right] \tag{7.30}$$

where $\bar{a}_r, \bar{a}_\theta, \bar{a}_\phi$ are the unit vectors in the r, θ and ϕ directions, respectively.

Substituting from Eqn (7.6) into Eqn (7.30), the $\bar{a}_r$ and $\bar{a}_\theta$ components are zero, leaving

$$\nabla \times \bar{A} = \bar{a}_\phi \frac{1}{r}\left[\frac{\partial}{\partial r}\left(-\frac{Il}{4\pi}\sin\omega[t-(r/c)]\sin\theta\right) - \frac{\partial}{\partial\theta}\left(\frac{Il}{4\pi r}\sin\omega[t-(r/c)]\cos\theta\right)\right]$$

$$= \bar{a}_\phi \frac{1}{r}\left[\frac{Il}{4\pi}\frac{\omega}{c}\cos\omega[t-(r/c)]\sin\theta + \frac{Il}{4\pi r}\sin\omega[t-(r/c)]\sin\theta\right] \tag{7.31}$$

and, knowing that $\bar{H} = \bar{a}_r H_r + \bar{a}_\theta H_\theta + \bar{a}_\phi H_\phi$, we have, from Eqns (7.5) and (7.31),

$$H_\phi = \frac{Il}{4\pi}\sin\theta\left[\frac{\omega}{cr}\cos\omega[t-(r/c)] + \frac{1}{r^2}\sin\omega[t-(r/c)]\right] \tag{7.32}$$

For the space outside the current element, the relationship between the H field and the E field is given by one of Maxwell's equations [Eqn (9.10) with $J = 0$], i.e.

$$\nabla \times \bar{H} = \varepsilon\frac{\partial \bar{E}}{\partial r} = \varepsilon\left[\bar{a}_r\frac{\partial E_r}{\partial t} + \bar{a}_\theta\frac{\partial E_\theta}{\partial t} + \bar{a}_\phi\frac{\partial E_\phi}{\partial t}\right] \tag{7.33}$$

Thus, if we expand the left-hand side of this equation, then with the help of Eqn (7.32) we can find the electric field components. Since there is only one component of the $\bar{H}$ field, H_ϕ, we have by analogy with Eqn (7.30)

$$\nabla \times \bar{H} = \bar{a}_r\frac{1}{r\sin\theta}\frac{\partial}{\partial\theta}(H_\phi\sin\theta) - \bar{a}_\theta\frac{1}{r}\cdot\frac{\partial}{\partial r}(rH_\phi)$$

Substituting from Eqn (7.32), this becomes

$$\nabla \times \bar{H} = \bar{a}_r\frac{1}{r\sin\theta}\frac{\partial}{\partial\theta}\left\{\frac{Il\sin^2\theta}{4\pi}\left[\frac{\omega}{cr}\cos\omega[t-(r/c)] + \frac{1}{r^2}\sin\omega[t-(r/c)]\right]\right\} - \bar{a}_\theta\frac{1}{r}\frac{\partial}{\partial r}\left\{\frac{Il\sin\theta}{4\pi}\left[\frac{\omega}{c}\cos\omega[t-(r/c)] + \frac{1}{r}\sin\omega[t-(r/c)]\right]\right\}$$

After performing the differentiations, and rearranging the terms,

$$\nabla \times \bar{H} = \bar{a}_r\frac{Il\cos\theta}{2\pi}\left[\frac{\omega}{cr^2}\cos\omega[t-(r/c)] + \frac{1}{r^3}\sin\omega[t-(r/c)]\right] + \bar{a}_\theta\frac{Il\sin\theta}{4\pi}\left[-\frac{\omega^2}{c^2 r}\sin\omega[t-(r/c)] + \frac{\omega}{cr^2}\cos\omega[t-(r/c)] + \frac{1}{r^3}\sin\omega[t-(r/c)]\right] \tag{7.34}$$

Equating the $\bar{a}_\theta$ terms from the right-hand sides of Eqns (7.33) and (7.34),

$$E_\theta = \frac{1}{\varepsilon}\frac{Il\sin\theta}{4\pi r}\int\left[-\frac{\omega^2}{c^2}\sin\omega[t-(r/c)] + \frac{\omega}{cr}\cos\omega[t-(r/c)]\right.$$
$$\left.+\frac{1}{r^2}\sin\omega[t-(r/c)]\right]\mathrm{d}t$$

which, after integrating, becomes

$$E_\theta = \frac{Il\sin\theta}{4\pi\varepsilon}\left[\frac{\omega}{rc^2}\cos\omega[t-(r/c)] + \frac{1}{cr^2}\sin\omega[t-(r/c)] - \frac{1}{\omega r^3}\cos\omega[t-(r/c)]\right] \tag{7.35}$$

By a similar argument, the $\bar{a}_r$ terms on the right-hand sides of Eqns (7.33) and (7.34) are used to find the radial component of the E field:

$$E_r = \frac{Il\cos\theta}{2\pi\varepsilon}\left[\frac{1}{cr^2}\sin\omega[t-(r/c)] - \frac{1}{\omega r^3}\cos\omega[t-(r/c)]\right] \tag{7.36}$$

We have shown, therefore, that the non-zero field components in the space surrounding the current element are

$$H_\phi = \frac{Il\sin\theta}{4\pi}\left[\frac{\omega}{cr}\cos\omega[t-(r/c)] + \frac{1}{r^2}\sin\omega[t-(r/c)]\right] \tag{7.37}$$

$$E_\theta = \frac{Il\sin\theta}{4\pi\varepsilon}\left[\frac{\omega}{c^2 r}\cos\omega[t-(r/c)] + \frac{1}{cr^2}\sin\omega[t-(r/c)]\right.$$
$$\left.-\frac{1}{\omega r^2}\cos\omega[t-(r/c)]\right] \tag{7.38}$$

$$E_r = \frac{Il\cos\theta}{2\pi\varepsilon}\left[\frac{1}{cr^2}\sin\omega[t-(r/c)] - \frac{1}{\omega r^3}\cos\omega[t-(r/c)]\right] \tag{7.39}$$

corresponding with Eqns (7.7)–(7.9) in Section 7.3.

By substituting $\omega = 2\pi f$, $c = f\lambda$ and $\eta = \sqrt{(\mu/\varepsilon)}$ (where η is the free-space impedance), these equations can be expressed in a form which explicitly involves λ:

$$H_\phi = \frac{Il\sin\theta}{2\lambda}\left[\frac{1}{r}\cos\omega[t-(r/c)] + \frac{\lambda}{2\pi r^2}\sin\omega[t-(r/c)]\right] \tag{7.40}$$

$$E_\theta = \frac{\eta Il\sin\theta}{2\lambda}\left[\frac{1}{r}\cos\omega[t-(r/c)] + \frac{\lambda}{2\pi r^2}\sin\omega[t-(r/c)]\right.$$
$$\left.-\frac{\lambda^2}{4\pi^2 r^3}\cos\omega[t-(r/c)]\right] \tag{7.41}$$

$$E_r = \frac{\eta Il\cos\theta}{\lambda}\left[\frac{\lambda}{2\pi r^2}\sin\omega[t-(r/c)] - \frac{\lambda^2}{4\pi^2 r^3}\cos\omega[t-(r/c)]\right] \tag{7.42}$$

Problems

7.1 Draw the polar diagram produced by two isotropic antennas, which are fed with equal currents if:
(a) the separation between the antennas is (i) 0.25λ; (ii) 0.7λ; (iii) 1.1λ;
(b) the phase difference between the currents is (i) 0°; (ii) 30°; (iii) 90°.

7.2 Two isotropic elements, 0.4λ apart, are fed with currents of equal magnitude and a phase difference of 90°. Find the number of nulls in the polar diagram, and their direction, and the magnitude of the field normal to the line joining the elements.

Answer: 2 nulls, $\pm 51.2°$.

7.3 Find the position of the main beam, relative to the line of the array, if five elements, 0.5λ apart, are fed with currents of equal magnitude and phase.

Answer: Main beam normal to the line of elements.

7.4 Repeat problem 7.3 for separations between the elements of (i) 0.25λ and (ii) 0.625λ, and sketch the radiation pattern in all three cases, noting the effect of increasing the spacing between the elements.

7.5 Plot the radiation pattern for an array of ten identical isotropic sources in line, and spaced at intervals of 0.375λ, if the phase difference between each source is $-135°$.

7.6 Use pattern multiplication to find the polar diagram of the array of isotropic radiators shown, assuming that the current magnitude is the same in each radiator.

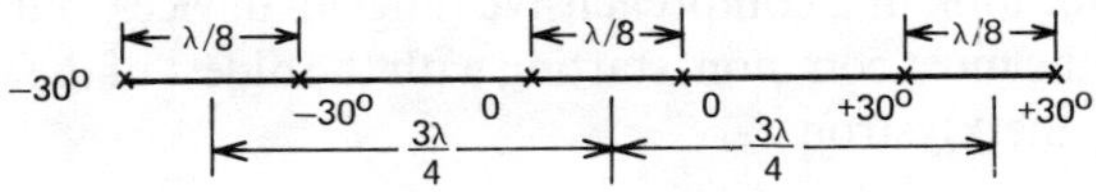

7.7 Use pattern multiplication to find the radiation pattern of an array of three isotropic elements in line, spaced at half-wavelength intervals, if the currents are in phase, but the magnitude of the current in the middle element is twice that in the outer ones.

7.8 Sketch the polar diagram of a six-element linear array in which the elements are fed with currents of equal magnitude, and a constant phase difference of 30°. The distance between adjacent elements is 0.6λ.
Compare this polar diagram with that produced by the array shown:

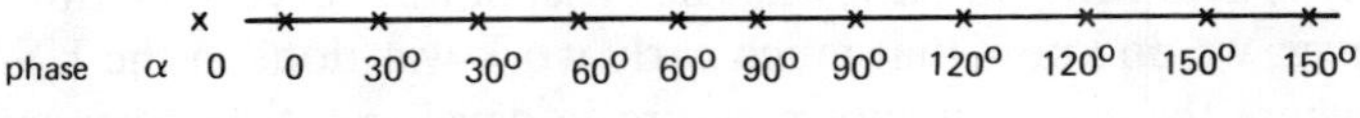

Spacing between adjacent elements is 0.3λ.

8 ACTIVE MICROWAVE DEVICES

We are considering here systems that operate at frequencies above about 1 GHz, covering the bands used for industrial and domestic heating, microwave communications, satellite communications and radar.

A notable feature of this range of frequencies is that some of the active devices are vacuum tubes. The power levels required for many systems cannot be achieved with solid-state techniques and this has meant a continuous development and refinement of the vacuum tubes that were first introduced many years ago.

Whether or not a solid-state device can be used in a particular application will depend almost entirely on the power levels involved. If these levels do not exceed 100 W, solid-state will be a natural choice. For higher powers, however, vacuum tubes must be used; which type will depend on other considerations such as bandwidth and noise level.

Here we will not look at a comprehensive range of devices, but we will study the operation of the most common, starting with the oldest of the vacuum tubes currently in use, the klystron.

8.1 Klystron

The origins of this tube are not clear. In the late 1930s, workers in several countries wrote about the possibility of using velocity modulation to generate microwaves; the space-charge modulation approach, which formed the basis of operation of lower frequency valves such as the triode, could not operate at such high frequencies. Which group should be credited with actually making the first prototype klystron is not known. The onset of war veiled such things in secrecy. However, we do know that much early work was done in the USA and in the UK, where there was intense pressure to develop satisfactory oscillators and amplifiers for use in the new detection system called radar.

8.2 Two-cavity Klystron

The basic form of the klystron amplifier is shown in Fig. 8.1. As we can see, it consists of a cathode, electron gun, two cavities and a collector. Electrons are

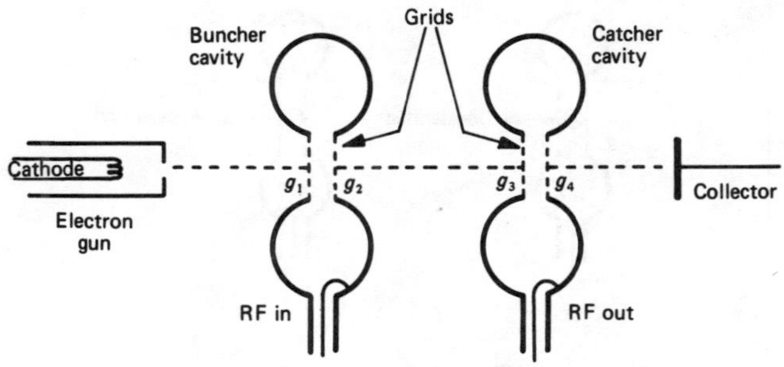

Fig. 8.1 Diagram of a two-cavity klystron.

generated at the cathode, directed into a fine stream along the axis of the tube by the electron gun, and, after passing through two cavities, they leave the tube via the collector. The cavity nearer to the cathode is called the buncher, and the other is called the catcher. The microwave signal to be amplified is coupled into the buncher cavity by either a waveguide or a coaxial loop. A similar device is used to abstract the amplified signal from the catcher cavity. The shape of the cavities will depend on the particular design; it could be inside the vacuum tube surrounding the electron gun, electron beam and collector, with the coupling by waveguide designed to launch the signal into the cavity, or it could be partly outside the tube, which would facilitate easier coupling of the microwaves, but would also require glass-to-metal seals between the cavities and the vacuum tube. Whatever the method used, the basic shape of the cavities is toroidal with two fine mesh grids across the axis to allow the electron beam to penetrate. The grids contain the cavity fields between them. There are some klystrons in which grids are not used; then there is no impairment to the passage of the electrons, but there are fringe fields around the axial gaps in the cavities.

The term "velocity modulation" describes the action of the microwave signal on the electron beam. In our discussion we are talking about the steady-state operation of the valve; that is, the way in which the valve works after it has warmed up and any transient effects have died away.

In all microwave tubes there is a transfer-of-energy mechanism that we will accept without attempting to explain how it takes place. If an electron passes through an electric field it experiences a force that causes it to change velocity. If the electron is speeded up, its kinetic energy is increased, and that increase must be taken from the field. If the electron is decelerated it loses energy, and the field gains an equivalent amount, so that energy in the system is conserved.

Now we consider the passage of an electron between the grids of a cavity (Fig. 8.2). In the steady state, the potential across the grids will change at the microwave frequency of the field within the cavity itself. On one half-cycle [Fig. 8.2(a)] grid 1 will be positive with respect to grid 2. The electron passing through the grids in the direction shown will experience a decelerating force and will therefore slow down, giving up energy to the field in the cavity. The degree of

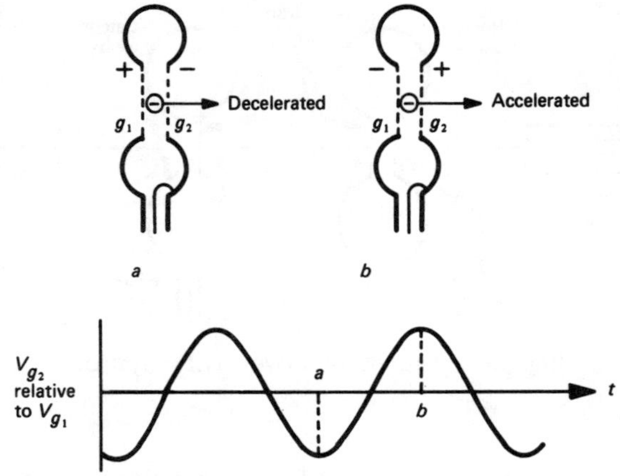

Fig. 8.2 Effect of grids on the electron beam in a two-cavity klystron.

deceleration will depend on the field across the grids at the time of the passage of the electron. We assume in our discussion that the grids are close together and the field remains constant during the transit of the electron. On the other half-cycle [Fig. 8.2(b)] the electron sees an accelerating force and is therefore speeded up, absorbing energy from the cavity field. When there is no potential difference between the grids the electron will pass through with velocity unchanged.

To clarify this velocity modulation mechanism further we can represent the movement of electrons, relative to the microwave field across the cavity grids, on a distance–time diagram, sometimes called the Applegate diagram after an early American worker on klystron characteristics (Fig. 8.3).

The sinusoid at the bottom of the diagram represents the fields across the

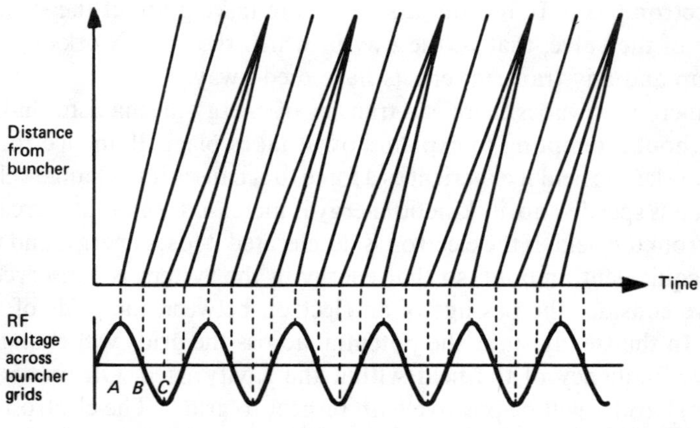

Fig. 8.3 Distance–time diagram for electrons passing through buncher.

grids. The lines above it represent the passage of an electron; the angle of each line is a measure of its velocity. Consequently, an electron passing through the grids when the field is maximum accelerating (point *A*) will have a larger slope (further distance in a given time) than one that leaves when there is no potential difference between the grids (point *B*); and that in turn has a larger slope than an electron suffering maximum deceleration (point *C*). As we might expect the faster electrons catch up the slower ones, and bunches of electrons are formed. The diagram shows that these bunches form around the electrons that pass through with unchanged velocity as the field goes from decelerating to accelerating.

In the two-cavity klystron it is the function of the first cavity to create these bunches – hence its name. The cavity is tuned, by either adjusting a screw in the wall or deforming the wall itself in some way, to the incoming microwave frequency. When the electron stream, which we assume to be approaching the buncher at uniform velocity, passes through the grids, the velocity modulation causes bunches to form at some point along the axis of the tube in the direction of the collector. The catcher is placed at this point, allowing the tight bunches to arrive between the grids. The distance between the buncher and catcher will be fixed once the valve is constructed. However, the distance taken to form the bunches will depend on their initial velocity, and this allows the collector voltage to control the bunching distance. In the steady state, the bunches arrive between the catcher grids when grid 3 is positive with respect to grid 4 (Fig. 8.1). Then the bunch is decelerated, and the cavity field absorbs the loss in kinetic energy, thus enhancing the strength of the field in the catcher cavity.

We can see from Fig. 8.3 that, although bunches are formed, if there is a continuous stream of electrons from the cathode, not all electrons are involved in the bunching process. Those that are not will not add to the catcher field and may in fact take energy from it. This mechanism reduces the efficiency of the valve.

The electron gun, consisting of a complex arrangement of electrodes, is designed to focus the electron steam into a filament along the axis of the tube. There will be some blurring at the edges of the beam, because of the space-charge forces that tend to spread the beam out.

The collector is designed to perform two functions. It must collect the electrons from the beam, without allowing any reflections or secondary emissions, and it must conduct away the heat generated by the impact of the electron beam. The first is achieved by shaping the collector in such a way that any secondary electrons do not pass back down the tube. The second requires that either a large structure of good conducting material is used to allow natural convection cooling, or that forced-air blowing be included in the overall system design.

It is possible to feed some of the catcher field back into the buncher by providing a suitable coupling mechanism between them, and so create a two-cavity klystron oscillator, but it is not much used in practice.

8.3 Reflex Klystron

The commonest klystron oscillator, particularly for laboratory work or for use as a local oscillator, is the reflex klystron. It uses only one cavity. A schematic representation is shown in Fig. 8.4. An electron gun directs electrons from the cathode through the grids of the cavity. The velocity of the electron beam is determined by the resonator–cathode voltage. As the beam passes through the resonator cavity grids, it is velocity-modulated, as we discussed in the last section with the aid of Fig. 8.2. Passing from the resonator grids the beam travels towards the reflector which is at a large negative potential. The beam is therefore in a retarding field, which it is eventually unable to resist, and it is forced to reverse direction and travel back down the tube towards the cathode. It therefore passes through the grids a second time and, on emerging from the

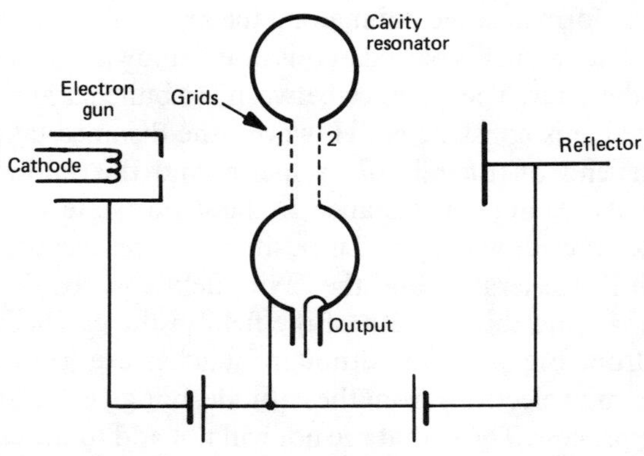

Fig. 8.4 Diagram of a reflex klystron.

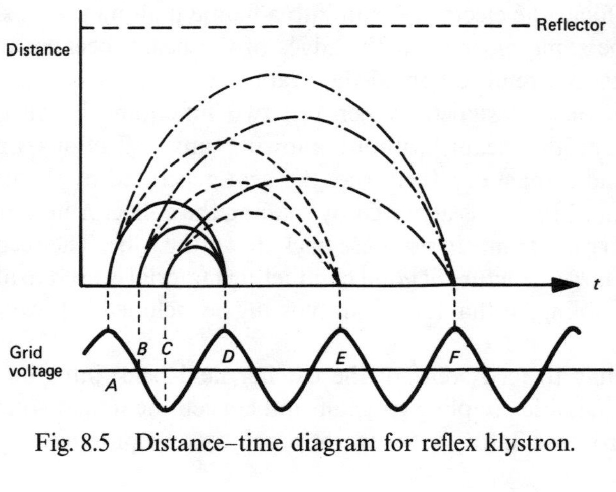

Fig. 8.5 Distance–time diagram for reflex klystron.

grids, the beam is collected on the body of the resonator. The technical problem of providing adequate heat dissipation limits the power output of this tube.

The influence of the field in the resonator on the electron beam is similar to that of the buncher and catcher in the two-cavity klystron. The modulated beam travelling through the cavity for the first time gradually breaks up. The faster electrons travel further towards the reflector, before having to turn back, than those electrons that have already been retarded by the action of the resonator field. The depth of penetration of the electrons into the reflector region is governed by the reflector voltage. To make the valve oscillate, the returning beam must arrive back between the grids in such a way that (a) the bunches are completely formed, and (b) the field is such that it provides maximum retardation, i.e. grid 1 is maximum negative with respect to grid 2. We can see this condition by referring to Fig. 8.5, which shows a distance–time diagram for the reflex klystron. The bunches form around electrons that pass through the grids for the first time as the field is moving from accelerating to decelerating (point *B*), and to achieve maximum transfer of energy from the beam to the resonator field the bunch must arrive back between the grids when the field is at *D*, *E*, *F*, etc. The transit time for the electrons in the resonator–reflector region, to give maximum output, is $(n-\frac{1}{4})T$, where T is one cycle of the field and $n=1, 2, 3 \ldots$.

Evidently, from the diagram, point D is at $n=1$, E at $n=2$, etc. Two other conditions are of interest. If the bunches arrive back between the grids when grid 1 is positive with respect to grid 2, they will be in an accelerating field and will therefore take energy from the field in the cavity, and there will be no output. Second, the bunches could arrive between the grids when there is a retarding field, but it is not at a maximum. Then the amount of energy taken from the bunches will be reduced, and the output from the klystron will fall.

When using a reflex klystron it is normal practice to adjust the resonator and reflector voltages to achieve maximum output. The frequency can normally be varied by a tuning screw, which slightly deforms the shape of the resonator.

The variation of output power with reflector voltage is shown in Fig. 8.6. The maxima are clearly defined. Each hump is called a mode and, as we can see, there may be several. In general, the amplitude of each maximum increases as the

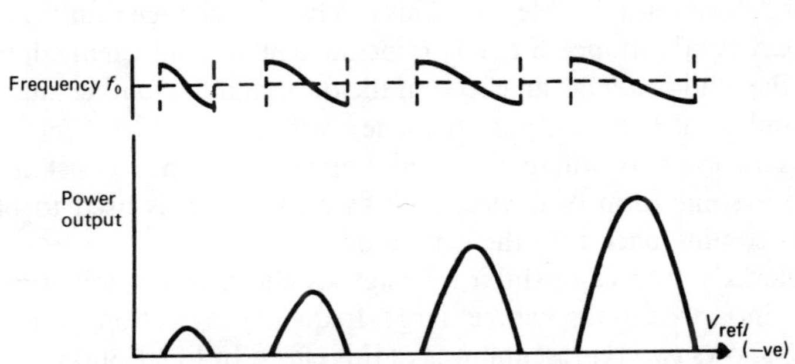

Fig. 8.6 Output characteristics of reflex klystron.

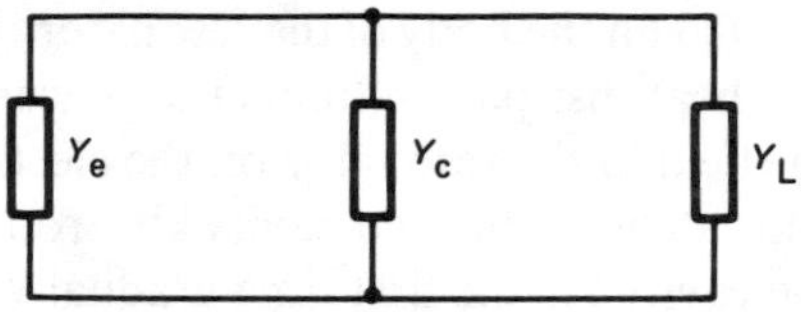

Fig. 8.7 Simplified equivalent circuit for reflex klystron.

reflector voltage increases; thus the largest output is achieved in mode 1, when $n = 1$ in the above formula, corresponding to a high voltage on the reflector. Occasionally, however, mode 1 is not the largest; second-order effects, such as multiple transits through the grids, reduce its level below that of mode 2.

Although the resonant frequency of the cavity itself is determined entirely by its geometry and size, when the valve is operating other factors affect the output frequency. There are three main components that combine to form the total impedance of the system; in their simplest form they can be considered as three parallel admittances, as shown in Fig. 8.7.

Y_e is the admittance of the electron beam, Y_c is the admittance of the cavity, and Y_L is the load, or output, admittance. Variation of the reflector or resonator voltage will change the bunching and velocity of the electron beam, thus varying the beam current and hence Y_e. The output frequency will therefore be varied. Figure 8.6 shows a typical example of the variation of frequency with reflector voltage. We can see that the output frequency varies across each mode; the frequency is the same at each mode maximum, but the frequency range of the variation across the mode may not be constant. This frequency dependence on the reflector voltage is often called electronic tuning, and it is this characteristic that made the reflex klystron attractive as a local oscillator in a radar receiver. By means of an automatic frequency-control circuit, the frequency of the received microwave signal was compared with that from the klystron, and if the difference was not equal to the required intermediate frequency (typically 45 or 60 MHz) the reflector voltage could be adjusted to bring the klystron output frequency to an appropriate value.

We have already noted that the cavity can be tuned mechanically by adjusting a screw or deforming a flexible wall. This mechanical change results in a change in Y_c, the cavity admittance. Since Y_L is included in the equivalent admittance of the tube, the load can have an effect on the output; if the load changes, Y_L will change, and hence the output frequency will alter. This effect can cause difficulties in systems where the load impedance is not constant, and to overcome it some form of device, such as an isolator, is used to block any reflections coming back into the generator.

The reflex klystron can exhibit a hysteresis effect, which sometimes causes problems in applications where large frequency variations are required. Figure 8.8 shows two representations of this effect. In Fig. 8.8(a) we see that as the reflector voltage is increased the output follows the normal mode shape, but

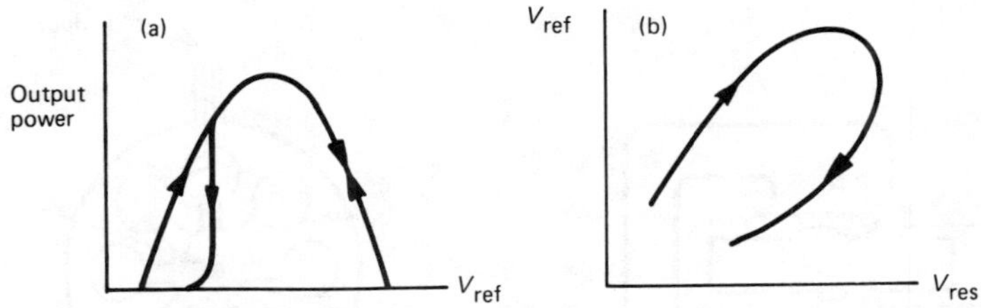

Fig. 8.8 Hysteresis effect (a) across a mode; (b) constant power curve.

as the reflector voltage is decreased the characteristic follows a different path. Figure 8.8(b) shows how hysteresis affects the curve of resonator voltage against reflector voltage for a fixed output power. Instead of being a single curve it is double valued. The characteristic is produced by following one side of the curve, by increasing both resonator and reflector voltages, and then following the other side as those voltages are decreased.

Hysteresis is caused by second-order effects that we have not taken into account.

8.4 Magnetron

The reflex klystron is not able to deliver large quantities of power. For applications such as industrial heating, domestic cooking and radar, from 1 kW up to several MW of pulsed power may be required. In the early days of radar there was intense pressure to produce an oscillator that would generate frequencies in the centimetre wavelength range for use in aircraft tracking systems, and in the early 1940s the magnetron was introduced. Although its design has been refined since then, there have been very few significant changes from the crude prototypes that demonstrated that the first magnetrons would work.

Figure 8.9 shows the essential features of the valve. It has five major components:

(i) an axial magnetic field, usually produced by a permanent magnet;
(ii) a cathode that is heated to generate electrons;
(iii) an anode block that has several slots cut into it (slow-wave structure);
(iv) a coupling device, either a loop or a hole, to couple energy from a slot in the anode block to external circuitry;
(v) the space between the cathode and the anode.

The operation of the magnetron is dependent on the interactions between the electrons travelling in the anode–cathode space and the static and r.f. electromagnetic fields in that region. The analysis is complex and will not be considered here.

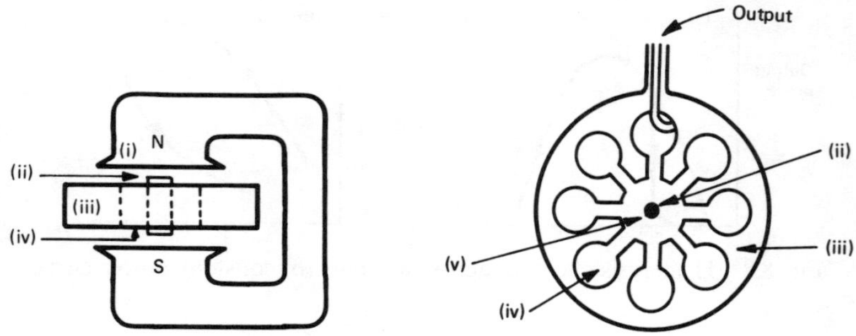

Fig. 8.9 Diagram of magnetron.

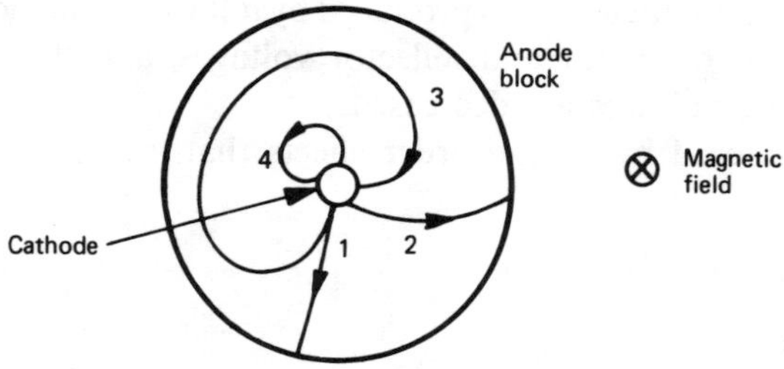

Fig. 8.10 Electron-paths in anode–cathode space.

We will consider briefly the action of the dc fields on electrons leaving the cathode, and we shall also look at the anode block and consider it as a slow-wave structure, whose purpose is to slow down a wave circulating in the cathode–anode space to a velocity approximately the same as that of the electrons moving in that space. Energy can then be transferred from the electron beam to the electromagnetic wave. The frequency of the output depends on the design of the slow-wave structure, as we shall see.

Between the anode and the cathode there is a strong electric field, produced by a high negative dc voltage on the cathode. In the absence of a magnetic field, an electron leaving the cathode moves radially to the anode (path 1 in Fig. 8.10). In this diagram we assume that the anode block has no indentations. An axial magnetic field applied to the structure will create an additional, rotational force on an electron in the anode–cathode space. The force is a function of both the electron velocity (determined by the anode–cathode voltage) and the strength of the magnetic field. The combination of radial and rotational forces on the electron can produce different trajectories, as shown in Fig. 8.10. If the electron velocity is not up to some critical value, the electron will follow a curved path on to the anode – path 2 in the diagram. If the velocity of the electron is very high, the rotational force of the magnetic field will draw the electron back into the cathode (path 4). Between these two extremes an electron, by a suitable combination of fields, will follow a curved trajectory almost to the anode, and then follow a wide loop back to the cathode (path 3). These paths are those that would occur in the concentric arrangement shown in Fig. 8.10. However, the function of the magnetron is to produce microwave oscillations of an electromagnetic field. To do that, the field must be able to exist inside the tube, and it must also be capable of interacting with the electron beam, from which it will receive its energy. The basic mechanism involved is of an electromagnetic wave travelling around the space between cathode and anode and, during its progress, absorbing energy from the electron cloud in that space. For satisfactory interaction between electrons and the beam to occur, the electron velocity and the electromagnetic field travelling-wave velocity must be substantially the same, as we noted earlier. The slow-wave structure used to support a travelling wave, whose phase velocity is about the same as that of the electron beam (i.e. about one tenth of the velocity of light), consists of a series of cavities cut into the anode block. The simplest shape is an even number of circular cavities evenly distributed around the block, and connected to the cathode–anode space by a short coupling slot. The cavities are as similar as they can be made, and we can assume that they resonate at identical frequencies. They behave in the same way as the cylindrical cavity described in section 9.21. Normally they would be operated in the fundamental TE_{101} mode.

The fields at the coupling holes will fringe into the cathode–anode space and influence any electron passing by. If the polarity of the field is such that an electron is accelerated it will take up energy from the cavity. The consequence of the increase in electron velocity is to move the electron back towards the cathode, and there it will create secondary electrons. (In the design of the cathode, this secondary emission, due to the return of high-velocity primary electrons, is a significant source of electrons, and allows the valve to run at a lower cathode temperature than would otherwise be possible.)

On the other hand, if the polarity of the field is in the opposite direction, the electrons passing through it will be slowed down and lose energy to the field. This is the condition required for successful operation. The loss of energy causes the electron to move into a wider orbit, and as it leaves the influence of that cavity it is affected by the field from the next. If by the time the electron has moved from the field due to the first cavity to that due to the second, the field polarities have changed, then it will be decelerated by that field, move out into a wider orbit, and continue to the next cavity where the field will again be a retarding one, and so on. Eventually the electron orbit will take it on to the anode. The trajectory just described is shown in Fig. 8.11.

In fact, the structure of the anode block can support a large number of travelling wave modes. The one we have described, in which the fields at alternate slots are 180° out of phase, is called the π mode. It is the one most commonly used in practice. Because a large number of modes are possible there is an inherent instability as the travelling wave can slip from one mode to another. Early in the development of the magnetron it was appreciated that

Fig. 8.11 Path of electron undergoing deceleration at successive slots.

Fig. 8.12 Strapping between alternate slots.

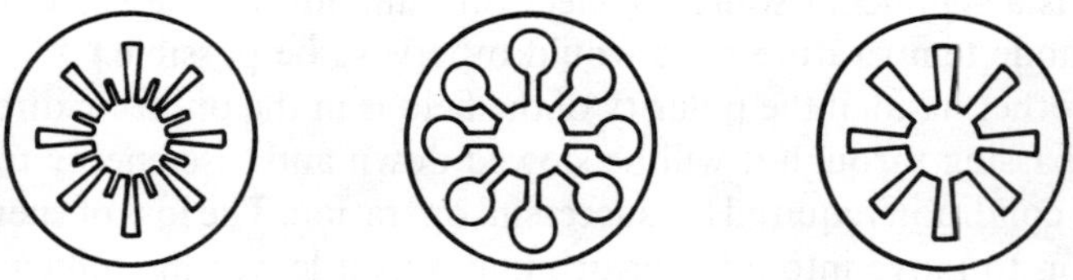

Fig. 8.13 Various anode block shapes. (a) Rising sun, (b) hole and slot, (c) slot.

better stability could be achieved if the π relationship could be maintained. A simple, but effective, method was "strapping", in which alternate "pole pieces" are interconnected by metal straps over the anode block, as shown in Fig. 8.12. This device was very successful in reducing the valve instabilities.

The coupling loop or waveguide aperture used to take energy from the valve (Fig. 8.9) can be placed in any one of the cavities. Work performed by some of the original designers showed that there was no advantage in coupling from more than one cavity; that chosen will depend on which is the most convenient mechanically.

Many different configurations have been used for the anode block, and some are shown in Fig. 8.13. Of particular significance is the "rising sun" arrangement of Fig. 8.13(a), which is designed for use at high frequencies, where the slot and cavity shape are more difficult to manufacture. The alternate short and long slots support field patterns that stabilize the π mode without the need to use strapping.

8.5 Travelling-wave Tube (TWT)

The klystron amplifier has the drawback of being a narrowband device. For some applications, particularly in communications, wideband operation is essential but the resonant cavities of the klystron limit its response to a few megahertz. The purpose of the klystron cavities is to provide an RF field that can velocity-modulate an electron beam, in the case of the buncher, or respond to such modulation in the case of the catcher. However, Kompfner[1] reported that another method of allowing the RF field to interact with the electron beam could be used without relying on resonant cavities. He introduced the travelling-wave tube in which the electromagnetic signal to be amplified is transmitted close to an electron beam for a distance of several wavelengths; their proximity allows an interaction to take place in which the beam loses energy to the RF wave. We have noted several times in the preceding sections that, before a constructive interaction can occur, the electron beam must be travelling slightly faster than the electromagnetic wave. This means that, for the TWT to function as an amplifier, the wave must be slowed down to about the same velocity as the beam, which for normal operating conditions is about one tenth of the velocity of light. The wave velocity of interest is that along the axis of the tube. To achieve this reduction, a slow-wave structure is used. For low-power tubes, the slow-wave structure is a helix. A diagram of a helix TWT is shown in Fig. 8.14. The axial velocity component of the wave propagating along the helix is $c \sin \alpha$, where α is the helix pitch and c is the velocity of light. For small α this velocity is approximately $c\alpha$, where α is in radians.

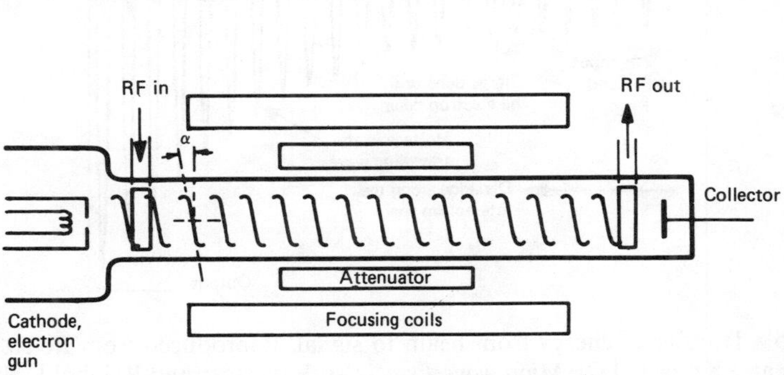

Fig. 8.14 Diagram of travelling-wave tube.

Other components of the TWT can be seen in the diagram. The electron beam is generated by a cathode, and focused by an electron gun. Additional focusing is necessary to confine the beam to the axis over the length of the tube, so it is provided by an axial magnetic field, produced by a coil surrounding the tube. The beam passes through the helix and then strikes a collector, where it is absorbed. The collector is a robust piece of copper, shaped to reduce secondary emission, and attached to a large heat sink.

The helix is made from fine wire and supported by ceramic rods. Its diameter is substantially greater than that of the beam. At each end of the helix some coupling device must be used to interconnect the electromagnetic wave with the external circuitry.

There are two inter-related mechanisms involved in the operation of the TWT:

(i) the creation of travelling waves on the electron beam, due to the action of the axial field produced by the signal on the helix;
(ii) the effect of these electron beam travelling waves on the propagation characteristics of the helix.

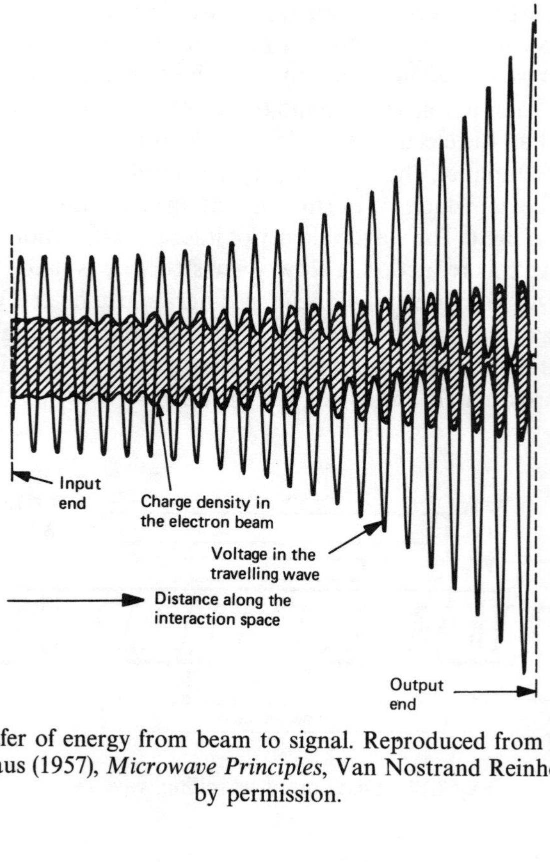

Fig. 8.15 Transfer of energy from beam to signal. Reproduced from Reich, Skalnik, Ordung and Kraus (1957), *Microwave Principles*, Van Nostrand Reinhold, New York, by permission.

The slow-wave on the helix produces an axial electric field that, for a single-frequency signal, is sinusoidal in the z direction. Some sections of the beam will be accelerated by this axial field, and some decelerated. The natural charge-repulsion inside the beam will cause slight bunching to occur, indicative of velocity modulation as we discussed in earlier sections. Figure 8.15 shows schematically the relationship between the axial field and the beam. The diagram represents one instant of time. However, the electric field will travel along the helix, and that induces two travelling waves on the beam. A theoretical description of this mechanism of setting up travelling waves on the beam by slight perturbations of the electron flow is given in several texts, e.g. Atwater[2] and Collin.[3]

The existence of the two travelling waves on the beam has an influence on the characteristics of the slow-wave structure. In Fig. 8.16 the beam current, and a transmission line representation of the slow-wave structure, are shown. There will also be a small leakage current δi between the beam and the structure. This leakage current, which is directly proportional to the current on the beam, will contain components of the two-beam travelling waves. The combination of the normal transmission line travelling waves on the helix when no beam is present and the travelling waves on the electron beam, is to produce four travelling waves on the helix in the presence of the beam. One is a backward wave and the other three are forward waves. The backward wave, if allowed to travel to the input coupler, would be reflected there and then travel forward, thus causing interference with the forward waves and possibly generating oscillations. An attenuator (or an isolator) is placed at some point along the tube to prevent such

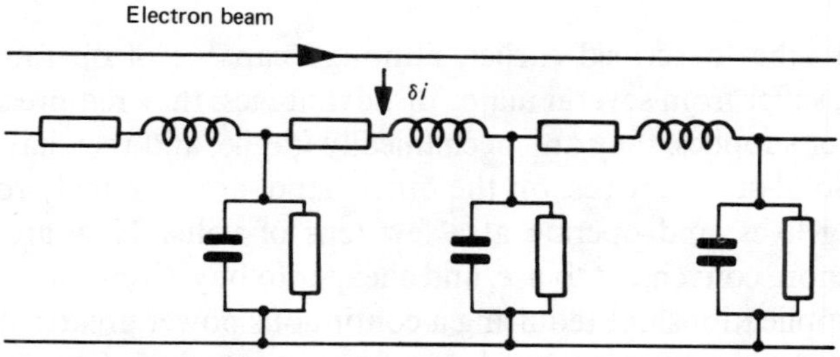

Fig. 8.16 Transmission line representation of travelling-wave tube.

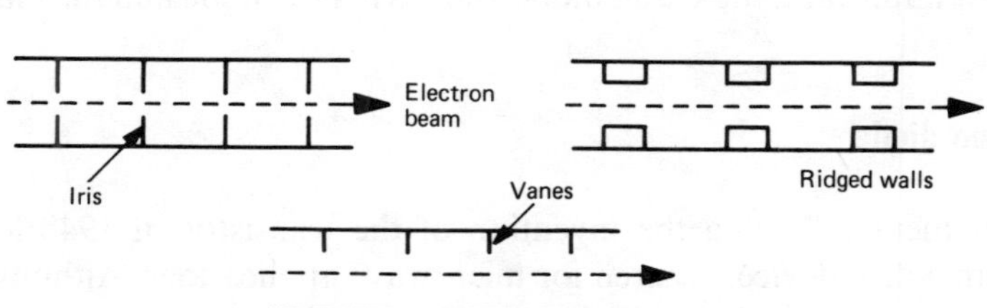

Fig. 8.17 Waveguide slow-wave structures.

interference. Of the three forward waves, one has no amplitude variation (the propagation constant is imaginary), another is attenuated, but the third grows exponentially with z, the distance from the input end of the helix. It is this last wave that gives the amplification required. The actual gain achieved will depend on how efficiently the energy in the beam can be transferred to the field on the helix. The growth wave is continuously amplified as it passes along the tube, and at the output end it is coupled to the external circuit.

Typically, a helix type TWT will operate over an octave bandwidth, and have a gain of 30 dB or more. Uniform gain across the band may be essential for many applications, and this will limit the bandwidth available.

The helix slow-wave structure is convenient, and easy to manufacture, but it can be used only for low-power tubes. The description above assumes that the electron beam consists of electrons moving in straight lines from cathode to anode. In practice the electron flow is much more complicated because of the mutual repulsion between the electrons, and this causes the beam to be blurred at the edges. For high power, the electron beam is of large diameter and it could easily damage the helix if it came in contact with it. The helix itself is not able to withstand bombardment by stray electrons; hence other structures are used in high-power systems. Figure 8.17 shows examples of some slow-wave structures based on waveguide propagation.

One of the most important civil applications of the TWT is as a high power amplifier (HPA) in a satellite transponder (see Section 14.4).

8.6 Solid-state Devices

The vacuum tubes described earlier, although capable of operating at high power levels, suffer from several major disadvantages; they require stable, high voltage, power supplies; they are mechanically fragile, and they have relatively short lives. Solid-state devices, on the other hand, are extremely robust, have long working lives, and operate at a few tens of volts. They are inherently smaller and more convenient to use, and cheaper to buy. Consequently they are used for all applications not requiring a continuous power greater than several tens of watts. As has happened at lower frequencies, there have been several different devices proposed and developed, but here we shall look only at the three most important; the Gunn diode, the IMPATT diode and the GaAsFET.

8.7 Gunn diode

The excitement following the invention of the transistor in 1948 led to an upsurge in active device research for microwave applications. Although many modifications were made to bipolar transistor designs, in an attempt to make them suitable for operation at microwave frequencies, there were clearly some

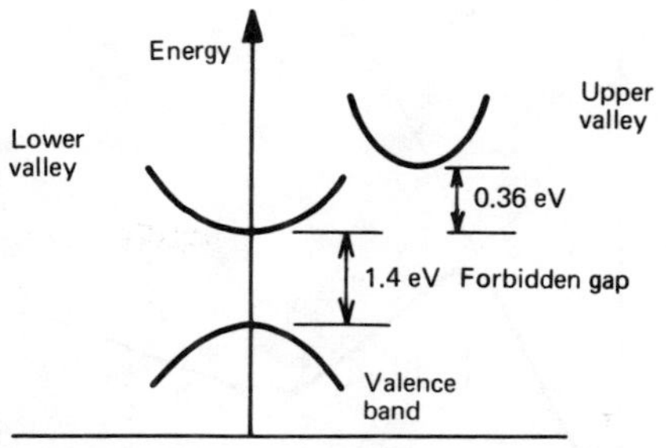

Fig. 8.18 Energy-level diagram for GaAs.

intrinsic limitations that prevented lower-frequency techniques from being applied successfully. From 1961 to 1963, however, there was a major research breakthrough. In 1961 Ridley and Watkins,[4] and in 1962 Hilsum,[5] wrote theoretical papers predicting that, given certain conditions that are satisfied by a few compounds, microwave oscillations could be produced directly from a thin sample of such a material. In 1963, Gunn,[6] while examining the noise properties of a slice of GaAs, observed a regular pulse of current from his sample when the applied voltage exceeded some level; the Gunn device was discovered. Gunn followed up this work with a set of very precise and difficult measurements in which he investigated the variations of electric field across the device when it was operating in an oscillatory mode, and as a result he confirmed many of the earlier theoretical predictions.

We can see the requirements of a suitable material from the energy level diagram of GaAs shown in Fig. 8.18. The conduction band is separated from the valence band by what is called the "forbidden gap", which is 1.4 eV wide. When there is no field across the diode (so called because it has two terminals, a cathode and an anode, and not because it exhibits any rectifying properties) the free conduction electrons reside in the lowest conduction state, shown as the lower valley in the diagram. As the electric field across the device increases, the drift velocity also increases; the straight line OA in Fig. 8.19 shows the relationship between them. The gradient of OA is known as the electron mobility in the lower valley, say μ_L. As the electric field is increased, it eventually achieves a value E_{Th} at which the linear relationship between electron velocity and electric field breaks down, and the velocity begins to decrease, along CD. There is now a negative gradient until the field reaches E_v at point D, after which the velocity again increases linearly with electric field. Beyond D the characteristic follows the line OB which represents an electron mobility of μ_u. Clearly, μ_u is smaller than μ_L.

Along DB the conduction electrons, which were in the lower valley, have now moved to the upper valley where their mobility is reduced, and their mass is increased. Between C and D the transition over the 0.36 eV gap between the valleys is taking place. To avoid breakdown of the diode material, which must

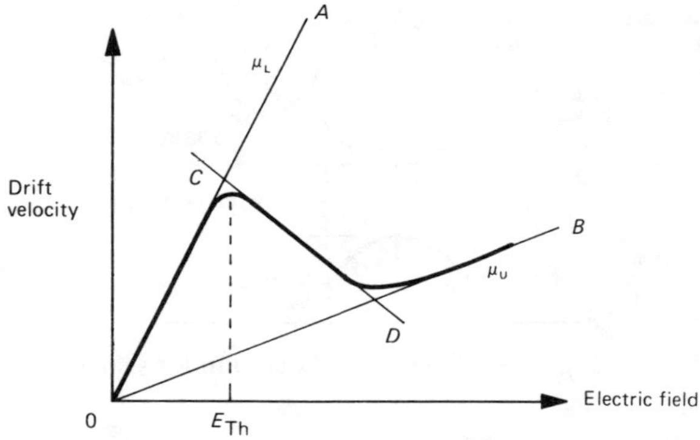

Fig. 8.19 Variation of drift velocity with electric field in a Gunn device.

have suitable conduction band energy levels, the difference between them must be much smaller than the forbidden gap. There are some III–V compounds that satisfy this requirement. GaAs is the best understood because, since the Gunn experiments, it has been used widely as a diode material. Another compound of interest is InP which may eventually prove to be even more appropriate. Intensive work is being done in several countries to overcome some of the technical problems associated with its production.

The distribution of electric field across the diode is constant until E_{Th} is exceeded. Thereafter the diode is operating in the unstable *CD* region. Gunn showed that in this region a domain of high-intensity field, and low-mobility electrons, forms at the cathode, and gradually drifts across to the anode, where it leaves the device and produces a pulse of current in the external circuit. The variation of output current with time is shown in Fig. 8.20. Once a domain forms at the cathode no further domain can be sustained, so there is only one domain present at any time. Some properties of the cathode region ensure that the domain begins there consistently, and therefore the periodicity of the pulses is determined by the transit time of the domain across the device, which in turn is related to the device thickness. For operation in the range 5–10 GHz the material would be about 10 μm across. To sustain oscillations in a circuit, the circuit resistance and the device resistance must cancel, implying that at the

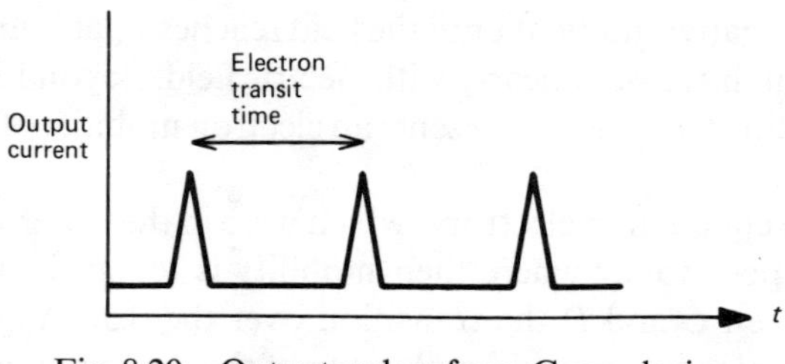

Fig. 8.20 Output pulses from Gunn device.

frequency of interest the device resistance must be negative. That this is so can be seen from Fig. 8.19. The electron drift velocity is related directly to current flow through the device, and the electric field is proportional to the terminal voltage; hence the line *CD* represents a negative resistance. In practice it will have a value of no more than 10Ω. Apart from this negative resistance, there will be reactive components associated with the device, its leads and the packaging, all of which will affect the operating frequency. A simplified equivalent circuit is shown in Fig. 8.21.

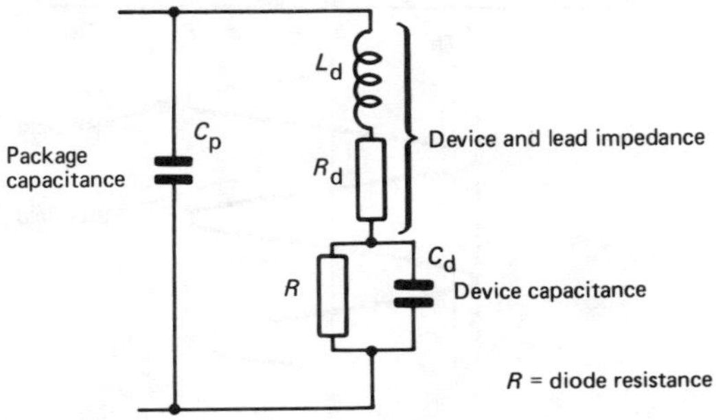

Fig. 8.21 Simplified equivalent circuit of Gunn device.

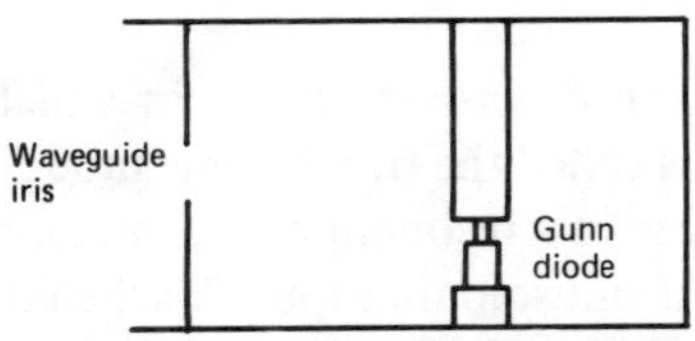

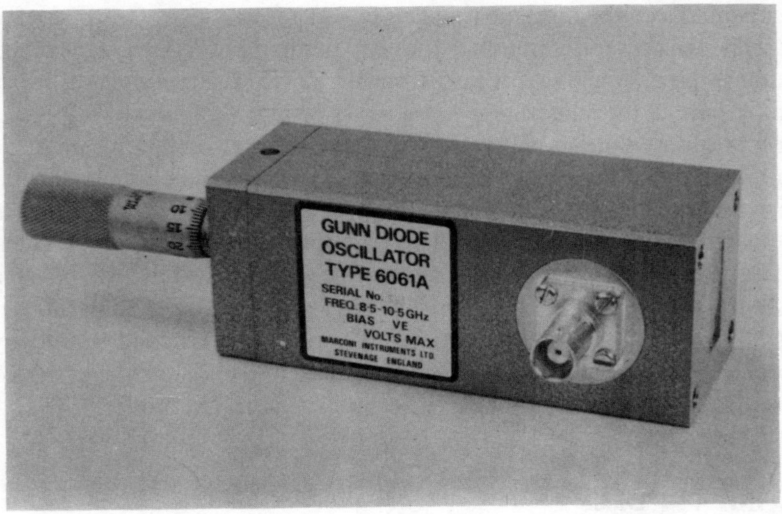

Fig. 8.22 Waveguide mounting for Gunn device. Reproduced from Marconi catalogue by permission of Sanders Division, Marconi Instruments Ltd.

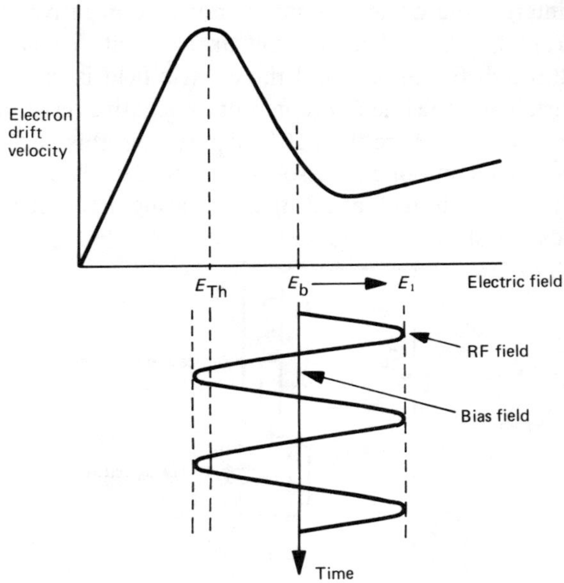

Fig. 8.23 Biasing for the LSA operation mode.

The mode of operation described above, in which the applied voltage is maintained above E_{Th}, is called the transit-time mode. To produce oscillations, the diode is usually placed in a resonant microwave circuit which acts as a high Q tuned system, giving a sinusoidal output. A schematic of a mounting for the Gunn diode is shown in Fig. 8.22. The electrodes are produced by crystal growth on to a conducting substrate, or by ion implantation. One of the factors that limits the power obtainable is the heat dissipation that can be achieved. To maximize the conductivity of heat from the device a large integral heat sink, forming part of the substrate, is used.

In the transit-time mode we saw that the natural oscillating frequency is determined by the length of the sample: the thinner the sample, the shorter the transit time, and hence the higher the frequency. However, the transit time mode is not the only one that is possible.[7] By applying a dc bias which takes the electric field just below E_{Th}, the diode can be made to switch into the negative resistance region as the alternating voltage takes the field value above E_{Th} (Fig. 8.23). Then the frequency of oscillation can be determined entirely by the external circuit in which the device is mounted. This mode of operation is known as the limited space–charge accumulation (LSA) mode.[8,9]

8.8 IMPATT Diode

The IMPATT diode takes its name from IMPact Avalanche and Transit Time diode. It is one of several microwave devices based on the avalanche growth of carriers through a diode junction placed in a very high electric field. Sometimes called the Read diode after the man who proposed it, this device was first produced experimentally in 1965 by Johnston, de Loach and Cohen.[10] Since then IMPATT oscillators have been made for frequencies throughout the microwave spectrum, from 300 MHz to 300 GHz. They are more efficient (typical value about 15%), and can produce significantly more power than the Gunn device, but at the expense of greater noise and higher power supply voltage.

As the name implies, there are two factors that are particularly important in the operation of IMPATT diodes. First is avalanche multiplication. This phenomenon can occur at electric fields above about 10^5 V/cm. Carriers receive sufficient energy to produce electron–hole pairs by impact ionization. The ionization rate is strongly, but not linearly, related to the electric field; a five-fold increase in ionization rate can be produced by doubling the electric field.

The other important factor is the finite carrier drift velocity, and its variation with electric field strength. The electric fields used in IMPATT devices are much higher than the threshold fields discussed in the previous section on Gunn devices. Operating fields are above 10^5 V/cm and at such levels the electron drift velocity variation with electric field suffers a saturation effect and remains substantially constant. The value of the drift velocity depends particularly on the device material; 6×10^6 cm/s and 8×10^6 cm/s are typical for Si and GaAs, respectively, when used at an operating temperature of about 200 °C.

The basic form of the IMPATT is shown in Fig. 8.24; electrons are released at the p – n junction and then drift through an intrinsic n region. There are two significant widths associated with the device; W_A, the avalanche region, in which ionization takes place and a concentration of electrons is formed, and W_D, the drift region through which the electron bunch moves at velocity v_s.

The field across the p^+–n junction is well in excess of the breakdown value, with the maximum field E_p being approximately given by

$$E_p = \frac{q}{\varepsilon} N W$$

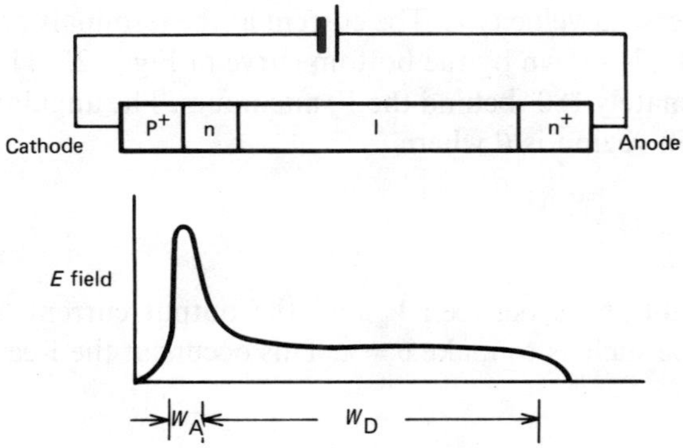

Fig. 8.24 IMPATT diode structure.

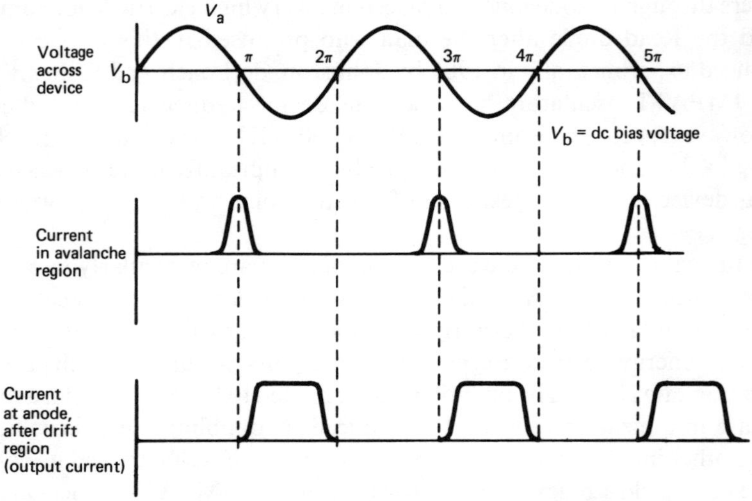

Fig. 8.25 Delay between output current and applied voltage in IMPATT diode.

where W is the depletion width $= W_A + W_D$, N is the doping concentration in the n-type material, q is the electron charge and ε is the permittivity of the material.

When an IMPATT is used in an oscillator circuit it is reverse biased by a fixed voltage V_b, and superimposed on V_b is the voltage due to the microwave field in the oscillator, V_a. If V_b is chosen such as to just support avalanche breakdown, we can see from Fig. 8.25 how the diode sustains oscillations in the circuit.

The ionization mechanism has an inherent inertia; consequently the ionization rate does not respond instantaneously to changes in junction voltage. As V_a increases, the ionization rate increases, and continues to increase when V_a falls, until V_a reaches zero. The ionization then decreases also. The resulting electron density is shown as a function of time, relative to V_a in Fig. 8.25. The avalanche current peak is approximately 90° behind the maximum of V_a. The electrons leave the avalanche region after distance W_A and enter the drift region which they traverse at velocity v_s. The current at the terminals of the IMPATT, in response to V_a, is shown by the bottom curve of Fig. 8.25. The fundamental will be approximately 180° behind the V_a maximum. The angular displacement through the drift region is θ where

$$\theta = \omega \frac{W_D}{v_s} \tag{8.1}$$

For complete antiphase between V_a and the output current, the oscillation frequency will be such as to make $\theta = \pi$. This occurs at the Read frequency, f_r where

$$f_r = \frac{v_s}{2W} \tag{8.2}$$

The width of a Si-based device, which has $v_s \simeq 6 \times 10^6$ cm/s for an operating frequency of 10 GHz, would therefore be about 3 μm.

Read's Equation

The theoretical basis for the IMPATT device was provided by Read in 1958. Following his work, several refinements have been made to his analysis to account for the manner of realizing the device. The variation of avalanche current density J_a with time is given by the Read equation:

$$\frac{dJ_a(t)}{dt} = \frac{3J_a(t)}{\tau_a}\left[\int_0^W \alpha(E)\,dx - 1\right] \tag{8.3}$$

where x is the distance through the device from the junction and α is the ionization coefficient, which is the probability that a carrier will experience an ionizing collision in a unit length. τ_a is the time taken for an electron travelling at the saturation velocity v_s to travel over the avalanche region, i.e.

$$\tau_a = W_A/v_s$$

Associated with the avalanche region is an inductance L_A and a capacitance C_A:

$$L_A = \frac{W_A}{AJ_{DC}k} \tag{8.4}$$

$$C_A = \frac{\varepsilon A}{W_A} \tag{8.5}$$

A is the device area, J_{DC} the mean current density and k is a factor relating the characteristics of the semiconductor material to the ionization coefficient, at maximum field:

$$k = \frac{3}{\tau_a}\frac{\varepsilon}{qN}\alpha(E_p)$$

From Eqns (8.4) and (8.5) the resonant frequency of the avalanche region is f_{ra} where

$$f_{ra}{}^2 = \frac{1}{(2\pi)^2 L_A C_A} = \frac{J_{DC}k}{\varepsilon} \tag{8.6}$$

At frequencies above f_{ra} the device resistance becomes negative, and the reactance changes from inductive to capacitive.

Power frequency limitations

There is a maximum voltage V_m, which may be applied to the diode, corresponding to the maximum field E_m which the diode can withstand without breakdown over the depletion layer W. When the electron velocity is at its saturation level v_s, the maximum current I_m is

$$I_m = \frac{E_m \varepsilon v_s A}{W} \tag{8.7}$$

where A is the cross-sectional area of the device and ε the permittivity of the semiconductor material.

The maximum power input

$$P_m = I_m V_m \tag{8.8}$$

$$= \frac{E_m \varepsilon v_s A}{W} E_m W$$

$$= E_m{}^2 \varepsilon v_s A$$

The depletion layer capacitance $C = \varepsilon A/W$. Therefore

$$P_m = \frac{E_m{}^2 v_s{}^2}{\Delta \pi f^2 X_c} \tag{8.9}$$

where

$$X_c = \frac{1}{2\pi f C} \quad \text{and} \quad f = \frac{v_s}{2W}$$

Hence if X_c is limited, P_m is proportional to $1/f^2$ in the millimetre wave range.

Thermal limitation

The major limiting factor other than the intrinsic limit implied above is the amount of heat that can be conducted away from the junction. The dissipation of heat can be improved by mounting the device on a high conductivity material such as copper or diamond. Most of the heat is generated in the avalanche region where, for GaAs or Si, the operating temperature may be about 200 °C. The thermal conductivity of the diode depends on the thermal conductivity of the semiconductor material, the thermal resistance between the semiconductor and the heat sink, and the thermal conductivity of the sink material. For a particular system, the power dissipation decreases linearly with frequency.

8.9 Field-effect Transistors

Probably the most promising type of microwave solid-state device is the field-effect transistor in one or other of its many forms. Based on GaAs, it is used for various purposes: high power generation, low noise and high frequency amplification in satellite and microwave terrestrial links. GaAs has a higher electron mobility than Si, and therefore a shorter transit time, allowing better high-frequency performance.

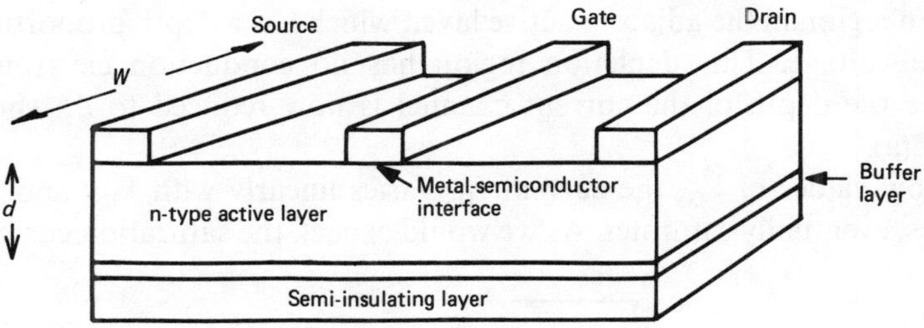

Fig. 8.26 Microwave FET.

The basic form of the device is shown in Fig. 8.26. The buffer layer isolates the n-type active layer from the semi-insulating substrate. The active layer, which is less than half a micrometre deep, has a doping level of some 10^{17} donors/cm^3. Source and drain ohmic contacts are placed either directly on to the active layer, or on to another, more heavily doped, layer in the contact area. Electrons travel from the source to the drain, which is positive relative to the source, via the active layer. The path available for the electrons is called the channel. With this arrangement the current from source to drain, I_{DS}, increases as the applied voltage V_{DS} increases. The relationship between I_{DS} and V_{DS} is linear at low voltage as the electron velocity v varies linearly with the electric field produced by V_{DS}. As the voltage increases, v increases less quickly than the electric field, and eventually when a critical value of E is reached, $E_c \simeq 3 \times 10^3$ V/cm, the electron drift velocity saturates at v_s. This characteristic is shown in Fig. 8.27.

The current I_{DS} is given by

$$I_{DS} = wdvqn \tag{8.10}$$

where wd = width × depth of the current channel, and $v.q.n.$ = velocity, unit charge, and density of the electron flow, respectively.

In the case just discussed, where the field just reaches E_c, the minimum depth of the channel is d_0. Clearly I_{DS} varies directly with v and therefore reaches a saturation value when $v = v_s$.

By introducing an electrode (the gate) between source and drain, an additional control on I_{DS} is possible. The gate, which is reverse biased, creates a

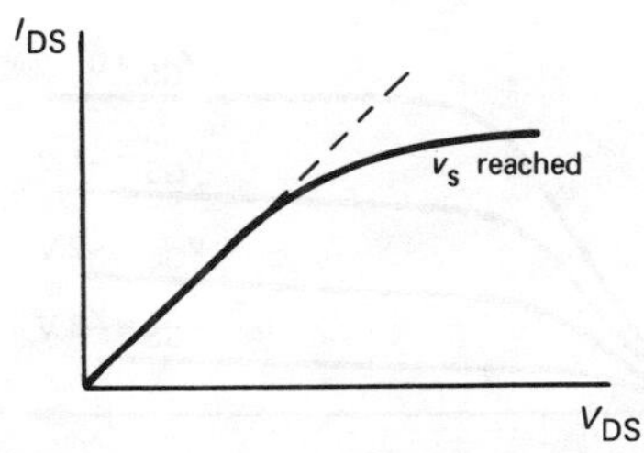

Fig. 8.27 Source–drain current–voltage characteristic for FET.

depletion region in the adjacent active layer, which has a depth proportional to the gate voltage. This depletion region has no conduction electrons, and therefore the depth of the current channel is now reduced to d_1, shown in Fig. 8.28(a).

For low values of V_{GS} the current I_{DS} varies linearly with V_{DS}, and, as V_{DS} increases, eventually saturates. As we would expect, the saturation current and

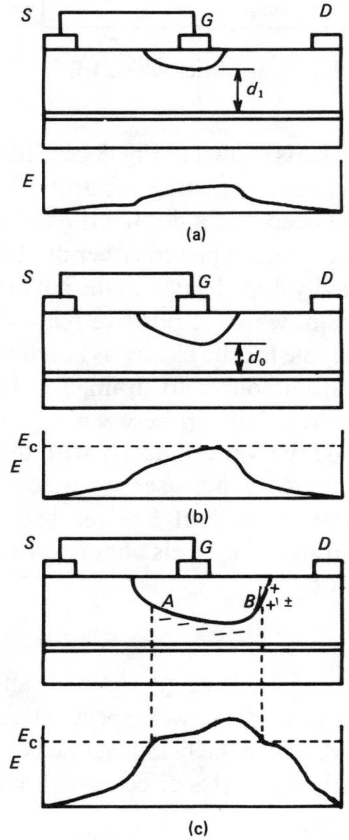

Fig. 8.28 Effect of applied voltage on current channel in a FET.

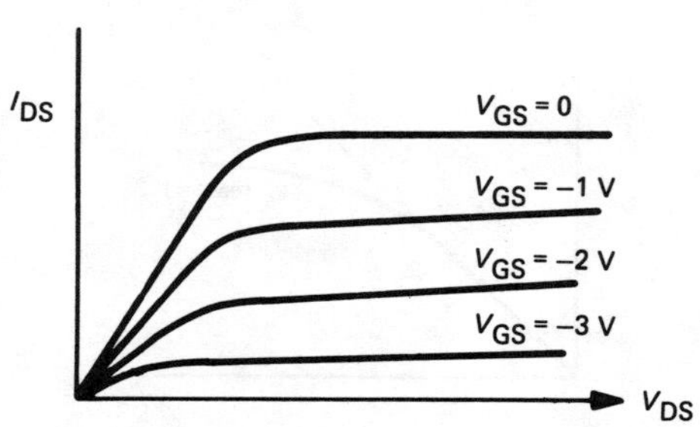

Fig. 8.29 Variation of output current with applied voltage.

the value of V_{DS} at which it is reached decrease as V_{GS} becomes more negative. A typical family of curves is shown in Fig. 8.29.

When V_{GS} is zero the gate still has an effect: the field beneath the gate produces a depletion layer. The reduction in channel depth produced by the depletion layer induces an increase in electron drift velocity as the field beneath the active layer increases [Fig. 8.28(b)]. When the field reaches the critical value E_c the velocity saturates, the depth of the channel is d_0, and the current through the device is given by

$$I_{DS} = wqnv_s d_0 \tag{8.11}$$

Any further increase in V_{DS} deepens the depletion layer, hence reducing d still further. For I_{DS} to remain constant in the area of saturated velocity, the electron density increases. Figure 8.28(c) shows the effect of this deeper depletion layer. The point at which the field becomes equal to E_c moves towards the source, and therefore is reduced in voltage (point A), producing a deeper channel than d_0, and therefore a larger current. This suggests that the current characteristic, after the saturation effect in Fig. 8.28(c), increases again with V_{DS}. GaAs demonstrates such a gentle positive slope beyond the initial saturation.

Another feature of the depletion layer on the drain side of A, in Fig. 8.28(c), is that an accumulation of electrons develops over the distance AB where the field is greater than E_c; point B is that position where the channel depth reaches the value it had at A. The value of n increases to maintain I_{DS} since d is reduced, and the electron velocity is saturated. Beyond B, a positive charge layer is formed which helps to maintain constant I_{DS}.

The bottom edge of the depletion layer is not parallel to the surface of the device, but is deeper on the drain side. This is due to the voltage gradient over the width of the gate producing a wedge shaped layer, with rounded edges. As we can see from the diagrams, the precise shape varies with V_{DS}.

This qualitative description shows the effect of the gate electrode when it is connected directly to the source. By its influence, the I_{DS}/V_{DS} characteristic is modified. However, the device is designed to be used as a transistor with a control voltage on the gate. As we noted in Fig. 8.29, by negatively biasing the gate the saturation effect occurs at a lower V_{DS} than for $V_{GS} = 0$. The reverse biased gate junction deepens the depletion layer and hence restricts I_{DS}. Eventually the depletion layer reaches the buffer layer and the channel is closed. The gate voltage at which this occurs, known as the pinch-off voltage V_p, is determined by the electrical properties, and the depth, of the active layer:

$$V_p = \frac{qnd^2}{2\varepsilon\varepsilon_0} \tag{8.12}$$

where q, n and d are as we defined them earlier, and ε, ε_0 are the relative permittivity of GaAs and permittivity of free space respectively.

A lumped component equivalent circuit for the device is shown in Fig. 8.30.

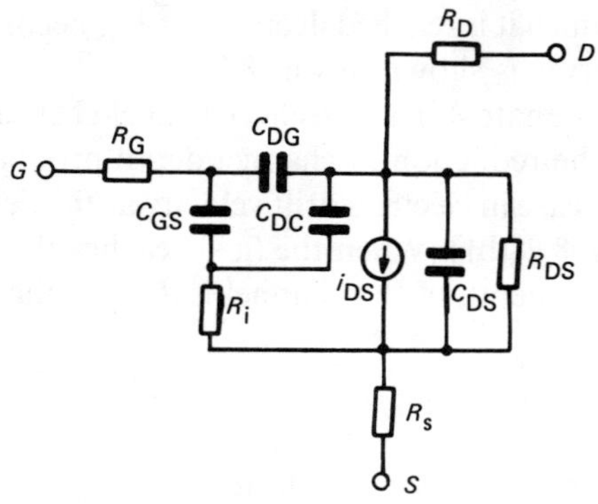

Fig. 8.30 FET equivalent circuit.

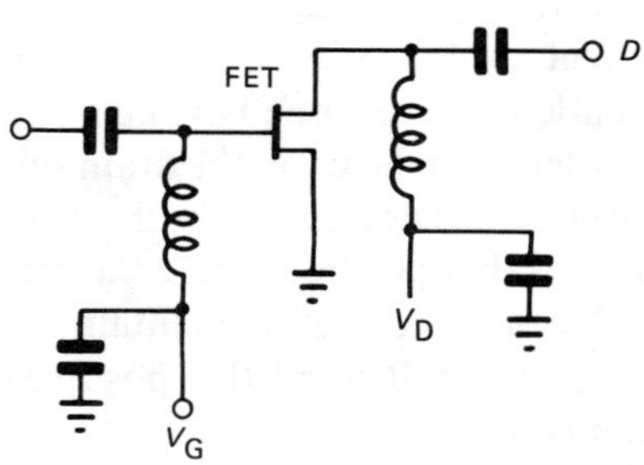

Fig. 8.31 Biasing network for an FET.

From this circuit, the maximum frequency of oscillation is

$$f_m = \frac{g_m}{4\pi(C_{GS} + C_{DG})}\left[\frac{R_i + R_s + R_G}{R_{DS}} + \frac{g_m}{(C_{GS} + C_{DG})} C_{DG} R_G\right]^{1/2} \tag{8.13}$$

The performance of the device is related to the bias circuit used to establish and maintain the dc gate voltage; variations in this voltage induce noise and therefore low values of source and gate resistance are desirable. To minimize the variation in bias point, a two-source biasing configuration is preferred, such as that shown in Fig. 8.31.

This brief description refers to the MESFET (metal semiconductor field-effect transistor) which has a metal gate (Schottky) contact. It has a very low noise figure, and is capable of operating at high millimetre wave frequencies. Many other FETs exist, or are being developed; by varying the geometry of the device and the nature of the gate, particular features are enhanced. For further information, see Lamming[11] and Ha.[12]

8.10 Conclusion

From the point of view of the early 1980s it seems that the FET in one form or another, probably built on GaAs, will become the stock active microwave semiconductor device. Suitable for integrated circuit design, it can be clustered to produce large amounts of power, or low noise amplification, and it is the subject of intense research and development. Although the high saturation velocity of GaAs makes it very attractive, other III–V compounds could turn out to be even better.

As we noted earlier, there are many applications that require power levels above those which can be achieved from solid-state devices. Consequently, there continues to be significant development in microwave vacuum tube design, in an attempt to improve the efficiency and performance of those tubes that have been established for nearly half a century, and to discover entirely new ways of generating high power.

References

1. Kompfner, R., "The travelling-wave tube", *Wireless World*, **52**, 349 (1946).
2. Atwater, H.A., *Introduction to Microwave Theory*, McGraw-Hill, Maidenhead, 1962.
3. Collin, R.E., *Foundation for Microwave Engineering*, McGraw-Hill, Maidenhead, 1966.
4. Ridley, B.K. and Watkins, T.B., "The possibility of negative resistance effects in semiconductors", *Proceedings of the Physical Society*, **78**, 293–304 (Aug 1961).
5. Hilsum, C., "Transferred electron amplifiers and oscillators", *Proceedings of the Institute of Radio Engineers*, **50**(2), 185–9 (Feb 1962).
6. Gunn, J.B., "Microwave oscillations of current in III–V semiconductors", *Solid-state Communications*, **1**, 88–91 (Sept 1963).
7. Pattison, J., "Active microwave devices", *Microwave Solid-state Devices and Applications* (ed. Morgan, D.V. and Howe, M.J.), Pergamon, Oxford, 1980, Chapter 2.
8. Hobson, G.S., *The Gunn Effect*, Clarendon, Oxford, 1974.
9. Bosch, B.G. and Engelmann, R.W., *Gunn-effect Electronics*, Pitman, London, 1975.
10. Johnston, R.L., de Loach, B.C. and Cohen, B.G., "A silicon diode microwave oscillator", *Bell System Technical Journal*, **44**, 2 (Feb 1965).
11. Lamming, J., "Active microwave devices – FET's and BJT's", *Microwave Solid-state Devices and Applications* (ed. Morgan, D.V. and Howe, M.J.), Pergamon, Oxford, 1980, Chapter 4.
12. Ha, Tri T., *Solid-state Microwave Amplifier Design*, Wiley, Chichester, 1981.

9 PASSIVE MICROWAVE DEVICES

In the last chapter we considered some common microwave active devices. When microwave energy is generated it is used by either radiating it into space via an antenna or passing it along a transmission system. In either case some circuitry will be involved at both the transmitting and receiving ends. This circuitry, and the transmission system, have features that are special to signals of microwave frequencies. In this chapter we will consider first the commonest microwave component, the rectangular waveguide, and go on to look at several of the passive devices that comprise microwave circuitry.

9.1 Waveguides

Waveguides are distinguished by having only one guiding surface. They are of two main types: (a) metal tubes of any cross-section, and (b) dielectric rods. In the first case the wave travels down the inside of the tube and there is no leakage to the outside. In the second, the wave is propagated over the outer surface of the rod and there is a considerable field in the surrounding medium.

In this chapter we shall be looking at the first type only.

For best transmission, tubular waveguides should be made of a low-loss, conducting material. If runs are short, and cost is important, aluminium or brass will be used, but good quality waveguides having low attenuation are made from copper which, for some applications where low loss is paramount, might be plated with silver or gold. The energy is propagated along the inside of the tube so the inner surface should be smooth and the cross-section must be uniform throughout the length of the guide.

9.2 Rectangular Waveguide

Power is transmitted down a waveguide by electric and magnetic fields which travel in a manner similar to that of travelling voltage waves in a transmission line.

Figure 9.1 shows a rectangular waveguide of internal cross-section $a \times b$, in relation to a right-handed system of coordinates. This means that the broad face

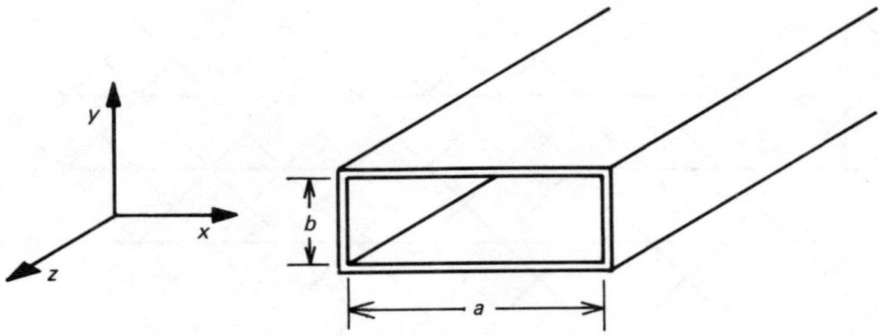

Fig. 9.1 Rectangular waveguide dimensions.

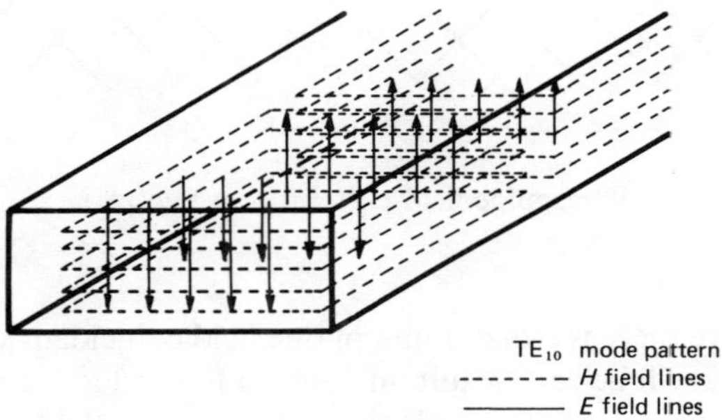

Fig. 9.2 Field patterns of TE_{10} rectangular waveguide mode.

of the guide is in the x direction, the narrow face in the y direction, and the wave is propagated in the z direction.

The wave inside the guide is described in terms of a field pattern. Figure 9.2 shows the pattern for the fundamental rectangular waveguide mode, TE_{10} (defined below). The way in which this field pattern is set up can be considered from two points of view, which will be considered in turn in the following sections.

9.3 Interference of Two Plane Waves

In Figure 9.3 two uniform plane waves of identical amplitude and phase are shown travelling in different directions. In each, the electric field vector is normal to the plane of the paper. As the waves travel across each other there will be constructive and destructive intereference, resulting in a field pattern that is the algebraic sum of the component fields at each point. Where a maximum of one of the waves crosses a maximum of the other there will be a maximum in the resultant field, and conversely there will be a minimum where the minima of the two component fields are coincident. Examples of these points are at E for the minimum and F for the maximum in Fig. 9.3.

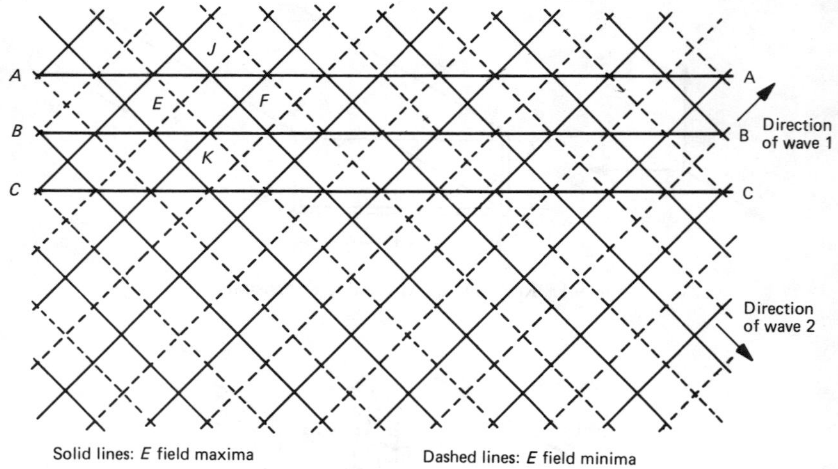

Fig. 9.3 Waveguide field pattern from two plane waves.

At points where there is a maximum of one field coincident with a minimum of the other there will be zero resultant field. In Fig. 9.3 a line is drawn joining such points, shown as *AA*. Since there is zero electric field along this line, a conducting surface could be inserted parallel to the *E* field lines along *AA* without perturbing the field. Similarly, a conducting surface could be placed along *BB*. These two conducting planes could thus form the side walls of a rectangular waveguide. Top and bottom conducting walls, parallel with the plane of the paper, can be used without altering the field pattern since the *E* field lines will meet them normally.

The resultant field pattern between *AA* and *BB* in Fig. 9.3 is therefore a representation of one that can exist inside a rectangular waveguide. As we shall see later the pattern between *AA* and *BB* represents the fundamental rectangular waveguide mode. If, instead of the side walls being along *AA* and *BB*, they were placed along *AA* and *CC*, a different pattern, representing a second-order mode, would exist. The question of modes is discussed in Section 9.7.

9.4 Cut-off Wavelength

The pattern shown in Fig. 9.3 between lines *AA* and *BB* can be used to quantify the cut-off wavelength in terms of the free space wavelengths and the guide width, i.e. the distance between *AA* and *BB*. An expanded representation of the pattern between a maximum and a minimum is given in Fig. 9.4. We can see that:

λ_g = wavelength of the field pattern inside the guide, i.e. the distance

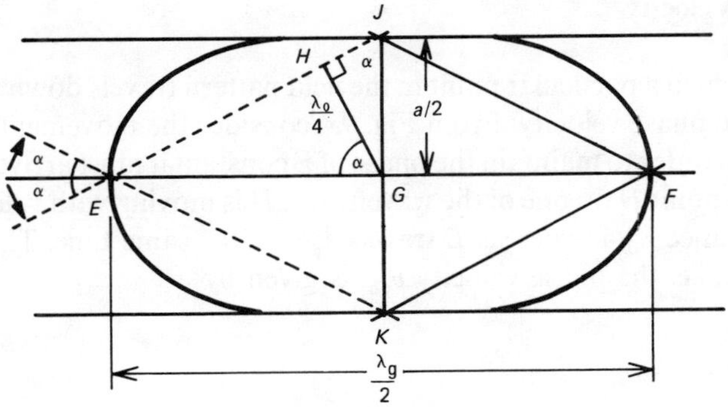

Fig. 9.4 Details at field pattern.

between two adjacent points of similar phase in the direction of propagation

λ_0 = wavelength of the wave in the medium filling the guide, say air

$$= \frac{\text{velocity of light in air}}{\text{frequency of the wave}} = \frac{c}{f}$$

From the diagram, if the angle between the direction of propagation of the two component waves is 2α (i.e. each wave propagates at an angle α to the axis of the guide),

$$\sin\alpha = \frac{\lambda_0}{2}\frac{1}{a} \quad \text{and} \quad \cos\alpha = \frac{\lambda_0}{\lambda_g}$$

Using the identity $\cos^2\alpha + \sin^2\alpha = 1$

$$\frac{\lambda_0{}^2}{(2a)^2} + \frac{\lambda_0{}^2}{\lambda_g{}^2} = 1$$

or

$$\frac{1}{\lambda_0{}^2} = \frac{1}{\lambda_g{}^2} + \frac{1}{(2a)^2} \tag{9.1}$$

where a is the width of the guide. $2a$ is called the cut-off wavelength λ_c for the TE_{10} mode; hence Eqn (9.1) becomes

$$\frac{1}{\lambda_0{}^2} = \frac{1}{\lambda_g{}^2} + \frac{1}{\lambda_c{}^2} \tag{9.2}$$

The cut-off wavelength is so called because it is the maximum value that λ_0 can have. If λ_0 becomes greater than λ_c the wave will not propagate down the guide. Thus the waveguide behaves as a high-pass filter: only waves of frequency greater than the cut-off value $f_c (= c/\lambda_c)$ can be carried. This aspect of waveguide behaviour will be considered in further detail in later sections.

9.5 Phase Velocity

The rate at which a particular point in the field pattern travels down the guide is known as the phase velocity. From Fig. 9.4 consider the movement of point *E* towards *F*. In order to maintain the phase of *E* constant it must arrive at *G* at the same time as point *H* on one of the wavefronts. *H* is moving with velocity *c*, and covers a distance $\lambda_0/4$, whereas *E* travels $\lambda_g/4$ in the same time. Therefore the velocity of *E*, i.e. the phase velocity v_{ph}, is given by

$$v_{ph} = c\frac{\lambda_g}{\lambda_0} \tag{9.3}$$

9.6 Group Velocity

It would violate the theory of relativity if energy were transported down the guide at the phase velocity, and in fact that does not happen. The velocity of energy movement is known as the group velocity, and in terms of our plane-wave model in Fig. 9.4 it is equal to the component of the plane wave velocity in the *z* direction. The plane wave travels at velocity *c* in air so the component in the *z* direction is

$$v_g = c\cos\alpha = c\frac{\lambda_0}{\lambda_g} \tag{9.4}$$

It is worth noting from the last two equations that

$$v_{ph} \cdot v_g = c^2 \tag{9.5}$$

9.7 Rectangular Waveguide Modes

The above discussion has been concerned with a particular field pattern inside the waveguide—that given by Figs. 9.2 and 9.3 and the section between *AA* and *BB* of Fig. 9.4. As we noted earlier, it is the TE_{10} mode, often called the fundamental rectangular waveguide mode because it has the lowest cut-off frequency.

In the designation TE_{10} the letters stand for Transverse Electric, indicating that the field pattern has no electric component in the *z* direction, it being contained entirely in the *x–y*, transverse, plane. The numerals 1, 0 are used to indicate the number of half-sinusoidal variations of the *E* field in the *x* and *y* directions, respectively. For any value of *y* the electric field goes through one half-sinusoidal variation with *x*, i.e. from zero at one wall, through a maximum in the guide-centre, to zero at the other wall (Fig. 9.2). For any value of *x* there is no variation of field with *y*, i.e. the field is constant in the *y* direction. Thus TE_{10} designates a transverse electric mode having one half-sinusoidal variation in the

x direction (across the broad side of the guide) and no variation in the y direction.

In fact, there are many possible TE modes. The general designation is TE_{mn} where m is the number of half-sinusoidal variations in the x direction, and n the number of similar variations in the y direction. We shall see later that the cut-off wavelength for the TE_{mn} mode is given by

$$\frac{1}{(\lambda_{c_{mn}})^2} = \left(\frac{m}{2a}\right)^2 + \left(\frac{n}{2b}\right)^2 \tag{9.6}$$

where a and b are the dimensions shown in Fig. 9.1.

There is another set of modes possible in a rectangular waveguide; the Transverse Magnetic, or TM, modes, in which there is no component of magnetic field in the direction of propagation. The general mode is designated TM_{mn}, and it too has a cut-off wavelength given by Eqn (9.6). The suffices mn have the same meaning as in TE_{mn}.

Unlike coaxial transmission line, there is no TEM mode in a waveguide: a mode cannot exist in which both the electric and magnetic vectors have no component in the z direction.

9.8 Field Theory of Propagation Along a Rectangular Waveguide

Although the above discussion, in which the fields inside the waveguide are considered to result from the intereference of two plane waves, gives a very useful insight into the mechanism of waveguide propagation, it has some disadvantages. In particular, it does not allow an evaluation of the field components inside the guide. A deeper appreciation can be obtained by using Maxwell's equations. These equations state in generalized form the relationship that must exist between the electric and magnetic fields in any medium, provided the characteristics of the medium and the initial conditions of the fields are known. It would be inappropriate to develop Maxwell's equations here, but several texts discuss their derivation in some detail, and reference should be made to one of these, e.g. Bleaney and Bleaney[1] or Paul and Nasar,[2] for background information.

We start by stating the equations in generalized vector form:

$$\nabla \cdot \bar{E} = \frac{\rho}{\varepsilon} \tag{9.7}$$

$$\nabla \cdot \bar{B} = 0 \tag{9.8}$$

$$\nabla \times \bar{E} = -\frac{\partial \bar{B}}{\partial t} \tag{9.9}$$

$$\nabla \times \bar{H} = \bar{J} + \varepsilon \frac{\partial \bar{E}}{\partial t} \tag{9.10}$$

$$\bar{B} = \mu\bar{H} \tag{9.11}$$

$\bar{E}$ and $\bar{H}$ are the electric and magnetic field vectors, respectively, $\bar{B}$ is the magnetic flux density, and $\bar{J}$ is the displacement current density. All four vectors are related to the space inside the guide; for air the value of $\bar{J}$ is zero.

The other symbols represent the electric characteristics of the material filling the guide. Normally the charge density (ρ) is zero, and for air the permittivity (ε_0) and the permeability (μ_0) are those for free space, i.e.

$$\varepsilon_0 = \frac{1}{36\pi \times 10^9} \quad \text{Farad/m}$$

$$\mu_0 = 4\pi \times 10^7 \quad \text{henry/m}$$

Although the following analysis is lengthy, it allows us to build up an awareness of an analytical description of waveguide propagation. By considering the argument in some detail, rather than merely isolating the important equations, we can appreciate more fully the underlying theory, and that will be of value in discussing several of the devices considered later in the chapter.

If we assume immediately that there is a sinusoidal signal being transmitted down the guide, then the time derivatives in Eqns (9.9) and (9.10) can be replaced by $j\omega$, and they become, when $\bar{B}$ is replaced by μH,

$$\nabla \times \bar{E} = -j\omega\mu\bar{H} \tag{9.12}$$

$$\nabla \times \bar{H} = j\omega\varepsilon\bar{E} \tag{9.13}$$

and, since $\rho = 0$,

$$\nabla \cdot \bar{E} = 0 \tag{9.14}$$

Now we must solve these equations for waveguide having a rectangular cross-section.

Taking the curl of Eqn (9.12), and substituting from Eqn (9.13),

$$\nabla \times \nabla \times \bar{E} = \omega^2 \varepsilon\mu\bar{E} \tag{9.15}$$

Using Eqn (9.15) in the vector identity,

$$\nabla \times \nabla \times \bar{E} = -\nabla^2\bar{E} + \nabla(\nabla \cdot \bar{E})$$

and combining the result with Eqn (9.14), we have

$$\nabla^2\bar{E} + \mu\varepsilon\omega^2\bar{E} = 0 \tag{9.16}$$

This is a vector representation of the E field wave equation, and to solve it for the waveguide it must be put into an appropriate coordinate system. In rectangular coordinates

$$\nabla^2\bar{E} = \bar{x}\nabla^2 E_x + \bar{y}\nabla^2 E_y + \bar{z}\nabla^2 E_z \tag{9.17}$$

From Eqns (9.16) and (9.17) there are three equations to be solved, one for each coordinate direction:

$$\nabla^2 E_x = -\mu\varepsilon\omega^2 E_x \tag{9.18}$$

$$\nabla^2 E_y = -\mu\varepsilon\omega^2 E_y \tag{9.19}$$

$$\nabla^2 E_z = -\mu\varepsilon\omega^2 E_z \tag{9.20}$$

By developing Eqns (9.16) and (9.17) in terms of $\bar{H}$ rather than $\bar{E}$, a similar set of equations is produced for the $\bar{H}$ components. The six equations resulting are related by the boundary conditions at the waveguide walls, and these conditions must be used when the equations are solved.

We will now solve Eqn (9.19), which can be expanded to

$$\frac{\partial^2 E_y}{\partial x^2} + \frac{\partial^2 E_y}{\partial y^2} + \frac{\partial^2 E_y}{\partial z^2} = -\mu\varepsilon\omega^2 E_y \tag{9.21}$$

In principle, E_y may vary with x, y and z, and variation in one direction will be independent of variation in another, so the technique of separation of variables can be used.

Let the solution of E_y be

$$E_y = XYZ \tag{9.22}$$

where XYZ are functions of only x, y and z, respectively.

Then, substituting Eqn (9.22) into Eqn (9.21),

$$\frac{\partial^2 X}{\partial x^2} YZ + X\frac{\partial^2 Y}{\partial y^2} Z + XY\frac{\partial^2 Z}{\partial z^2} = -\mu\varepsilon\omega^2 XYZ$$

or

$$\frac{1}{X}\frac{\partial^2 X}{\partial x^2} + \frac{1}{Y}\frac{\partial^2 Y}{\partial y^2} + \frac{1}{Z}\frac{\partial^2 Z}{\partial z^2} = -\mu\varepsilon\omega^2 \tag{9.23}$$

At any specific frequency the right-hand side of Eqn (9.23) is constant, so for this expression to be valid each of the terms on the left-hand side must also be constant; hence we can write

$$\frac{1}{X}\frac{\partial^2 X}{\partial x^2} = k_x^{\,2} \tag{9.24}$$

$$\frac{1}{Y}\frac{\partial^2 Y}{\partial y^2} = k_y^{\,2} \tag{9.25}$$

$$\frac{1}{Z}\frac{\partial^2 Z}{\partial z^2} = k_z^{\,2} \tag{9.26}$$

The significance of k_x, k_y and k_z will be clear later.

Introducing $k^2 = \mu\varepsilon\omega^2$, Eqn (9.23) gives

$$k_x^{\,2} + k_y^{\,2} + k_z^{\,2} = -k^2 = -\mu\varepsilon\omega^2 \tag{9.27}$$

We can distinguish between the transverse and z direction components; letting

$$k_x{}^2 + k_y{}^2 = -k_c{}^2 \tag{9.28}$$

we have

$$k_z{}^2 = -k^2 + k_c{}^2 \tag{9.29}$$

or

$$k_z = \pm j\sqrt{(k^2 - k_c{}^2)} \tag{9.30}$$

The solution of Eqn (9.26), which gives the variation in the z direction, is, by analogy with voltage waves on a transmission line (following Eqn (6.11))

$$Z = \exp(\pm k_z z) \tag{9.31}$$

and from Eqn (9.30), when $k^2 > k_c{}^2$,

$$k_z = \pm j\beta_g \tag{9.32}$$

where

$$\beta_g = \sqrt{(k^2 - k_c{}^2)} \tag{9.33}$$

hence

$$Z = \exp(\pm j\beta_g z) \tag{9.34}$$

Then E_y is given by

$$E_y = XY\exp[j(\omega t \pm \beta_g z)] \tag{9.35}$$

where here the time dependence implicit in assuming a sinusoidal signal is included in the term $\exp(j\omega t)$.

Equation (9.35) represents two travelling waves, one in the forward and one in the backward direction, in a similar way to the waves on a transmission line (see Section 6.1).

This method of solution could be used to develop expressions for the other components in the E and H fields.

For normal propagation $k^2 > k_c{}^2$, making β_g real and k_z imaginary, and therefore $\exp(\pm j\beta_g z)$ is a phase term. However, if $k_c{}^2 > k^2$, β_g is imaginary and k_z is real, indicating attenuation. In that case there is rapid attenuation. The transition between propagation and attenuation occurs at that value of $\omega(=\omega_c)$ for which

$$k_c{}^2 = k^2$$

i.e.

$$k_c{}^2 = \mu\varepsilon\omega_c{}^2$$

or

$$\omega_c = k_c/\sqrt{(\mu\varepsilon)} \tag{9.36}$$

k_c has dimensions of 1/length and it can be defined by

$$k_c = 2\pi/\lambda_c \tag{9.37}$$

We can rewrite Eqn (9.29), i.e.

$$k_z^2 = -k^2 + k_c^2$$

as

$$-\left(\frac{2\pi}{\lambda_g}\right)^2 = -\left(\frac{2\pi}{\lambda_0}\right)^2 + \left(\frac{2\pi}{\lambda_c}\right)^2$$

or, rearranging terms,

$$\frac{1}{\lambda_0^2} = \frac{1}{\lambda_g^2} + \frac{1}{\lambda_c^2} \tag{9.2}$$

where

$$\lambda_g = \frac{2\pi}{\beta_g} \qquad \text{(cf. Eqn (6.18))}$$

and

$$\lambda_0 = \frac{2\pi}{k}$$

These wavelengths λ_0, λ_g and λ_c have been discussed in Section 9.4. Here we are not yet able to specify λ_c, because we have not considered a particular mode, but that will be done shortly.

9.9 Transverse Electric (TE) Modes

In the last section we derived some general expressions for modes in a rectangular waveguide, but we cannot go further without specifying the class of mode in which we are interested. As we saw in Section 9.7 there are two classes, TE and TM. In this section we consider TE modes in a rectangular guide, and we can use our earlier equations to set up the field equations in the waveguide; we then introduce the effect of having no component of the E field in the z direction, thus allowing a solution to Maxwell's equations to be obtained.

Returning to Eqn (9.12),

$$\nabla \times \bar{E} = -j\omega\mu\bar{H}$$

and expressing it in cartesian coordinates,

$$\bar{x}\left(\frac{\partial E_z}{\partial y} - \frac{\partial E_y}{\partial z}\right) + \bar{y}\left(\frac{\partial E_x}{\partial z} - \frac{\partial E_z}{\partial x}\right) + \bar{z}\left(\frac{\partial E_y}{\partial x} - \frac{\partial E_x}{\partial y}\right)$$
$$= -j\omega\mu(\bar{x}H_x + \bar{y}H_y + \bar{z}H_z) \tag{9.38}$$

For TE modes there is no E field in the z direction, and hence $E_z = 0$. From Eqn (9.31) the dependence of the other fields in the z direction is $\exp(\pm k_z)$, giving $\partial/\partial z = k_z$.

Using these two conditions in Eqn (9.38),

$$\bar{x}(-k_z E_y) + \bar{y}(k_z E_x) + \bar{z}\left(\frac{\partial E_y}{\partial x} - \frac{\partial E_x}{\partial y}\right)$$

$$= -j\omega\mu(\bar{x}H_x + \bar{y}H_y + \bar{z}H_z)$$

Equating x, y and z components,

$$k_z E_y = j\omega\mu H_x \tag{9.39}$$

$$k_z E_x = -j\omega\mu H_y \tag{9.40}$$

$$\frac{\partial E_y}{\partial x} - \frac{\partial E_x}{\partial y} = -j\omega\mu H_z \tag{9.41}$$

By a similar development from Eqn (9.13), additional relationships between the E and H fields can be found, viz.

$$\frac{\partial H_z}{\partial y} - k_z H_y = j\omega\varepsilon E_x \tag{9.42}$$

$$k_z H_x - \frac{\partial H_z}{\partial x} = j\omega\varepsilon E_y \tag{9.43}$$

$$\frac{\partial H_y}{\partial x} - \frac{\partial H_x}{\partial y} = 0 \tag{9.44}$$

To find the field components we can first use these equations to express the transverse field components E_x, E_y, H_x, H_y in terms of the single component in the z direction, H_z.

From Eqns (9.40) and (9.42),

$$\frac{\partial H_z}{\partial y} + \frac{k_z^2}{j\omega\mu} E_x = j\omega\varepsilon E_x$$

$$j\omega\mu \frac{\partial H_z}{\partial y} + k_z^2 E_x = -\omega^2 \mu\varepsilon E_x$$

or

$$j\omega\mu \frac{\partial H_z}{\partial y} = -(k_z^2 + \omega^2\mu\varepsilon) E_x$$

i.e.

$$E_x = -\frac{j\omega\mu}{k_z^2 + \omega^2\mu\varepsilon} \frac{\partial H_z}{\partial y}$$

$$= -\frac{j\omega\mu}{k_c^2} \frac{\partial H_z}{\partial y} \tag{9.45}$$

where

$$k_c^2 = k_z^2 + \omega^2 \mu \varepsilon$$

Substituting Eqn (9.45) back into Eqn (9.40), and rearranging terms gives

$$H_y = \frac{k_z}{k_c^2} \frac{\partial H_z}{\partial y} \tag{9.46}$$

In a similar way we can find expressions for the other transverse components:

$$E_y = \frac{j\omega\mu}{k_c^2} \frac{\partial H_z}{\partial x} \tag{9.47}$$

and

$$H_x = \frac{k_z \partial H_z}{k_c^2 \partial x} \tag{9.48}$$

Thus in order to find the transverse field components we only need to determine H_z. It can be found as follows.

By analogy with the discussion on E_y in Section 9.8, H_z must be a solution of

$$\nabla^2 H_z + \mu\varepsilon\omega^2 H_z = 0 \tag{9.49}$$

and in the same way it will have the form

$$H_z = XY \exp\left[j(\omega t \pm \beta_g z)\right] \tag{9.50}$$

We remember that X and Y must each satisfy second-order differential equations, so they must each have two solutions. Two satisfactory solutions are

$$X = A \sin jk_x x + B \cos jk_x x \tag{9.51}$$

$$Y = C \sin jk_y y + D \cos jk_y y \tag{9.52}$$

Then the full solution for H_z is

$$H_z = (A \sin jk_x x + B \cos jk_x x)(C \sin jk_y y + D \cos jk_y y) \exp\left[j(\omega t \pm \beta_g z)\right] \tag{9.53}$$

The unknowns in this expression (A, B, C, D, k_x, k_y) are found from either the boundary or the initial conditions. The boundary conditions imposed by the walls of the guide are as follows:

(i) there can be no normal component of magnetic field at the guide walls;
(ii) there can be no tangential component of electric field at the guide walls.

Condition (i) means that

$$\frac{\partial H_z}{\partial x} = 0 \quad \text{at} \quad x = 0, a$$

and

$$\frac{\partial H_z}{\partial y} = 0 \quad \text{at} \quad y = 0, b$$

thus making $A = C = 0$ in Eqn (9.53), leaving

$$H_z = H_1 \cos jk_x x \cos jk_y y \exp(\pm j\beta_g z) \tag{9.54}$$

(omitting, but not forgetting, the time dependent term $\exp[j\omega t]$).

$H_1 = BD$, and is determined by the initial field applied to the guide. If we use this value of H_z in Eqn (9.47) we have

$$E_y = \frac{\omega\mu k_x H_1}{k_c^2} \sin jk_x x \cos jk_y y \exp(\pm j\beta_g z)$$

and from condition (ii) $E_y = 0$ at $x = 0$, a, which is satisfied by making

$$jk_x = m\pi/a \tag{9.55}$$

By a similar argument, consideration of E_x imposes the relationship

$$jk_y = n\pi/b \tag{9.56}$$

Then, substituting for jk_x and jk_y from Eqns (9.55) and (9.56) into Eqn (9.54) gives

$$H_z = H_1 \cos\frac{m\pi}{a} x \cos\frac{n\pi}{b} y \exp(\pm j\beta_g z) \tag{9.57}$$

where m and n can be any integer and indicate the number of half-sinusoidal variations of field in the x and y directions, respectively, i.e. they refer to the TE_{mn} mode.

By inserting Eqn (9.57) into Eqns (9.45), (9.46) and (9.48), the full set of field components for the TE_{mn} mode are obtained, i.e.

$$E_x = \frac{j\omega\mu n\pi H_1}{bk_c^2} \cos\frac{m\pi x}{a} \sin\frac{n\pi y}{b} \exp(\pm j\beta_g z) \tag{9.58}$$

$$E_y = -\frac{j\omega\mu m\pi H_1}{ak_c^2} \sin\frac{m\pi x}{a} \cos\frac{n\pi y}{b} \exp(\pm j\beta_g z) \tag{9.59}$$

$$E_z = 0 \tag{9.60}$$

$$H_x = -\frac{j\beta_g m\pi H_1}{ak_c^2} \sin\frac{m\pi x}{a} \cos\frac{n\pi y}{b} \exp(\pm j\beta_g z) \tag{9.61}$$

$$H_y = -\frac{j\beta_g n\pi H_1}{bk_c^2} \cos\frac{m\pi x}{a} \sin\frac{n\pi y}{b} \exp(\pm j\beta_g z) \tag{9.62}$$

$$H_z = H_1 \cos\frac{m\pi x}{a} \cos\frac{n\pi y}{b} \exp(\pm j\beta_g z) \tag{9.63}$$

where, from Eqns (9.27) and (9.28) as before,

$$k_c{}^2 = \mu\varepsilon\omega^2 + k_z{}^2$$

From Eqn (9.28)

$$k_c{}^2 = -(k_x{}^2 + k_y{}^2)$$

Substituting for k_x and k_y from Eqns (9.55) and (9.56), respectively,

$$k_c{}^2 = \left[\left(\frac{m\pi}{a}\right)^2 + \left(\frac{n\pi}{b}\right)^2\right]$$

Therefore

$$\mu\varepsilon\omega^2 + k_z{}^2 = \left(\frac{m\pi}{a}\right)^2 + \left(\frac{n\pi}{b}\right)^2$$

From $\omega = 2\pi f$, $\quad \omega^2 = \left(\frac{2\pi c}{\lambda_0}\right)^2 \quad$ and $\quad c = 1/\sqrt{(\mu\varepsilon)}$

Therefore

$$\mu\varepsilon\omega^2 = \left(\frac{2\pi}{\lambda_0}\right)^2$$

and

$$k_z{}^2 = (j\beta_g)^2 = -\left(\frac{2\pi}{\lambda_g}\right)^2$$

Hence

$$\left(\frac{2\pi}{\lambda_0}\right)^2 - \left(\frac{2\pi}{\lambda_g}\right)^2 = \left(\frac{m\pi}{a}\right)^2 + \left(\frac{n\pi}{b}\right)^2$$

Comparing this equation with Eqn (9.6)

$$\left(\frac{2\pi}{\lambda_c}\right)^2 = \left(\frac{m\pi}{a}\right)^2 + \left(\frac{n\pi}{b}\right)^2$$

Therefore

$$\frac{1}{\lambda_c{}^2} = \left(\frac{m}{2a}\right)^2 + \left(\frac{n}{2b}\right)^2 \tag{9.64}$$

This expression, which for the TE_{10} mode gives $\lambda_c = 2a$ as in Section 9.4, allows us to find the cut-off wavelength for any TE mode, given the cross-sectional dimensions of the waveguide. We can see that λ_c will vary with the size of guide, and therefore with its aspect ratio a/b.

9.10 Transverse Magnetic (TM) Modes

As we noted in Section 9.7, the other class of modes in a rectangular waveguide is that in which the magnetic field has no component in the z direction (i.e. $H_z = 0$). Modes of this type are called Transverse Magnetic because the magnetic field is contained entirely in the xy plane. The general mode designation is TM_{mn} where m and n have the same meaning as in Section 9.7.

By an analysis similar to that used in considering TE modes, the six field components of the TM_{mn} mode are obtained:

$$E_x = \frac{j\beta_g m\pi E_1}{a k_c^2} \cos\frac{m\pi x}{a} \sin\frac{n\pi y}{b} \exp(\pm j\beta_g z) \tag{9.65}$$

$$E_y = \frac{j\beta_g n\pi E_1}{b k_c^2} \sin\frac{m\pi x}{a} \cos\frac{n\pi y}{b} \exp(\pm j\beta_g z) \tag{9.66}$$

$$E_z = E_1 \sin\frac{m\pi x}{a} \sin\frac{n\pi y}{b} \exp(\pm j\beta_g z) \tag{9.67}$$

$$H_x = \frac{j\omega\varepsilon n\pi E_1}{b k_c^2} \sin\frac{m\pi x}{a} \cos\frac{n\pi y}{b} \exp(\pm j\beta_g z) \tag{9.68}$$

$$H_y = -\frac{j\omega\varepsilon m\pi E_1}{a k_c^2} \cos\frac{m\pi x}{a} \sin\frac{n\pi y}{b} \exp(\pm j\beta_g z) \tag{9.69}$$

$$H_z = 0 \tag{9.70}$$

Equations (9.64) and (9.33) for the cut-off wavelength and the phase constant are also valid for TM modes. If we examine Eqn (9.64) again, and express it in terms of cut-off frequency rather than wavelength, it becomes

$$f_c = c\left[\left(\frac{m}{2a}\right)^2 + \left(\frac{n}{2b}\right)^2\right]^{1/2} \tag{9.71}$$

and it is clear that there is a dependence on a and b as well as on m and n. Taking as an example $b = 2a$ we can show the relationship between the cut-off frequencies of several modes on the simple diagram of Fig. 9.5. The cut-off

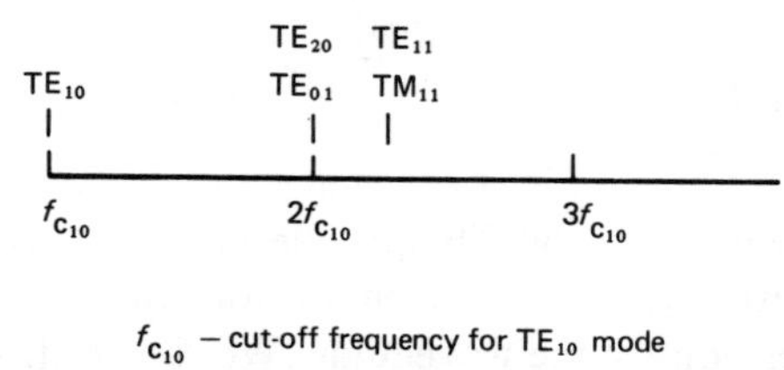

Fig. 9.5 Cut-off frequencies for low-order rectangular modes.

values are given relative to that for the TE_{10} mode, and it can now be seen why it is called the fundamental. Some modes (e.g. TE_{11}, TM_{11}) have the same value of f_c; they are called degenerate because propagation that starts in one mode can easily change (i.e. degenerate) into the other. Such things as bends, roughness in the guide or poorly fitted flanges can cause this mode conversion and because waveguide detectors are mode dependent it results in a loss of power.

9.11 Field Equations for the Fundamental TE_{10} Mode

Putting $m = 1$, $n = 0$ into Eqns (9.58) to (9.63) and adjusting the coefficients, we have the following field components:

$$E_x = 0 \tag{9.72}$$

$$E_y = E_1 \sin\frac{\pi x}{a} \exp(\pm j\beta_g z) \tag{9.73}$$

$$E_z = 0 \tag{9.74}$$

$$H_x = \frac{\beta_g}{\omega\mu} E_1 \sin\frac{\pi x}{a} \exp(\pm j\beta_g z) \tag{9.75}$$

$$H_y = 0 \tag{9.76}$$

$$H_z = \frac{jak_c^2 E_1}{\omega\mu\pi} \cos\frac{\pi x}{a} \exp(\pm j\beta_g z) \tag{9.77}$$

These are the equations of the field patterns considered in Section 9.9 and which are shown in Fig. 9.2. As well as being the fundamental, this is also the simplest rectangular waveguide mode; there is only one half-sinusoidal variation of electric field across the guide in the x direction and none in the y direction. Similar points of phase down the guide are separated by λ_g.

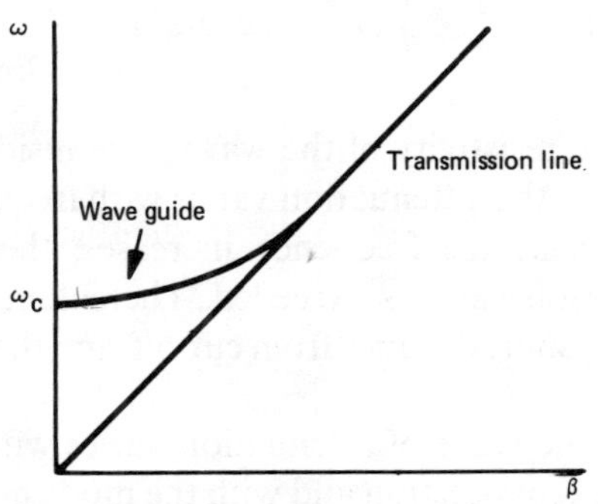

Fig. 9.6 Rectangular waveguide ω–β diagram.

From Eqns (9.72) to (9.77) the field patterns for the TE_{10} mode can be drawn, giving the diagram we discussed earlier (Fig. 9.2). The introduction of a cut-off frequency implies that the phase constant β_g is frequency dependent, unlike the phase constnat β discussed in Chapter 6 on transmission lines. The variation of β_g with frequency can be shown on an $\omega-\beta$ diagram. Figure 9.6 shows a typical characteristic for rectangular waveguide with β_g falling to zero (no propagation) when $\omega=\omega_c$, the cut-off frequency.

At high frequencies the $\omega-\beta$ diagram is linear. For comparison, the $\omega-\beta$ characteristic of a transmission line is also shown. The equations relating ω and β are

for waveguide $$\omega\sqrt{(\mu\varepsilon)}=\sqrt{(k_c^{\,2}+\beta_g^{\,2})} \qquad (9.78)$$

where, for a given guide, μ, ε and k_c are constant, and

for transmission line $$\omega\sqrt{(\mu\varepsilon)}=\beta \qquad (9.79)$$

9.12 Attenuation in Rectangular Waveguide

In the above analysis we have assumed implicitly that the walls of the guide are perfect conductors. Although the conductivity may be very high, there will be some loss at the waveguide walls caused by two factors. First, the finite conductivity of the wall material gives rise to ohmic loss and, second, surface roughness, which becomes more important as the frequency increases. The second cause can be reduced at the cost of a high-quality polished finish on the inside wall surface. The ohmic loss is a function of the wall itself and therefore it is intrinsic once the wall material has been chosen.

An expression for the attenuation along a rectangular air-filled waveguide, due to ohmic loss, can be obtained by finding the wall currents and then calculating the power loss into the guide wall. In this way the attenuation is shown[5,7] to be given by

$$\alpha=\frac{R_s}{120\pi b}\frac{2\pi}{\lambda_0\beta_g}\left[1+2\frac{b}{a}\left(\frac{f_c}{f}\right)^2\right] \quad \text{nepers/m} \qquad (9.80)$$

where R_s is the surface resistivity of the waveguide inside wall.

Figure 9.7 shows how this attenuation varies with frequency. Each curve has a characteristic shape; as the frequency increases, there is a rapid fall in attenuation as the cut-off value is exceeded. Thereafter, the attenuation falls gently to a minimum, a short distance from cut-off, and then gradually increases with frequency.

It can be seen that the value of attenuation varies with the size and aspect ratio of the waveguide cross-section and with the mode number, as well as with the guide conductivity.

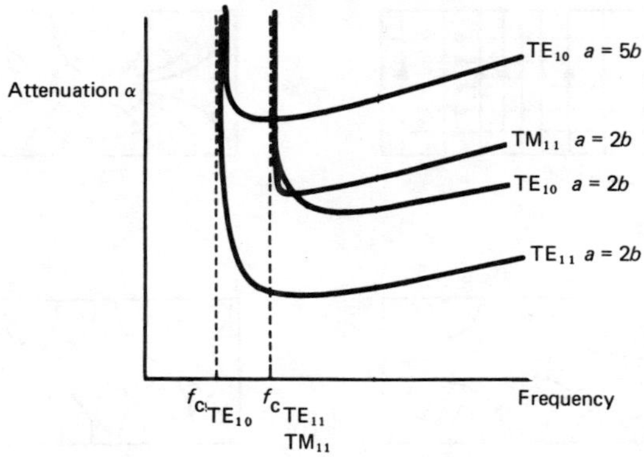

Fig. 9.7 Rectangular waveguide attenuation characteristics.

9.13 Evanescent Modes

Although we have discussed the cut-off mechanism as an effect that occurs suddenly as the frequency is reduced, in fact there is some propagation below cut-off, but very little. Below cut-off the propagation coefficient is real, and therefore the wave attenuates exponentially with distance from the source. The modes that are present in the region of the source end of the waveguide when it is operating below cut-off are called evanescent modes.

For transmission purposes, evanescent modes cannot be used, but there are some applications where the rapid attenuation can be useful. Because the forward wave attenuates very rapidly, there is little or no reflection from a short circuit or other junction placed in the guide, and any reflected wave is also heavily attenuated. This means that the reflected wave can be ignored allowing the guide input impedance to appear to be purely reactive, a characteristic that finds application in a number of measuring techniques.

9.14 Rectangular Mode Patterns

We discussed the existence of higher-order modes in an earlier section, but the variation of fields in both the x and y directions makes the shapes of the field patterns for the general mn mode difficult to visualize.

In Fig. 9.8 we present patterns for some of the lower-order modes alongside diagrams showing methods of exciting them. Notice that the E and H field lines cross at right-angles, that the magnetic field lines do not terminate on the walls of the guide but form closed loops, and that the positions of maximum field in the cross-section of the guide vary with the mode number.

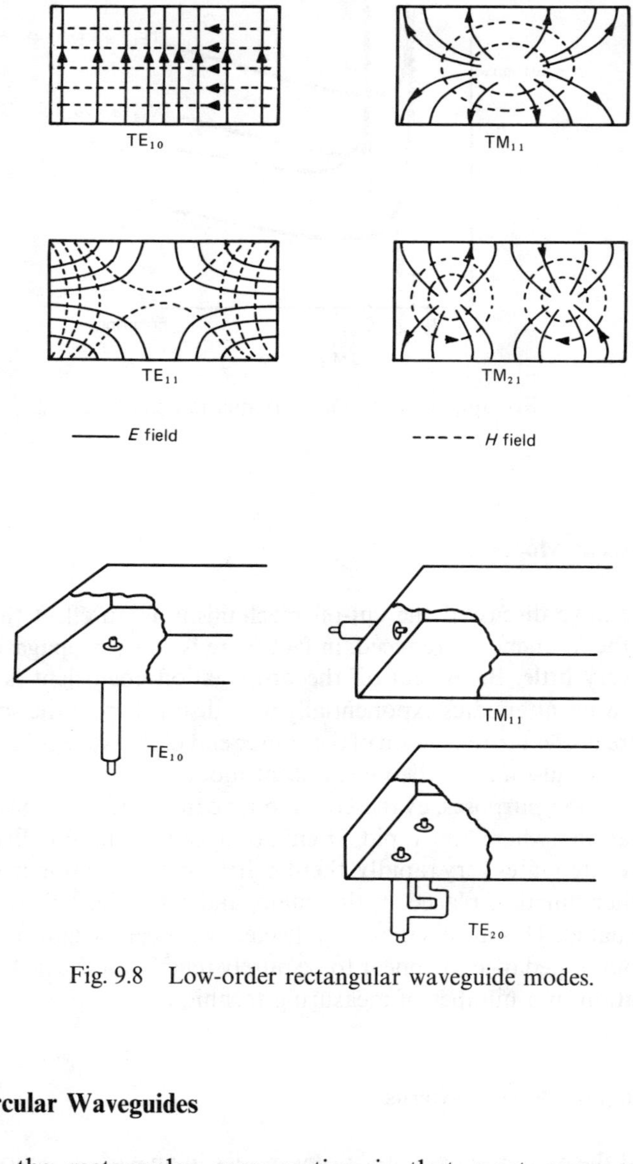

Fig. 9.8 Low-order rectangular waveguide modes.

9.15 Circular Waveguides

Although the rectangular cross-section is that most commonly used in waveguides, there are several others that are used from time to time, some of which are shown in Fig. 9.9. The field patterns can be found by solving Maxwell's equations and applying the appropriate boundary conditions, which usually means that a coordinate system allied to the shape of the cross-section must be introduced. As an example, we shall look briefly at the solution for a circular waveguide, which is used in several special devices that need the mode-pattern symmetry which is characteristic of some of the low-order modes.

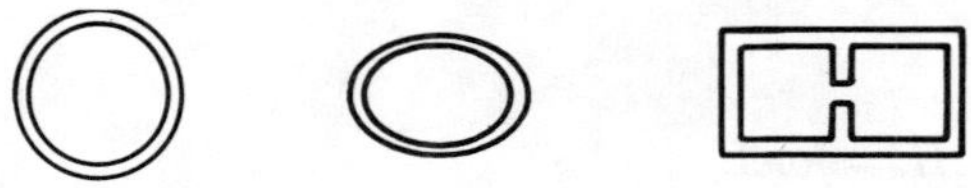

Fig. 9.9 Waveguide cross-sections.

It is not necessary to go through all the steps we followed earlier; we can assume that the space inside the guide has the same effect on Maxwell's equations as in the rectangular waveguide case.

Starting with Eqns (9.12) and (9.13),

$$\nabla \times \bar{E} = -j\omega\mu\bar{H} \tag{9.12}$$

$$\nabla \times \bar{H} = j\omega\varepsilon\bar{E} \tag{9.13}$$

we can use a similar argument to that of Section 9.7 Equation (9.12), expressed in cylindrical coordinates, which has the form

$$\bar{r}\frac{1}{r}\left[\frac{\partial E_z}{\partial \theta} - \frac{\partial (rE_\theta)}{\partial z}\right] + \bar{\theta}\left[\frac{\partial E_r}{\partial z} - \frac{\partial E_z}{\partial r}\right] + \bar{z}\frac{1}{r}\left[\frac{\partial (rE_\theta)}{\partial r} - \frac{\partial E_r}{\partial \theta}\right]$$

$$= -j\omega\mu[\bar{r}H_r + \bar{\theta}H_\theta + \bar{z}H_z] \tag{9.81}$$

Separating the three components, and equating left and right sides of Eqn (9.81),

$$\frac{1}{r}\left[\frac{\partial E_z}{\partial \theta} + \frac{\partial (rE_\theta)}{\partial z}\right] = -j\omega\mu H_r \tag{9.82}$$

$$\frac{\partial E_r}{\partial z} - \frac{\partial E_z}{\partial r} = -j\omega\mu H_\theta \tag{9.83}$$

$$\frac{1}{r}\left[\frac{\partial (rE_\theta)}{\partial r} - \frac{\partial E_r}{\partial \theta}\right] = -j\omega\mu H_z \tag{9.84}$$

As before, we will assume that the z dependence of the fields is $\exp(k_z)$, and therefore we can replace $\partial/\partial z$ in Eqns (9.82) and (9.83) by k_z. Then

$$\frac{1}{r}\frac{\partial E_z}{\partial \theta} - k_z E_\theta = -j\omega\mu H_r \tag{9.85}$$

$$k_z E_r - \frac{\partial E_z}{\partial r} = -j\omega\mu H_\theta \tag{9.86}$$

$$\frac{1}{r}\left[\frac{\partial (rE_\theta)}{\partial r} - \frac{\partial E_r}{\partial \theta}\right] = -j\omega\mu H_z \tag{9.87}$$

Similarly, starting with Eqn (9.13), the following additional equations are developed:

$$\frac{1}{r}\frac{\partial H_z}{\partial \theta} - k_z H_\theta = j\omega\varepsilon E_r \tag{9.88}$$

$$k_z H_r - \frac{\partial H_z}{\partial r} = j\omega\varepsilon E_\theta \tag{9.89}$$

$$\frac{1}{r}\frac{\partial}{\partial r}(rH_\theta) - \frac{1}{r}\frac{\partial H_r}{\partial \theta} = j\omega\varepsilon E_z \tag{9.90}$$

Again, two families of modes are possible, transverse electric (TE) and transverse magnetic (TM).

9.16 Circular Waveguide (TE) Modes

By definition, the z component of electric field is zero, and putting $E_z = 0$ into Eqns (9.85) to (9.86) reduces them to

$$k_z E_\theta = j\omega\mu H_r \tag{9.91}$$

$$k_z E_r = -j\omega\mu H_\theta \tag{9.92}$$

$$\frac{1}{r}\frac{\partial}{\partial r}(rE_\theta) - \frac{1}{r}\frac{\partial E_r}{\partial \theta} = -j\omega\mu H_z \tag{9.93}$$

$$\frac{1}{r}\frac{\partial H_z}{\partial \theta} - k_z H_\theta = j\omega\varepsilon E_r \tag{9.94}$$

$$k_z H_r - \frac{\partial H_z}{\partial r} = j\omega\varepsilon E_\theta \tag{9.95}$$

$$\frac{1}{r}\frac{\partial}{\partial r}(rH_\theta) - \frac{1}{r}\frac{\partial H_r}{\partial \theta} = 0 \tag{9.96}$$

The only z component is H_z, and by algebraic manipulation of the above equations the other components can be obtained in relation to it. For example, from Eqn (9.88),

$$H_\theta = -\frac{1}{k_z}\left[j\omega\varepsilon E_r - \frac{1}{r}\frac{\partial H_z}{\partial \theta}\right]$$

and putting that into Eqn (9.92) gives

$$E_r = \frac{j\omega\mu}{k_z^{\,2}}\left[j\omega\varepsilon E_r - \frac{1}{r}\frac{\partial H_z}{\partial \theta}\right]$$

Hence

$$E_r = \frac{-j\omega\mu}{r(k_z^{\,2} + \omega^2\varepsilon\mu)}\frac{\partial H_z}{\partial \theta}$$

or, if k_c is introduced from Eqn (9.45),

$$E_r = \frac{-j\omega\mu}{k_c^2}\frac{1}{r}\frac{\partial H_z}{\partial \theta} \qquad (9.97)$$

By a similar sort of argument the other three components are

$$E_\theta = \frac{j\omega\mu}{k_c^2}\frac{\partial H_z}{\partial r} \qquad (9.98)$$

$$H_r = \frac{k_z}{k_c^2}\frac{\partial H_z}{\partial r} \qquad (9.99)$$

$$H_\theta = \frac{k_z}{k_c^2}\frac{1}{r}\frac{\partial H_z}{\partial \theta} \qquad (9.100)$$

Thus if we can find an expression for H_z we can then find the values of the other components.

By analogy with the arguments leading to Eqn (9.20), H_z must satisfy the vector wave equation

$$\nabla^2 H_z = -\omega^2 \varepsilon\mu H_z \qquad (9.101)$$

which has a solution of the form

$$H_z = R\theta Z \qquad (9.102)$$

where R, θ and Z refer to the three cylindrical coordinates of Fig. 9.10.

In cylindrical coordinates

$$\nabla^2 H_z = \frac{\partial^2 H_z}{\partial r^2} + \frac{1}{r}\frac{\partial H_z}{\partial r} + \frac{1}{r^2}\frac{\partial^2 H_z}{\partial \theta^2} + \frac{\partial^2 H_z}{\partial z^2} \qquad (9.103)$$

Putting Eqns (9.103) and (9.102) into Eqn (9.101), and remembering that the z component solution is

$$Z = H_z \exp(k_z z) \qquad (9.104)$$

Eqn (9.101) becomes

$$\theta Z\frac{\partial^2 Z}{\partial r^2} + \frac{\theta Z}{r}\frac{\partial R}{\partial r} + \frac{RZ}{r^2}\frac{\partial^2 \theta}{\partial \theta^2} + R\theta k_z^2 Z = k^2 R\theta Z$$

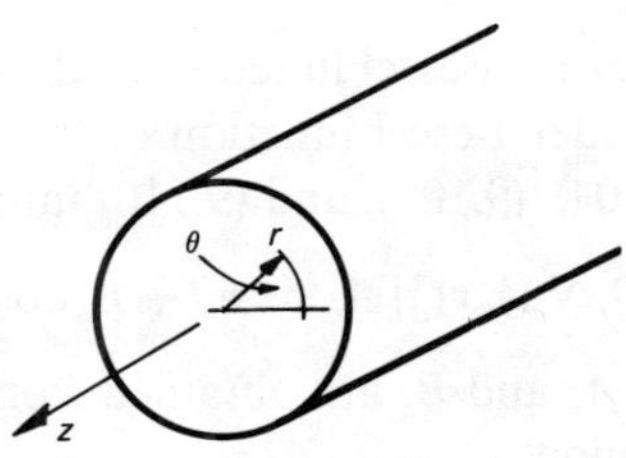

Fig. 9.10 Circular waveguide.

and dividing by $R\theta Z$

$$\frac{1}{R}\frac{\partial^2 R}{\partial r^2}+\frac{1}{Rr}\frac{\partial R}{\partial r}+\frac{1}{\theta r^2}\frac{\partial^2\theta}{\partial\theta^2}+k_z{}^2=-k^2$$

or

$$\frac{1}{R}\frac{\partial^2 R}{\partial r^2}+\frac{1}{r}\frac{1}{R}\frac{\partial R}{\partial r}+\frac{1}{r^2}\frac{1}{\theta}\frac{\partial^2\theta}{\partial\theta^2}+k_c{}^2=0 \tag{9.105}$$

where, as before [Eqn (9.29)],

$$k^2+k_z{}^2=k_c{}^2$$

From this we can establish two independent expressions:

$$\frac{r^2}{R}\frac{\partial^2 R}{\partial r^2}+\frac{r}{R}\frac{\partial R}{\partial r}+k_c{}^2=n^2 \tag{9.106}$$

and

$$\frac{1}{\theta}\frac{\partial^2\theta}{\partial\theta^2}=-n^2 \tag{9.107}$$

where n^2 is a constant. The solution for θ is straightforward:

$$\frac{\partial^2\theta}{\partial\theta^2}+n^2\theta=0 \tag{9.108}$$

Hence

$$\theta=A_n\sin n\theta+B_n\cos n\theta \tag{9.109}$$

However, the solution for R is less simple. Equation (9.106).

$$\frac{\partial^2 R}{\partial r^2}+\frac{1}{r}\frac{\partial R}{\partial r}+\left(k_c{}^2-\frac{n^2}{v^2}\right)R=0$$

does not have a simple closed-form solution. It appears in several problems and it is called the Bessel equation. Solutions are given in tabular form and designated by the symbols $J_n(k_c r)$ and $N_n(k_c r)$. Thus the solution of Eqn (9.105) is written

$$R=C_nJ_n(k_cr)+D_nN_n(k_cr) \tag{9.110}$$

$J_n(k_cr)$ is called the nth order Bessel function of the first kind, of argument k_cr, and $N_n(k_cr)$ is the nth order Bessel function of the second kind.

Combining Eqns (9.104), (9.109), and (9.110) into Eqn (9.102),

$$H_z=[C_nJ_n(k_cr)+D_nN_n(k_cr)][A_n\sin n\theta+B_n\cos n\theta]\exp(k_zz) \tag{9.111}$$

The coefficients C_n, D_n, A_n and B_n are obtained from the boundary conditions or initial fields in the guide.

The nth order Bessel function of the second kind, $N_n(k_c r)$, increases to infinity as r approaches zero. The effect of having this term in the expression for H_z would therefore be to have a field of infinite magnitude along the axis of the waveguide. This is not possible and hence, on physical grounds, $D_n = 0$, leaving

$$H_z = C_n J_n(k_c r)[A_n \sin n\theta + B_n \cos n\theta] \exp(k_z z) \tag{9.112}$$

The terms giving the variation of field with θ depend on the positioning of the $\theta = 0$ reference. By choosing it correctly, A_n can be made zero without imposing additional constraints on the field. Then

$$H_z = B_n \cos n\theta \cdot C_n J_n(k_c r) \exp(k_z z)$$

or

$$H_z = H_1 \cos n\theta \cdot J_n(k_c r) \exp(k_z z) \tag{9.113}$$

where H_1 is determined by the initial field applied to the guide. This solution is not yet complete: H_z must satisfy the conditions at the internal surface of the waveguide. At the guide wall, radius a, $J_n(k_c a)$ must be zero since H_z must be zero there. In addition, also at the guide walls, radius a, $\partial H_z/\partial r = 0$ using boundary conditions (ii) following Eqn (9.53), which implies that E_θ is zero from Eqn (9.113), [Eqn (9.98)]. For $\partial H_z/\partial r = 0$,

$$\frac{\partial}{\partial r}[H_1 J_n(k_c a) \cos n\theta \exp(\pm j\beta_g z)] = 0$$

giving

$$\frac{\partial}{\partial r} J_n(k_c a) = 0 \tag{9.114}$$

$J_n(k_c a)$ does not have a regular periodicity so there is no simple expression for k_c. Instead, tabulated values of the roots of Eqn (9.114) must be used. These depend on n and they are denoted by p'_{nm}. The prime indicates the roots of the first derivative of $J_n(k_c a)$, n is the Bessel function order, and m is the number of the root or the number of zeros of the field [i.e. of $J'_n(k_c a)$] in the range 0 to a.

So now we have developed a full solution of H_z, i.e.

$$H_z = H_1 J_n\left(\frac{p'_{nm}}{a}, r\right) \cos n\theta \exp(\pm j\beta_g z) \tag{9.115}$$

Returning to the other field components, Eqn (9.115) can be used in Eqns (9.97), (9.99) and (9.100), giving

$$E_r = -\frac{j\omega\mu}{k_c^{\,2}} \frac{1}{r} \frac{\partial}{\partial\theta}\left[H_1 J_n\left(\frac{p'_{nm}}{a}, r\right) \cos n\theta \exp(\pm j\beta_g z)\right]$$

$$= \frac{j\omega\mu n}{r k_c^{\,2}} H_1 J_n\left(\frac{p'_{nm}}{a}, r\right) \sin n\theta \exp(\pm j\beta_g z) \tag{9.116}$$

$$E_\theta = \frac{j\omega\mu}{k_c^{\ 2}} \frac{\partial}{\partial r}\left[H_1 J_n\left(\frac{p'_{nm}}{a}, r\right) \cos n\theta \exp(\pm j\beta_g z)\right]$$

$$= \frac{j\omega\mu}{k_c^{\ 2}} H_1 \left(\frac{\partial}{\partial r}\left[J_n\left(\frac{p'_{nm}}{a}, r\right)\right]\right) \cos n\theta \exp(\pm j\beta_g z) \tag{9.117}$$

$$H_r = \frac{k_z}{k_c^{\ 2}} H_1 \left[\frac{\partial}{\partial r} J_n\left(\frac{p'_{nm}}{a}, r\right)\right] \cos n\theta \exp(\pm j\beta_g z) \tag{9.118}$$

$$H_\theta = \frac{k_z}{k_c^{\ 2}} \frac{1}{r} n H_1 J_n\left(\frac{p'_{nm}}{a}, r\right) \sin n\theta \exp(\pm j\beta_g z) \tag{9.119}$$

As before, we have obtained the transverse field components in terms of the single longitudinal component H_z.

9.17 Circular Mode Cut-off Frequency

The nomenclature used to describe the transverse electric fields in a circular waveguide is similar to that in a rectangular waveguide. TE_{nm} designates the nm mode. The cut-off wavelength in the circular case is given by

$$\lambda_{c_{nm}} = \frac{2\pi a}{p'_{nm}} \tag{9.120}$$

and the cut-off frequency by

$$f_{c_{nm}} = \frac{p'_{nm} C}{2\pi a} \tag{9.121}$$

Values of p'_{nm} are given in Table 9.1. We can see from the table that the lowest

Table 9.1 First ten TE modes in circular waveguide, and corresponding values of p'_{nm}

Mode	p'_{nm}
TE_{11}	1.841
TE_{21}	3.054
TE_{01}	3.832
TE_{31}	4.201
TE_{41}	5.318
TE_{12}	5.331
TE_{51}	6.416
TE_{22}	6.706
TE_{02}	7.016
TE_{32}	8.015

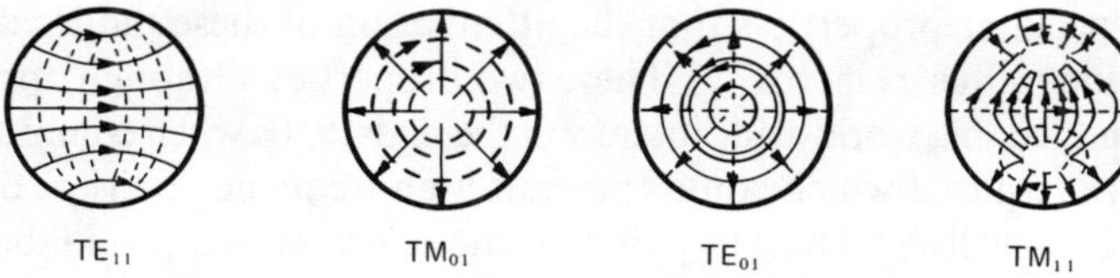

Fig. 9.11 Low-order circular waveguide modes.

cut-off frequency is for the TE_{11} mode, when $p'_{11} = 1.841$. This is called the fundamental or dominant mode. The field patterns across the guide for some of the lowest order modes are shown in Fig. 9.11.

9.18 Attenuation in Circular Waveguide

The waveguide walls have finite conductivity and so give rise to some ohmic loss. The attenuation of a TE_{nm} wave caused by the wall losses is given by

$$\alpha = \frac{R_s}{a\eta} \frac{1}{(1-(f_c/f)^2)^{1/2}} \left[\left(\frac{f_c}{f} \right)^2 + \frac{n^2}{(p'_{nm})^2 - n^2} \right] \tag{9.122}$$

We can see that $n = 0$ is a special case, for then

$$\alpha = \frac{R_s}{a\eta} \frac{1}{\sqrt{[1-(f_{c01}/f)^2]}} \tag{9.123}$$

and the variation of α with frequency is negative, i.e. the attenuation decreases as the frequency increases.

Typical curves of TE mode attenuation are shown in Fig. 9.12 and the difference between the TE_{0m} modes and the others is clear.

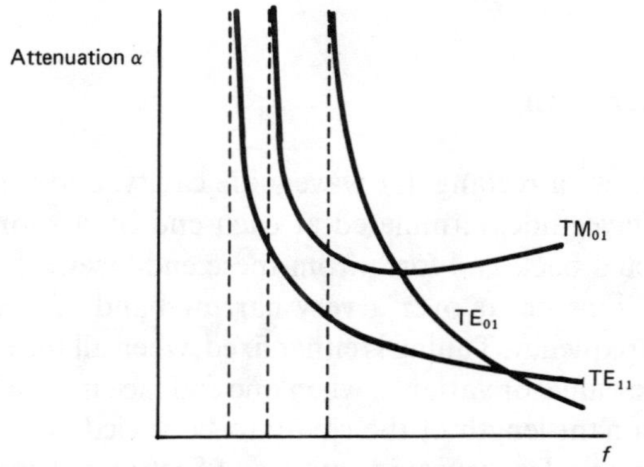

Fig. 9.12 Circular waveguide attenuation characteristics.

Another peculiar property is that the attenuation of these modes decreases as a, the guide radius, is increased. These two properties, although appreciated for some time, led to a world-wide interest in the mid-1950s when communication systems were required which could transmit high-frequency carriers over long distances with low loss.[3] The TE_{01} circular mode was chosen, and laboratories in the UK, USA and Japan spent the next 20 years in developing production systems. Several were used successfully. From our discussion there would appear to be few problems, but in practice there are several which produced major difficulties. The TE_{01} mode is not the fundamental, and the waveguide is overmoded because of its large diameter. These factors mean that several modes can propagate down the guide, and at bends and corners, which are essential parts of a realistic system, the TE_{01} mode can be easily converted into other modes, with a consequent loss of power. These other spurious modes may have different propagation velocities from the TE_{01}, and that can have the effect of degrading the shape of the signal if there is a double mode conversion – from TE_{01} to spurious mode at one point along the guide and then back from spurious mode to TE_{01} at a further point. The performance of the guide can be improved by using something more complex than a simple circular cylindrical tube. By putting a helix of fine wire, set in resin, on the inner surface, many of the spurious modes can be filtered out since the helix will attenuate all non-TE_{0m} modes. Of course, the helix adds significantly to the cost of making the guide, but if the spacing between repeaters along the guide can be reduced thereby, a net saving may result. The precision quality of the waveguide requires that it is inserted into some protective jacket and laid in such a way that there are no tight bends. The latter considerations again add to the cost. Until the early 1970s it seemed that, on balance, circular waveguide would be an attractive long-haul transmission medium. Then optical fibres came along, and although for a few years both systems were developed in parallel in the large telecommunications laboratories, it soon became clear that fibre systems (see Chapter 12) could be better, and now their performance is such that the waveguide approach is no longer viable.

9.19 Rectangular Cavity

Figure 9.13(a) shows a rectangular waveguide cavity, and we can recognise it as a length of waveguide, terminated at each end by a short-circuiting face. Waves are reflected back and forth from these end faces. The cavity is highly resonant; oscillations occur over a very narrow band of frequencies centred on the resonant frequency. Tuning is either fixed, when all the cavity dimensions can have just one value, or variable, when one end face is attached to a plunger, allowing the resonant length of the cavity to be varied.

Any of the rectangular waveguide modes of Section 9.7 can exist within the cavity. We shall consider in detail the fundamental TE_{10} mode. The boundary

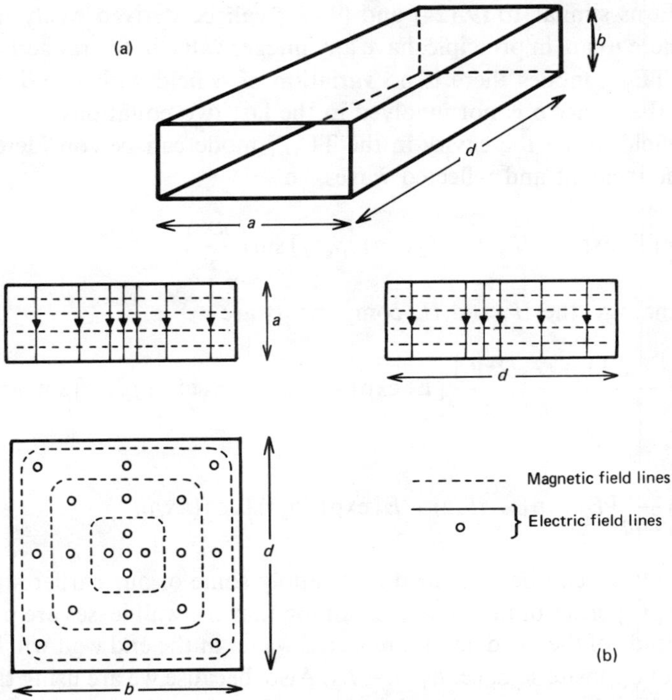

Fig. 9.13 (a) Rectangular waveguide cavity; (b) field pattern for TE_{101} mode.

conditions used in the development of the solutions to the field equations within the waveguide also apply in the cavity, which puts an additional constraint on the field pattern. The conducting walls at $z=0$ and $z=d$ (the end walls) impose the condition that at these walls the tangential electric field, E_y, must be zero. Thus d must be an integral number of half guide-wavelengths long. The lowest order cavity mode is when $d=\lambda_g/2$. Then the E field is zero at the end walls and has one maximum, in the centre of the cavity.

If the width of the guide is a and the free-space wavelength is λ_0, the first resonant length can be found from Eqn (9.2) to be

$$d=\frac{\lambda_g}{2}=\frac{\lambda_0}{2[1-(\lambda_0/2a)^2]^{1/2}} \tag{9.124}$$

The resonant frequency corresponding to a cavity of width a and length d in this fundamental (TE_{101}) mode is found directly from

$$f_{r_{101}}=c/\lambda_{0_{101}}$$

where

$$\lambda_{0_{101}}=2ad/(a^2+d^2)^{1/2} \tag{9.125}$$

Clearly, Eqns similar to (9.124) and (9.125) can be derived easily for TE_{10p} modes where p can, in principle, have any integer value in the range $1 \leqslant p \leqslant \infty$.

In the TE_{10p} modes there is no variation of E field with y and hence the height of the cavity b is not involved in the last two equations.

The E field inside the cavity in the TE_{101} mode can be considered as the sum of the incident and reflected waves, i.e.

$$Ey = [E_1 \exp(-j\beta_g z) + E_2 \exp(j\beta_g z)] \sin\left(\frac{\pi x}{a}\right) \tag{9.126}$$

and, by analogy, the H_x and H_z components are

$$H_x = -\frac{\sqrt{[1-(\lambda_0/2\alpha)^2]}}{\eta}[E_1 \exp(-j\beta_g z) - E_2 \exp(j\beta_g z)] \sin(\pi x/a) \tag{9.127}$$

$$H_z = \frac{\lambda_0}{2a\eta}[E_1 \exp(-j\beta_g z) - E_2 \exp(j\beta_g z)] \cos(\pi x/a) \tag{9.128}$$

These equations can be simplified if we apply some of our earlier statements about the properties of the cavity. Assuming that the wall losses are negligible, the magnitude of the incident and reflected waves at the end wall will be equal, although of opposite sign, i.e. $E_1 = -E_2$. Also, because we are using the TE_{101} mode, in which $E_y = 0$ at $z = 0$ and $z = d$, and there is only one E field maximum between the end walls, $\beta_g = \pi/d$. Then Eqn (9.126) can be written

$$\begin{aligned} E_y &= E_1\left[\exp\left(-j\frac{\pi}{d}z\right) - \exp\left(j\frac{\pi}{d}z\right)\right]\sin\frac{\pi x}{a} \\ &= -2jE_1 \sin\frac{\pi}{d}z \sin\frac{\pi}{a}x \end{aligned} \tag{9.129}$$

and using this expression in Eqn (9.127) and (9.128) we have

$$H_x = \frac{2E_1}{\eta}\frac{\lambda}{2d}\sin\frac{\pi x}{a}\cos\frac{\pi z}{d} \tag{9.130}$$

$$H_z = -\frac{\lambda}{2a}\frac{2E_1}{\eta}\cos\frac{\pi x}{a}\sin\frac{\pi z}{d} \tag{9.131}$$

where

$$\eta = \sqrt{(\mu/\varepsilon)}$$

From these three expressions the field patterns inside the cavity can be plotted for the fundamental mode. Figure 9.13(b) shows a representation of these fields.

Higher-order Modes

We can extend our discussion of the TE_{101} mode to the more general case of the TE_{mnp} mode where m, n and p represent the number of half-sinusoidal variations of field in the x, y and z directions, respectively. In theory, each parameter can have an infinite number of values.

The rectangular waveguide TE_{mn} mode has an H_z field of the form

$$H_z = [A\exp(-j\beta_g z) + B\exp(j\beta_g z)]\cos\frac{m\pi}{a}x\cos\frac{n\pi}{b}y \tag{9.132}$$

As in our discussion on the TE_{101} mode, we can assume that $A = -B$ and that $H_z = 0$ at $z = 0, d$ making $\beta_g = p\pi/d$.

Then Eqn (9.133) can be written

$$H_z = -2jA\sin\frac{p\pi}{d}z\cos\frac{m\pi}{a}x\cos\frac{n\pi}{b}y \tag{9.133}$$

For the general TE_{mnp} mode the relationship between the transverse field components E_x, E_y, H_x and H_y, are the same as those given by Eqns (9.58) to (9.62) in Section (9.9).

Substituting Eqn (9.133) into those equations gives

$$H_x = \frac{2jA}{k_c^2}\left(\frac{p\pi}{d}\right)\left(\frac{m\pi}{a}\right)\sin\frac{m\pi}{a}x\cos\frac{n\pi}{b}y\cos\frac{p\pi}{d}z \tag{9.134}$$

$$H_y = \frac{2jA}{k_c^2}\left(\frac{p\pi}{d}\right)\left(\frac{n\pi}{b}\right)\cos\frac{m\pi}{a}x\sin\frac{n\pi}{b}y\cos\frac{p\pi}{d}z \tag{9.135}$$

$$E_x = \frac{2A\omega\mu}{k_c^2}\left(\frac{n\pi}{b}\right)\cos\frac{m\pi}{a}x\sin\frac{n\pi}{b}y\sin\frac{p\pi}{d}z \tag{9.136}$$

$$E_y = -\frac{2A\omega\mu}{k_c^2}\left(\frac{m\pi}{a}\right)\sin\frac{m\pi}{a}x\cos\frac{n\pi}{b}y\sin\frac{p\pi}{d}z \tag{9.137}$$

where as before

$$k_c^2 = \left(\frac{m\pi}{a}\right)^2 + \left(\frac{n\pi}{b}\right)^2 \tag{9.138}$$

For the TE_{mnp} cavity mode

$$\beta_g = \frac{p\pi}{a} = \left[\left(\frac{2\pi}{\lambda_0}\right)^2 - k_c^2\right]^{1/2} \tag{9.139}$$

Re-arranging Eqns (9.138) and (9.139) we can derive the resonant free-space wavelength for a cavity of dimensions a, b, d operating in the TE_{mnp} mode, viz.

$$\lambda_0 = 2\pi\Bigg/\left[\left(\frac{m\pi}{a}\right)^2 + \left(\frac{n\pi}{b}\right)^2 + \left(\frac{p\pi}{d}\right)^2\right]^{1/2} \tag{9.140}$$

It is worth noting that, for a particular resonant frequency, the resonant length of the cavity, d, depends also on the cross-sectional dimensions of the waveguide appropriate to the mode of interest [Eqns (9.125) and (9.140)].

One of the main applications of multimode cavities is in microwave heating;[4] for example, in microwave ovens, where the object is to heat the sample uniformly, there is a need for a uniform field over the central space to be occupied by the sample, and this uniformity can be achieved by a combination of high-order modes and the use of a mode-stirrer, which is a metal plate, made to change its position and so vary the resonant length, and therefore the position of field maxima in the oven.

9.20 Cavity $Q - \mathrm{TE}_{101}$ Mode

As in all resonant systems, there are some losses, and eventually the oscillations in the cavity will decay. It is shown in other texts[5,6] that the cavity Q, which is a measure of the sharpness of the resonance curve, can be expressed as a ratio of the energy stored in the cavity to the mean loss at the walls, or

$$Q = \omega_r \frac{W_s}{W_L} = \omega_r \times \frac{\text{energy stored in the cavity}}{\text{average power loss}} \tag{9.141}$$

where ω_r is the angular resonant frequency of the cavity.

W_s is the energy stored in the fields within the cavity. It is shared between the electric and magnetic fields, and it can therefore be found by calculating the maximum of one, when the other is zero.

The energy in an electric field is $\frac{1}{2}\varepsilon E^2$ integrated over the volume of the cavity:

$$W_s = \frac{\varepsilon}{2} \int_0^d \int_0^b \int_0^a |E_y|^2 \,\mathrm{d}x\,\mathrm{d}y\,\mathrm{d}z \tag{9.142}$$

since there is only one field component, E_y in the TE_{101} mode.

Substituting for E_y from Eqn (9.129),

$$W_s = \frac{\varepsilon}{2} \int_0^d \int_0^b \int_0^a \frac{\varepsilon}{2} E_0{}^2 \sin^2 \frac{\pi x}{a} \sin^2 \frac{\pi z}{d} \,\mathrm{d}x\,\mathrm{d}y\,\mathrm{d}z$$

where E_0 depends on the initial field in the cavity.

Remembering to replace $\sin^2\theta$ by $\frac{1}{2}[1 - \cos 2\theta]$, this expression is integrated and reduces to

$$W_s = \frac{\varepsilon abd E_0{}^2}{8} \tag{9.143}$$

The average power loss at the walls is $\frac{1}{2}I^2 R_s$ where I is the peak wall current induced by the fields in the cavity, and R_s is the resistivity of the walls. The current on any wall is directly related to the tangential magnetic field at the wall surface. This field will vary with position in the cavity and therefore integration is required over each of the wall surfaces. The symmetry of the system means that opposite walls will have identical losses, hence

$$W_L = \frac{R_s}{2}\left\{2\int_0^b\int_0^a |H_x|^2_{z=0}\,dx\,dy + 2\int_0^d\int_0^b |H_z|^2_{x=0}\,dy\,dz + 2\int_0^d\int_0^a (|H_x|^2 + |H_z|^2)\,dx\,dz\right\} \tag{9.144}$$

i.e.

W_L = loss in (end walls + side walls + top and bottom walls)

If we now develop Eqn (9.144) using

$$H_x = -j\frac{E_0}{\eta}\frac{\lambda_g}{2d}\sin\frac{\pi x}{a}\cos\frac{\pi z}{d}$$

and

$$H_z = j\frac{E_0}{\eta}\frac{\lambda_g}{2a}\cos\frac{\pi x}{a}\sin\frac{\pi z}{d}$$

We have

$$W_L = \frac{R_s\lambda_g{}^2 E_0{}^2}{8\eta^2}\left[\frac{ab}{d^2} + \frac{bd}{a^2} + \frac{1}{2}\left(\frac{a}{d} + \frac{d}{a}\right)\right] \tag{9.145}$$

Then, from Eqns (9.143) and (9.145) and after some rearranging, the Q factor becomes

$$Q = \frac{\pi\eta}{R_s}\left[\frac{2b(a^2 + d^2)^{3/2}}{ad(a^2 + d^2) + 2b(a^3 + d^3)}\right] \tag{9.146}$$

9.21 Circular Cavity

By analogy with the previous section we can see that a length of circular waveguide with short-circuited ends will also form a cavity. The separation of the ends, d in Fig. 9.14(a), determines its resonant frequency; resonance occurs when d is an integral number of half-guide-wavelengths long. For each of the circular waveguide modes (Section 9.17) there is a corresponding family of cavity resonating modes, e.g. the fundamental TE_{11} circular waveguide mode will give rise to the TE_{11l} family of cylindrical cavity modes, where l indicates the number of half-guide-wavelengths in the length of the cavity. The fundamental cavity mode corresponds to $d = \lambda_g/2$, and it is designated the TE_{111} mode. In terms of the dimensions of the cavity, the resonant frequency of the TE_{111} mode is given by

$$f_{111} = \frac{c}{2\pi}\left[\left(\frac{\pi}{d}\right)^2 + \left(\frac{p'_{11}}{a}\right)^2\right]^{1/2}$$

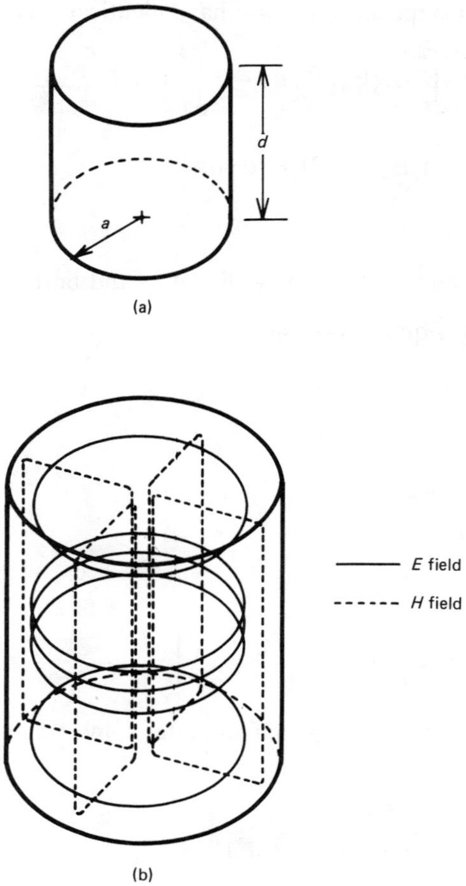

Fig. 9.14 (a) Circular waveguide cavity; (b) field pattern for TE_{011} mode.

where p'_{11} is the Bessel function coefficient of Section 9.17.

Another mode which has several applications is the TE_{011}. Its field pattern, shown in Fig. 9.14(b), produces no current flow between the cylindrical wall and the cavity ends. This means that the cavity can be constructed to allow the distance d to be varied (in applications where the cavity is to be tuned, as in a wavemeter), without imposing very rigorous constraints on its fabrication; good contact between the wall and the end faces is not essential.

The Q factor for TE modes has a form which is a function of the dimensions of the cavity, the mode number, and the skin depth of the wall material, δ_s.

From Collin,[5] for the TE_{mnl} mode

$$Q = \frac{\lambda_0}{2\pi\delta_s}\left[1-\left(\frac{n}{p'_{nm}}\right)^2\right]\left[(p'_{nm})^2+\left(\frac{l\pi a}{d}\right)^2\right]^{3/2} \Bigg/ \left[(p'_{nm})^2+\frac{2a}{d}\left(\frac{l\pi a}{d}\right)^2+\left(1-\frac{2a}{d}\right)\left(\frac{nl\pi a}{p'_{nm}d}\right)^2\right]$$

where δ_s is related to the conductivity of the walls, σ, by

$$\delta_s = \sqrt{(2/\omega\mu\sigma)}$$

Coupling into the cavity depends on the mode required in the particular application. The aperture between the cavity and the waveguide feed has to be so placed that, if possible, only the mode of interest is generated. Some modes are degenerate and special techniques may be necessary to absorb or avoid those which are unwanted.

A full discussion on circular waveguide cavities can be found in Harvey[7] and Marcuvitz.[8]

9.22 Rectangular Waveguide Components–TE_{10} Mode

In any practical application, several components are essential to provide a useable microwave system. A typical lay-out is shown in Fig. 9.15. In this section we will discuss briefly the passive devices shown; generators have been considered separately in Chapter 8.

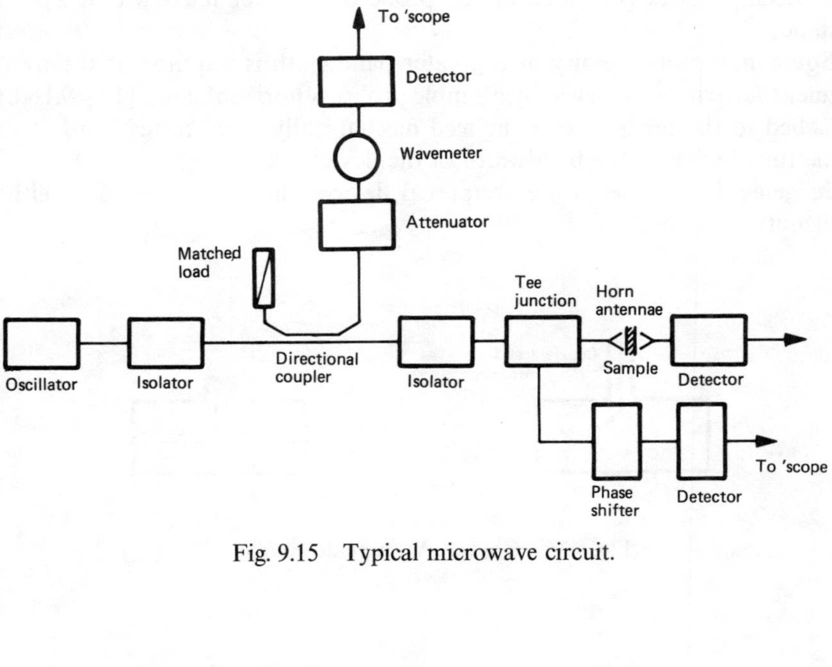

Fig. 9.15 Typical microwave circuit.

Nearly all components are designed to operate in a particular mode. In our discussion we will assume that the fundamental TE_{10} mode is being transmitted. The purpose of most passive microwave devices is self-evident from their names, but their operation is not always quite so obvious. In most cases there has been considerable analytical work done to evaluate how these devices work, and their operating performance characteristics. Here we shall not consider any analysis, and if the reader would like to explore this area, reference should be made to the specialist texts available.[5,7]

9.23 Waveguide–Coaxial Transformer

To launch a wave into the guide some mechanism is required that will transform the rf energy on the feed from the generator (often coaxial cable) into microwave energy. Alternatively, and usually at higher frequencies, the generator is mounted directly on to the waveguide and indeed may be integral with it. Examples of a coaxial–waveguide transformer and a directly coupled feed are shown in Fig. 9.16.

In Fig. 9.16(a), energy is coupled from the generator via a simple probe, the centre of a coaxial feed. The probe penetrates into the guide through a broad face, and acts like an aerial. The electric fields induced between the probe and the waveguide propagate in all directions. By placing a short-circuit behind the probe, energy reflected from the short is propagated in the forward direction, in phase with that leaving the probe. The degree of coupling between the probe and the guide depends on the depth of penetration of the probe into the guide. The guide itself acts as a filter, which causes the many spurious (often evanescent) modes generated at the probe to be attenuated within a short distance.

Sometimes more rigidity or a greater bandwidth is required at the transformer than can be achieved by a simple probe. A horizontal bar [Fig. 9.16(b)] attached to the probe makes the feed mechanically more robust, and at the same time increases the bandwidth of the device.

In general, these feeds are reciprocal devices that can be used as either transmitters or receivers.

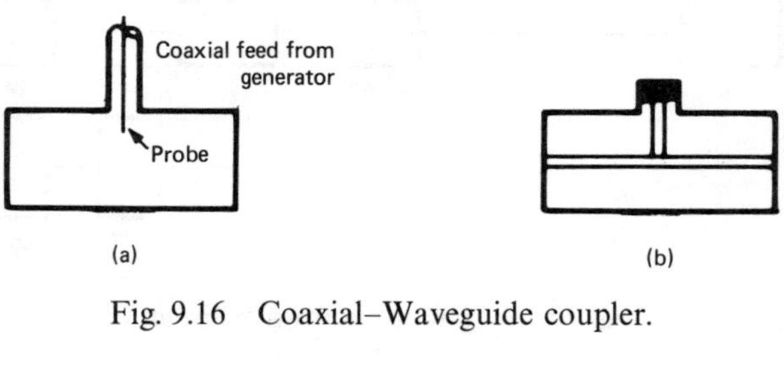

Fig. 9.16 Coaxial–Waveguide coupler.

In the above paragraphs we have looked only at electric field probes, but the TE_{10} mode can also be generated by inducing a magnetic field in the guide. This is done by placing a loop through an end wall. The current variation produced in the loop by a signal from the generator induces a magnetic field, and again the filtering action of the guide creates the conditions for propagation of the TE_{10} mode.

9.24 Attenuator

If the amplitude of the wave in the guide is too high, it must be reduced by using an attenuator, which consists of a vane made of a lossy material. The degree of attenuation depends on the strength of field at the vane; by varying the position of the vane, different levels of attentuation are possible. Two common forms of attenuator are shown in Fig. 9.17. In Fig. 9.17(a) the vane is attached to a micrometer through the sidewall of the guide. Attenuation is a minimum when the micrometer is fully extended, and the vane is against the sidewalls of the waveguide, where the electric field is at a minimum. Maximum attenuation occurs when the micrometer is screwed in and the vane lies along the axis of the guide in a position of maximum electric field. Good quality attenuators are tested in the manufacturer's laboratory and a calibration chart is issued to relate the reading on the micrometer screw to the attenuation introduced by the vane. For some applications, an attenuator is used to reduce the level of the signal in the guide to a workable value, and in that case a simple ungraduated screw will suffice. In order to avoid reflections the ends of the attenuator vane may be either stepped or tapered. The attenuator in Fig. 9.17(b) shows a different type of mechanism – often used in the high-frequency microwave, or millimetre, range. The vane has a cycloid profile and it is introduced into a slot along the centre of a broad face

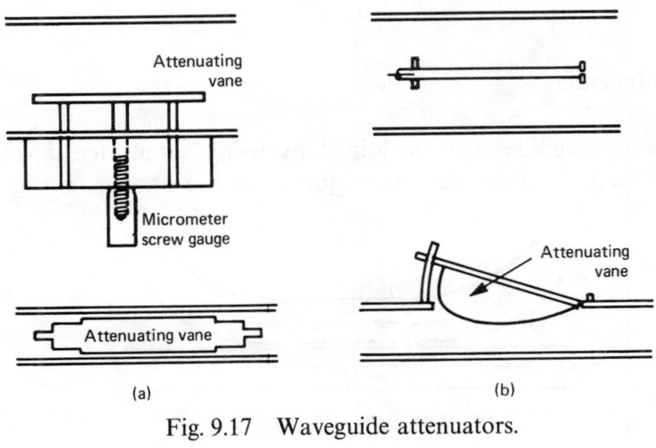

Fig. 9.17 Waveguide attenuators.

of the guide. The degree of attenuation increases with the depth of penetration into the waveguide, achieving a maximum when the vane is fully inserted.

9.25 Directional Coupler

In many systems a proportion of the field is taken from the main waveguide to be measured or sampled in some way. A directional coupler allows a known fraction of the power in the primary guide to be filtered off. It consists of a second waveguide, contiguous with the first, and coupled to it by one or more holes, which may be circular, rectangular or cruciform in shape. The degree of coupling depends on the size of the hole and its position relative to the axis of the waveguide, if it is in the broad face. A double (four-port) coupler is often used in which energy is passing from port 1 to port 2 in the primary guide. Port 3 couples out a fraction of the forward wave and port 4 couples out a similar proportion of the backward wave. In this type of coupler (Fig. 9.18) the coupling holes are separated by $\lambda_g/4$. Hence, part of the forward wave from port 1 to port 2 is coupled through hole *A* and an equal part through hole *B*. The coupled fields add and are detected at port 3. Power from the forward wave coupled through hole *B* and travelling towards port 4 is cancelled by the power from hole *A* travelling in that direction. The effect of the coupling holes is given by two parameters:

(i) Coupling factor. With matched loads at ports 2 and 3, any power at port 4 is due to that coupled from the primary guide. The coupling factor is given by $10\log(P_1/P_4)$ dB. The typical values are 3, 10, 12, 20 dB.
(ii) Directivity. The coupling holes will themselves produce reflections, and there will not be total isolation between ports 3 and 4. The directivity measures the ratio of the power in these ports for a matched load at port 2, i.e. directivity $= 10\log(P_4/P_3)$ dB and this value would not be expected to be less than 30 dB.

9.26 Tee Junctions

Rectangular waveguides can be joined by using simple tee devices. Three commonly used, fundamental mode, junctions are shown in Fig. 9.19. To

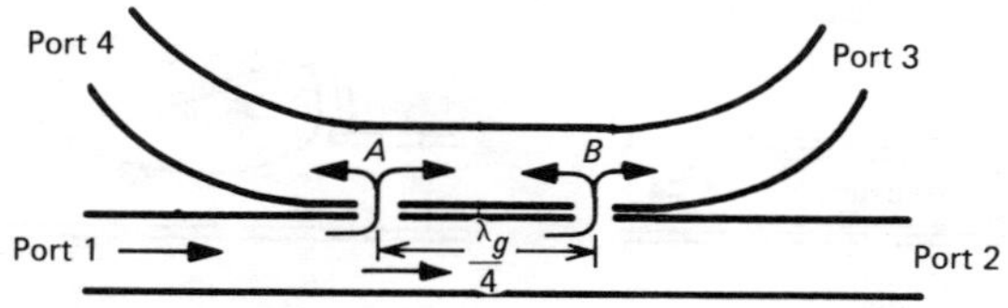

Fig. 9.18 Directional coupler.

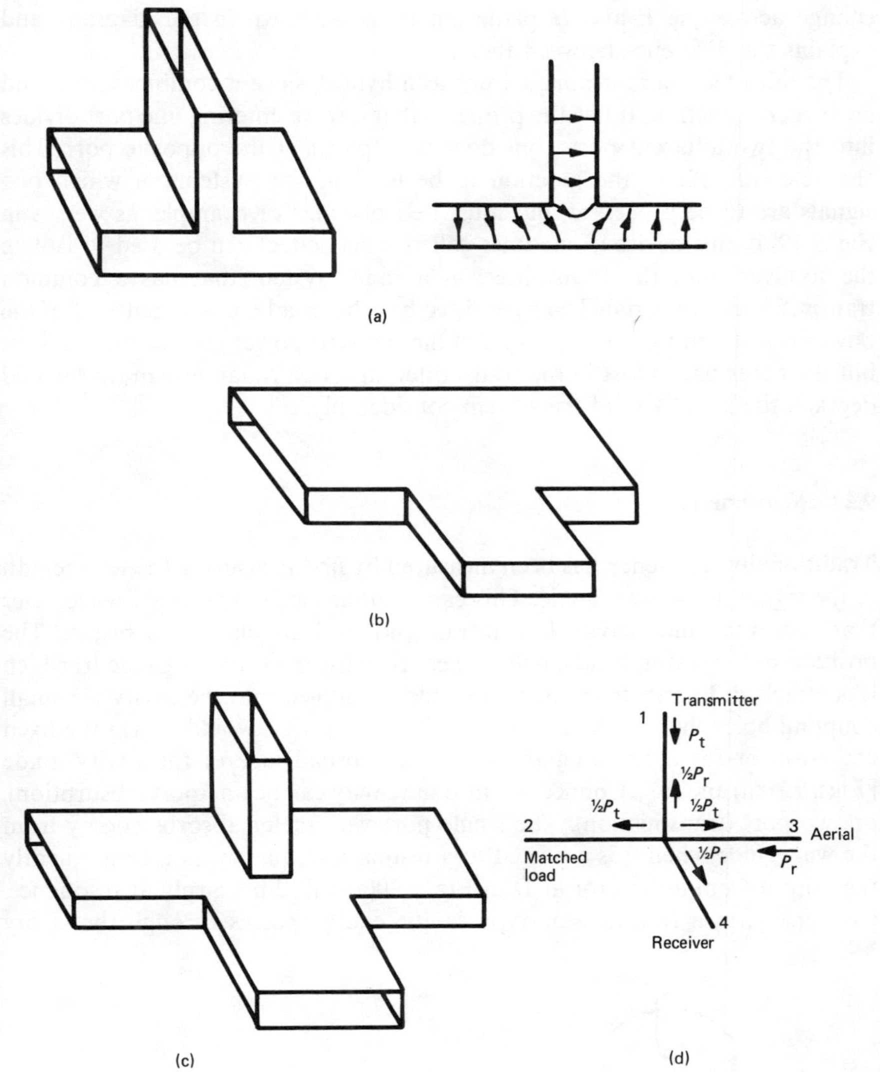

Fig. 9.19 Rectangular waveguide Tee junctions: (a) *E* plane; (b) *H* plane; (c) hybrid.

distinguish between junctions into the broad or narrow side of the guide, the first [Fig. 9.19(a)] is called an *E* plane tee, and the second [Fig. 9.19(b)] is called an *H* plane tee. Apart from the difference in geometry, these tees have different characteristics. In the *E* plane tee a wave entering port 1 divides equally into ports 2 and 3, but in phase opposition, so that at a given distance from the junction the signal out of port 2 is 180° out of phase with that from port 3. The *H* plane junction does not exhibit such phase reversal. The *E* field change across the E and H plane junctions is shown in the diagram, and explains the difference between them.

The third tee is a four-port, known as a hybrid, since it combines an E and an H plane junction. It has the property that a wave entering one port divides into the two adjacent ports, but does not appear at the opposite port. This characteristic allows the junction to be used in any systems in which two signals are to be processed, but must be isolated. For example, as we see in Fig. 9.19(d), this device (sometimes called a magic-tee) can be used to isolate the receiver from the transmitter, in a radar system that has a common transmit – receive aerial. The hybrid-tee has the disadvantage that half of the power is lost – in this example half of the received power goes to the receiver, but the other half is lost in the transmitter, and in a radar system that would degrade the sensitivity of the system considerably.

9.27 Wavemeter

Traditionally, frequency has been measured by first measuring the wavelength of the signal in the waveguide. This can be done easily by using a wavemeter that is a waveguide cavity having one end wall attached to a piston. The position of the piston can be determined from the micrometer gauge to which it is attached. Energy from the waveguide is coupled into the cavity via small coupling holes that, in the case of a cylindrical cavity, would be via the fixed end wall, or for a rectangular cavity via a broad face of the cavity guide [Figs. 9.20(a) and (b)]. Connection into the cavity can be one-port (absorbtion), or two-port (transmission). The single port wavemeter absorbs energy from the waveguide when it is adjusted to a resonant frequency, and consequently the output from a detector at D in Fig. 9.20(a) will dip sharply at resonance. Conversely, in a transmission-type cavity, energy passes through the cavity

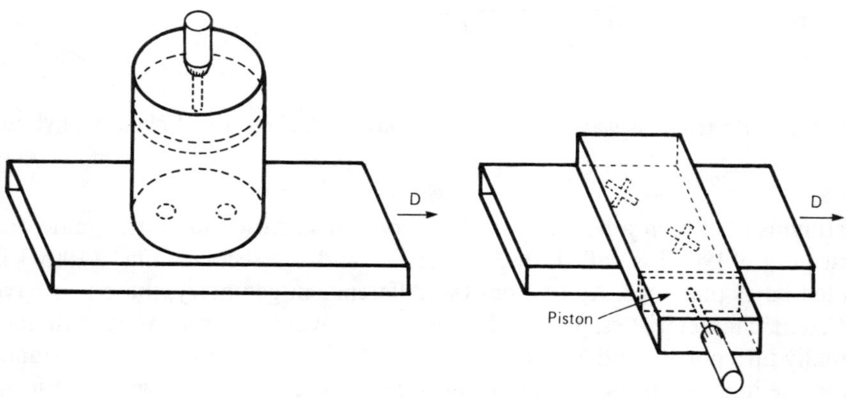

Fig. 9.20 Wavemeters: (a) cylindrical cavity; (b) rectangular cavity.

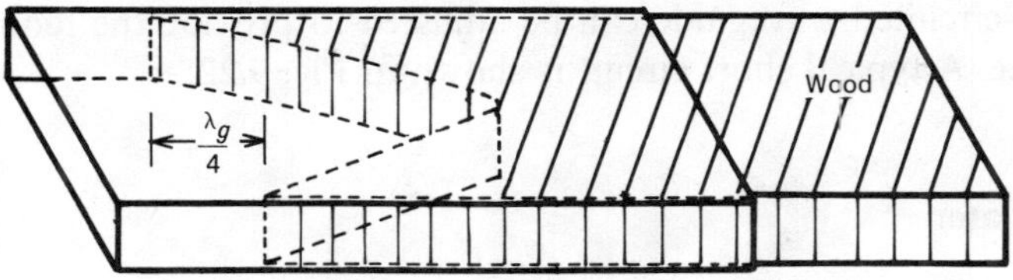

Fig. 9.21 Matched load.

most effectively at resonance; hence the detector in Fig. 9.20(b) will detect a surge of current at that frequency. In either case, the Q of the wavemeter will normally exceed 3000, and for high quality wavemeters it may be greater than 10 000.

9.28 Matched Load

There are several positions in a microwave circuit, e.g. at one side of a directional coupler, beyond a wavemeter, or at the opposite end of a sample to be measured, when a matched termination is required. A termination provides a match if it does not induce a reflected wave. Hence most matched loads operate on the principle of completely absorbing incident energy so that no reflections can take place. Figure 9.21 shows a typical matched load. The absorbing material is wood, or some other lossy dielectric, and it is shaped to offer a gradual taper to the incident wave, and therefore no reflecting surface. To further reduce the possibility of reflection, the ends of the legs of the double wedge are displaced by $\lambda_g/4$, so that any reflections from these ends cancel. The dielectric insert is several wavelengths long to ensure that any energy reflected from the distant end is absorbed on the return path.

9.29 Short Circuit

In Section 6.10 the use of stubs in transmission lines was discussed and there it was shown that the input impedance of a short-circuited stub can be altered by varying the stub length. A short circuit can be used in the same way in microwave systems. By attaching a short circuit to a plunger, the length of

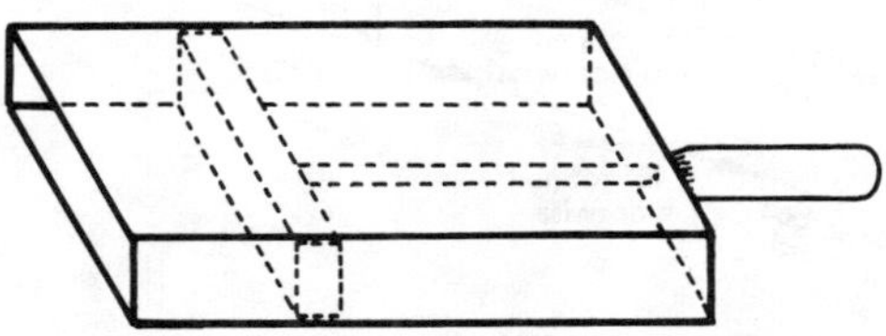

Fig. 9.22 Variable position short-circuit.

the short-circuited waveguide can be adjusted to provide the required input impedance. A typical short circuit is shown in Fig. 9.22.

9.30 Isolator

Here we are not able to develop the theoretical basis for ferrites; we will acknowledge that a ferrite is a material that, when subjected to a magnetic field, will cause a polarized electromagnetic wave passing through it to rotate. The direction of rotation is not dependent on the direction of propagation of the wave. Hence a device can be constructed such that in one direction the rotation assists in the transmission of the wave, whereas in the opposite direction the rotation blocks propagation. Reference to Fig. 9.23 will explain how this can happen. Let us assume that the ferrite and magnetic field are so chosen that the electric field vector is rotated clockwise by 45°. By using a 45° anticlockwise twist in the waveguide input before the ferrite, the rotation by the ferrite restores the wave to its original polarization and it can leave the output guide in the normal way. For a wave travelling in the opposite direction, however, the rotation through the ferrite is still clockwise, but the rotation through the 45° twist is clockwise also, resulting in the wave approaching the input guide horizontally polarized, and thus unable to pass through the system. There would be a reflection in such a case, but the resistive vane absorbs the energy in this backward wave and prevents reflections occurring. The loss in the forward direction is very small, whereas that in the backward direction might be as high as 30 dB. This device is used chiefly to isolate a generator from a system, to prevent mis-matches in the system from causing reflections which might interfere with the performance of the generator.

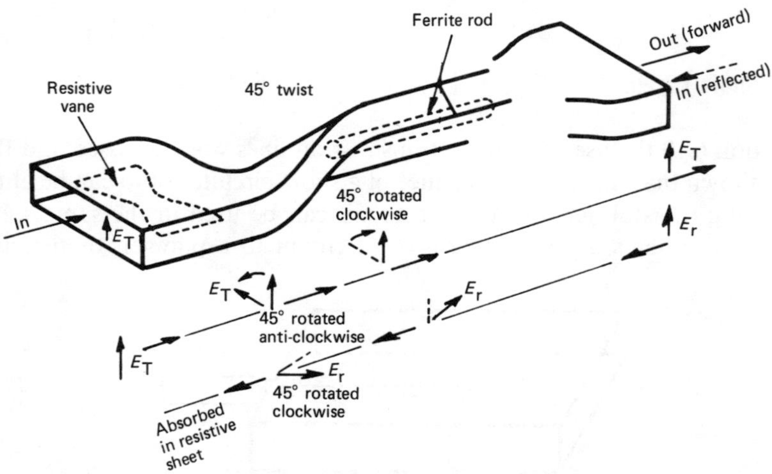

Fig. 9.23 Principles of operation of isolator.

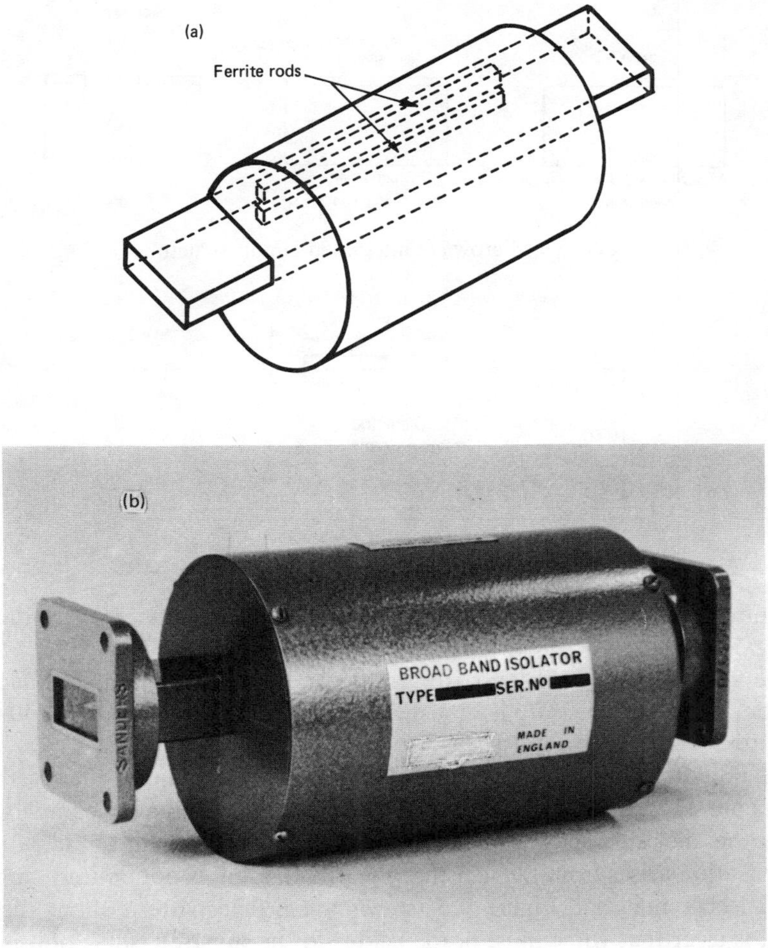

Fig. 9.24 Ferrite isolator. Reproduced from Marconi catalogue by permission of Sanders Division Marconi Instruments Ltd.

Another type of isolator is shown in Fig. 9.24. Here the magnetic field is across the guide, produced by a substantial permanent magnet.

9.31 Microstrip

The development of microwave integrated circuits, involving substrate based sources, detectors and passive components, has necessitated the development of waveguiding techniques suitable for integrated circuit applications. Some of the structures investigated for guiding waves on substrates are shown in Fig. 9.25. The most commonly used is the microstrip line and that is the

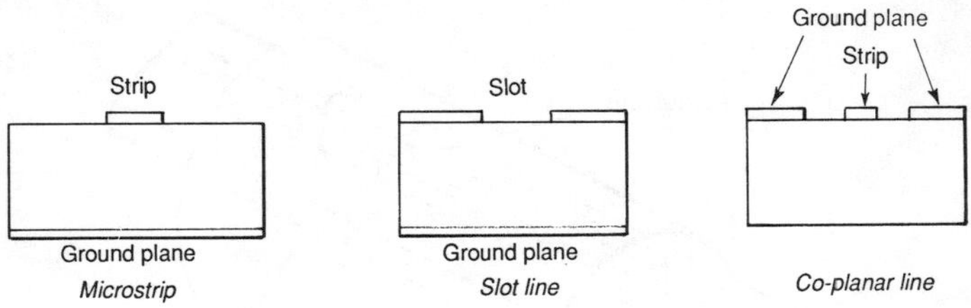

Fig. 9.25 Microwave integrated circuit structures.

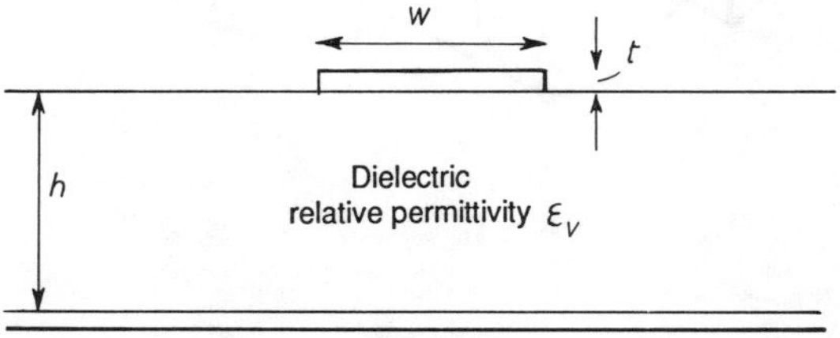

Fig. 9.26 Microstrip dimensions.

structure that we will consider here. The dimensions of interest are indicated in Fig. 9.26.

The ground plane and the strip are of good conducting material such as copper, while the substrate is of a dielectric material with low loss and high permittivity, typically $\varepsilon_r \approx 10$.

We noted earlier, in Section 9.7, that in pipe waveguides, the operating modes are TE and TM. In microstrip, the modes are extremely complex, but when the dielectric has a high permittivity they are almost TEM. The complication arises from the fact that the propagation is not entirely through the dielectric material. Figure 9.27 shows the *E* field pattern about the line and it is clear that while most of the field is in the substrate some of it is not. The analysis is therefore difficult.

Here we will not develop the underlying theory of propagation on a microstrip system; if the reader wishes to do so excellent treatments are found in references 9 and 10.

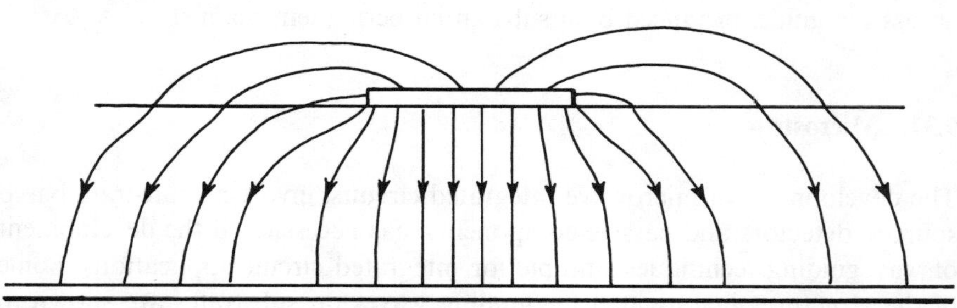

Fig. 9.27 *E*-field distribution in microstrip.

If the energy propagated along the microstrip were confined entirely to the dielectric, the analytic would be relatively straightforward, consisting of calculating the velocity along the microstrip from

$$v = c/\sqrt{\varepsilon_r}$$

and from that would follow good approximations for the other parameters of interest, in particular the capacitance and characteristic impedance.

Given that only a small amount of the energy propagates through air, the line can be modelled by a system in which all the energy travels in the substrate, but for which the dielectric permittivity is modified from its real value to an effective value ε_e. The difficulty is then to devise an expression ε_e in terms of w, h, and ε_r which can be verified by experiment. Many workers have tackled the problem of providing a suitable expression for ε_e and in the literature there is a variety of different equations available. One such equation is the following for the characteristic impedance, Z_0:

$$Z_0 = \frac{42.4}{\sqrt{(\varepsilon_r + 1)}} \ln\left\{1 + \left(\frac{4h}{w}\right)\left[\left(\frac{14 + 8/\varepsilon_r}{11}\right)\left(\frac{4h}{w}\right) + \sqrt{\left(\left(\frac{14 + 8/\varepsilon_r}{11}\right)^2\left(\frac{4h}{w}\right)^2 + \left(\frac{1 + 1/\varepsilon_r}{2}\right)\pi^2\right)}\right]\right\} \quad (9.147)$$

Clearly Eqn (9.148) is an empirical relationship. From an analytical point of view it gives an expression for Z_0 which can be used to derive other parameters. However a designer is usually required to determine the dimensions of the strip line given specified values for ε_r and Z_0. To do that Eqn (9.147) would be rearranged with w/h as a subject. The equations used for ε_r can vary with the value of w/h and therefore care must be taken in their application. Reference 11 gives one of the clearest expositions of the basic features of microstrip design.

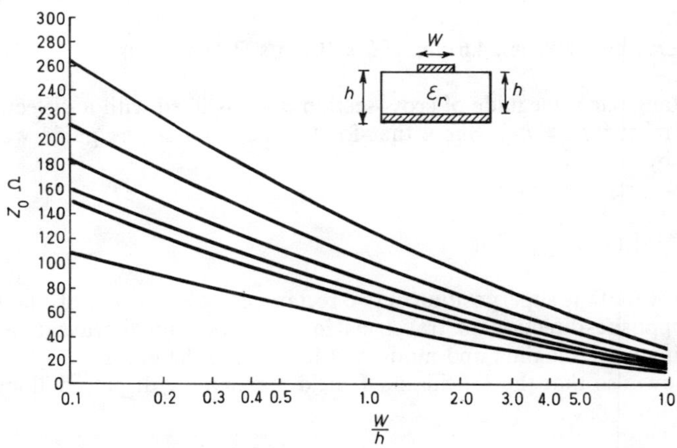

Fig. 9.28 Characteristic impedance of microstrip line.

In order to design a microstrip line it is necessary usually to use design curves of the type shown in Fig. 9.28. There it can be seen that the characteristic impedance of the microstrip is portrayed as a function of the dimensions w and h for a variety of values of dielectric permittivity. Hence the curves can be used either to find Z_0 for a particular strip line or to determine the ratio of w/h once the required characteristic impedance is known.

References

1. Bleaney, B.I. and Bleaney, B., *Electricity and Magnetism*, Oxford University Press, Oxford, 1978.
2. Paul, C.R. and Nasar, S.A., *Introduction to Electromagnetic Fields*, McGraw-Hill, Maidenhead, 1982.
3. Karbowiak, A.E., *Trunk Waveguide Communication*, Chapman & Hall, London, 1965.
4. Okress, E.C., *Microwave Power Engineering*, Vol. 2, Academic Press, London, 1968.
5. Collin, R.E., *Foundation of Microwave Engineering*, McGraw-Hill, Maidenhead, 1966.
6. Atwater, H.A., *Introduction to Microwave Theory*, McGraw-Hill, Maidenhead, 1962.
7. Harvey, A.F., *Microwave Engineering*, Academic Press, London, 1963.
8. Marcuvitz, M., *Waveguide Handbook*, McGraw-Hill, Maidenhead, 1951.
9. Gupta, H.C. and Sing, A., *Microwave Integrated Circuits*, Chapter 3, Wiley Eastern, 1974.
10. Edwards, T.C., *Foundations for Microstrip Circuit Design*, Wiley, Chichester, 1981.
11. Combes, P.F., Graffeuil, J. and Santerean, J.F., *Microwave Components, Devices and Active Circuits*, Wiley, 1987.

Problems

9.1 Find the values of $\lambda_0, \lambda_g, \lambda_c, v_{ph}$ and v_g for a rectangular waveguide of inner cross-section 2.3 cm × 1.0 cm when it carries a signal of 10 GHz in the TE_{10} mode.

Answer: 3 cm, 3.96 cm, 4.6 cm, 3.96×10^8 m/s, 2.27×10^8 m/s.

9.2 A rectangular waveguide of cross-section $a \times b$ is filled with a dielectric material of permittivity $\varepsilon = k\varepsilon_0$. Show that for the TE_{10} mode the guide wavelength is given by

$$\lambda_g = \frac{\lambda_0}{[k - (\lambda_0/\lambda_g)^2]^{1/2}}$$

9.3 (a) Show that the superposition of two rectangular TE_{10} modes of equal amplitude and opposite direction of travel (as in a short-circuited waveguide) produces standing wave maxima and minima of field along the guide.
(b) Show also that the E field and H field maxima are displaced along the guide by $\lambda_g/4$.

9.4 Calculate the minimum value of a, in a rectangular waveguide of width a, operating in the TE_{10} mode over the frequency range from 10 to 11.5 GHz if the variation in v_g is not to exceed 20% of its value at 10 GHz.

Answer: 2.34 cm.

9.5 A TE_{10} signal at 10 GHz is propagating down a waveguide of width $a = 2$ cm. What is the change in phase velocity if the width is increased to 2.4 cm?

Answer: 0.70×10^8 m/s.

9.6 Find the distance between two adjacent wave minima in a 2.3 cm × 1.0 cm rectangular air-filled waveguide which is propagating in the TE_{10} mode at the frequency at which the TE_{21} mode could just begin to propagate.

Answer: 0.799 cm.

9.7 Sketch the electric and magnetic field patterns, in both the cross-sectional and longitudinal planes of a rectangular waveguide operating in the TE_{20} mode. From the model of two interfering plane waves, find the cut-off wavelength for that mode in terms of the cross-sectional dimensions $a \times b$ ($\lambda_c = a$).

9.8 An air-filled rectangular waveguide has a TE_{10} mode cut-off frequency of 8 GHz. The signal frequency is 15 GHz, and the ratio $a/b = 2$. If the guide is now filled with perspex of relative permittivity 2.5, find the guide wavelength and phase velocity of the signal in the guide.

Would you expect higher order modes to be propagated in the dielectric-filled guide?

Answer: 1.34 cm, 2.02×10^8 m/s; yes.

9.9 A wave of frequency 9.95 GHz is travelling in the TE_{10} mode down a rectangular waveguide of width 1.5 cm. How far along the guide will the wave travel before it is attenuated to 1/e of its initial values?

Answer: 2.76 cm.

10 TELEPHONY

The telephone system is the largest integrated system in the world. Individual subscribers, separated by half the globe, can, in many cases automatically, make contact using this system. There is very rapid progress towards a complete direct-dialling international network, and it now seems very likely that the development of microcircuits, telecommunications satellites and new transmission media will allow a telephone link to be established to an individual in the remotest part of the earth via his own radio transceiver, identified by his unique world telephone number and accessed via satellite.

As a subject, telephony is usually restricted to considerations of the way in which subscribers are inter-connected, and an analysis of the traffic they generate. Traditionally, voice traffic, limited to a 3.4 kHz bandwidth, is assumed to be the nature of the signal sent over the telephone system, and we shall take that view in this chapter. However, the rapid growth of data traffic, generated by offices, banks, computers, airlines, etc., and the introduction of digital telephone systems in many countries, have raised the question of putting data over the telephone network and thus providing an integrated service. We will take a brief look at this area at the end of the chapter.

Telephony introduces several new concepts that appear strange at first, and tele-traffic theory (the study of telephone traffic and its relation to the design of switching and network systems) uses probabilistic models that need to be studied carefully in order to be understood, although the level of difficulty is well within the scope of an undergraduate statistics course, which most engineers follow.

We start with a section on traffic theory to establish a mathematical foundation. Systems were developed without the use of such theory, and indeed a mathematical analysis for many aspects of modern computer-controlled exchanges has not yet been developed, but if analysis can be applied it allows designers to make full use of their system, or an administration to predict how a manufacturer's equipment will perform.

10.1 Telephone Traffic

The use made of a telephone exchange, or a group of trunk circuits, is determined by both the rate at which calls arrive and the length of time for which they are held. The term "traffic" takes these two quantities into account. Two units of traffic are employed: the CCS (hundred-call seconds) which is used principally in the USA; and the erlang which is used in Europe and other parts of the world.

If a circuit carries one call continuously it is said to carry one erlang (= 360 CCS) of traffic. The erlang, named after the Danish pioneer of teletraffic theory, A.K. Erlang,[1] is expressed in mathematical terms as

$$A = ys \tag{10.1}$$

where A = traffic in erlangs, y = mean call arrival rate (= calls per unit time) and s = mean call holding time measured in the same time units as y.

10.2 Inter-arrival and Call-holding Times

Usually the quantities y and s in Eqn (10.1) are mean values because in practice calls arrive at random and they last for a random length of time. The associated inter-arrival and holding time distributions could be determined by taking measurements over a long period, and building up a pattern for these parameters. However, apart from being a lengthy process, this approach has the disadvantage that the results would be accurate only for the time and place of the measurements; it would not necessarily follow that such distributions were applicable in general. To have some idea of the shape of the inter-arrival time and holding time distributions is, however, most important from the traffic theorist's point of view, and if they can be approximated by well-understood mathematical models the development of a theoretical foundation to the subject can follow.

Over the years there have been many series of measurements on traffic from subscribers, and there is widespread agreement that the negative exponential distribution, with an appropriate choice of mean value, is satisfactory for both parameters. Figure 10.1 compares the negative exponential holding time curve

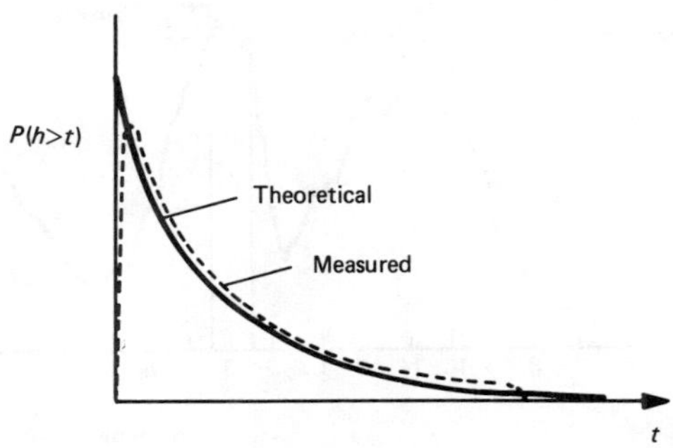

Fig. 10.1 Call holding time distribution.

with the results of a typical measurement. We notice that the ends of the distribution depart from reality, but, taking into account some of its attractive features, it is sufficiently accurate to stand as a valuable theoretical model.

The negative exponential distribution can be handled with comparative ease, but the property that is most appealing is that it is "without memory", i.e. the probability that a call arrives in an interval dt is not dependent on the time that has elapsed since the last call or, for the holding time distribution, the probability that a call releases in a given interval dt is not related to the time for which the call has been in progress. Some thought will satisfy you that analysis would be significantly complicated if the history immediately before dt had to be taken into account.

The numerical values used for the mean inter-arrival time and the mean holding time depend on the application. The mean inter-arrival time is the inverse of the mean calling intensity, and therefore depends on the level of offered traffic. For the mean holding time, telephone administrations assume some reasonable value, typically 120 or 180 seconds.

10.3 Traffic Variation

The amount of telephone traffic passing through a particular exchange depends on several factors: the nature of the subscribers on the exchange (business, residential or mixed), the time of day, the month of the year, the occurrence of holidays and catastrophies, and the tariff in operation.

Figure 10.2 gives a typical distribution of traffic on a weekday in an exchange with a preponderance of business subscribers.

We see from the diagram that the night-time traffic is low, that there is a morning peak between 9.30 am and noon and that there is an afternoon peak between 2.00 pm and 4.30 pm. There is a slight increase in traffic between

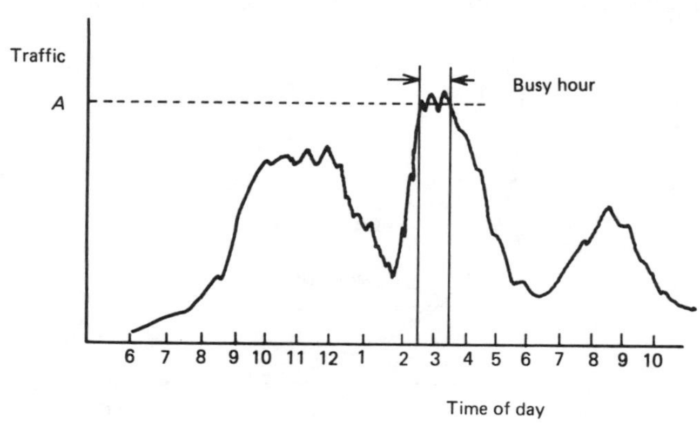

Fig. 10.2 Typical daily traffic distribution.

6.00 pm and 10.00 pm owing to cheap-rate trunk calls from the residential subscribers. The heavy afternoon peak is typical of a system that offers the business subscriber a financial incentive for leaving trunk calls to the afternoon – by charging more for morning calls. The difference between the morning and afternoon peaks could be reduced by lowering the tariff differential. From the point of view of efficient use of the system, the distribution of traffic should be as uniform as possible throughout the day, and one of the functions of having a variable tariff scheme is to induce subscribers to use parts of the day that would not be their first choice.

10.4 Busy Hour

In the last two sections we have discussed the variation of traffic and its dependence on several factors. However, the designer must be able to use some traffic value against which he can design his exchange or determine the number of trunk circuits required. The traffic level is an average, taken over several days, and over the busiest period. The period is usually one hour, and the average traffic over that hour is called the "busy hour" traffic. In Fig. 10.2 the busy hour is 2.30 pm to 3.30 pm, and the mean traffic over that period, A, would be used as the value of the busy hour traffic. In general, when traffic levels are quoted for routes or switching systems, it is implied that they are busy hour traffics. For example, if a circuit is said to carry 0.6 erlangs, it will be busy, on average, for 0.6 h (36 min) during the busy hour (not necessarily consecutive periods of time).

10.5 Lost-calls-cleared Systems

There are two major types of system used in telephony – lost-calls-cleared, and delay; although in a delay system a waiting subscriber may become impatient and leave, or the system may run out of holding positions. In a lost-calls-cleared system, if a call arrives to find there is no free path available to its destination it is cleared down. Once the system has tried to set up the call and has been unable to do so, there is no mechanism for putting the call in a queue to await a free path. Most of the theory developed in this chapter will concern lost-calls-cleared systems.

10.6 Equations of State

Calls arrive at the network and depart from it at random, and the number of calls in progress, that is the number of devices busy, will vary in a random fashion. In the following we use N to be the maximum number of devices in

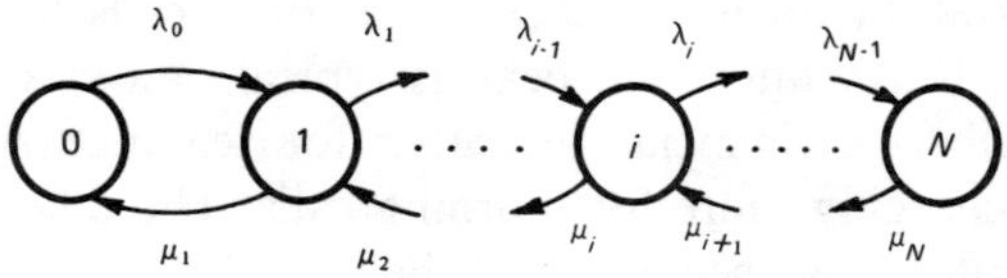

Fig. 10.3 State transition diagram.

the system, and i to indicate the number of calls in progress. Clearly, i is equal to the number of busy devices, and it is often called the state of the network. When $i = N$ the network is full and we use the term "congested" to describe it; any further calls arriving when the network is in state N will be unable to find a free device.

In developing a mathematical model to allow us to analyse the properties of the system (the most important being the probability that the network is congested), we make the following assumptions:

(i) the calls are independent;
(ii) the rate of call arrivals when the network is in state i is λ_i;
(iii) the rate of call departure when the network is in state i is μ_i;
(iv) only one event (departure or arrival) can occur at a particular instant of time.

One common method of representing the states of a system, and the transitions between them, is by the state transition diagram of Fig. 10.3. There are $N+1$ states, going in unit steps from no calls in progress (state 0) to a congested system (state N), i.e.

$$0 \leqslant i \leqslant N \qquad i = 0, 1, 2 \ldots$$

The parameters λ_i and μ_i are the rates of call arrival and departure respectively, when the network is in state i. As we saw earlier, λ_i is related directly to the traffic offered to the system, and μ_i is determined by the nature of that traffic. Our objective is to develop an expression for the probability that the system is in state i, i.e. we want to find the state probabilities.

The probability that the system is in state i at some time $t + \mathrm{d}t$ is equal to the sum of the following:

(i) the probability that the system was in state i at time t and no call arrived or departed in time $\mathrm{d}t$;
(ii) the probability that the system was in state $i-1$ at time t and a call arrived in time $\mathrm{d}t$;
(iii) the probability that the system was in state $i+1$ at time t and a call released during time $\mathrm{d}t$.

As before, we are assuming here that no more than one transition can take place in the interval $\mathrm{d}t$.

In symbols we can write this as

$$[i]_{t+\mathrm{d}t}=[i]_t(1-\lambda_i\,\mathrm{d}t-\mu_i\,\mathrm{d}t)+[i+1]_t(\mu_{i+1}\,\mathrm{d}t)+[i-1]_t(\lambda_{i-1}\,\mathrm{d}t) \quad (10.2)$$

where $[x]\equiv$ probability that the network is in state x.

Rearranging this expression

$$\frac{[i]_{t+\mathrm{d}t}-[i]_t}{\mathrm{d}t}=-(\lambda_i+\mu_i)[i]_t+\mu_{i+1}[i+1]_t+\lambda_{i-1}[i-1]_t \quad i=1,2,3,\ldots \quad (10.3)$$

$i=0$ is a special case, and

$\mu_0=0$ (no departure from state 0)

$\lambda_{-1}=0$ (state -1 does not exist)

Then Eqn (10.3) becomes

$$\frac{[0]_{t+\mathrm{d}t}-[0]_t}{\mathrm{d}t}=-\lambda_0[0]_t+\mu_1[1] \quad (10.4)$$

As $\mathrm{d}t\to 0$ Eqns (10.4) and (10.3) can be written in differential form:

$$\frac{\mathrm{d}[0]_t}{\mathrm{d}t}=-\lambda_0[0]_t+\mu_1[1]_t \quad (10.5)$$

and

$$\frac{\mathrm{d}[i]_t}{\mathrm{d}t}=-(\lambda_i+\mu_i)[i]_t+\mu_{i+1}[i+1]_t+\lambda_{i-1}[i-1]_t \quad (10.6)$$

10.7 Statistical Equilibrium

In Section 10.4 we discussed busy hour traffic. If the mean traffic over that period is used for design purposes we can assume that there will be no long-term change in the state probabilities with time. Therefore we can assume that during the busy hour the time dependence of the state probabilities is zero, and $\mathrm{d}[i]/\mathrm{d}t=0$.

Applying this condition to Eqns (10.5) and (10.6) gives

$$\lambda_0[0]=\mu_1[1] \quad (10.7)$$

$$(\lambda_1+\mu_1)[1]=\lambda_0[0]+\mu_2[2] \qquad \text{etc.}$$

In general

$$(\lambda_i+\mu_1)[i]=\lambda_{i-1}[i-1]+\mu_{i+1}[i+1] \quad (10.8)$$

Substituting successive equations gives

$$\lambda_1[1]=\mu_2[2]$$

$$\lambda_2[2]=\mu_3[3] \qquad \text{etc.}$$

The general expression becomes

$$\lambda_{i-1}[i-1] = \mu_i[i] \tag{10.9}$$

or, in words,

(probability of being in state $i-1$) × (rate of call arrivals in state $i-1$)
= (probability of being in state i) × (rate of call departure in state i)

In the long run, then, we can say that the number of calls leaving the system is equal to the number of calls arriving. If this were not so, as time passed the state of the system would gradually creep towards 0 or N and stay there.

Statistical equilibrium is a concept that lies at the heart of the models we are going to develop in the next few pages.

10.8 State Probability

We have made considerable use of the notion of state probability $[i]$ in the last section, and it is worthwhile remembering, before developing the theory further, that this probability is equivalent to the proportion of the busy hour that the system is in state i.

In Fig. 10.4 the call arrivals and departures are plotted over the busy hour for a system having $N = 5$. This diagram shows the way in which the system changes state; note that only one transition can occur at any instant.

Taking state 4 as an example:

$$[4] = \frac{t_1 + t_2 + t_3 + t_4 + t_5}{3600} = \frac{\sum \text{Times spent in state 4}}{\text{busy hour}}$$

which is the fraction of the busy hour during which exactly four devices are busy.

We can also see from that diagram that

$$[0] + [1] + [2] + [3] + [4] + [5] = 1$$

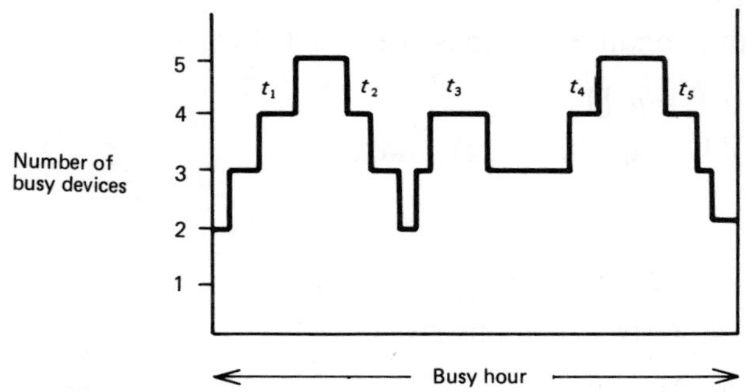

Fig. 10.4 Changes of state for a 5-device system.

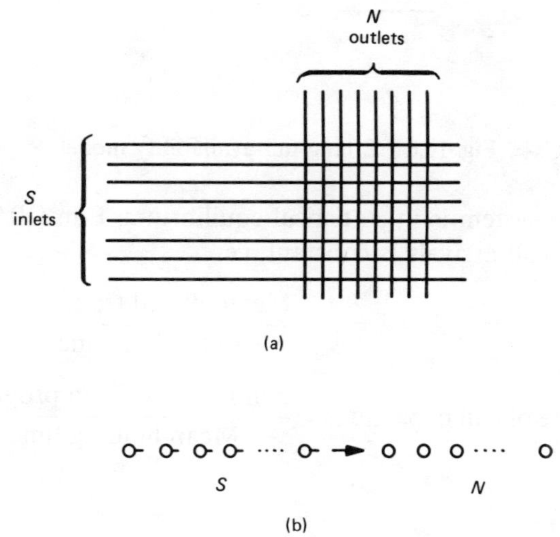

Fig. 10.5 Full availability system: (a) switch representation; (b) theoretical model.

In general, if there are N possible states

$$\sum_{i=0}^{N} [i] = 1 \tag{10.10}$$

Although this may appear to be rather obvious, it is a most important equation, as we shall see.

10.9 Full-availability Models

A full-availability system is one in which there are S sources of traffic and N outlets and any free source can seize any free outlet, regardless of the state of the system. The simplest example is a matrix switch with $S \times N$ cross-points as shown in Fig. 10.5(a). A straightforward representation, which we shall use, is shown in Fig. 10.5(b).

In order to evaluate $[i]$ for a full-availability group we need to have information about the offered traffic and how it depends on the state of the system; that in turn depends on the size of S in comparison with N.

10.10 Erlang Distribution: $S \gg N$

If the number of sources is very much greater than the number of devices, it can be assumed that, no matter how many devices are busy, the rate of call arrivals will be constant. We can therefore ignore the number of sources, and our model becomes that shown in Fig. 10.6.

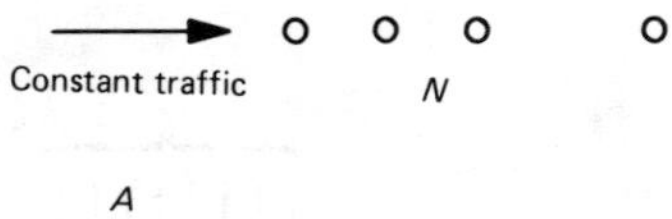

Fig. 10.6 Erlang full-availability model.

Assuming the system is in statistical equilibrium, Eqn (10.9) holds. In this case the rate of call arrivals is constant, i.e.

$$\lambda_i = \lambda = \frac{A}{s} = \frac{\text{Mean offered traffic}}{\text{Mean holding time}}$$

$$\text{and the rate of call departures} = \frac{\text{Number of calls in progress}}{\text{Mean holding time}}$$

i.e.

$$\mu_i = \frac{i}{s}$$

Then from Eqn (10.9)

$$[i]\frac{i}{s} = [i-1]\frac{A}{s}$$

Putting in values for i:

$$[1] = A[0]$$

$$[2] = \frac{A^2}{2}[0]$$

$$\vdots$$

$$[i] = \frac{A^i}{i!}[0] \tag{10.11}$$

$$\vdots$$

$$[N] = \frac{A^N}{N!}[0]$$

Using Eqn (10.10)

$$[0] + A[0] + \cdots + \frac{A^N}{N!}[0] = 1$$

Hence

$$[0] = \frac{1}{\sum_{v=0}^{N} \frac{A^v}{v!}}$$

and using this in Eqn (10.11), the probability of being in state i is

$$[i] = \frac{A^i}{i!} \bigg/ \sum_{v=0}^{N} \frac{A^v}{v!} \tag{10.12}$$

Thus we can find $[i]$ for all values of i if we know A and N.

Equation (10.12) is called the probability distribution of the states of the system, and it is a truncated form of the Poisson distribution (section 13.3).

10.11 Time Congestion

Time congestion E is defined as the proportion of the busy hour for which the network is in state N, i.e.

Time congestion $E = [N]$

It follows directly from equation (10.12) that

$$E = [N] = \frac{A^N}{N!} \bigg/ \sum_{v=0}^{N} \frac{A^v}{v!} = E_N(A) \tag{10.13}$$

This is known as Erlang's loss formula. It has been tabulated extensively and it can be evaluated simply on a computer by using the iteration

$$E_i(A) = \frac{AE_{i-1}(A)}{i + AE_{i-1}(A)} \tag{10.14}$$

and the initial condition that $E_0(A) = 1$.

10.12 Call Congestion

Another type of congestion is call congestion, B. It is the probability that a call arrives to find the system fully occupied, i.e.

$$B = \frac{\text{Number of calls arriving when the system is in state } N}{\text{Total number of calls arriving in the busy hour}}$$

In the Erlang case,

$$B = \lambda \cdot [N] \bigg/ \sum_{i=0}^{N} \lambda \cdot [i]$$

and since λ is constant

$$B = [N] \bigg/ \sum_{i=0}^{N} [i]$$

Again applying Eqn (10.10), we have

$$B = [N]$$

Thus for the Erlang case

call congestion = time congestion

This model is used extensively in practice. It is applied whenever the assumption can be made that the calling rate is independent of the number of calls in progress.

10.13 Bernoulli Distribution: $S \leqslant N$

If the number of sources is less than the number of devices, a different model has to be used because we cannot assume that the traffic intensity is indepenent of the number of calls in progress. The model must now take account of the number of sources (Fig. 10.7). The simplest way of taking account of the state of the system is to relate the traffic offered to the traffic offered per free source and the number of free sources available, i.e.

$$\lambda_i = (S - i)\alpha/s \tag{10.15}$$

where α is the traffic per free source, $S - i$ is the number of free sources and, as before, the mean holding time is s.

Applying the statistical equilibrium Eqn (10.9), once more

$$\lambda_{i-1}[i-1] = \mu_i[i]$$

$\mu_i = i/s$ as in the Erlang case, and if this is used, along with λ_{i-1} from Eqn (10.15), we have

$$(S - i + 1)[i-1]\frac{\alpha}{s} = \frac{i}{s}[i]$$

or

$$[i] = \frac{S - i + 1}{i}\alpha[i-1]$$

giving

$$[1] = S \cdot \alpha \cdot [0]$$

$$[2] = \frac{S-1}{2} \cdot \alpha \cdot [1] = \frac{S(S-1)\alpha^2}{2}[0] \qquad \text{etc.}$$

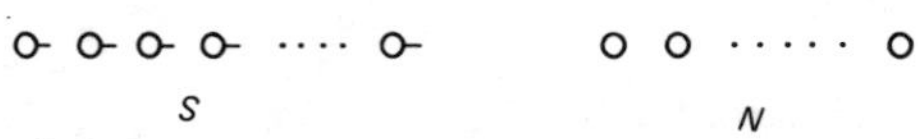

Fig. 10.7 Limited source full-availability model.

and generally

$$[i] = \frac{S!}{(S-i)!i!}\alpha^i[0] = \binom{S}{i}\alpha^i[0]$$

Summing over all i and applying Eqn (10.10),

$$\sum_{i=0}^{S} [i] = 1 = \left[1 + S\alpha + \frac{S(S-1)\alpha^2}{2} + \frac{S(S-1)(S-2)\alpha^3}{3!} + \cdots + \alpha^S\right][0]$$

(Note that the summation is from $0 \leqslant i \leqslant S$ since $S \leqslant N$)

$$[0] = \frac{1}{1 + S\alpha + S(S-1)\frac{\alpha^2}{2!} + \cdots + \alpha^S} = \frac{1}{(1+\alpha)^S}$$

and the general term can now be written

$$[i] = \binom{S}{i}\alpha^i \frac{1}{(1+\alpha)^S}$$

$$= \binom{S}{i}\left(\frac{\alpha}{1+\alpha}\right)^i\left(1 - \frac{\alpha}{1+\alpha}\right)^{S-i}$$

or

$$[i] = \binom{S}{i}a^i(1-a)^{S-i} \tag{10.16}$$

where

$$a = \frac{\alpha}{1+\alpha}$$

If $S < N$ there will be no time congestion because the state N is never reached; there will always be at least $N - S$ free devices.

If $S = N$ then the time congestion is

$$[N] = a^S = a^N \tag{10.17}$$

But the call congestion will be zero because there are no calls arriving when the network is in state N, all the sources being busy.

The quantity a in Eqn (10.16) is the carried traffic per device.

10.14 Engset Distribution: $S > N$

This case models the situation in which the number of sources is greater than the number of devices, but not so large that the traffic offered is constant. As in the Bernoulli case we shall assume here that the traffic offered is related directly to the number of free sources, allowing us to start with the same state equation, viz.

$$(S-i+1)[i-1]\frac{\alpha}{S}=\frac{i}{S}[i] \tag{10.18}$$

from which

$$[1]=S\alpha[0]$$

$$[2]=\frac{S(S-1)\alpha^2}{2}[0] \qquad \text{etc.}$$

and generally

$$[i]=\binom{S}{i}\alpha^i[0] \tag{10.19}$$

We can now sum $[i]$ over all states $0 \leqslant i \leqslant N$ and apply Eqn (10.10), i.e.

$$\sum_{i=0}^{N}\binom{S}{i}\alpha^i[0]=1$$

or, changing the variable to avoid confusion later,

$$[0]=1\Bigg/\sum_{v=0}^{N}\binom{S}{v}\alpha^v$$

Then, inserting this expression in Eqn (10.19) gives

$$[i]=\binom{S}{i}\alpha^i\Bigg/\sum_{v=0}^{N}\binom{S}{v}\alpha^v \tag{10.20}$$

Time congestion

From Eqn (10.20)

$$[N]=\binom{S}{N}\alpha^N\Bigg/\sum_{v=0}^{N}\binom{S}{v}\alpha^v \tag{10.21}$$

which is denoted by $E(N,S,\alpha)$. Values of E for a given N, S and α can either be calculated from Eqn (10.21) or be obtained from tabulated results.

Call congestion

From the previous definition

$$B=(S-N)\frac{\alpha}{S}[N]\Bigg/\sum_{i=0}^{N}[i](S-i)\frac{\alpha}{S}$$

Substituting from Eqn (10.21) for $[N]$ and rearranging gives

$$B = \binom{S-1}{N}\alpha^N \Big/ \sum_{v=0}^{N} \binom{S-1}{v}\alpha^v$$
$$= E(N, S-1, \alpha) \tag{10.22}$$

10.15 Probability of Occupancy of Particular Devices

We have just developed the three most important full-availability models, and in each case the general term has been $[i]$, the probability that any i devices are busy.

There are some applications where a distinction must be made between the two probabilities of i busy devices:

(i) that any i devices are busy, denoted by $G(i)$;
(ii) that a particular i devices are busy, denoted by $H(i)$.

The first is what we have calculated so far. To find the second, $H(i)$, we consider the situation shown in Fig. 10.8.

The problem is, given that j out of N devices are busy, what is the probability, $H(i)$, that a particular i are busy?

We can argue as follows:

Given that j devices in total are busy, there are $j-i$ busy devices not included in our i, and these $j-i$ can be distributed over the $N-i$ positions available for them in $\binom{N-i}{j-i}$ ways. Then

$$\frac{\text{Number of ways the particular } i \text{ are busy}}{\text{Total number of ways of distributing } j \text{ over } N} = \frac{\binom{N-i}{j-i}}{\binom{N}{j}}$$

Hence

$$H(i) = \sum_{j=i}^{N} \left[\binom{N-i}{j-i} \Big/ \binom{N}{j}\right] P(j) \tag{10.23}$$

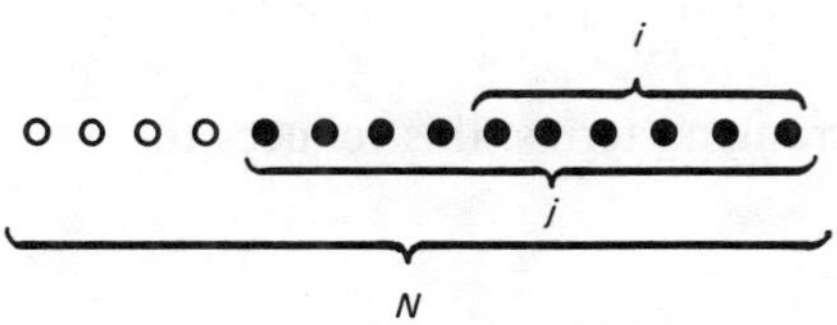

Fig. 10.8 Model to find $H(j)$.

noting that the summation can only be over the range for which $j \geqslant i$. The value of $H(i)$ depends of course on the quantity $P(j)$ which, in the terms introduced at the beginning of this section, is equivalent to $G(j)$. Applying the models of Section 10.2, there are three cases.

Erlang

$$H(i) = \sum_{j=i}^{N} \left[\binom{N-i}{j-i} \Big/ \binom{N}{j} \right] P(j)$$

$$= \sum_{j=i}^{N} \left[\binom{N-i}{j-i} \Big/ \binom{N}{j} \right] \cdot \left[\frac{A^j}{j!} \Big/ \sum_{\nu=0}^{N} \frac{A^\nu}{\nu!} \right] \tag{10.24}$$

from Eqn (10.13)

$$= \sum_{j=i}^{N} \frac{(N-i)!}{(j-i)!(N-j)!} \cdot \frac{j!(N-j)!}{N!} \cdot \left[\frac{A^j}{j!} \Big/ \sum_{\nu=0}^{N} \frac{A^\nu}{\nu!} \right]$$

$$= \left[A^N \Big/ N! \sum_{\nu=0}^{N} \frac{A^\nu}{\nu!} \right] \cdot \frac{(N-i)!}{A^N} \sum_{j=i}^{N} \frac{A^j}{(j-i)!}$$

$$= \left[\frac{A^N}{N!} \Big/ \sum_{\nu=0}^{N} \frac{A^\nu}{\nu!} \right] \cdot \frac{(N-i)!}{A^N} \left[\frac{A^i}{1} + \frac{A^{i+1}}{1} + \cdots + \frac{A^N}{(N-i)!} \right]$$

$$= E_N(A) \cdot \frac{A^i (N-i)!}{A^N} \left[1 + A + \cdots + \frac{A^{N-i}}{(N-i)!} \right]$$

$$= \frac{E_N(A)}{E_{N-i}(A)} \tag{10.25}$$

Showing that $H(i)$ in the Erlang case can be found from two Erlang functions.

Bernoulli

In this case

$$P(j) = \binom{N}{j} a^j (1-a)^{N-j}$$

and

$$H(i) = \sum_{j=i}^{S} \left[\binom{N-i}{j-i} \Big/ \binom{N}{j} \right] \binom{N}{j} a^j (1-a)^{N-j}$$

Expanding and rearranging terms, this reduces to

$$H(i) = \frac{\binom{S}{i}}{\binom{N}{i}} a^i \qquad \text{for } S < N \tag{10.26}$$

Table 10.1

	$G(i)$	$H(i)$
Erlang	$E_i(A) = (A^i/i!) \Big/ \sum_{\nu=0}^{N} (A^\nu/\nu!)$	$\dfrac{E_N(A)}{E_{N-i}(A)}$
Bernoulli	$\binom{N}{i} a^i (1-a)^{N-i}$	$\dfrac{\binom{S}{i}}{\binom{N}{i}} a^i$ (a^i for $S = N$)
Engset	$E(N, S, \alpha) = \binom{S}{i} a^i \Big/ \sum_{\nu=0}^{N} \binom{S}{\nu} \alpha^\nu$	$\dfrac{E(N, S, \alpha)}{E(N-i, S-i, \alpha)}$

In the special case when $S = N$, which is common in practice,

$$H(i) = a^i \tag{10.27}$$

Engset

From the earlier analysis

$$P(j) = \binom{S}{j} \alpha^j \Big/ \sum_{\nu=0}^{N} \binom{S}{\nu} \alpha^\nu$$

so

$$H(i) = \sum_{j=1}^{N} \left[\binom{N-i}{j-i} \Big/ \binom{N}{j} \right] \cdot \left[\binom{S}{j} \alpha^j \Big/ \sum_{\nu=0}^{N} \binom{S}{\nu} \alpha^\nu \right] \tag{10.28}$$

which becomes

$$H(i) = \frac{E(N, S, \alpha)}{E(N-i, S-i, \alpha)} \tag{10.29}$$

Summarizing the results obtained for the full availability models considered above, we can tabulate the formulae we have derived as shown in Table 10.1.

10.16 Link Systems

A link system is a multi-stage switching network that consists of a set of first-stage switches, connected to a set of second-stage switches, which in turn are connected to a set of third-stage switches, and so on. The number of stages of switching used depends on the purpose of the system. We will consider first a two-stage system as shown in Fig. 10.9(a). There we see that there are four

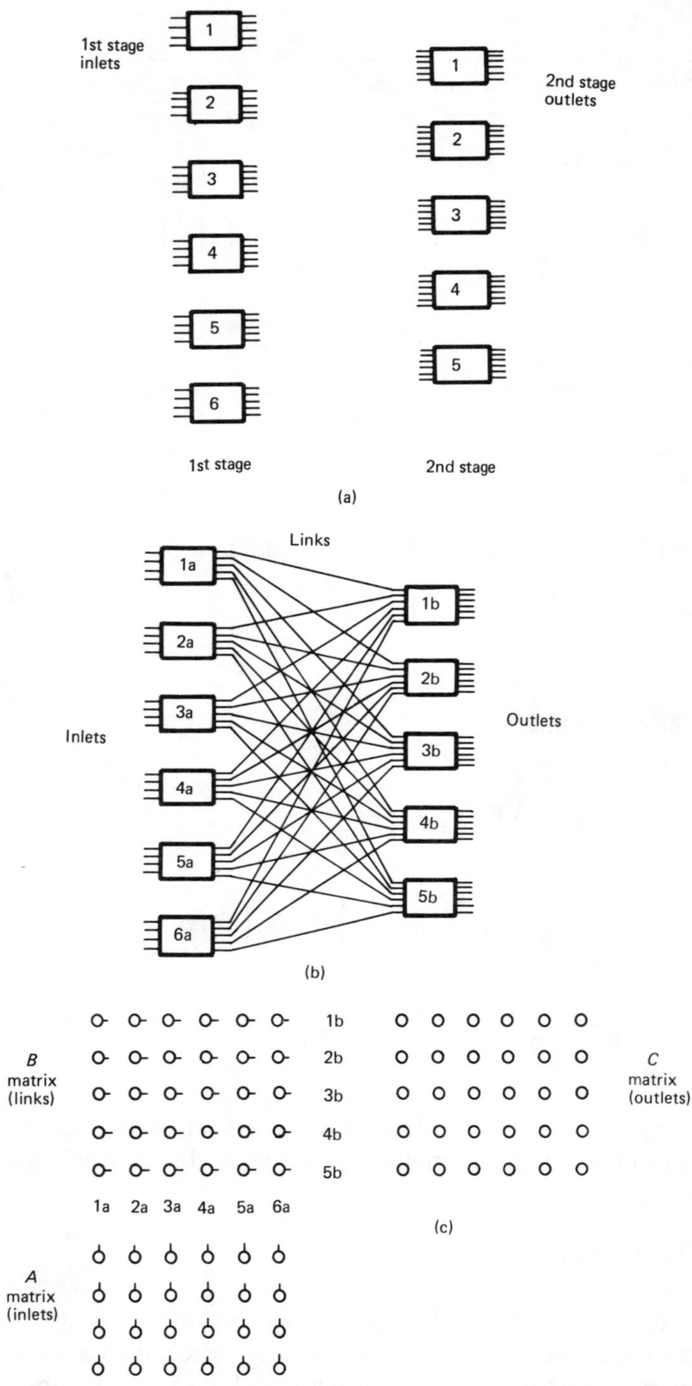

Fig. 10.9 Two-stage links system; (a) switches; (b) complete with links; (c) chicken diagram equivalent.

inlets and five outlets on each of the six first-stage switches, and six inlets and six outlets on each of the five second-stage switches.

The connections, or links between the two switching stages are shown in Fig. 10.9(b). We see that the diagram is very messy, and even if the links are drawn only when there is a call in progress along it, the lines indicating a particular call will be difficult to trace when the number of calls is high. A different type of representation suits our purposes better, i.e. one that was first

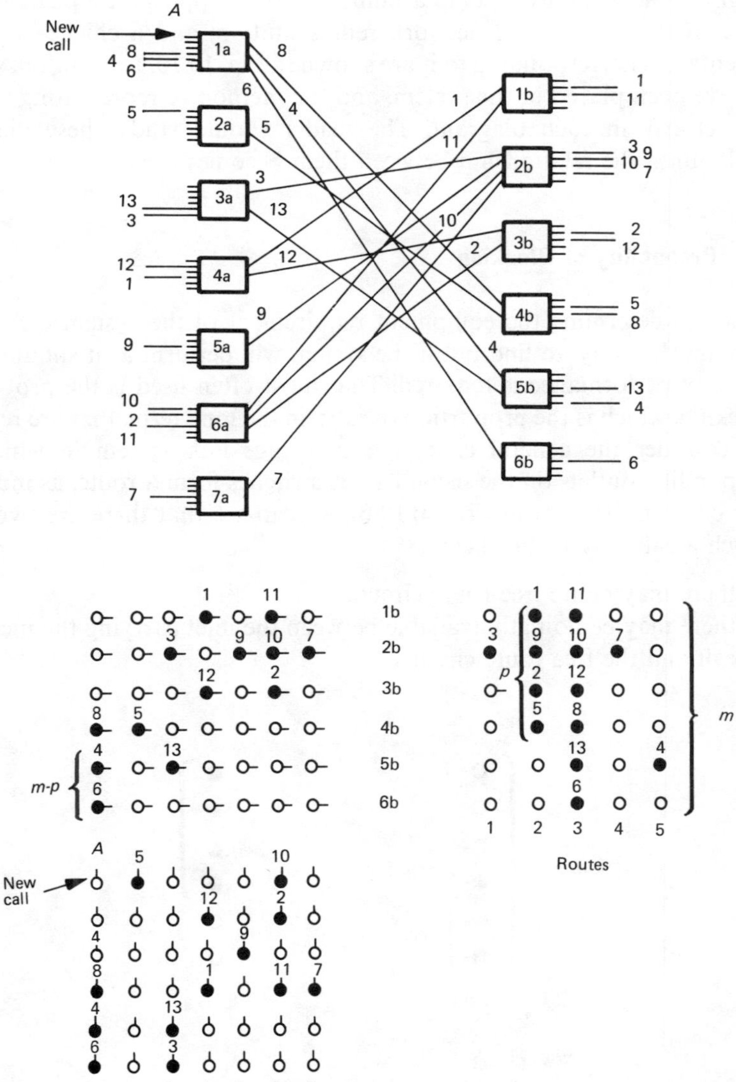

Fig. 10.10 Blocking in a two-stage link system.

introduced by workers at the Swedish company L.M. Ericsson, and is called a chicken diagram. The chicken diagram representation of the network in Fig. 10.9(a) is shown in Fig. 10.9(c). Now the link connections are not shown, but are implied by the diagram. A column on the A matrix represents the inlets of a first-stage switch; the corresponding column on the B matrix represents the outlets of the first-stage switch. A row on the B matrix corresponds to the inlets on a second-stage switch, and a row on the C matrix corresponds to the outlets on the same second-stage switch. Calls can be traced through the network by placing a number on the appropriate place in each matrix. An example of the network representation, and a chicken diagram representation corresponding to it are shown in Fig. 10.10. In addition, several calls have been placed in the system, and the method of representing them is shown clearly in each diagram. The reader should study these diagrams carefully until the relationship between them becomes clear.

10.17 Probability of Blocking

In order to determine the equipment requirement of the system or, putting that in another way, to find out if the system will perform as it should, some measure of performance is required. That most often used is the probability of blocking, which is the proportion of calls, in the long term, that are rejected. If we consider the general case of a two-stage link system in which the corresponding outlets on the second-stage switches form a route, as indicated by the chicken diagram in Fig. 10.10(b), we can see that there are two ways in which a call may be blocked:

(i) there may be no free route circuit;
(ii) there may be no path available between the inlet carrying the incoming call and the free route circuits.

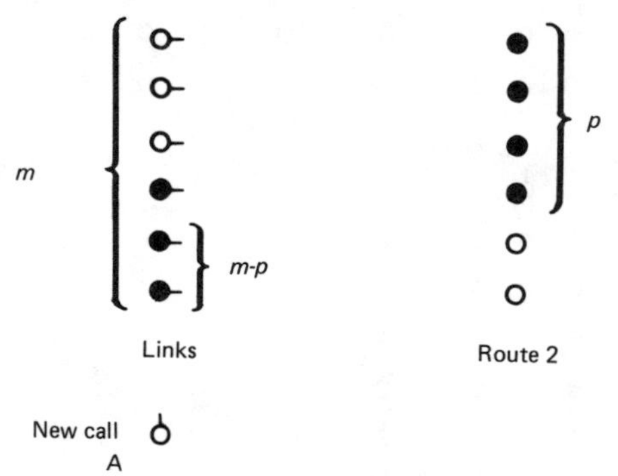

Fig. 10.11 Paths from a particular inlet to a route.

The diagram shows both of these conditions; route 3 has no free circuits available, whereas if inlet A on first-stage switch 1 is wanting to be connected to route 2, on which there are free route circuits, it cannot be, because there is no available link.

If we now look at this particular connection between inlet A and route 2 further, we can develop from it an analytical approach to the probability of blocking. The part of the network involved in the connection from A to route 2 is shown in Fig. 10.11.

There are m links to a route from any first-stage switch, since any route has one circuit on each of the second-stage switches. Thus a route consists of m outlet circuits, one on each second-stage switch.

Let us assume that there are p busy circuits on the route, then a call will not be able to be connected if, at the same time, the particular $m-p$ links, having access to the free $m-p$ route circuits, are busy, as they are in the diagram.

The first probability, that there are any p route circuits busy, we denote by $G(p)$, and the second probability, that $m-p$ particular devices are busy is denoted by $H(m-p)$. The probability of blocking, that is the probability that a call cannot be connected, is the joint probability of these two events, summed over all possible values of the variable p. Denoting the probability of blocking by E, we have then that

$$E = \sum_{p=0}^{m} G(p)H(m-p) \tag{10.30}$$

This apparently simple expression, known as the Jacobaeus equation, after its development by Christian Jacobaeus of L.M. Ericsson[2] in 1950, has been the cornerstone of blocking calculations on link systems for many years. There are some assumptions implicit in its form which become important, and indeed limiting, under certain circumstances: the most significant is that the two probabilities $H(m-p)$ and $G(p)$ are independent. Provided that the traffic level in the network is low, this independence assumption can be made without concern, but at higher traffic levels the analysis must be modified to take the conditional dependence of $G(p)$ on $H(m-p)$ into account.

The evaluation of Eqn (10.30) depends on the probability distributions appropriate to $G(p)$ and $H(m-p)$. In most cases one of those summarized in Table 10.1 is chosen. Often $G(p)$ will be Erlang because the number of route circuits (m) will be much smaller than the number of potential traffic sources (all the inlets on the first stage). The distribution used for $H(m-p)$ will depend entirely on the relationship between the inlets and outlets of the first-stage switches. An example will demonstrate how the Jacobaeus equation can be applied to a two-stage system.

EXAMPLE: In a two-stage link system the outlet circuits are combined into routes – each route has one circuit from each outlet switch. Using a chicken diagram representation, find the internal link congestion if the total traffic offered to the network is 42 erlangs, and if the outlets are formed into routes, as explained in the first sentence. Assume that there is equal demand for each route.

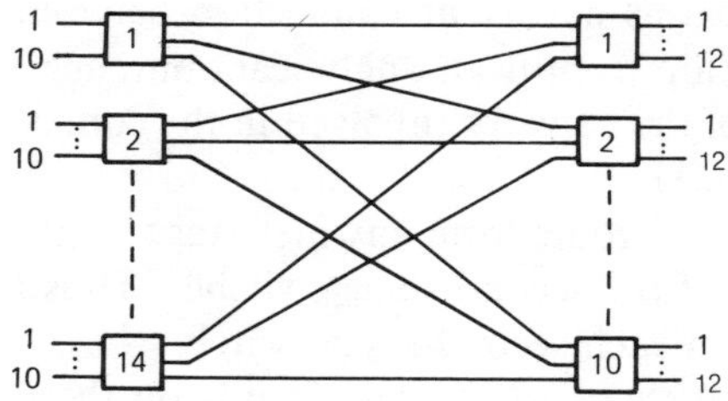

The chicken diagram representation is

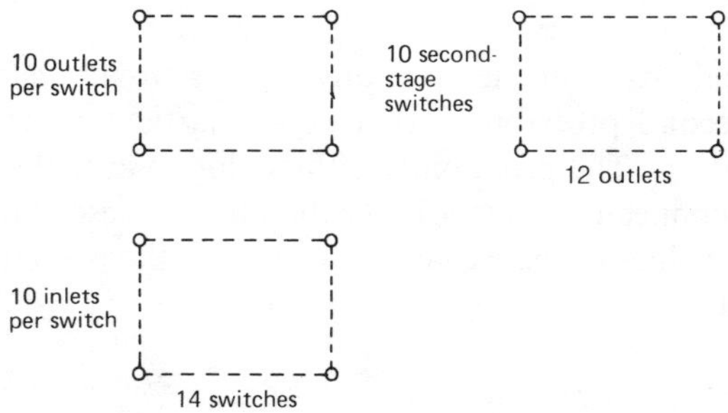

There are 12 routes, and 42 erlangs of total offered traffic; therefore

each route is offered $42/12 = 3.5$ e

The Jacobaeus equation gives the probability of blocking as

$$\sum_{p=0}^{m} G(p)H(m-p)$$

where $G(p) =$ probability that p devices are busy on the route and $H(m-p) =$ probability that $m-p$ particular links are busy. Since there are 140 inlets to the network, and each route consists of 10 devices, we can assume an Erlang distribution on the route, i.e.

$$G(p) = \frac{A^p}{p!} \bigg/ \sum_{v=0}^{m} \frac{A^v}{v!}$$

There is the same number of inlets as outlets on a first-stage switch, and therefore the Bernoulli distribution is appropriate for $H(m-p)$, giving, from Table 10.1,

$$H(m-p) = a^{m-p}$$

where $a =$ traffic carried per link.

There are 140 links in total; therefore, assuming the congestion to be small,

$$a = 42/140 = 0.3\text{ e}$$

Hence the total blocking, E, is

$$E = \sum_{p=0}^{10} \frac{(3.5)^p/p!}{\sum_{v=0}^{10} 3.5^v/v!}(0.3)^{10-p}$$

Evaluating this expression for each value of p, and summing, gives the total

$$E = 0.0082$$

The internal link congestion = total congestion − route congestion.

In any route there are 10 circuits, and we calculated earlier that the traffic offered per route is 3.5 erlangs. Therefore the route congestion $= E_{10}(3.5) = 0.0023$. Hence internal congestion $= 0.0082 - 0.0023 = 0.0059$.

Although we shall not do more here, the Jacobeaus approach can be used for a wide variety of link system structures, but they do not involve any change in the principles we have developed.

10.18 Delay System

In the foregoing sections we dealt exclusively with lost-calls-cleared systems in which a call is lost if it arrives when all the devices are occupied. In such systems there is no mechanism for allowing calls to wait until a free device becomes available.

If queueing is allowed, however, the call can be delayed, or held within the system until it has access to a free device. In this section we shall briefly examine the simplest delay model. It has the following attributes:

(i) calls arrive at random with constant mean traffic and negative-exponential distribution of inter-arrival times (Poisson offered traffic);
(ii) the holding times follow a negative-exponential distribution;
(iii) there are n devices (servers);
(iv) the queue can accept an infinite number of waiting calls:
(v) the service discipline is first in – first out, i.e. calls are passed to devices in order of arrival.

Queues satisfying (i), (ii) and (iii) are usually classified as $M/M/n$ in books on queueing theory, after the nomenclature of Kendall[3] who proposed a systematic classification of queues.

The queue we will now study is idealized through the assumption (iv), which could not be achieved in practice, but there are many real situations which are sufficiently close to it to make the analysis applicable.

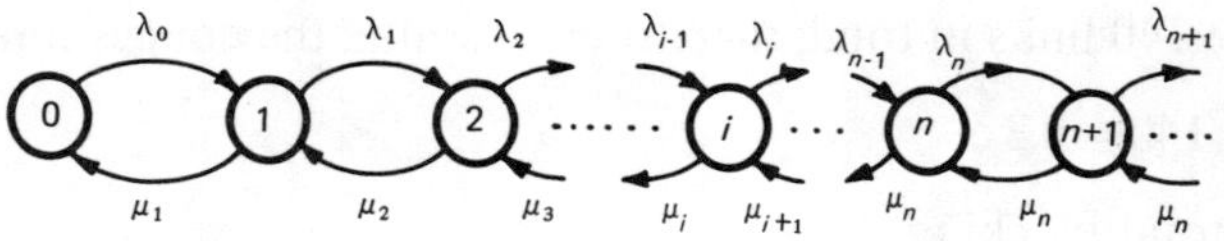

Fig. 10.12 State transition diagram for a delay system.

If, as before, the mean calling rate and the mean departure rate at any instant are given by λ_i and μ_i, where i is the number of calls is the system, we can draw a transition diagram as shown in Fig. 10.12.

If we compare this diagram with that of Fig. 10.3 we can see two differences. In the delay system there is an infinite number of states, and the departure coefficient μ_i is dependant on the number of calls in progress when $i \leqslant n$, and thereafter it is equal to $n\mu$ for all i.

We can use the same approach as we did in the analysis of the Erlang loss formula, since here too we have assumed that the mean traffic offered is constant.

Starting with the relevant birth and death equation, assuming that we have statistical equilibrium (which in a queueing system requires that the mean traffic offered per device is less than unity, i.e. $A/n < 1$), we have

$$\mu_i[i] = \frac{A}{s}[i-1]$$

and

$$\mu_i = \frac{i}{s}$$

Therefore

$$[i] = \frac{A}{i}[i-1]$$

Putting in values

$$[1] = A[0]$$

$$[2] = \frac{A}{2}[1] = \frac{A^2}{2}[0]$$

$$[3] = \frac{A}{3}[2] = \frac{A^3}{3!}[0]$$

etc. to

$$[n-1] = \frac{A^{n-1}}{(n-1)!}[0]$$

$$[n] = \frac{A^n}{n!}[0] = \frac{A}{n}[n-1]$$

For $i > n$ the factor A/n will remain constant, so we have

$$[n+1] = \frac{A}{n}[n]$$

$$[n+2] = \left(\frac{A}{n}\right)^2 [n]$$

and in general

$$[i] = \left(\frac{A}{n}\right)^{i-n} [n] \tag{10.31}$$

Again we use the normalizing condition that the sum of all state probabilities is unity:

$$\sum_{i=0}^{\infty} [i] = 1 = [0]\left\{\sum_{i=0}^{n-1} \frac{A^i}{i!} + \sum_{k=0}^{\infty}\left(\frac{A}{n}\right)^k \frac{A^n}{n!}\right\}$$

where the variable k is used instead of i in the range $i \geqslant n$.

$\sum_{k=0}^{\infty}(A/n)^k$ is the infinite sum of a geometric series of ratio A/n, which has the value $1/[1-(A/n)]$

Therefore

$$1 = [0]\left\{\sum_{i=1}^{n-1} \frac{A^i}{i!} + \frac{A^n}{n!} \cdot \frac{1}{1-(A/n)}\right\}$$

or

$$[0] = 1 \Bigg/ \left[\sum_{i=0}^{n-1} \frac{A^i}{i!} + \frac{A^n}{n!} \frac{n}{n-A}\right] \tag{10.32}$$

The probability that a call is delayed is the same as the probability that all devices are busy, which is the sum of the state probabilities $[i]$ in the range $i \geqslant n$, i.e.

$$\begin{aligned}
\text{Prob (call is delayed)} &= \sum_{i=n}^{\infty} [i] \\
&= \sum_{i=n}^{\infty} \left(\frac{A}{n}\right)^{i-n} \frac{A^n}{n!}[0] \\
&= \frac{A^n}{n!}\frac{n}{n-A}[0] \\
&= \frac{(A^n/n!)n/(n-A)}{\sum_{i=0}^{n-1} A^i/i! + (A^n/n!)\cdot n/(n-A)}
\end{aligned}$$

This expression, symbolized by $E_{2,n}(A)$, is known as the Erlang delay formula. For purposes of computation it can be expressed in terms of the Erlang loss formula as

$$E_{2,n}(A) = \frac{nE_n(A)}{n - A + AE_n(A)} \tag{10.33}$$

where $E_n(A)$ is given by Eqn. (10.13).

In queueing systems, several parameters other than the probability that a call is delayed may be of interest, e.g. the mean queue length, the mean delay of all calls, the mean delay of delayed calls, maximum delay, and so on. Further treatment of queues related to telephony can be found in Cooper[3] and Fry.[4]

10.19 Simulation

Modern systems involve a considerable amount of control by microprocessors dedicated to a particular small task and operating on an individual bit or word basis. The sequence of tasks involved in passing a call through the system can be represented as a series of queues, and at each queue there is the possibility of delay. If the traffic is very high, the total delay may become so great that the processors cannot cope and the throughput falls dramatically. Teletraffic theory has to provide the tools to predict this degradation in performance. However, sufficient analytical power is not available in queueing theory. It provides the facility to study fairly simple problems, but when there are several queues interacting the models can become too difficult to analyse. Resort must then be made to simulation. A computer program is written to represent the behaviour of the system to be studied and it is used in conjunction with some form of simulation package which generates calls in a prescribed way (e.g. with negative exponential inter-arrival and call holding time distributions), provides random number streams and organizes the simulation of the system operation. Simulation is a most powerful tool and it can be used for any system. It relies on:

(i) sufficient computer power;
(ii) a satisfactory software model of the system being studied;
(iii) the correct choice of the behaviour of the traffic offered;
(iv) a proper evaluation of which results to take;
(v) sufficient length and frequency of runs to account for the effects of correlation.

The last condition is one that must be in mind when a simulation is being organized; correlation can produce errors in results if it is not taken into account. In particular, the urge to reduce cost by making runs short should not inhibit the need for long runs to make the results reliable.

Simulation also has disadvantages. It is costly because it requires long runs and therefore considerable computer time, it is time consuming for the analyst because the programs to model the system being studied are often very complex, and the results are not of general applicability since they are related to a particular set of initial conditions such as system dimensions and traffic level.

We shall not consider simulation in detail here, but there are some useful texts available on the subject. Until there is a major breakthrough in the analysis of delay systems (equivalent to the Jacobaeus approach to link systems) simulation will be an essential weapon in the armoury of the teletraffic engineer.[5]

10.20 Switching and Signalling

When a subscriber makes a telephone call a series of events takes place which have the following sequence, illustrated in Fig. 10.13.

1. Calling subscriber lifts handset.
2. Exchange detects demand for call.
3. Check is made to find free equipment in the exchange.
4. When equipment becomes free, dial tone is returned to the caller. If equipment is not immediately available at the exchange there may be a long dial-tone delay.
5. Subscriber dials digits.
6. Exchange interprets the digits and routes call to its destination.
7. Check is made to determine whether or not the called subscriber is free.
8. If called subscriber is busy, busy tone is returned to caller.
9. If called subscriber is free, ringing tone is returned to caller and ring current is sent forward to ring called subscriber's bell.
10. If called subscriber is unavailable, caller hangs up and equipment clears down.
11. If called subscriber answers, ring tone and ring current cease.
12. Speech path is set up and metering commences.
13. When conversation ceases called subscriber hangs up and metering ceases.
14. Calling subscriber hangs up and system clears down.

Note that if an unallowed number is dialled, or service to the number dialled has been discontinued, number-unobtainable tone is returned to the caller.

From the above description we can see that the process of signalling has at least three functions:

(i) To indicate the state of the call to each subscriber by
  (a) dial tone,
  (b) ring tone and ring current,
  (c) busy tone or number-unobtainable tone.

(ii) To tell the system what to do next, by indicating the path for the call for example.

(iii) To initiate a billing procedure – usually by tripping the calling subscriber's meter at the correct charging rate – to enable the administration to gather the revenue needed to provide the service.

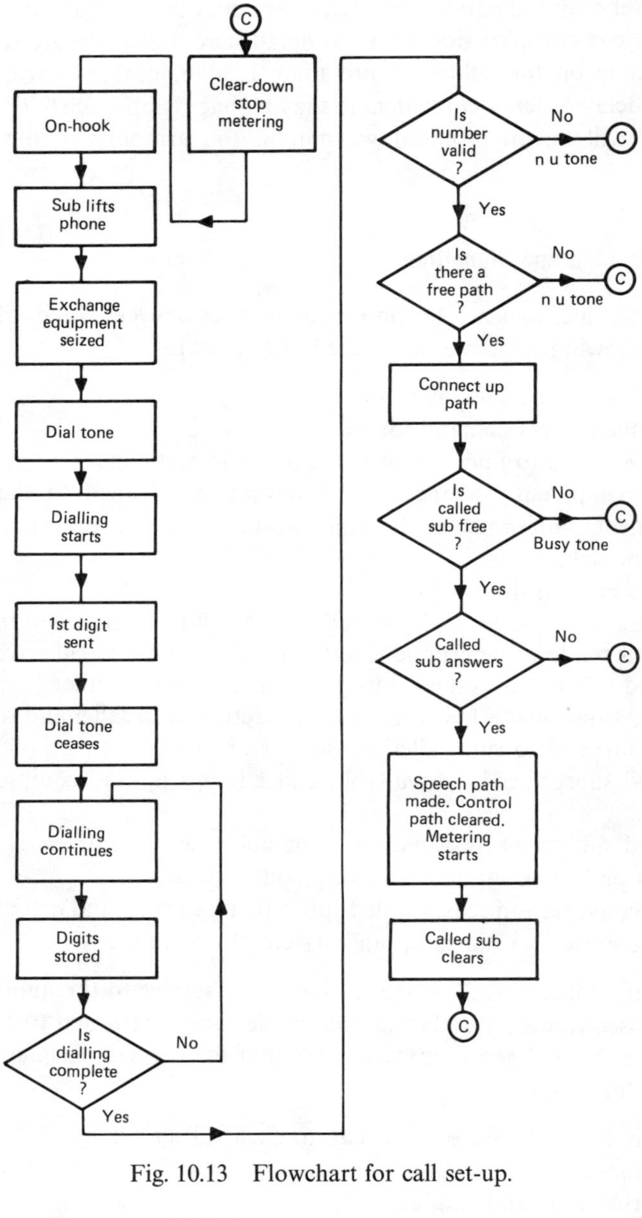

Fig. 10.13 Flowchart for call set-up.

In the early days of telephony when calls were handled by an operator, these signalling functions were done manually. Indeed in small communities having up to two hundred lines, or so, the operator provided an automatic call-transfer facility of some sophistication: the more interest the operator had in local affairs the better that part of the service could be!

The flow-chart of Fig. 10.13 applies particularly to the setting up of calls that share the same exchange. Although most calls are still of that type, in many large cities an increasing number of calls involve inter-exchange working, and the use of direct-dialling systems has created a greater demand for trunk calls. Inter-exchange signalling is necessary to allow the calling subscriber's local exchange to keep control of the call, and later we shall look in some detail at the signalling systems used for the purpose. Evidently, inter-exchange systems are not restricted in speed as are the subscriber's signalling systems because of the slow-moving dial, and they are not required to produce the tones which the local exchanges must produce to keep the subscriber informed of the progress of the call.

10.21 Types of Signalling System

We shall not look at any of the historical systems, but we shall look at those which are commonly in use. They fall into four main categories:

(i) loop disconnect – dc signalling;
(ii) multi-frequency – ac signalling;
(iii) voice frequency – ac signalling;
(iv) common channel signalling.

Loop Disconnect Signalling

Until recently, the universal means at the subscriber's disposal for indicating the number he wished to call was the telephone dial, and it is still present on many handsets. It was developed to its present state many years ago. It has several advantages that make it difficult to replace; it is economical, relatively cheap to make and virtually trouble free, it is very robust, and simple to use. But it has a major disadvantage: it operates very slowly by the standards of modern electronics, and its slow motion places a definite limit on the speed at which signals can be sent to the exchange.

When the dial is rotated the finger hole corresponding to the required digit is pulled round to the finger guard and then the dial is released. A governor inside the dial causes it to rotate back automatically at a fixed speed, causing a series of pulses to be sent down the subscriber's line. After the dial has returned

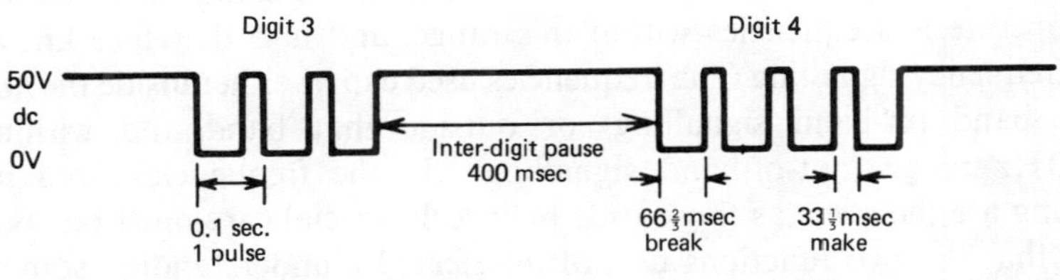

Fig. 10.14 Ideal output from telephone dial.

to its rest position the next digit is dialled. The time between the last of the pulses for one digit, and the first pulse of the next is called the inter-digit pause. It is this pause that allows the exchange to recognize the end of a digit. A typical sequence, with average operating times (nominally the pulses are sent at a rate of 10 per second) is shown in Fig. 10.14. Although the dials are designed to operate at a nominal 10 pulses per second, the exchange equipment will still function reliably if the pulse speed is between 7 and 12 pulses per second and the break is between 63 and 72 per cent of the pulse period.

Multi-frequency (mf) Signalling

Modern subscribers-handsets are fitted with key-pads instead of dials, and this development allows signals between subscriber and exchange to be made more quickly. Generally the keypads send out frequencies instead of pulses to represent a digit. Most systems use two frequencies to represent a particular digit. Whether or not the much increased speed of mf signalling can be realized depends on the local exchange. An mf key-phone working into a Strowger-type exchange gives no overall improvement, it just changes the position in the process at which the caller must wait: when a dial is used there is a long inter-digit pause and short post-dial delay, whereas in an mf system there is little inter-digit pause because the numbers can be keyed in very quickly, but since the exchange equipment operates at 10 pulses/s the mf signal must be converted into normal digits producing a lengthy post-dial delay.

However, in modern exchanges the equipment will respond directly to the mf signals and the call can be set up very quickly.

Voice-frequency (vf) Signalling

The use of dc signalling is limited by several factors. In particular, over long lines the variation of equipment performance, and the effect of the characteristics of the lines used to transmit between exchanges, degrade the pulse shape of the dialled digits to such an extent that errors can be caused. In addition, the receiving equipment requires a higher voltage than would be necessary if the system used alternating current signals. Apart from these considerations multi-channel working makes dc signalling inappropriate.

The normal telephone channel occupies a bandwidth of 300–3400 Hz, from the range of 0–4000 Hz which it is allocated. If ac signalling is to be used it must operate at frequencies within this range, and it is therefore known as voice-frequency signalling. The frequencies used can be either inside the normal speech band (in-band signalling) or outside that band and within the 0–4000 Hz range (out-of-band signalling). As the frequencies used for vf signalling are the same as those used in speech, special care must be taken to ensure that the two functions do not interfere. By understanding something of the characteristics of speech it is possible to specify signalling systems that will not be operated erroneously by speech frequencies. Since signalling is done by tones within the baseband of the telephone channel it is possible to use vf signalling when the channels are multiplexed on to a common carrier – either line or radio.

Common Channel Signalling

An alternative to a signalling system being associated individually with a speech path is to couple the signals for a large number of calls together and send them on a separate signalling channel, i.e. send the signals on a common channel. This method of signalling is particularly useful for interexchange working, when the signals will be sent over a data-type high speed link, and in PCM systems. In the latter, one channel is dedicated to signalling in a 32 channel system. The signals must be coded so that they can be related to a specific speech channel. There is an inherent security problem in common channel signalling, so provision should be made for a second signalling channel to be used in the event of failure, and an error-detection facility must be provided.

10.22 Voice-frequency Signalling

As noted in Section 10.21, vf signalling can be either in-band or out-of-band (although some writers refer to in-band as vf signalling, and treat out-of-band separately).

In-band Signalling

Since the signalling frequency is within the 300–3400 Hz speech bandwidth there are obvious problems associated with this signalling method. It cannot be operated during speech and the equipment must be able to distinguish between a speech pattern and a signal. There are two parameters available for variation: the signal frequency and the signal recognition time. Other considerations that will assist in distinguishing between speech and signal are:

(i) speech at the signal frequency is accompanied by other frequencies;
(ii) more than one signal frequency could be used;
(iii) the signals could be coded bursts of the signal frequency.

Choice of Frequency

This is related to the frequency characteristics of speech. As shown in Fig. 10.15, the energy in English is predominant at lower frequencies (maximum at 500 Hz or thereabouts) and it falls gradually over the rest of the band. This suggests that a high signal frequency should be used to reduce the possibility

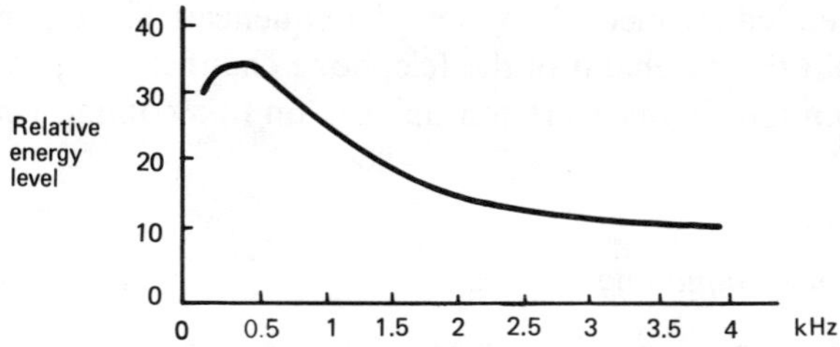

Fig. 10.15 Variation of energy with frequency in English speech.

of imitation by speech frequencies. However, there are other considerations that suggest that frequencies at the upper end of the band should not be used. At the higher frequencies there is an increase in cross-talk, so low-level transmission is necessary. If the amplitude is low the receiver must be very sensitive, raising the prospect of imitation signalling by low-level speech. The channel characteristics may vary at the upper end of the band on different links so that on some older links the cut-off may be below 3400 Hz. In addition, the variation of amplitude with frequency is quite marked at the high frequencies so any change in signal frequency will result in a significant change in signal amplitude. Between these considerations some compromise is necessary and in practice the frequency chosen lies in the range 2040–3000 Hz.

Signal Duration

By delaying the recognition of the signal until it has persisted for some time the chance of signal imitation is reduced significantly. Using a guard circuit (see below) and a 40 ms recognition delay the probability of the signal receiver responding to a speech frequency is reduced to a low level. Figure 10.16(a) shows the degree of improvement that may be possible as the delay time is increased.

Guard Circuit

As can be seen in Fig. 10.16(a) the effect of signal recognition delay is greatly enhanced if a guard circuit is used in the receiver to increase the rejection of imitation signals. Figure 10.16(b) shows the basic features of such a system. The circuit shown there forms part of the vf receiver shown in Fig. 10.17.

The energy coupled from the speech channel is amplified and then passes along two paths:

(i) through a bandpass filter tuned to the signal frequency, forming part of the signal circuit;
(ii) through the guard circuit which includes a band-stop filter allowing all but the signal frequency to pass.

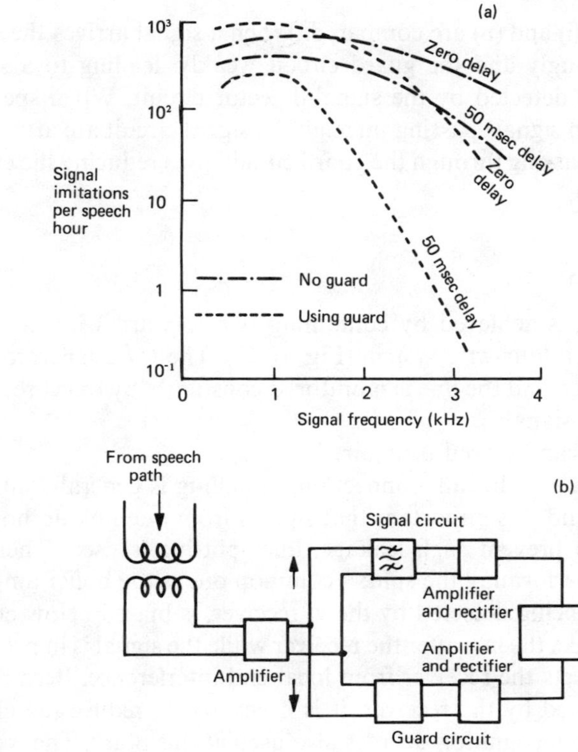

Fig. 10.16 (a) Effect of guard circuits, with and without delay; (b) block diagram of system using guard circuit.

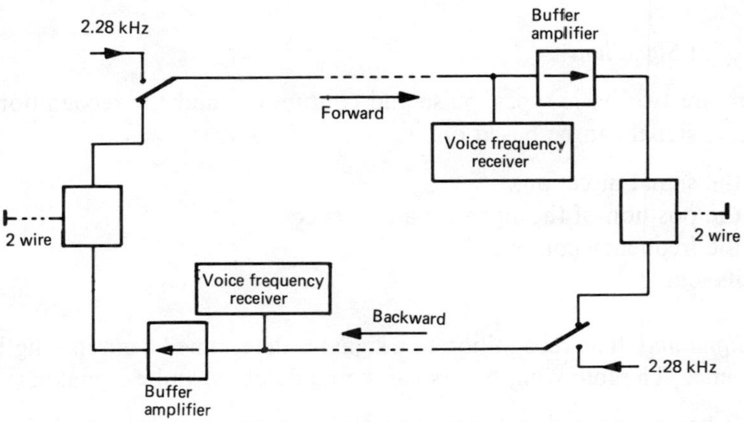

Fig. 10.17 Voice frequency signalling system.

The outputs from (i) and (ii) are compared. When a signal arrives the signal circuit responds strongly and the guard circuit weakly, leading to a strong positive signal being detected by the signal-detector circuit. When speech is present, any imitation signals passing through the signal circuit are attenuated by the strong signal passing through the guard circuit, thus reducing the chance of spurious imitation.

Basic Layout of the System

Two-way working is achieved by combining two forward wires and two backward wires into a four-wire system (Fig. 10.17). The tone generators are switched in as required and the line is monitored constantly by the vf receivers for the presence of a signal.

Echo suppression can be used if required.

When employed in a multi-link connection, signalling is generally done on a link-by-link basis and it is important that signals from one link do not leak over to the next. To prevent such leakage, line splitting is used. There are various methods of performing the split; a common one is the buffer amplifier which, on the signal being detected by the vf receiver, is biased below cut-off, which effectively breaks the line after the receiver while the signal is in progress. The buffer also protects the receiver from local-end interference. Because the line splitting is initiated by the receiver, it is necessary to reduce the chance of signal imitation, so recognition delay is also used at this point. The receiver on detecting a signal frequency, waits for 20 ms before initiating the line split. During that 20 ms there will be leakage into the next link, but it will not be a problem as it much less than the normal recognition delay of 40 ms used in the receiver.

Line splitting may also be inserted at the transmit end of each direction to prevent switching transients passing down the line at the same time as the signal.

Types of vf Signalling

There are two basic types, pulse and continuous, and the recognition of a particular signal can be based on:

(i) the signal direction;
(ii) the position of the signal in a sequence;
(iii) the frequency content;
(iv) its length.

Pulse Signalling. The recognition of a signal is determined from its length and its sequence. The following points can be made about pulse signalling:

(i) It has a higher signal repertoire than continuous signalling.
(ii) It can be transmitted at a higher voltage level and therefore provides a better SNR.

Table 10.2

Signal	Duration
Seizure (forward)	65 ms
Dial pulsing (forward) (loop disconnect)	57 ms
Called subscriber answers (backward)	250 ms
Called subscriber clears (backward)	250 ms
Calling subscriber clears (forward)	900 ms
Release-guard (backward)	1000 ms
Busy (backward)	Continuous

(iii) It is less influenced by interference.
(iv) It complicates the dc/ac and ac/dc conversions because the pulses have to be carefully timed.
(v) It requires a memory facility at the receiver for pulse recognition.

EXAMPLE: The UK AC9 signalling system is typical of that used in many countries. The signal frequency is 2280 Hz and the pulses used are shown in Table 10.2.

The calling subscriber's clear signal should initiate the backward release-guard signal. If the release guard is not received, the outgoing end sends seizure followed by forward clear until release guard is received.

A continuous tone busies the sending end relay set (busy back signal) and raises an alarm if it continues for several minutes.

The release guard facility ensures that calls are not passed down to the receiving end if there is a fault. When a fault occurs and the forward-clear signal is sent, the outgoing equipment is guarded from accepting another call until the release-guard signal is received. The repeated use of the seizure and calling-subscriber-clear signal will initiate the release-guard signal once the fault has cleared.

Note that the backward answer signal and the called-subscriber-clears signal have the same pulse length, but they are distinguishable by their place in the sequence.

Metering is stopped by either the forward clear or backward clear signals.

Continuous Signalling. There are two types of continuous vf signalling: (i) compelled and (ii) two-state non-compelled. Both are used, (i) in CCITT system 5 and (ii) in Bell SF. In (i) a signal ceases when an acknowledgement is received, whereas in (ii) the signal information is carried by the change in state, and acknowledgement is not used.

In terms of reliability (i) is preferable to pulse, which is preferable to (ii), but (i) is much slower than (ii) because it depends on acknowledgement signals, and hence on propagation time. This slowness of (i) means that it is used only in special applications. In the second continuous mode, (ii), the tone is on for much of the non-speech time, which can lead to overload on the transmission system if the signal level is not limited.

10.23 Outband Signalling

This term is generally applied to a signalling system in which the signal frequency is in the range 3400–4000 Hz. The CCITT recommended frequency is 3825 Hz, although 3700 and 3850 Hz are also used. This method is applicable only to carrier systems because the equipment used for baseband transmission may attenuate the signal frequency.

Compared with in-band signalling there are two advantages:

(i) There is no need to take steps to avoid signal imitation from speech – guard circuits and line splits are not required.
(ii) Signals and speech can be transmitted simultaneously.

The general layout of an out-band signalling system is shown in Fig. 10.18.

Signalling across the switching stage is usually dc, hence out-band signalling is link by link with ac/dc, dc/ac converter at each end. Filtering is used as shown to ensure that there are no spurious frequencies and to isolate the signal before the receiver.

This type of signalling has the advantage of simplicity and consequent cheapness. It can be either two-stage continuous or pulse. There are inherent

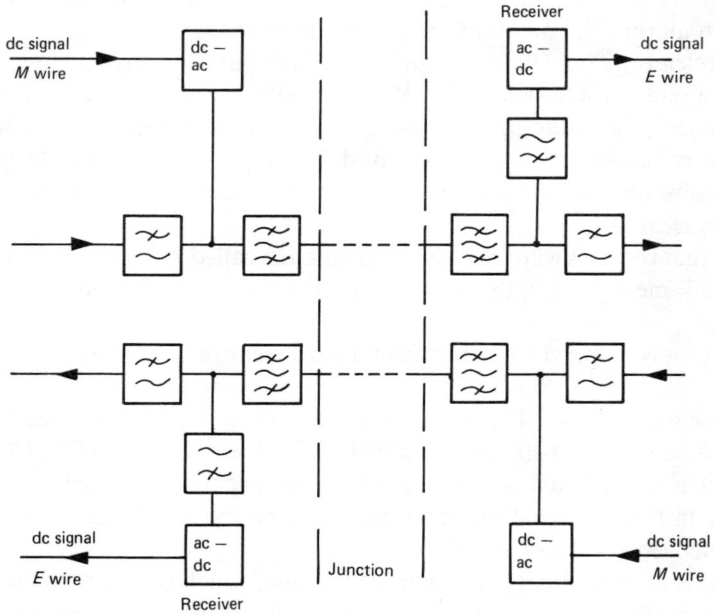

Fig. 10.18 Transmitter and receiver for outband signalling.

Table 10.3

Signal	Tone	
	Forward	Backward
Idle	Off	Off
Seizure	On	Off
Decadic address	Off (pulse breaks)	Off
Release	Off	On when clear forward first
Answer	On	On
Clear Back	On	Off
Release guard	Off	On
Blocking	Off	On

difficulties in each mode; continuous cannot be at too high a level because of the risk of overload, so the receiver needs to be sensitive, whereas pulse requires more sophisticated circuitry to perform the memory function. On balance, the continuous mode is preferred because of its simplicity (compared with in-band signalling where the pulse mode has advantages).

Out-band signalling is more attractive than in-band, but cannot always be applied without considerable expense to existing transmission plant; consequently it is commonly used in new fdm systems, but vf is the most common on existing systems.

There are two modes of continuous signalling, tone-off idle and tone-on idle, and there is little to choose between them for many applications, although tone-on idle would be preferred if the system were in general use because tone-off idle would tend to overload the transmission system.

A typical signal code for a tone-off idle system is shown in Table 10.3.

10.24 Switching Networks

In switching systems the number of cross-points is a good measure of cost, and part of the effort involved in switch design is concerned with reducing the number.

An idea of the way in which the number of cross-points can be reduced,

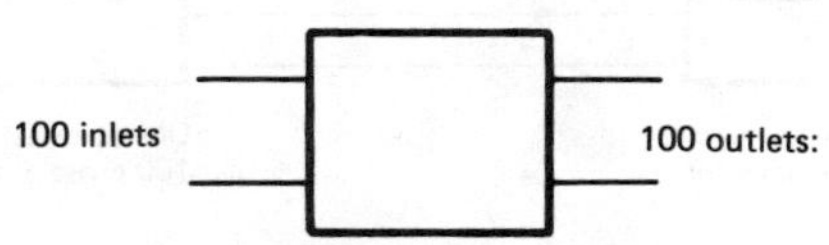

Fig. 10.19 Full-availability switch.

and the effect of that reduction on network performance, can be obtained from the following simple approach.

If there are N lines into and N lines from a single switch the number of cross-points required is N^2, e.g. the matrix shown in Fig. 10.19 requires 10 000 cross-points.

This large number of cross-points allows any free inlet to be connected to any free outlet regardless of the connections made between other inlets and outlets – a full availability loss-less system.

However, each inlet, if it is a subscriber's line, will carry a very small load, probably less than 0.1 erlangs, so it will be busy for only 6 min on average during the busy hour. This means that there will be fewer than 100 calls in progress simultaneously and therefore the system could operate with many fewer cross-points.

As a first step towards reducing the size of the network the matrix can be split into two parts; the first part having 100 inlets and 25 outlets and the second part having 25 inlets and 100 outlets [Fig. 10.20(a)]. This reduces the number of cross points by 50%.

A further saving can be made if the inlet and outlets switches are further subdivided, as in Fig. 10.20(b). However, this last division raises serious problems. Each inlet switch has access to only one outlet switch so, for example, inlet switch 3 cannot route a call to outlet switch 5, etc. One way round this difficulty would be to interconnect the links between the inlet and outlet

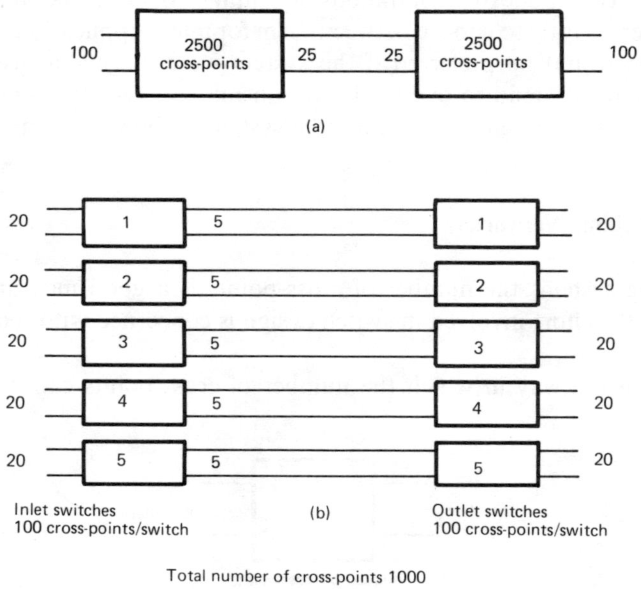

Fig. 10.20 (a) Cross-point reduction by vertical partitioning, (b) Cross-point reduction by vertical and horizontal partitioning.

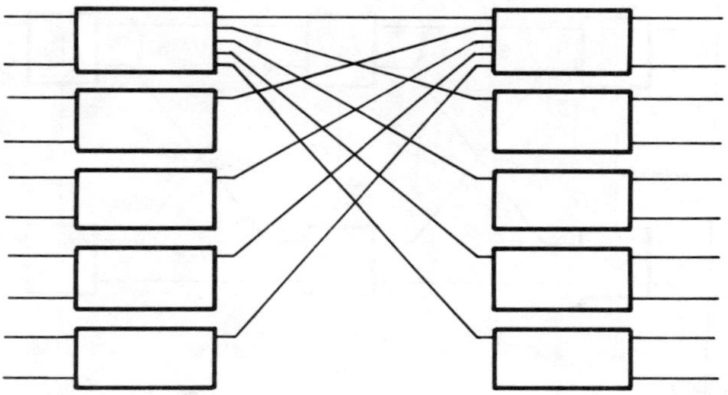

Fig. 10.21 Diagonal link connections to increase availability.

switches in the manner shown in Fig. 10.21, but that itself has the drawback that only one of the calls from an inlet switch can be connected with a particular outlet switch.

It seems that the reduction in availability has been too great and that more cross-points need to be used to make the system less restrictive. A middle stage can be included to make the distribution of paths between the inlet and outlet stages more satisfactory. This distributor stage will, in this example, have 25 × 25 cross-points as shown in Fig. 10.22.

In the same way as was done with the inlet and outlet stages, the distributor stage can be sub-divided into five 5 × 5 switches and the link pattern between first and second stages, and second and third stages arranged to provide five paths through the network from any first stage (inlet) switch to any third stage (outlet) switch (Fig. 10.23).

Clearly there has been a very large reduction in the number of cross-points. The chance that a call is lost because of the reduced availability is a question that can be answered by calculating the internal blocking of the network at given traffic levels, using the Jacobaeus equation, or some other suitable method.

Considering once again the three stages of the network in Fig. 10.23, we can see that the first stage has many more inlets than outlets (100 and 25, respectively), the third stage has many more outlets than inlets, and the second stage has an equal number of inlets and outlets. Figure 10.24 represents this

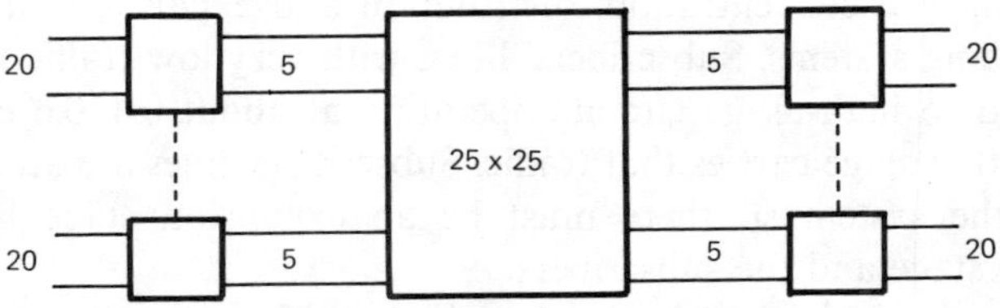

Fig. 10.22 Three-stage link system.

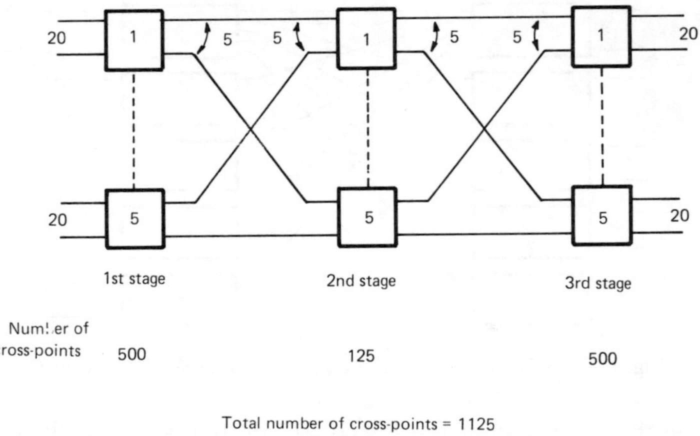

Fig. 10.23 Cross-point reduction on middle stage. Total number of cross-points = 1125.

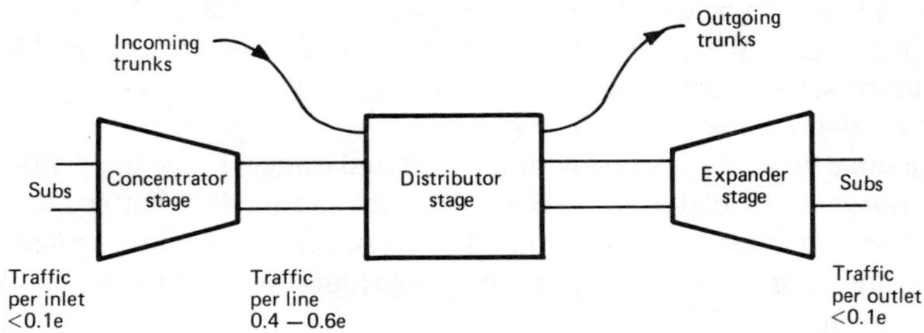

Fig. 10.24 Functional diagram of three-stage link system.

aspect of the three stages diagrammatically. For obvious reasons the stages are named:

First stage many more inlets than outlets – concentrator
Second stage same number of inlets and outlets – distributor
Third stage many more outlets than inlets – expander

This division into concentration, distribution and expansion stages typifies many switching systems. Subscribers' lines, with very low traffic per line, are concentrated to high-usage circuits operating at about 0.4–0.6 erlangs, and the distribution stage carries that traffic. Subscribers' lines are attached to the outlets of the system so there must be an expansion stage between the distribution stage and the subscribers.

Trunk circuits, which are designed for high traffic, come directly to, and go directly from, the distribution stage, as shown.

10.25 Basic Analogue Switching Systems

Switching systems can be classified in several different ways, but here we shall restrict our interest to separating them on the basis of their method of control. Two methods are in common use: (i) step-by-step; or (ii) centralized control.

10.26 Step-by-Step

In step-by-step systems, the control path and speech path are the same. As each digit is dialled, the connection is made to one further stage, until the final selection when two digits are required.

Strowger-type systems operate in this way. As an example consider a local exchange call on a four-digit number system (Fig. 10.25).

When the caller goes off-hook, the line circuit connects his line to a free first numeral selector (group selector). This connection may take some time, and not until the first selector is seized is dial tone returned to the caller. The dial-tone delay is a measure of system capacity from the subscriber's point of view, and most administrations specify a maximum value for this time. If it is too long, the subscriber feels that the system is not responding to his demand for service and he may hang-up.

Having seized the first selector the control waits for the first digit to be dialled; on its receipt the first selector wipers are racked up to the corresponding level and hunt round for a free outlet to a second selector. The second digit causes the second selector to go through a similar process and seize a final selector. The final selector has subscribers' line circuits attached to it, and so each of its hundred contacts represents a different subscriber. Consequently it needs two digits for its operation.

If at some stage during the setting up of the call, after the dial tone has been

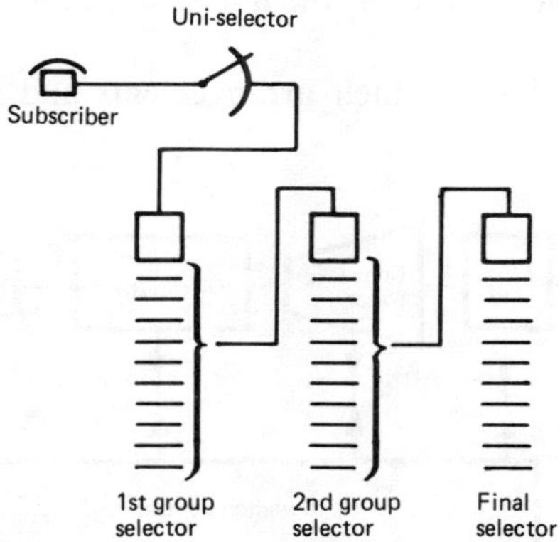

Fig. 10.25 Diagrammatic representation of a step-by-step system.

received a free selector is not available, a busy tone is returned to the caller; the wipers, having hunted over all the outlets to find a free selector at the next stage, without success, automatically return to their home position. In this method of control, part of the path is set up and then the system waits for more information before extending the path a little further and waiting again, thus setting up the call in a step-by-step manner.

10.27 Common Control

The step-by-step system described earlier operates on the principle that the control of the call follows the same path as the speech circuit. This means that all the control equipment is provided on a per-call basis. However, some of it need not be, and savings can be made by separating some of the control from the speech path. It can then be used only as required by a call, and can then become available for other calls. For example, the register–translator, which is only required during the routing, or path selection process, could be placed in a common-control area, used to set up a path between caller and called, and then released for use in setting up another call. Figure 10.26 shows a diagram of a common-control system.

The switching block will not be of the Strowger-type, but either a cross-bar switch or a reed-relay system. A cross-bar switch consists of a matrix of horizontal and vertical conductors that can be made to inter-connect at any required cross-point by the vertical bar trapping a metal finger attached to the horizontal bar when the latter is tilted by a relay operated by the marker-circuit in the common-control. Figure 10.27 shows a diagram of a typical cross-bar switch. The reed-relay switch is an "electronic" cross-point consisting of two contacts inside an evacuated envelope and surrounded by an inductive coil. When the coil is energized it induces a magnetic field which forces the contacts together. The reed-relay switch is also of the matrix type with horizontal and vertical wires, the interconnection again being controlled by the marker.

Although systems differ in their arrangements and in the names used to

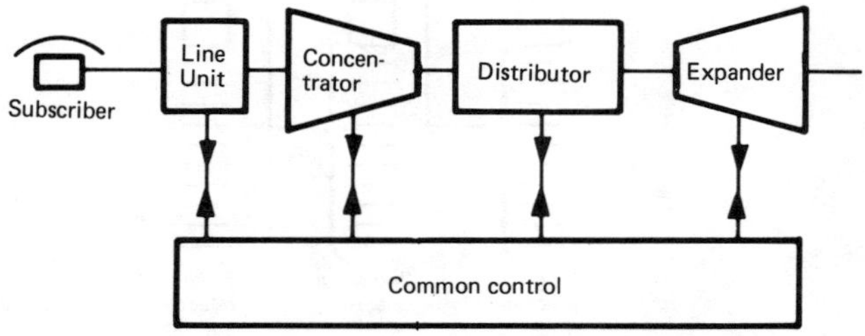

Fig. 10.26 Use of common control.

Fig. 10.27 Typical cross-bar switch. Reproduced from Standard Telephones & Cables Ltd (1973) *The BXB System Components, Equipment and Power* by permission of STC Telecommunications Ltd.

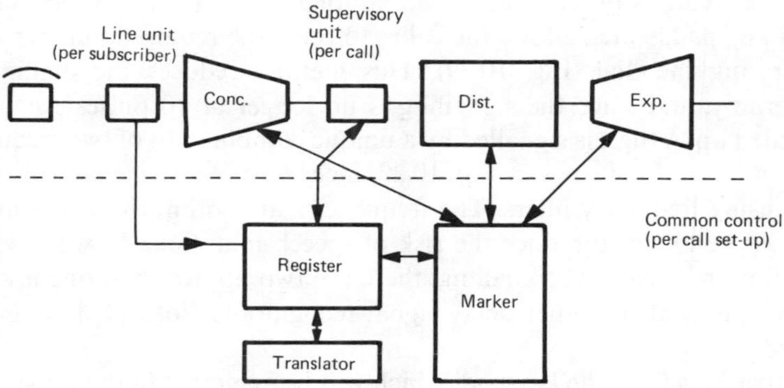

Fig. 10.28 Centralized control exchange.

identify particular units, Fig. 10.28 shows the basic form of a common-control exchange.

Subscribers are attached directly to a subscriber's line unit which recognizes a request-for-call condition. When the calling subscriber goes off-hook, the line unit indicates the request to the common control that seeks a free outlet from the first switching stage, which is a concentrator, and returns dial tone to the caller. Each outlet from this first stage of switching is connected to a call-monitoring (supervisory) unit that maintains an awareness of the call for metering purposes, and reacts to a clear-down signal. The calling subscriber dials and the digits are stored in a register in the common control. If translation is required, the register passes the digits to a translator when one becomes available, and the translation is sent to the marker. The marker examines the switching units for a free path to the called outlet and tests the outlet to find out if it is free. If it is not, busy tone is returned to the caller. If it is free, a path through the switches is set up, ring-current is sent to the called subscriber and ring-tone is sent back to the caller. The common control is released and the progress of the call is noted by the call-monitoring unit.

There will usually be fewer translators provided than registers and many fewer registers than call monitoring units, the provision being proportional to the relative holding times of these elements except that in an exchange which requires only one translator, two may be provided for security reasons.

There is considerable saving in expense if the most provided control equipment, the subscriber's line unit, is made as simple as possible, and more complex features of the control are placed in those units that are required on a per set-up basis only, like the registers and markers.

10.28 Multi-frequency Signalling

The rather tedious procedure of dialling long numbers can be alleviated by using a key-pad, which allows the caller to press the required number instead of operating the dial (Fig. 10.29). This method reduces the dialling time considerably, and since the signalling is no longer at 10 pulses/s it too can be speeded up. A digit is signalled by a unique combination of two frequencies, one from each of the rows in Fig. 10.29. Detection at the exchange is carried out by using frequency filters. The frequencies are within the voiceband and care must be taken to reduce the risk of speech imitation. As we saw in our discussion on in-band vf signalling, there are two approaches, one involves a guard circuit and the other delay signal recognition. Both of these methods are used.

Another benefit of the key-pad, which will be exploited in digital systems is that there are more combinations of frequencies than the ten required for

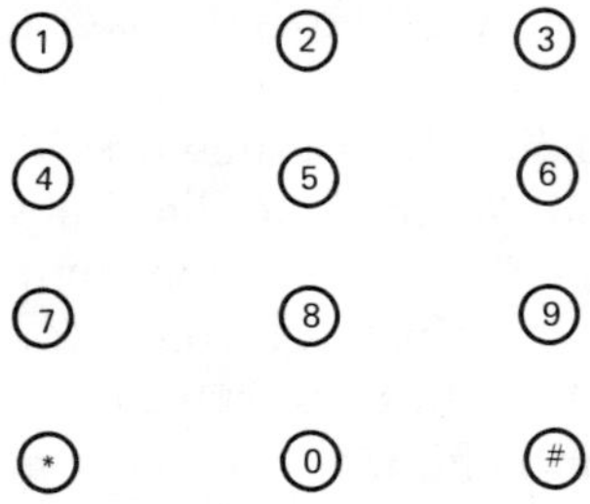

(a)

f_1 (Hz)	1209	1336	1477	
f_2 (Hz)	697	770	852	941

(b)

Fig. 10.29 (a) Keypad arrangement; (b) frequencies used in keypad.

digits. If there are 16 possible combinations – one frequency from four, twice – there are five potential combinations that could used to signal other information to the exchange. Some systems have only two additional unused combinations because the pair of frequencies are chosen from a group of four and a group of three, giving 12 in all.

The development of common control based on computer operation removed some of the constraints on inter-exchange signalling. Very rapid reception, detection and processing were possible, and for signalling between processors care need not be taken to counter speech imitation. CCITT No. 4 was developed as a fast-inter-exchange mf signalling method based on transmitting signals as a combination of two from five or two from six, depending on the repertoire required.

10.29 Common-channel Signalling

One of the benefits of the rapid development of digital systems has been the international commitment to a new signalling system. The communication between exchange processors could be more efficiently performed if it were in a format and on a link disassociated from the speech channels. In addition, it is no longer true that communications on "telephone" networks are restricted to telephony. Management and administration information within the system is transmitted best in digital form, passing from one computer to another, and digital systems will accommodate integrated services that combine data and voice. The conventional signalling systems of dc loop disconnect and vf have severe limitations in respect of digital working and demand for modern facilities. They are relatively slow, they have a limited repertoire and they are not in a form that can be readily dealt with by a computer processor.

There has been international agreement that the signalling method to be used in digital working will be CCITT No. 7. Many signals can be sent along one channel that will operate at a rate of 64 kbits/s. The channel will not follow the same path as the speech channels; i.e. it will be non-associated and it may go through some intermediate node, even if the speech paths do not (Fig. 10.30).

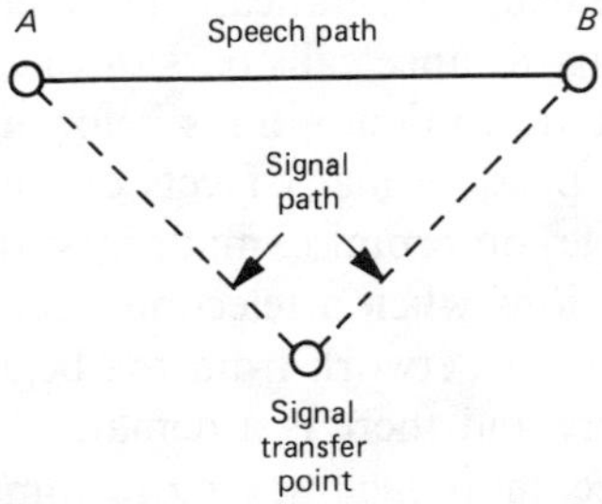

Fig. 10.30 Separate speech and signal paths in common-channel signalling.

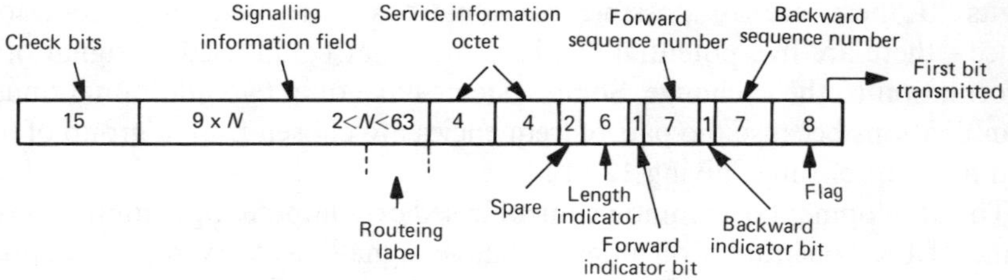

Fig. 10.31 CCITT No. 7 signalling packet.

The format can be considered to have two major parts, the message transmission part and the user part. Figure 10.31 shows the layout of a typical signalling message. The user part is that which contains the signalling information, which may be of various lengths depending on the particular message being transmitted. The rest is the vehicle that carries the user part. It consists of identity bits for the originating signalling node, for the destination signalling node and for the type of user (e.g. telephony or data). Bits to allow error detection are added and there are bits concerned with acknowledging that the signal has been correctly received. Finally there is a short code to indicate the start of the message. The implementation of CCITT No. 7 depends on the particular digital system. Basically, however, the message is formed by several processors at one signalling point and received and translated into tasks at the destination signalling point. If an error is detected, the signal is not acknowledged and it is retransmitted along with all subsequent signals to ensure that the correct sequence is maintained.

The format of CCITT No. 7 signals, and the common-channel working, mean that some of the analytical problems concerned with finding the capacity of a signalling channel, and the delay in passing through a signalling point are similar to those found in packet-switched networks.

10.30 The Telephone Network

The telephone network has developed dramatically in recent years so that it is now possible to make calls automatically between subscribers separated by thousands of miles. Long-distance calls pass through several stages of switching and several possible transmission links before reaching their destination, and to make such calls possible many facets of telecommunications must be integrated and reasonable compromises made by systems designers. In practice, there are very few occasions when a telecommunications administration has the opportunity to design a network from the beginning. Usually some sort of network exists already and there is a demand for increased service: more lines, better performance, more facilities, more comprehensive direct dialling, etc. This demand has to be satisfied by grafting on to the existing system new

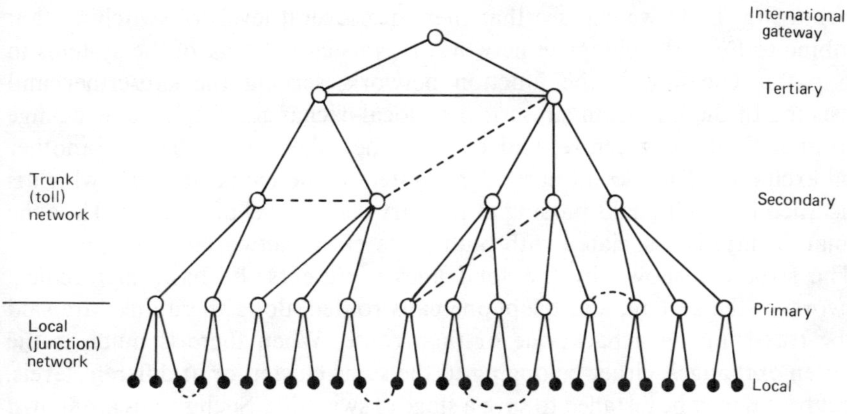

Fig. 10.32 Hierarchical routing system.

switching and transmission equipment that may be totally different in type and in its principles of operation. Recent years have seen two revolutionary developments: the integrated circuit leading to cheap, reliable, small and very powerful digital systems; and the optical fibre, which is much better in many respects than conventional transmission media. These two have led to an entirely new approach, and enormously enhanced facilities for future systems. One of the major planning problems is how to graft such new systems on to existing equipment. We shall return to this point later.

Although the structure of telephone networks has developed piecemeal as demand has increased, it still has some identifiable form. The passage of a call through a national network can be represented by the multi-level diagram shown in Fig. 10.32. Before going further, some comment is necessary on the terminology to be used. The words used in North America and the UK differ significantly in their meaning. Here we will refer to the various levels of switching by using the CCITT recommended terms. Table 10.4 relates them to those commonly used in North America and the UK.

Table 10.4 Terminology

CCITT	USA	UK
Primary centre	Toll centre	Group switching centre
Secondary centre	Primary centre	District switching centre
Tertiary centre	Sectional centre	Main switching centre
Trunk exchange	Toll office	Trunk exchange
Trunk network	Toll network	Trunk network
Trunk circuit	Trunk	Trunk (circuit)
Local exchange	End (central) office	Local exchange
Junction circuit	Inter-office trunk	Junction

From Fig. 10.32 we can see that there are several levels of switching that combine to form the complete network. It is usual to think of the systems in two parts. The first is the junction network, serving the subscriber and consisting of the link from subscriber to local exchange, from local exchange to primary switching centre, and back to the called subscriber via another local exchange. The second part of the system is the trunk network, which is concerned only with calls passing at primary centre level and above. Thus the primary centre is associated with both parts of the network.

The structure shown by the solid lines represents the basic hierarchical network of the system, and telephone calls routed along these links are said to be travelling on a backbone or final route. When there is much traffic between exchanges, either belonging to the same cluster, or at different levels, direct routes may be installed to save a stage of switching. Such routes are shown as broken lines in Fig. 10.32. These direct routes are dimensioned to a different grade of service from the final routes. Because a call finding all the direct-route circuits busy can be placed on the backbone route, the direct route can be dimensioned to a higher grade of service than the final route, and thus work more efficiently.

The number of exchanges at the various levels depends on several factors: the physical extent of the network, the number of subscribers, the amount of traffic, the forecast growth and the transmission methods used. Beyond the top level of the national system is a layer that gives access to the international network. This layer may consist of one or more international (usually called gateway) exchanges.

10.31 Numbering Schemes

In modern systems, the numbering scheme used by a telephone administration to allocate subscribers' numbers has an underlying plan, and there are a few constraints on the development of the plan that must be taken into account:

(i) It must provide each subscriber with a unique number within the national network.
(ii) The allocation to areas must be able to meet forecast growth for several decades.
(iii) The number of digits should not exceed that recommended by CCITT.

In principle, (i) is easy to satisfy if (ii) and (iii) have been met. The length of the number recommended by CCITT is $11 - n$ where n is the country code (see below). If, for example, n is 2, then the national number should not exceed nine digits in length. These digits are used to denote the subscriber's number on the local exchange, the exchange within a given area, and the area within the national numbering scheme. In many local exchanges there is a maximum capacity of 10 000 lines, thus the last four digits of the national number are

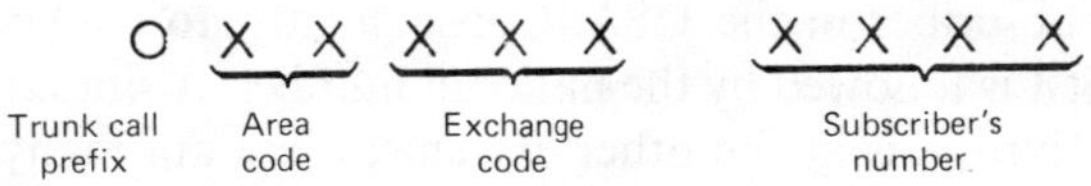

Fig. 10.33 National telephone number.

allocated to the subscriber's number in the exchange. Of the remaining five digits in our example, the first two would denote the area and the remaining three the exchange within the area. Thus the number has the form shown in Fig. 10.33.

The division of digits between area exchange and subscriber's code may not be the same as that shown – but all three components exist in every number.

For automatic long-distance dialling a prefix is necessary to indicate to the exchange equipment that a trunk call is being made. In many countries a "0" is used, but any other digit would do. In calculating the length of the national number the prefix is not included.

International Number

In the introduction to this chapter we imagined a future when person-to-person calls could be made very easily via individual instruments and satellite links. For such availability of connections to be possible each subscriber in the world must have a unique number. To achieve that, some agreement between countries is essential; each country must be identified by a number different from that of all other countries. That is achieved by agreement through CCITT on the way in which these codes, called country codes, are allocated.

The first digit of the country code is the zone code; the world is divided up into nine zones and each country belongs to a zone. The relationship between zone number and geographical area is shown in Table 10.5.

In all but two of the zones, one or two digits are added to the zone number to produce the country code. For example, Brazil has a zone number 5 and an additional country code digit 5 to give its country code 55. Brunei is in zone 6 and two further digits are added to give a country code 673.

The two exceptions are zones 1 (North America and the Caribbean) and 7 (USSR). Throughout each of these zones there is a linked numbering scheme that means, for example, that no subscriber in Canada has the same national

Table 10.5 Zone numbers

1 North America	6 Australasia
2 Africa	7 USSR
3 Europe	8 Eastern Asia
4 Europe	9 Far East and Middle East
5 South America	0 Spare

number as a subscriber in the USA. Consequently, to connect to anyone in zone 1 the digit 1 is followed by the national number. A similar situation exists in the USSR. Europe is at the other extreme; there are many countries with large national networks that have nine digits in their national numbers. For these, a two-digit country code is required, and that can only be achieved by having two zone numbers allocated to Europe.

The division of the world into the zones shown in Table 10.5 is intended to be satisfactory until early in the next century, but clearly, as some large countries develop their telephone networks, some adjustment will be necessary at some future time.

The national and international numbering schemes we have discussed above are the simplest. However, in several parts of the world there are small exceptions, particularly in regard to local calls. In the scheme where the national number is used for all calls within a country, it can lead to irritation on the part of the subscriber and long set-up times for the exchange equipment. Consequently, in many countries local calls use a shorter code. For calls within the same area, the area code is omitted, and for small single-exchange areas no exchange code is used for own-exchange calls. Coupled with this last arrangement will be a very short code for calls to adjacent exchanges; these arrangements are particularly well suited to rural areas. The disadvantage of short codes is that they change with the location of the calling subscriber, and therefore a short code directory must be available in each exchange area.

10.32 Routing Calls

The early type of switching equipment, called step-by-step or Strowger, operated by using the dialled pulses to move the selectors to the position corresponding to the digit dialled. In many ways this was an excellent system, but one major disadvantage was that it allowed no flexibility in the way calls were routed – the route was pre-determined by the dialled digits. Although some systems were modified to overcome this problem it was not until common-control equipment became widely used that the path a call took between calling and called subscribers could be chosen to allow the most efficient use of the available capacity in the system. The function of the common control in the routing process was to store the dialled digits in a register and then translate them into routing digits which would indicate to the switching system the path to take through the network. This register–translator combination is essential to automatic trunk and international dialling schemes; it allows the telephone administration to manage the system efficiently by changing routes as circumstances alter without having to change subscribers' numbers. This therefore separates the subscriber from the system. The subscriber dials the national number from any location and the register–translator automatically selects an appropriate route.

10.33 Digital Systems

Several factors have acted to push telephony from analogue to digital working. To make such a major change, telephone administrations and equipment manufacturers have been persuaded that it is in digital operation that the future lies, and the decade from the mid-seventies was characterized in all equipment producing countries by huge investments of manpower and plant in a race to manufacture an efficient digital telecommunications system. There are more manufacturers involved than at any other time and great financial commitments have been made in the hope that a market will be available for the many products being developed.

Traditionally the two basic elements of a telephone system were transmission and switching. However, with digital operation the whole system is considered as an entity.

Following the development of PCM several countries installed PCM links between analogue exchanges. These worked in a basic 24- or 32-channel format with an encoder and decoder. We can see how the 32-channel system is formed with reference to Fig. 10.34. Strictly speaking, it should be referred to as a 30-channel, 32-time-slot system, because two of the time slots contain signalling

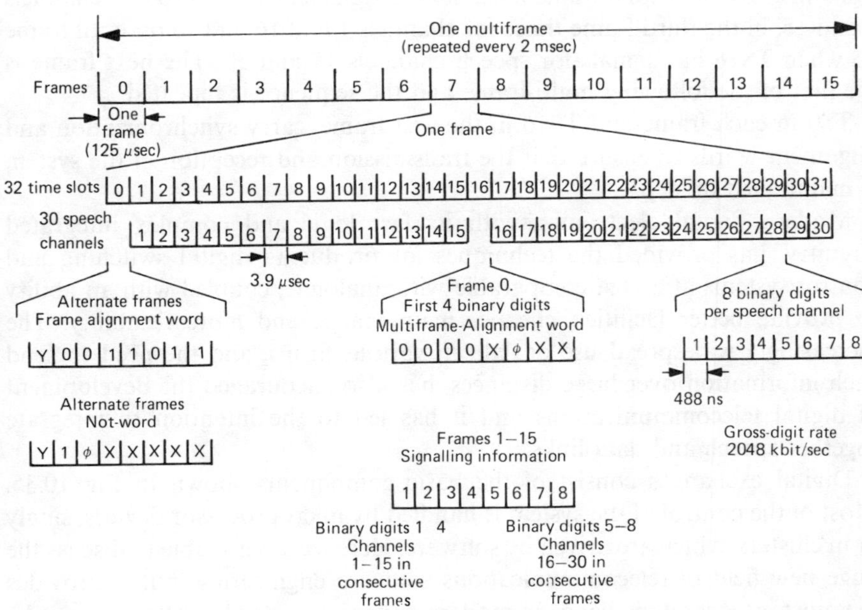

Fig. 10.34 32-frame PCM multiplex frame arrangement: *x*, digits not allocated to any particular function and set to state *one*; *y*, reserved for international use (normally set to state *one*); 0, digits normally *zero* but changed to *one* when loss of frame alignment occurs and/or system-fail alarm occurs (TS0 only) or when loss of multiframe alignment occurs (TS16 only).

and synchronizing information, not speech samples. The sampling rate of each speech channel determines the length of each time slot. For telephony the sampling rate used is 8 kHz, which means that the time between adjacent samples of the same channel is 125 μs. One frame, consisting of 32 time-slots lasts for this time. A time slot is therefore approximately 3.9 μs long. For reasons that we shall see in a moment, 16 frames are put together to form a multi-frame which has a time span of 2 ms. This is the basic unit of the PCM system applied to telephony.

Within a time slot there is the encoded information about the speech sample for that channel. In most systems it is coded into 256 ($=2^8$) levels and there are therefore 8 binary digits within each time slot; each digit must be less than 0.48 μs long. The technique of PCM is described in detail in Section 3.5.

The channel digits bearing speech information must be sent to the right destination and monitored for release and clear-down. Signals must therefore be associated with each channel and in PCM telephony that is done by allocating a signalling word to each channel once per multi-frame. Sufficient information for signalling can be contained in a 4-digit word so that an 8-digit word can contain two signals. Hence a signalling time slot contains information for two speech channels.

Figure 10.34 indicates that time slot sixteen (TS16) is used to carry the signalling. In the second frame it carries the signals related to speech channels 0 and 15, in the third frame those for channels 1 and 16 and so on until frame 15 when TS16 has signals for speech channels 15 and 30. The next frame is the first of the following multi-frame and the sequence is repeated.

TS0 in each frame and TS16 in the first frame, carry synchronization and alignment words to ensure that the transmission and reception of the system is maintained in step.

Modern digital electronics, with its fast logic and complex integrated circuitry, has provided the techniques for producing digital switching and control systems at a cost comparable with analogue, coupled with an ability to provide better facilities, cheaper maintenance and more flexibility. The increasingly widespread use of data in various forms, and the need to send such information over large distances, has also encouraged the development of digital telecommunications and it has led to the intention to integrate together speech and data links.

Digital exchanges consist of the basic components shown in Fig. 10.35. Most of the control of the system is handled by microprocessor devices, singly or in clusters, which are driven by software. Here we are not able to discuss the huge new field of telecommunications software engineering, but it provides the most important challenge in modern system design. The software must be efficient, reliable, secure, understandable and well documented. In theory it affords a degree of flexibility in the operation of the system which is much higher than is possible in hard-wire control. However, such is the complexity of large telephone networks that software development presents the most difficult problems to designers, for it must last for tens of years and although made up of very long programs, it must be able to cope with dramatic changes in hardware technology.

In analogue exchanges the usual figure of merit used is the grade of service, or probability of blocking, but in digital switches the blocking is virtually zero. In these systems the major problems concern delay in the processing of calls caused by the processor units becoming overloaded. The analysis of such problems is very difficult and if simple queueing theory models do not apply resort must be made to computer simulation. One of the difficult tasks of the software engineer is to produce a satisfactory compromise between short efficient programs and those that are longer, more complex and more reliable and secure.

10.34 Digital Switching

The digital switch can have many structural forms depending on the application, the number of connections required, and the technology used. The system shown schematically in Fig. 10.35 is a local exchange and it shows that there are two types of switch involved. A subscriber switch to act as a concentrator, and a central switch which has a distribution function.

The architecture of the subscriber switch will depend to a large extent on the number of subscribers to be attached to the exchange. If a small rural exchange is being connected the switch may act as no more than a multiplexor, being the mechanism whereby up to 30 speech channels are time division multiplexed on to a single PCM carrier. In such a case, since a group of 30 channels is the smallest PCM link available, it may be sensible to allow all subscriber lines full availability access. However if the number of subscribers is somewhat larger a concentrator will be necessary. As in analogue systems,

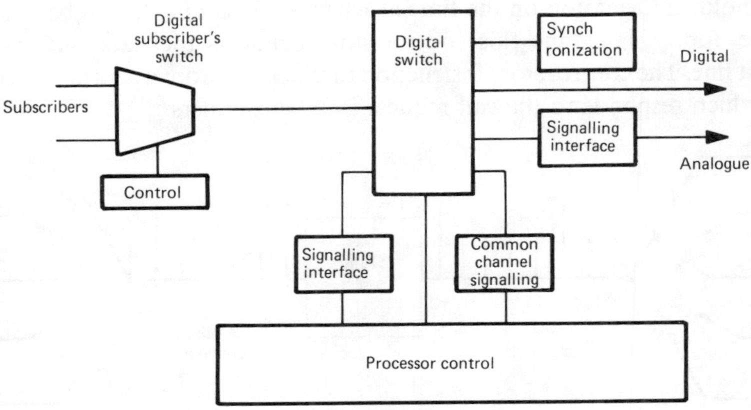

Fig. 10.35 Digital telephone exchange.

a concentrator provides significantly fewer lines on the outlet side than are attached at the input. The fact that the inlets are subscribers indicates that generally the traffic per inlet line will be less than 0.1 erlang thus a very low loss probability can be achieved even if the number of outlets is no more than a fraction of the number of inlets.

Time Switching and Space Switching

The last section dealt with the notion that subscribers lines could be connected to PCM links via a concentrator or a multiplexor. However, that is not the only function of digital switching. It is often used to interconnect a calling subscriber on an incoming line with a called subscriber on an outgoing line, or as a distribution stage as discussed earlier.

We saw in Section 10.33 that the channels on a PCM link are structured into a frame and multiframe arrangement so that any attempt to interconnect calling and called subscribers will require a timeslot adjustment; see Fig. 10.36 for an example: Five subscribers on the inlet side of a multiplexor are able to call subscribers on the output side; if, for example, subscriber C on inlet timeslot 3 wants to be connected to subscriber V on outlet timeslot 4, the inlet sample must be delayed in the switch for one timeslot before being sent out on timeslot 4 for the connection. The type of switch designed to provide the appropriate delay is called a Time Switch. In the early days of digital working, before the development of reliable and cheap computer technology, the delay would have been produced by acoustic delay lines by means of which the signal would have been physically slowed down in its transmission. In modern time switches the delay is much easier to produce. The sample from the inlet channel is stored in a buffer and then read out when the appropriate output timeslot arrives. Figure 10.37 illustrates the way in which this is done. Synchronization is obtained from the frame and multiframe alignment signals in the PCM format (Section 10.33). Each input timeslot word is stored in a buffer. A control store holds information on the time at which each sample has to be read out. At the appropriate time this control store connects the data buffer to the output line. The control store instructions are derived from a central processing unit which responds to the call request from subscribers.

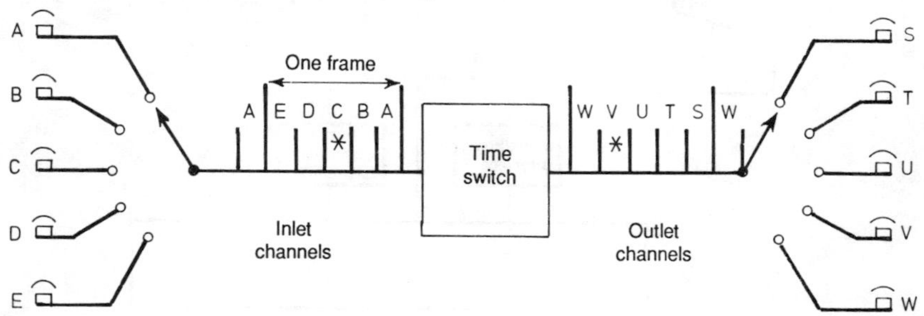

Fig. 10.36 Time switching.

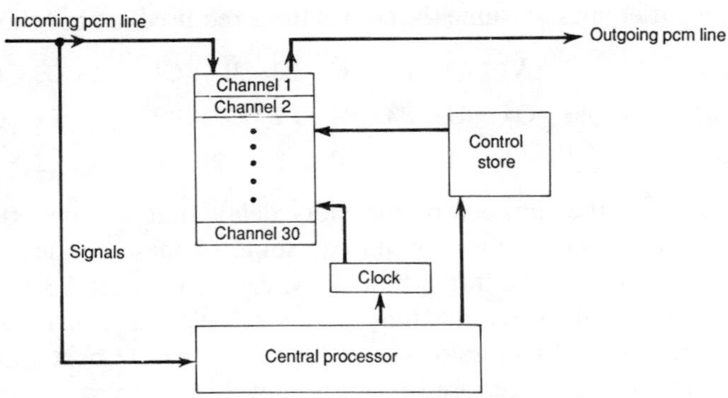

Fig. 10.37 Buffer delay for time switching.

Time switching is clearly not enough if there is more than one incoming and outgoing PCM line. Any channel on an inlet line must be able to obtain access to any channel on any outgoing line and to do this, space switching is necessary to provide the linking between the particular inlet and outlet lines.

For complete flexibility in the operation of the digital switch some combination of time and space switching is required. Although it is possible to achieve full switching with just one time switch (T) in combination with one space switch (S), such a switch would probably suffer internal blocking. More commonly a three-stage arrangement is used, either STS or TST. As an example we will consider the TST configuration, shown in Fig. 10.38. Each inlet time switch has one four-channel inlet line. A frame has four timeslots, and on the outlet side of the system there are three outlet time switches each connected to a PCM line of four channels. A, B and C are the inlet lines and

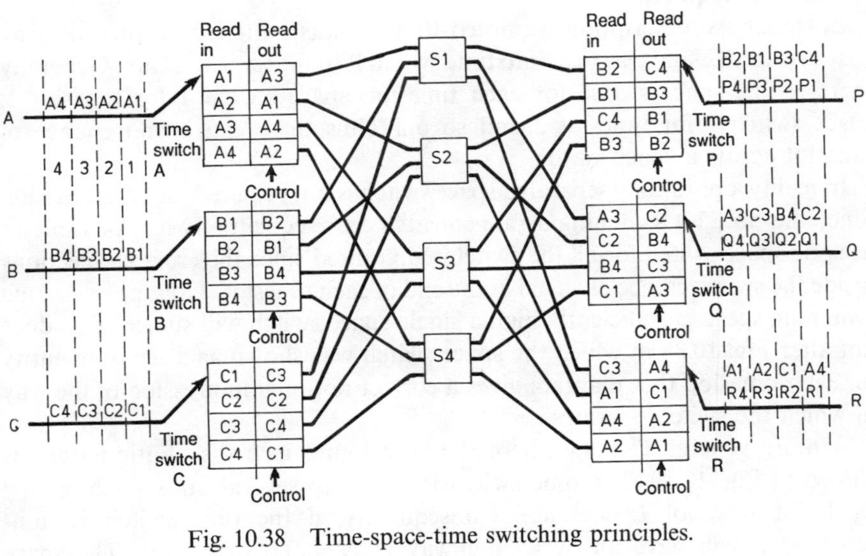

Fig. 10.38 Time-space-time switching principles.

P, Q and R are the outlet lines. Assume the connections required are as follows:

Inlet Channel	A1	A2	A3	A4	B1	B2	B3	B4	C1	C2	C3	C4
Outlet Channel	R4	R3	Q4	R1	P3	P4	P2	Q2	R2	Q1	P1	Q3
Delay (timeslots)	3	1	1	1	2	2	3	2	1	3	2	3

Also shown above is the number of timeslots delay that are necessary to make the input/output connections required. Some of these connections require a delay which crosses the frame boundary, e.g. to connect B3 to P2 requires a delay of three timeslots, meaning that a sample which arrives in timeslot 3 has to be held until timeslot 4 and timeslot 1 of the next frame, have passed; the sample is then released into timeslot 2 of the new frame to make the connection.

On the inlet side of the first-stage time switches, the samples in the speech channels are stored in cyclic order – as shown by the read-in column of the inlet time switches. Similarly the channels are read out from the outlet time switches in cyclic order, from timeslot 1 to timeslot 4 as shown. Consequently any time delay required must be produced by the combined delay between the two time switching stages.

Consider the call-in channel A1. It is to be connected to channel R4, requiring a delay of three timeslots between the input and output of the system. From Fig. 10.38 we can see that A1 is read into timeslot 1 and out in timeslot 2. (The choice of which timeslot to read it out is somewhat arbitrary, being determined to some extent by the calls already set up in the system.) A1 is switched in the second space switch to time switch R and is read into that switch in timeslot 2. The additional delay of two timeslots is provided in time switch R since A1 is stored until timeslot 4 arrives, when it is read out to line R as required.

In the above description we noted that A1 was switched in timeslot 2 by space switch S2, and was read into time switch R in the same timeslot. Generally there is one space switch for each timeslot; space switch 1 for timeslot 1, space switch 2 for timeslot 2 and so on. Thus there can be no change of timeslot across a space switch.

In reality a physically separated space switch is not required for each timeslot. Since the timeslot switching arrangements occur sequentially it is possible to use one space switch for all the switching required since the cross connections inside the switch can be changed at the end of each timeslot. Thus, provided the switching speed is sufficiently high a single space switch will suffice. This does not alter Fig. 10.38 in which the space switch was shown as a separate entity for each timeslot; that diagram gives a correct notional impression of the way in which the switch operates.

A more general TST switch of the type found in a distribution stage is shown in Fig. 10.39. The time switch is split into several units, each having M PCM links of L channels. Consequently, if the time switch is non-blocking it will have an outlet highway of $N = ML$ time slots. The space

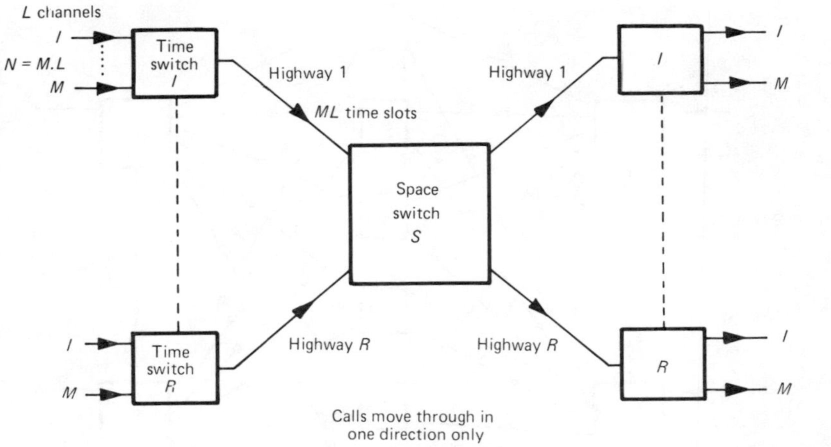

Fig. 10.39 Network representation of a digital switch.

switch is square with R inlet highways and R outlet highways. The purpose of the TST unit is to allow a particular call, which occupies a specific channel into one of the time switches, to be connected to a particular outlet channel. Basically the switching is between highways on either side of the space switch. Each highway has N time slots and in order for a particular call to be connected say from $H1$ to $H3$ it must find a time slot which is free in both highways. This slot may not be the same as the required incoming and outgoing slots for the call, and so some time delay, provided by the time switches, is necessary. The method used to produce the delay again depends on the technology employed, but the delay may be from 1 to 31 time slots depending on the relative positions of the time slots in and out of the unit, and the chosen free time slot in the space switch.

To understand the behaviour of the TST switch in terms of the link systems considered earlier it is important to appreciate that for each time slot the interconnections in the space switch will be different; at each time slot there will be a different set of calls in progress and the connections between the highways will last for only one time slot period then new connections will be established. This can be represented by having N space switches (Fig. 10.40), one for each time slot.

Whether or not blocking occurs in the TST unit depends entirely on the dimensions of the space switch, and since in modern systems switches are comparatively inexpensive they are usually large enough to make blocking negligible. For total non-blocking there must be at least as many outlets as inlets on the time switches, and the space switch highways must have $2N - 1$ time slots, where N is the number of time slots in a link to a time switch.

Digital switches are uni-directional, and that implies that two paths are required to connect two channels X and Y, one for conversation from X to Y and the other for conversation from Y to X. To reduce the control process,

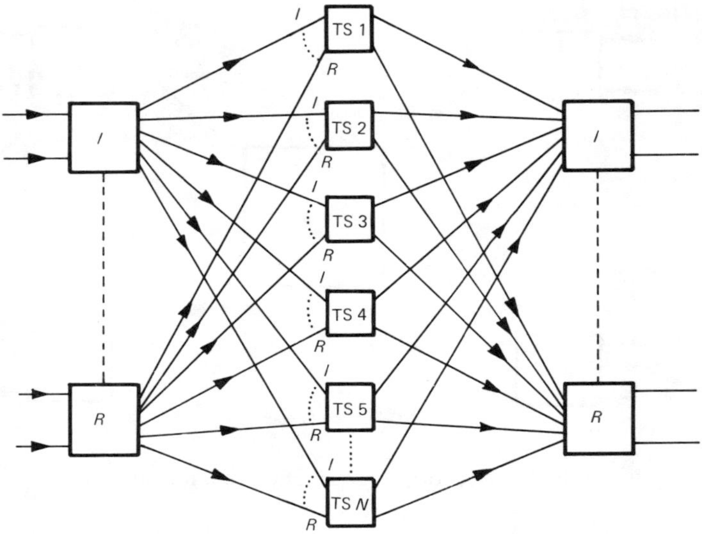

Fig. 10.40 Analogue equivalent of TST switch.

the *X* to *Y* slot is chosen according to whatever rules are used by the designer, and the *Y* and *X* interconnection is allocated a fixed number of time-slots from it, e.g. one, or half a frame. By this method, if the *X* to *Y* connection is available, the *Y* to *X* must also be free.

The digital switch just described is situated in a main exchange, forming part of the trunk network. Interconnections between exchanges are made, for the information paths, via PCM links. However, signalling is carried over a common channel, using signalling system CCITT No. 7 as described in Section 10.29. The inter-exchange signals will be concerned not only with setting up calls between exchanges, but with accounting, administration fault diagnosis and maintenance. As described earlier, the CCITT No. 7 system transmits signals as messages that can have variable length. Each message is preceded by labels that identify its origin and destination exchange, the type of message (call handling, fault, etc.) and includes error-detection and acknowledgement bits. If an error is detected in a message, that and all subsequent messages are retransmitted to ensure that the sequence received at the far end is in the correct order.

The error rate has an important bearing on the capacity of the signalling channel; retransmission of incorrectly received messages obviously takes up time that could be used for new signals and consequently slows down the overall process. The capacity is specified in terms of number of messages per busy hour, given that the delay from end to end is not greater than some predetermined value.

On the subscriber side of the digital switch there will be a local unit of some description. For large areas a local digital exchange would be used with mf signalling from subscriber to exchange where conversion to a PCM format would take place before concentration through a digital switch. Alternatively for very small units no exchange facility would be available, but a simple digital concentrator would be used to take in the analogue channels, convert them to PCM and multiplex them on to a single highway to the nearest local exchange. Calls between subscribers on the same concentrator would then have to pass through the local exchange. Signalling in these PCM links would be on TS16 and the control, software and firmware at the main exchange would convert it to common channel if a trunk call were required.

The introduction of digital systems is rarely a starting point for the telephone system. Usually a system exists and the digital equipment has to be grafted on to it. Whatever method is used, interworking between the old and new systems is required, and one area of difficulty is the interfacing of various analogue signalling systems with the new equipment designed to operate on TS16 or common channel. This interfacing can be a severe problem if there are many existing signalling schemes in a particular network, and the development of satisfactory units can add considerably to system costs.

10.35 Conclusion

In some respects, the whole concept of telecommunications is changing, and with it the services and functions provided by telecommunications operating companies. The very rapid growth of information technology, with its demand for high-speed, large-capacity, data links, with video output, and the introduction of new facilities on telephone exchanges, are both causing manufacturers to rethink the whole philosophy of how communication systems should best be provided. In the next few years the main point at issue will be to decide whether future systems should be fully integrated, or not. If so, should they be based on a digital telephone system, with its centralized switching units, or on one of the many data networks, with distributed control? Conversely, economy and efficiency may dictate an extension of the present hybrid scheme in which both methods co-exist, with perhaps an element of interworking between them. It is too early in the development of distributed systems to predict what changes will take place in the next decade, but that major redesign will occur is beyond doubt.

References

1. Brockmeyer, B., Halstron, H.L. and Jensen, A., *The Life and Works of A.K. Erlang*, Copenhagen Telephone Company, 1943.
2. Jacobaeus, C., 'A study on congestion in link systems', *Ericsson Technics*, No. 48, 1950.

3. Cooper, R.B., *Introduction to Queueing Theory*, Edward Arnold, London, 1981, Chapter 5.
4. Fry, T.C., *Probability and its Engineering Uses*, Van Nostrand Reinhold, Wokingham, 1965.
5. Fishman, G.S., *Principles of Discrete Event Simulation*, Wiley, Chichester, 1978.
6. Flood, J.E., *Telecommunication Networks*, Peter Perigrinus, London, 1975.
7. Bear, D., *Telecommunication Traffic Engineering*, Peter Perigrinus, London, 1976.
8. Hills, M.T., *Telecommunication Switching Principles*, George Allen & Unwin, London, 1979.

Problems

10.1 A traffic-recording machine takes measurements of the number of busy devices in a group every three minutes during the busy hour. If the sum of the devices busy over that period is 600, what is the value of traffic carried?

Answer: 30 erlangs.

10.2 A loss-system full availability group consists of five devices. If the mean call holding time is 180 seconds, and the call intensity is 80 calls/hour, what is the mean load per device?

Answer: 0.8 erlangs.

10.3 In a particular system, it was found that during the busy hour, the average number of calls in progress simultaneously in a certain full availability group of circuits was 15. All circuits were busy for a total of 30 seconds during the busy hour. Calculate the traffic offered to the group.

Answer: 15.13 erlangs.

10.4 A group of eight circuits is offered 6 erlangs of traffic. Find the time congestion of the group, and calculate how much traffic is lost.

Answer: 0.122; 0.73 erlangs.

10.5 A ninth circuit is added to the group in Question 10.1. What traffic will it carry?

Answer: 0.31 erlangs.

10.6 A system of six telephones has full availability access to six devices. Find the probability that $1, 2, \ldots, 6$ devices are busy. What is (a) the call, and (b) the time congestion of the system if the carried traffic is 2.4 erlangs?

Answer: (a) 0; (b) 0.004.

10.7 (a) Two erlangs of traffic are fed to three devices. What is the congestion, and how much traffic is lost?
(b) Two erlangs of traffic are fed to one device. The overflow is fed to a second device, and the overflow from that to a third. What is the overall congestion, and is the value of the traffic lost the same as in (a)?

Answer: (a) 0.2105, 0.421 erlangs; (b) 0.154, No.

10.8 Show that, if the assumption of statistical equilibrium is valid, the probability of a system being in state i is given in terms of the probability that it is in state 0 by

$$[i] = \frac{\prod_{j=0}^{i-1} \lambda_j}{\prod_{j=1}^{i} \mu_j}[0]$$

10.9 The state transition diagram below represents a system with an infinite number of devices subjected to calls arriving at random with fixed mean arrival rate, λ.

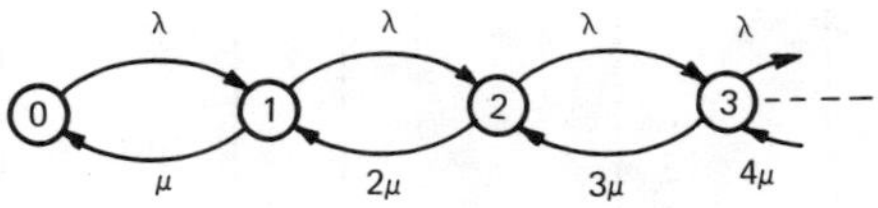

If $\lambda/\mu = A$, the mean offered traffic, use the birth and death equations, and assume statistical equilibrium, to show that

$$[i] = \frac{A^i \exp(-A)}{i!}$$

10.10 An infinite number of sources feed 5 erlangs of traffic into eight devices. Find the probability of the network being in each of its possible states, i, and check that $\sum_i [i] = 1$. Plot $[i]$ against i.

Answer: 0.0072, 0.0362, 0.0904, 0.1506, 0.1883, 0.1883, 0.1569, 0.1121, 0.0700.

10.11 Repeat question 10 for (i) 2 erlangs, and (ii) 7 erlangs of traffic, noting the variation in both the congestion and the distribution of $[i]$, with increasing traffic.

Answer: (i) 0.1354, 0.2707, 0.2707, 0.1805, 0.0902, 0.0361, 0.0120, 0.0035, 0.0009; (ii) 0.0013, 0.0088, 0.0306, 0.0715, 0.1251, 0.1751, 0.2044, 0.2044, 0.1788.

10.12 A telephone route of n circuits has to carry a normal load of 3 erlangs. If the grade of service must not exceed 0.03, what is the smallest value that n can have? In an emergency, there is a 20% increase in offered traffic. What will be the grade of service in this overload condition?

Answer: $n = 7$, 0.0438.

10.13 Draw the chicken diagram equivalent of the two-stage link network shown below, and calculate the internal blocking.

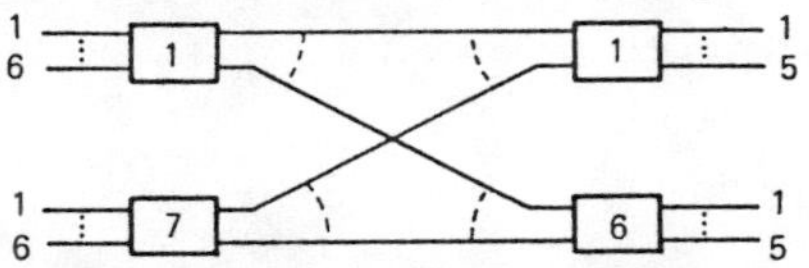

Corresponding outlets from second-stage switches form a route.

Traffic offered per route = 2.5 erlangs
Traffic carried per inlet switch = 1.8 erlangs

Answer: 0.0411

10.14 A time–space digital switching configuration is shown. What are the functions of the time and space elements in relation to the speech channels on the PCM inlets and outlets? Draw the analogue equivalent of this switch.

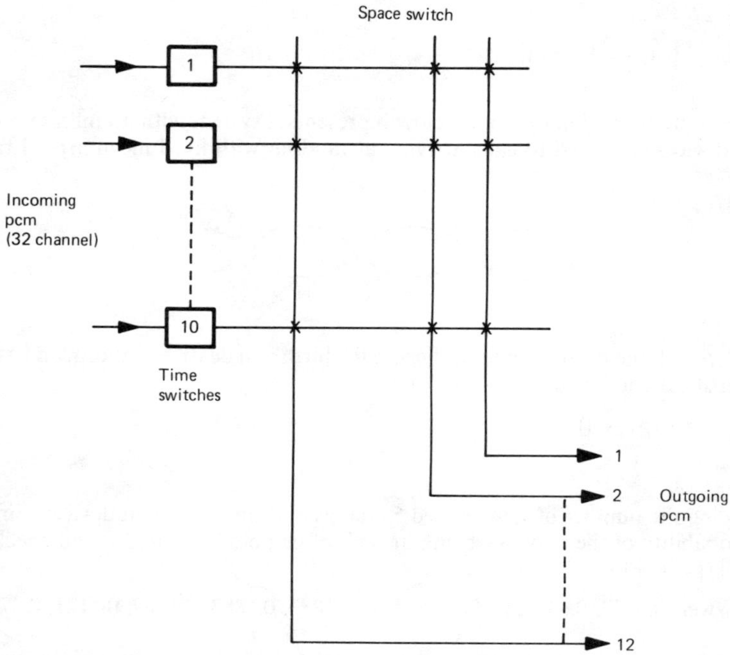

11 TELEVISION SYSTEMS

11.1 Introduction

In this chapter we shall be concerned primarily with the fundamentals of domestic broadcast television systems. There are, of course, many specialized closed-circuit television systems, but these are all based upon the same principles as the broadcast system and will not therefore be considered separately.

Television has progressed over a period of about 60 years from John Logie Baird's mechanical scan system to the modern highly sophisticated colour systems that can transmit both entertainment programmes and teledata information. During this period the quality of sound and visual image reproduction has increased steadily. The basic principles governing the transmission of visual information by electrical means have, however, remained unchanged. These principles rely upon a property of the human eye/brain combination known as "persistence of vision". This means that under certain conditions, the brain is unable to differentiate between a moving image focused on the retina of the eye and a rapid sequence of still images. Television is specifically designed for the human eye/brain combination, and we shall therefore begin by considering the subjective response of this combination to incident light energy.

11.2 Measurement of Light and the Response of the Eye

Light is a form of electromagnetic radiation and the power radiated from any source is measured objectively in watts. Light energy is visible only over a very restricted range of wavelengths, from 4×10^{-9} to 7×10^{-9} m. The subjective sense of "brightness" varies with the wavelength of incident light, where in this context brightness means the impression on an observer of whether a light source is bright or dim. The response of the eye varies between individuals but the variation in this response has been found to be small enough to allow a "standard observer response" to be defined. This standard response is shown in Fig. 11.1.

A black body radiator radiates equal energy at all wavelengths. The luminous

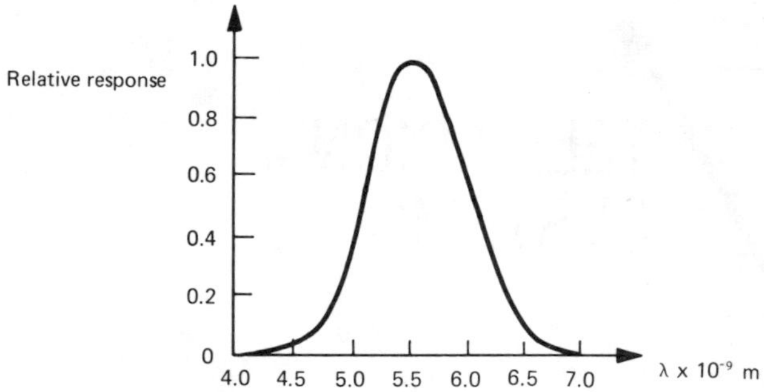

Fig. 11.1 The standard observer.

intensity of such a source is obtained by weighting the uniform spectral density of the source by the response of the standard observer. The resulting subjective unit is the candela. This is defined as 1/60th of the luminous intensity per cm^2 of a black body radiator maintained at a temperature of 2042K. The lumen is the amount of light passing through unit area at unit distance from a point source of 1 candela, i.e., the total emission from a source of 1 candela is 4π lumens. The lumen is the unit of luminous flux and, unlike the power output from a light source, depends upon the response of the standard observer.

It is clear from Fig. 11.1 that two light sources of equal power but different colour do not necessarily have the same luminous flux. Two light sources of different colour that appear equally bright to the standard observer do have the same value of luminous flux measured in lumens.

The significance of the last statement will become clear when we consider colour theory in Section 11.8. As far as monochrome television is concerned the response of the camera when viewing a coloured scene should closely resemble the response of the standard observer, i.e. the electrical output should be proportional to the brightness of the scene measured in lumens. Such an output signal is termed the luminance signal.

11.3 The Monochrome Television Waveform

In most of Europe, with the notable exception of France, television pictures are transmitted using the 625 line standard. We will consider the 625 line signal waveform as a means of explaining the transmission of a three-dimensional vision signal (i.e. one that is a function of time and both horizontal and vertical position) over a one-dimensional channel (i.e. a channel in which voltage is a function of time). The three-dimensional visual image to be transmitted is focused on to a photosensitive (in the electrical sense) surface by an optical lens system. The photosensitive surface may be considered as a large number of separate photoelectric transducers. Each of these transducers produces an

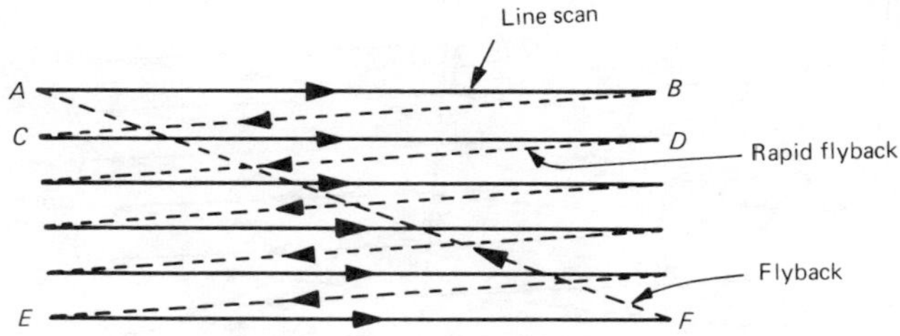

Fig. 11.2 Sequential scanning.

electrical output proportional to the intensity of that part of the image that falls upon it. The original image is thus decomposed into a large number of picture elements known as pels (or pixels). The electrical output corresponding to each pel is one-dimensional since it is a function only of time. The three-dimensional signal, which is the combined output of all pels, is actually transmitted as a rapid sequence of one-dimensional outputs. The original optical image is reconstructed by displaying each pel in its correct spatial position.

This image is produced, in a domestic television receiver, by modulating the beam current of a cathode ray tube (CRT). The screen of such a tube will have the same number of pels as the photosensitive plate in the television camera. The image is then reproduced by effectively connecting each pel in the camera to the corresponding pel in the CRT in turn. This is achieved by scanning the photosensitive plate in the camera with an electron beam synchronized to a second scanning beam in the CRT. Sequential scanning of the camera and CRT are shown diagrammatically in Fig. 11.2. The electron beam is deflected in both vertical and horizontal directions. The scan begins at point A in Fig. 11.2 and proceeds to point B. During flyback, which occurs very rapidly, no information is transmitted. The next line is scanned from C to D and the process is repeated until point F is reached. At this point flyback occurs to point A and the scanning of the complete picture is repeated.

We can see from Fig. 11.2 that each picture scan takes a finite time. If this time is longer than the persistence of vision period the eye will perceive flicker on the picture. If the scan time is very short the number of pels/second and hence the signal bandwidth will be large. The optimum scan rate is therefore one that is just fast enough to avoid flicker. The picture frequency at which flicker occurs is related to the luminance of the picture being viewed. For television systems, flicker is avoided when the picture rate is approximately 50 complete scans/second. In fact it is possible to avoid flicker when the picture rate is as low as 25 complete scans/second. This is made possible by use of interlaced scanning. This is shown in Fig. 11.3. The scan now occurs in two fields. The odd field starts at A and the line is scanned until point B. Flyback then occurs from B to C and so on. It will be seen from Fig. 11.3 that the last line is scanned from D to E, i.e. field flyback occurs from E to F. (In fact this flyback takes the equivalent of several line

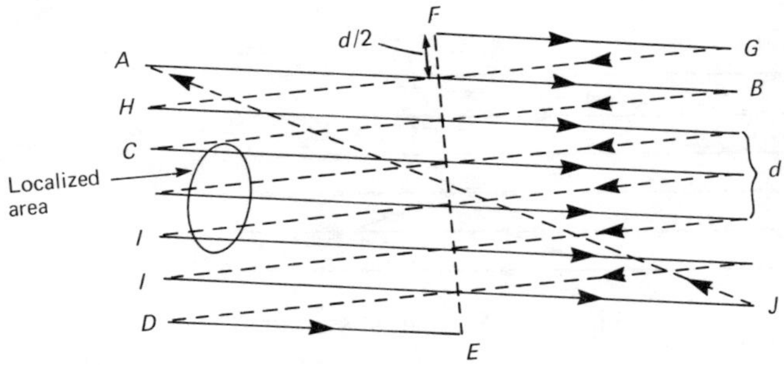

Fig. 11.3 Interlaced scanning.

scan periods so will not be a straight line as is shown in Fig. 11.3.) The even field starts half-way across the picture at *F* and the scan of the first line finishes at *G*, flyback occurring from *G* to *H*. The last line of the even field is scanned from *I* to *J*, flyback then occurs to point *A*, which is the start of the next odd field.

Interlaced scanning avoids flicker because although each complete picture is scanned 25 times/second, a localized area of the picture is scanned by both odd and even fields. Thus each such area is apparently scanned 50 times/second. This deception relies on the fact that the variation in brightness in a vertical direction usually occurs gradually. If there is a significant difference in the brightness between one line and the next, some localized flicker will be observed.

We pointed out earlier that it is necessary to synchronize the scans in the camera and the CRT. To achieve this, a line-synchronizing pulse is transmitted at the end of each line and a field-synchronizing pulse is transmitted at the end of each field. The synchronizing pulses trigger the flyback circuits in the television receiver. The transmitted video signal thus has distinct components, i.e., the picture signal that represents the variation in brightness of each line and the synchronizing pulses that are transmitted below black level. The composite waveform is shown in Fig. 11.4.

The 625 line transmissions in the UK use negative AM of the vision carrier, i.e., an increase in picture brightness produces a decrease in signal power. A portion of the envelope of the modulated signal is shown in Fig. 11.5. The advantage of negative modulaton is considered to be that the black spots

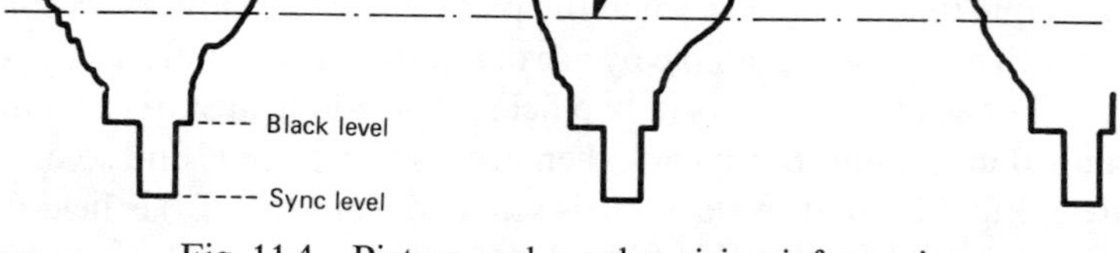

Fig. 11.4 Picture and synchronizing information.

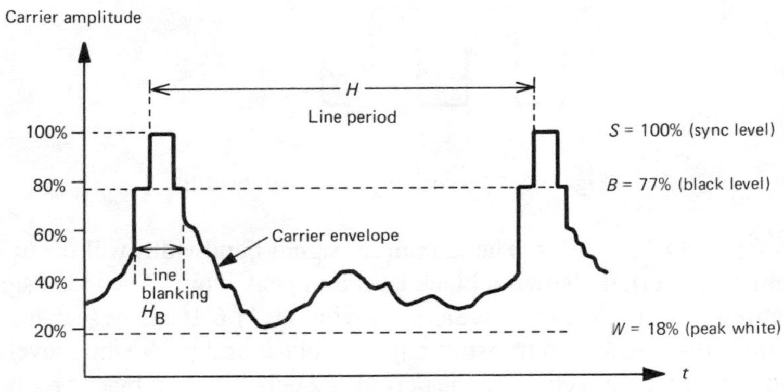

Fig. 11.5 Transmitted carrier envelope.

produced on the screen by some forms of ignition interference are less objectionable than the white spots that would be produced in the same circumstances if positive modulation were used. The ratio WB/BS shown in Fig. 11.5 is known as the picture/sync ratio and has a value of 1/3. This is a compromise figure that produces a reasonable picture quality and adequate synchronization in poor SNR conditions. Figure 11.5 shows that a portion of each line before, and just after, each line synchronizing pulse is below black level. This allows the line flyback to be suppressed and hence no visual output is produced during this interval. A similar provision is made at the end of each field, but the flyback interval is much longer, being equal to 25 line periods. Historically no information was transmitted during the field flyback period but in recent years part of this interval has been used for the transmission of teletext data signals. Teletext transmission is discussed in detail in Section 11.18.

11.4 Bandwidth of a Television Waveform

The bandwidth of a television waveform is directly related to the number of pels transmitted per second. In each 625 line picture 25 lines are blanked off at the end of each field to accommodate the field flyback. The number of lines seen by a viewer is thus $(L - L_B)$ where $L = 625$ and $L_B = 50$. The vertical resolution of the picture is equal to the number of lines seen. Assuming equal horizontal and vertical resolution the number of pels per line is $A(L - L_B)$ where $A(= 4/3)$ is the picture aspect ratio (width/height). Since each line is blanked out for a period of H_B the number of pels transmitted per second is

$$P_s = \frac{A(L - L_B)}{H - H_B} \tag{11.1}$$

For a 625 line system the line period $H = 64\,\mu s$ and $H_B = 12.05\,\mu s$ giving a

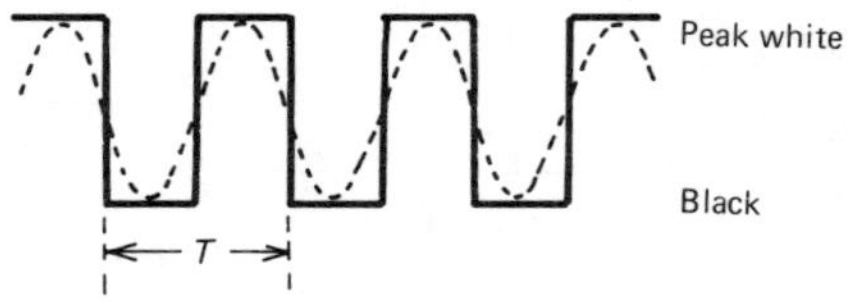

Fig. 11.6 Television picture bandwidth.

value of $P_s = 14.76 \times 10^6$/s. The maximum signal bandwidth will occur when adjacent pels alternate between black level and peak white. The video signal is then represented by a square wave as shown in Fig. 11.6. It can be seen from this figure that it is possible to transmit adjacent black and peak white levels by a sine wave with a period equal to the period of 2 pels. The bandwidth required is then $1/T$ which in the 625 line system is 7.38 MHz. It will be clear from Fig. 11.6 that when the signal bandwidth is restricted 7.38 M Hz the sharp edges of the square wave will be lost. This is virtually undetectable in practice because of the finite size of the scanning beam which does not allow detail of this resolution to be displayed in any case. Statistically speaking, very few pictures require the degree of resolution allowed by a bandwidth of 7.38 MHz. Use is made of this fact to reduce the signal bandwidth even further. The actual bandwidth allowed for 625 line transmissions is $K \times 7.38$ MHz. The constant K is known as the Kell factor. The value of K depends upon several factors, e.g. the width of each line on the CRT and the low pass filtering effect of the eye (which itself depends upon how close the viewer is to the CRT). The value of K used in Great Britain is 0.73. This gives an acceptable subjective picture quality and results in a video bandwidth of 5.5 MHz.

11.5 Choice of Number of Lines

There are various standards in operation throughout the world, but they all have one feature in common, each system has an odd number of lines. The reason for this can be explained by reference to Fig. 11.3. If each picture has $2n + 1$ lines, then each field will have $n + \frac{1}{2}$ lines. If we assume that the distance between adjacent lines of the same field is d the odd field scan finishes halfway across the last line. This means that field flyback (if assumed instantaneous) would cause the even field to start halfway across the first line and would therefore be a distance of $d/2$ above the first line of the previous field. Interlacing is thus achieved automatically with the same field deflection on both odd and even fields.

11.6 Synchronizing Pulses

These pulses are required to trigger both line and field timebase generators in the receiver. The detail of the line synchronizing pulse is shown in Fig. 11.7. The

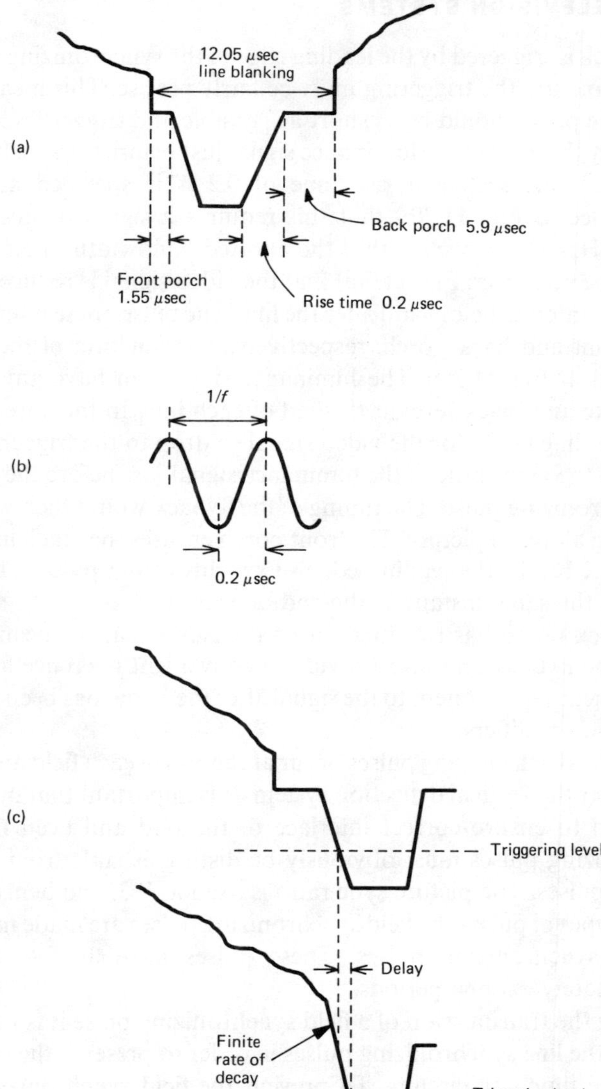

Fig. 11.7 Details of synchronizing pulses.

line flyback is triggered by the leading edge of the synchronizing pulse. To avoid a ragged picture the triggering must be fairly precise. This means that the rise time of the pulse should be as short as possible and trigger level should not be affected by the level of the luminance signal just before the synchronizing pulse. In the 625 line system a rise time of $0.2\,\mu s$ is specified and we can see by reference to Fig. 11.7(b) that this requires a signal bandwidth exceeding $f = 2.5$ MHz. This is well within the allotted bandwidth of 5.5 MHz.

It will be seen from Fig. 11.7(a) that the video signal is reduced to black level for a short interval before and after the line sync pulse; these intervals are known as the front and back porch, respectively. The function of the front porch is illustrated in Fig. 11.7(c). The luminance signal can have any value between peak white and black level at the end of each line. In the absence of the front porch the time taken for the video signal to drop to the triggering level would vary with the amplitude of the luminance signal just before the leading edge of the synchronizing pulse. The timing of the flyback would then vary from line to line giving a ragged picture. The front porch ensures that each line is reduced to black level before the leading edge of synchronizing pulse. The flyback then occurs at the same instant at the end of each line.

The back porch has two functions – it ensures that the beam is blanked off during line flyback and also provides a convenient reference for restoring the zero frequency component to the signal after the signal has been passed through ac coupled amplifiers.

The field synchronizing pulses occur at the end of each field and their purpose is to trigger the vertical deflection system. It is important that precise triggering is affected to ensure correct interlace of the odd and even fields. The field synchronizing pulses must obviously be distinguishable from the line synchronizing pulses. The picture sync ratio is fixed as 7/3, and hence to distinguish the two types of pulses the field synchronizing pulses are made much wider than the line synchronizing pulses. These pulses have, in fact, a duration of approximately 2.5 line periods.

During the transmission of a field synchronizing pulse it is also necessary to transmit the line synchronizing pulses in order to preserve the synchronization of the line timebase circuits. To prevent the field synchronizing pulses from obliterating the line synchronizing pulses they are split up into separate pulses equal in duration to half a line period. By dividing the field synchronizing pulses in this way the same synchronizing pulse may be transmitted at the end of the odd and even fields. The timing relationship between field and line synchronizing pulses is shown in Fig. 11.8. The negative going edge that normally triggers the line timebase is preserved at the correct instant during the field pulses. The negative going edge that occurs half-way across a line does not affect the line timebase circuit, which is halfway across its stroke, as such circuits are designed to ignore trigger pulses until they are close to the end of a stroke.

In the television receiver line and field synchronizing pulses are applied to both line and field timebase circuits. The field pulses trigger the line timebase

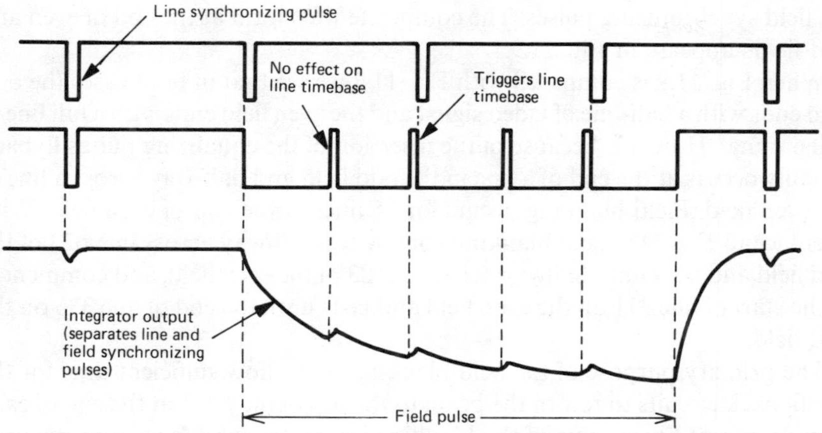

Fig. 11.8 Line and field synchronizing pulses.

circuits at the instants indicated in the Fig. 11.8. The duration of the line synchronizing pulse is too short to affect the field timebase circuits which are usually preceded by some form of integrator. A typical output is illustrated in Fig. 11.8. In this figure the field synchronizing pulse is shown starting at end of a line. In the next field, flyback, and hence the field synchronizing pulse, will occur half-way across a line. It is possible therefore for the residual output on the integrator to be slightly different on odd and even fields before the occurrence of a field pulse. If this is the case, incorrect interlacing will result. To eliminate this possibility 2.5 lines of narrow equalizing pulses are inserted before the start of

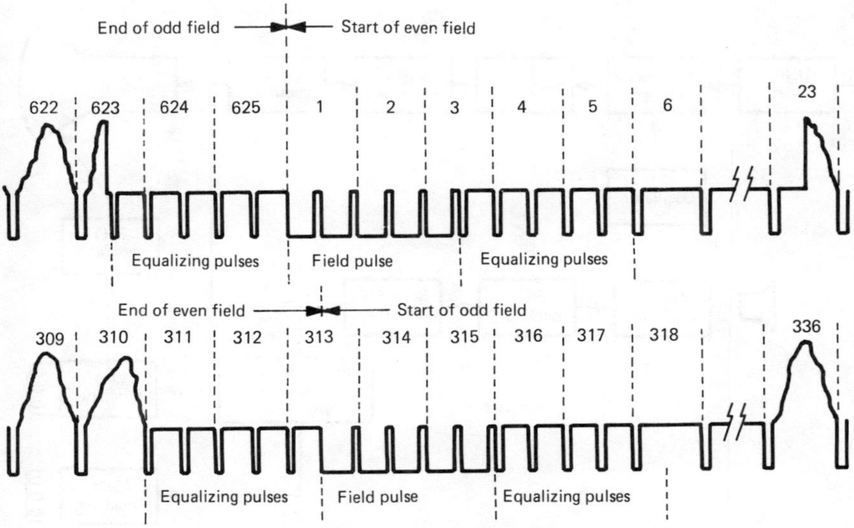

Fig. 11.9 Details of field and equalizing pulses.

the field synchronizing pulses. The composite waveform at the end of even and odd fields appears in Fig. 11.9.

When Fig. 11.9 is compared with Fig. 11.3 we see that in both cases the odd field ends with a half-line of video signal and the even field ends with a full line of video signal. However because of the insertion of the equalizing pulses flyback actually occurs at the end of a line in the odd field and half-way across a line in the even field. Field blanking occurs for 25 lines at the end of each field. With reference to Fig. 11.9, field blanking commences halfway across line 623 of the odd field and lasts until halfway across line 23 of the even field, and commences at the start of line 311 on the even field and lasts until the end of line 335 on the odd field.

The primary purpose of the field blanking is to allow sufficient time for the line flyback circuits to return the beam to the top of the CRT at the end of each scan. In recent times some of the blanking interval is used for transmission of teletext information. (This is discussed in detail in Section 11.18.) Clearly the deflection circuits of the CRT must be adjusted to exclude the blanking from the visible area of the CRT otherwise the transmitted data will be seen by the viewer as a series of variable length dashes.

11.7 The Television Receiver

The domestic television receiver is required to receive signals over a wide bandwidth and is based on the superheterodyne receiver. The television receiver is considerably more complicated than a broadcast radio receiver because the

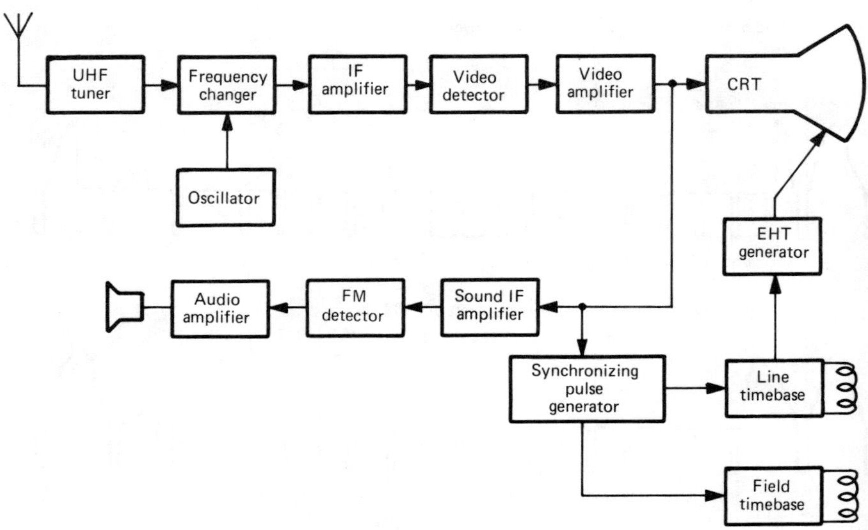

Fig. 11.10 Superheterodyne television receiver.

former is required to reproduce a video signal, synchronized scanning waveforms for the CRT, and a sound signal. The block diagram of a typical monochrome receiver is given in Fig. 11.10.

The British 625 line system transmits the video signal using AM and the sound signal is transmitted using FM. In the transmitted television signal the FM sound carrier is 6 MHz above the vision carrier. The local oscillator in the receiver operates at 39.5 MHz above the vision carrier, the difference frequency being used as the vision IF. The relative positions of sound and vision carriers will of course be reversed in the IF signal. The spectrum of the IF signal and the frequency response of the IF amplifier are shown in Fig. 11.11. The energy in the video spectrum is not, in fact, uniformly distributed as shown

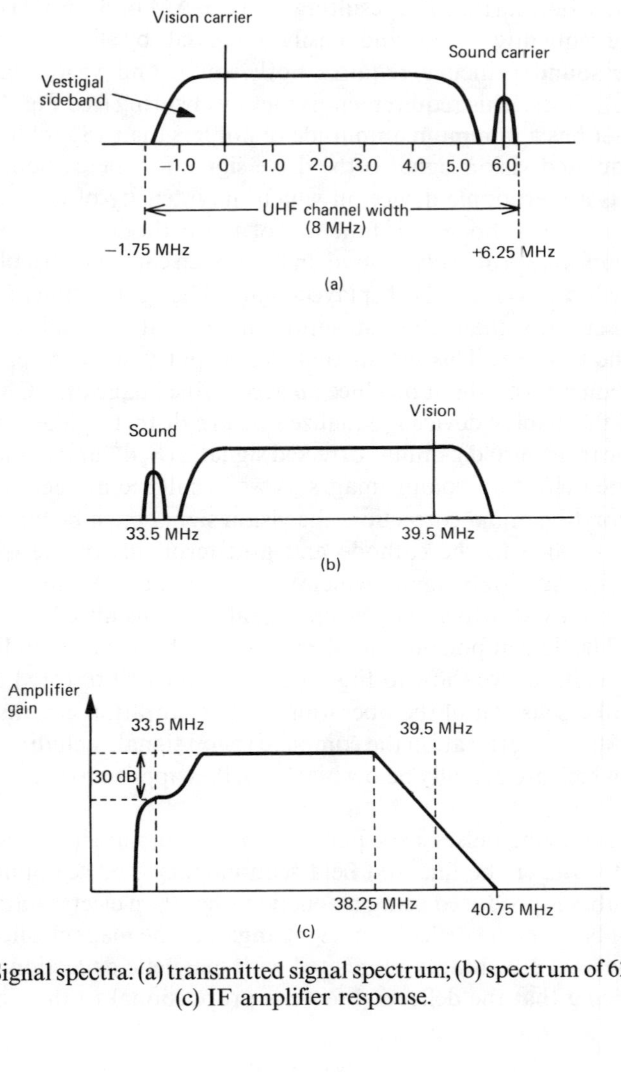

Fig. 11.11 Signal spectra: (a) transmitted signal spectrum; (b) spectrum of 625 IF signal; (c) IF amplifier response.

in this diagram, but tends to be concentrated towards the lower frequencies, i.e. frequencies close to the carrier.

The asymmetric response of the IF amplifier around the frequency of 39.5 MHz is required to compensate for the vestigial sideband transmission of the vision signal. The detection of VSB signals is discussed fully in Section 2.12. The IF amplifier response is 30 dB down at the sound carrier frequency. This is required because both sound and vision carriers are applied to the vision detector, which then produces sum and difference frequencies between these two carriers. The difference frequency of 6 MHz is used as the sound IF and this is known as the "intercarrier sound" principle. This component will be modulated in both amplitude and frequency. The 30 dB drop in the IF amplifier response at the sound carrier frequency ensures that the amplitude of the sound carrier is much less than the vision carrier. The resulting depth of AM of the 6 MHz component is then consequently small and easily removed by limiting. Clearly the intercarrier sound principle requires both sound and vision carriers to be present at all times. This requirement is met by ensuring (see Fig. 11.5) that the vision carrier has a minimum amplitude of not less than 18% of its peak value. The demodulated vision signal is the drive signal for the cathode ray display tube. This is a thermionic device, in which the intensity of the electron beam emitted from the cathode is a function of the cathode-to-grid voltage. The electron beam is electronically focused on to a screen coated with a photoemissive phosphor, which produces the displayed image. The light output of a CRT is not a linear function of the cathode-to-grid voltage, but is equal to this voltage raised to the power γ. This means that the output voltage E_y produced by a television camera would not produce an acceptable image on a CRT. The non-linearity of the display device is equalized by pre-distorting the camera output voltage. The transmitted gamma corrected signal is $(E_y)^{1/\gamma}$ and has an important but unwanted effect on colour images as well shall see in Section 11.12.

In addition to gamma correction, the vision signal must be further modified before it is applied to the cathode and grid terminals of the CRT. The dc component of the vision signal, which represents the average brightness of a picture, is removed when the vision signal is transmitted via ac-coupled amplifiers. The dc component also affects the synchronization of line and field timebase circuits (as we show in Fig. 11.12) and must be restored to the vision signal. A full discussion of the operation of dc restoration circuits is given by Patchett.[1] After dc restoration the complete vision signal, including synchronizing pulses (which are actually below black level), is applied to the control grid of the CRT.

The synchronizing pulses are separated from the picture information and are then used to trigger the line and field scan circuits. The scanning process in television tubes is produced using magnetic rather than electrostatic deflection, the electron beam being deflected at right angles to the magnetic field direction. There are several advantages associated with magnetic deflection systems, the main one being that the deflecting force is proportional to the electron beam

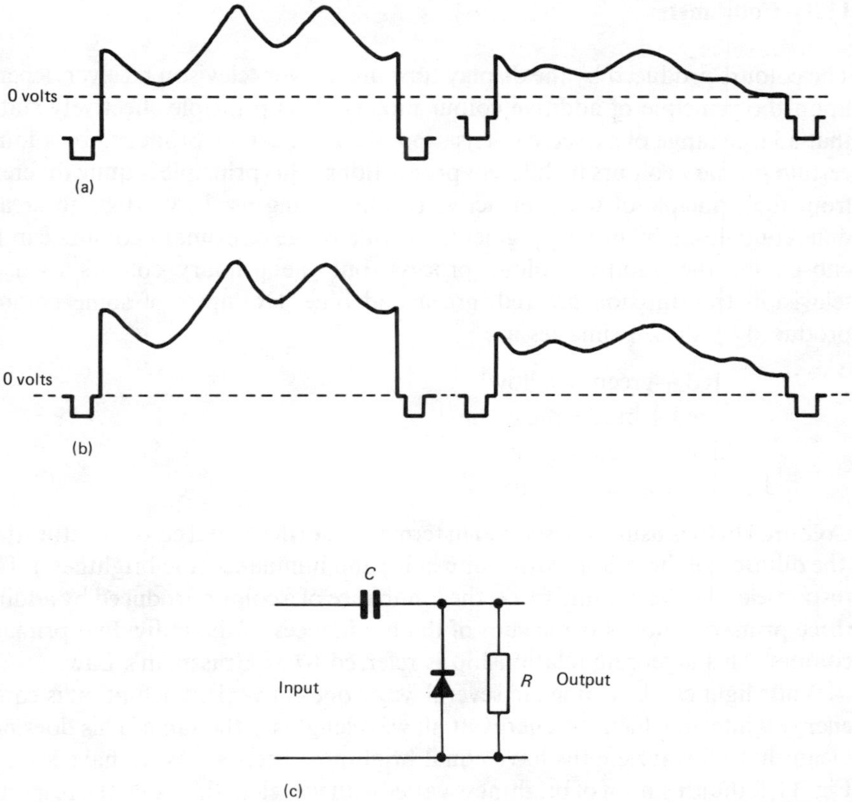

Fig. 11.12 DC restoration. (a) Before dc restoration (sync level varies with picture brightness); (b) after dc restoration (sync level is independent of picture brightness); (c) dc restoring circuit.

velocity and thus increases as the anode accelerating voltage increases. Electrostatic systems, which are commonly used for oscilloscope applications, produce a force that is independent of beam velocity. This means that a high accelerating potential, required to give a bright image, would be accompanied by a small deflection because individual electrons would be influenced by the deflecting force for a shorter interval. The current in the deflection coils has a sawtooth waveform, and during flyback the current changes very rapidly. This produces a considerable induced voltage, which is stepped up by transformer action and used to form the anode accelerating voltage. A typical accelerating voltage for a colour tube is 15 kV.

Colour television transmission uses the same standards as monochrome transmissions. There are, however, several specific requirements, which must be met to produce acceptable colour images, and in order to understand these requirements we must first consider the properties of coloured light.

11.8 Colorimetry

The colours produced by the display tube in a colour televison receiver depend upon the principle of additive colour mixing. This principle effectively states that a large range of subjective sensation of colour can be produced by adding certain primary colours in different proportions. This principle is quite different from the principle of the subtractive colour mixing used by artists to create different colours by mixing pigments. A wide range of primary colours can be chosen for the additive colour process, but the primary colours used in television transmission are red, green and blue. Examples of some colours produced by these primaries are:

red + green = yellow
red + blue = magenta
blue + green = cyan
red + blue + green = white

Coloured light is usually described in terms of hue (the actual colour), saturation (the dilution of the colour with white light) and luminance (the brightness). The response of the eye is additive, i.e. the luminance of a colour produced by adding three primary colours is the sum of the luminances of the individual primary colours. This algebraic relationship is referred to as Grassman's Law.

White light can be defined in several ways; one convenient definition is equal energy white, in which the energy at all wavelengths is the same. This does not mean that all wavelengths have equal brightness because, as we have seen in Fig. 11.1, the sensation of brightness varies with wavelength. Using the response of the standard observer, equal energy white can be expressed in terms of the three primaries, R, G, B as

$$1\,\text{lumen}\ W = 0.3\,\text{lumen}\ R + 0.59\,\text{lumen}\ G + 0.11\,\text{lumen}\ B \tag{11.2}$$

In colour television trichromatic units (T units) are used to simplify Eqn (11.2). White light is then said to be composed of equal quantities of red, green and blue light when the latter are expressed in T units:

$$1\,\text{lumen}\ W = 1T(R) + 1T(G) + 1T(B) \tag{11.3}$$

Where $1T$ unit of red = 0.3 lumen, $1T$ unit of green = 0.59 lumen, $1T$ unit of blue = 0.11 lumen.

In order for Eqn (11.3) to balance, evidently 1 lumen of white = $3T$ units. Equation (11.3) can be adapted to represent any colour in terms of R, G and B:

$$1T(C) = xT(R) + yT(G) + zT(B) \tag{11.4}$$

The coefficients x, y and z in Eqn (11.4) are the trichromatic coefficients of the colour C and it is evident that since this equation obeys Grassman's Law then $x + y + z = 1$. This means that the colour C is actually defined in terms of two trichromatic coefficients. If x and y are known then z can be obtained from

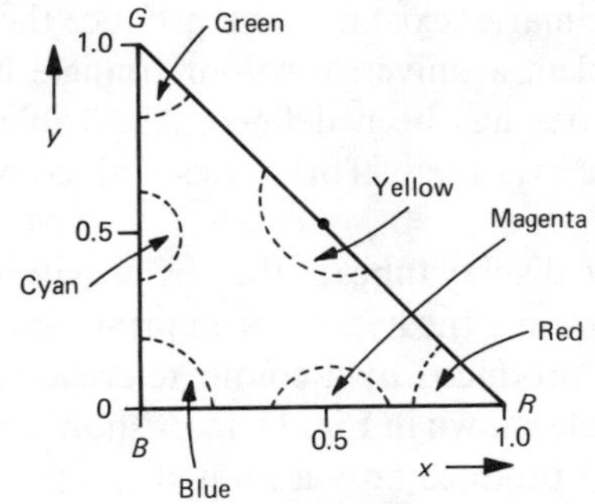

Fig. 11.13 Graphical representation of hue and saturation.

$z = 1 - (x + y)$. The representation of a colour is often done graphically in the form of a colour triangle as shown in Fig. 11.13, in which white is defined by the point $x = 0.33$, $y = 0.33$ (hence $z = 0.33$). The point $x = 0$, $y = 0$, on the other hand, represents saturated blue as in this case $z = 1$. The colour triangle therefore represents both the hue (the actual colour) and the saturation (dilution with white light), for example the point $x = y = 0.25$ represents 50% desaturated blue as $z = 0.5$ is the dominant component. The hue and saturation of any colour which can be produced using *RGB* primaries lies within the colour triangle of Fig. 11.13, but the range of colours produced in a colour television display system will ultimately depend upon the phosphors of the CRT.

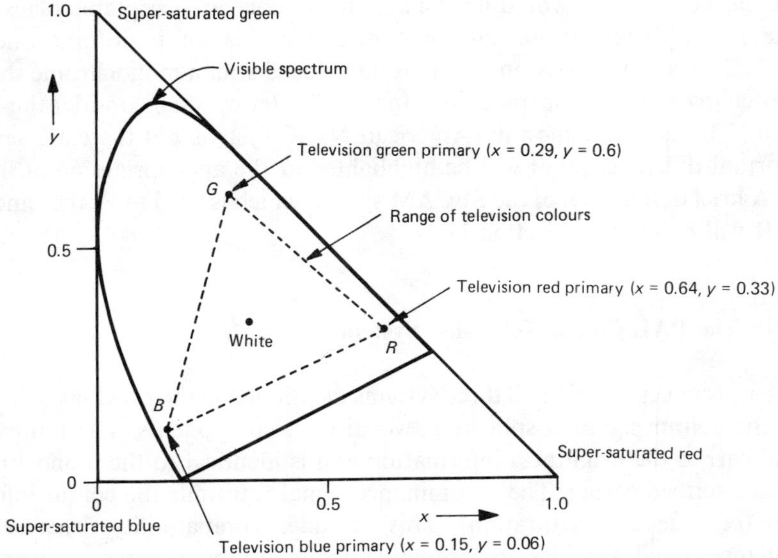

Fig. 11.14 Chromaticity diagram.

In reality no three primaries exist that can produce the full visible range, and to overcome this problem a universal colour triangle based on hypothetical super-saturated primaries has been defined. All visible colours can then be defined in terms of these hypothetical primaries. This colour triangle is shown in Fig. 11.14.

The primaries used in display tubes in the UK lie within the visible spectrum, which can be produced using the hypothetical primaries. The theoretical range of colours that can be produced by a colour television display tube then lies within the dotted triangle shown in Fig. 11.14. It should be noted that although the colour receiver can produce only a limited range of colour, this range is considerably greater than that of high-quality colour film. Equal energy white has the coordinates $x = 0.33$, $y = 0.33$. The white used for television is slightly different and has the coordinates $x = 0.313$ and $y = 0.329$, which is known as illuminant *D*. This in no way affects the validity of Eqn (11.2).

11.9 Requirements of a Colour Television System

We can conclude from Eqn (11.4) that in order to transmit colour pictures it is necessary to split the light from the scene being viewed into its red, green and blue components and to transmit these components, which are then recombined in the receiver. The bandwidth of the three signal components will be comparable to the figure of 5.5 MHz derived for monochrome signals in Section 11.4. Apart from the excessive bandwidth that such a transmission would require, it would not be compatible with monochrome receivers, i.e. the monochrome receiver would not be able to produce an acceptable black and white image. The requirement of compatibility is an important one, as invariably a colour television system is introduced after a monochrome service has been in operation for some time. In this chapter we shall consider the PAL system. This is based upon the American NTSC system, but there are several important differences that will be highlighted at the appropriate point in the text. A brief description of the SECAM system which is used in France and the USSR will be given in Section 11.11.

11.10 The PAL Colour Television System

The basic concept used in all three systems mentioned in the previous section is that the colour signal is split into two distinct components. The luminance signal carries the brightness information and is identical to the monochrome signal described earlier. The chrominance signal transmits the colour information (i.e. hue and saturation). This provides compatibility, because the luminance signal provides an acceptable image in monochrome receivers.

In a monochrome television system, the camera response closely approximates the response of the standard observer. In a colour receiver the red, green and blue signals should approximate to the energy in the primary colours. This is because the eye actually sees the colour image and performs its own relative attenuation on this image, hence when displaying white light the red, green and blue signals should be equal in magnitude. The luminance signal can be derived in two ways. One method is to use a standard monochrome camera tube; the other method is to combine the outputs of three separate camera tubes, each tube producing an output corresponding to a different primary colour. When viewing white light the outputs of these three cameras E_R, E_G and E_B are adjusted to be equal. The camera output voltages are therefore interpreted directly in trichromatic units. The luminance signal is derived by combining these outputs according to Eqn (11.2):

$$E_Y = xE_R + yE_G + zE_B \tag{11.5}$$

where

$x = 0.3$, $y = 0.59$ and $z = 0.11$.

The colour receiver requires a separate knowledge of the E_R, E_G and E_B signals. Ideally the chrominance signal should not transmit any brightness information, as this is already transmitted in the luminance signal. If we ignore the effects of gamma correction, we can show that this condition is satisfied when the chrominance signal is transmitted in the form of colour-difference signals. The colour-difference signals are

$$(E_R - E_Y) \qquad (E_G - E_Y) \qquad (E_B - E_Y)$$

where E_Y is the luminance signal defined by y Eqn (11.5).

$$(E_R - E_Y) = E_R - xE_R - yE_G - zE_B$$

i.e.

$$\begin{aligned} (E_R - E_Y) &= (1 - x)E_R - yE_G - zE_B \\ (E_G - E_Y) &= (1 - y)E_G - xE_R - zE_B \\ (E_B - E_Y) &= (1 - z)E_B - xE_R - yE_G \end{aligned} \tag{11.6}$$

When transmitting a black and white picture $E_R = E_G = E_Y$ and since $x + y + z = 1$, then

$$(E_R - E_Y) = (1 - x - y - z)E_R = 0$$

This result is true for the other colour-difference signals also, indicating that these signals do not contain any brightness information. The original E_R, E_G and E_Y can, of course, be reproduced at the receiver simply by adding the luminance signal to the colour-difference signals in turn.

The chrominance signal is in fact completely defined by any two of the colour difference signals. This can be shown as follows, from Eqn (11.5):

$$E_G = \frac{1}{y}E_Y - \frac{x}{y}E_R - \frac{z}{y}E_B$$

Therefore

$$(E_G - E_Y) = \frac{(1-y)}{y}E_Y - \frac{x}{y}E_R - \frac{z}{y}E_B$$

but $(1 - y) = x + z$. Hence

$$(E_G - E_Y) = -\frac{x}{y}(E_R - E_Y) - \frac{z}{y}(E_R - E_Y) \tag{11.7}$$

The given colour-difference signal can thus be derived from the other two colour-difference signals and need not be transmitted. In fact any one of the colour-difference signals could be reproduced from the other two. The values of $(E_R - E_Y)$, $(E_G - E_Y)$ and $(E_B - E_Y)$ can be calculated for a range of transmitted colours and the mean square value of the green colour difference turns out to be significantly less than the other two. The SNR performance will therefore be maximum when the $(E_R - E_Y)$ and $(E_B - E_Y)$ signals are transmitted.

In addition to providing compatibility, the colour television signal must be confined to the same bandwidth as a monochrome signal, otherwise adjacent channel interference would be a serious problem. The question then has to be answered of how it is possible to fit the chrominance signal into a bandwidth, which is already fully occupied by the luminance signal. Fourier analysis of the luminance signal shows that it does not completely occupy the bandwidth allocated to it. The scanning process can be regarded as a sampling operation, and the spectrum produced will be composed of sum and difference frequencies centred on harmonics of the sampling frequency (in this case the line frequency). Using the sampling analogy, if the sampling frequency is much larger than the maximum frequency of the signal being sampled, there will be periodic gaps in the spectrum of the sampled signal. This is the case with the luminance signal, the spectrum being of the form shown in Fig. 11.15.

The chrominance signals will have a spectrum of the same format; the actual amplitudes of the components will of course be different. If we assume that the red and blue colour difference signals can be combined it is apparent from Fig. 11.15 that the chrominance spectrum can be slotted into the gaps in the luminance spectrum. This requires the chrominance signal to modulate a sub-carrier of frequency equal to $\frac{1}{2}nf_L$ where n is an odd integer.

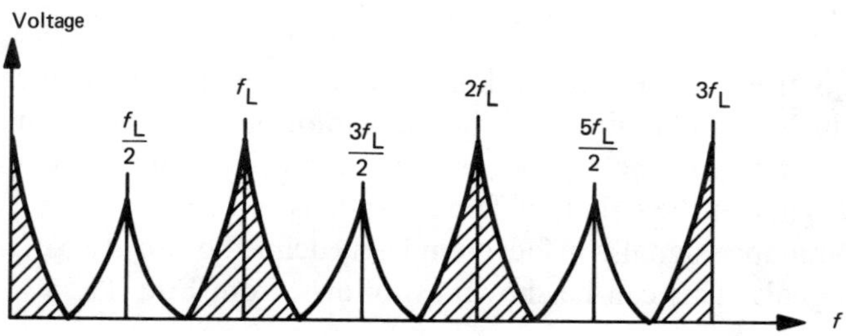

Fig. 11.15 Luminance amplitude spectrum.

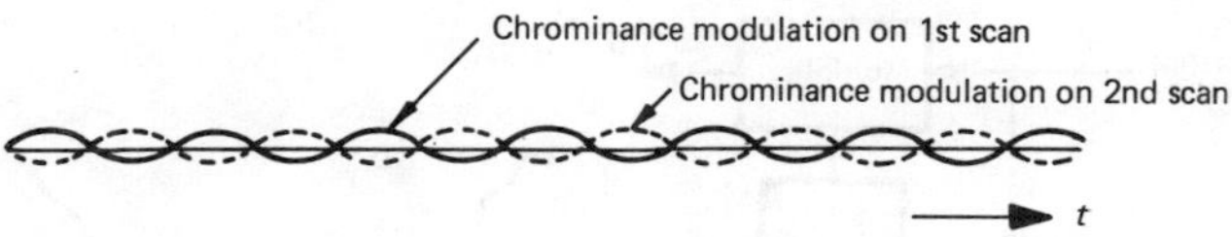

Fig. 11.16 Effect of chrominance signal on monochrome receiver.

It is important to consider the effect of such a modulated sub-carrier on a monochrome receiver (Fig. 11.16). This component will produce an unwanted modulation of the CRT in the receiver. As the chrominance sub-carrier is an odd multiple of the half-line frequency the unwanted modulation on any line should cancel when that line is scanned a second time. In fact this cancellation is not complete. There are several reasons for this, one of the more important being that the light output from a CRT is not linearly related to the grid–cathode voltage (i.e. the decrease in brightness of negative half cycles of modulation is not equal to the increase in brightness on positive half cycles of modulation). The result is that a very fine dot pattern is produced on the CRT of a monochrome receiver. This is only visible if the picture is viewed closely and it is usually ignored.

11.11 Transmission of the Chrominance Signals

There are of course two different colour-difference signals that must be fitted into the gaps in the luminance spectrum. The technique used is similar to the one described in connection with the transmission of data signals over telephone lines using multiphase PSK. Each of the colour-difference signals modulate orthogonal carriers of the same frequency, i.e. the carriers are shifted in phase relative to each other by 90°. The modulated chrominance sub-carrier is actually the sum of two DSB–SC–AM signals

$$h_c(t) = h_R(t)\cos(2\pi f_{\mathrm{sub}}t) + h_B(t)\sin(2\pi f_{\mathrm{sub}}t) \tag{11.8}$$

This composite signal may be conveniently slotted into the gaps in the luminance spectrum. Each of the two colour-difference signals is obtained from the chrominance signal at the receiver by coherent detection, which requires the local generation of $\cos(2\pi f_{\mathrm{sub}}t)$ and $\sin(2\pi f_{\mathrm{sub}}t)$ when components at twice the sub-carrier frequency are removed by filtering, i.e.

$$h_c(t)\cos(2\pi f_{\mathrm{sub}}t) = \tfrac{1}{2}h_R(t)[1 + \cos 4\pi f_{\mathrm{sub}}t)] + \tfrac{1}{2}h_B(t)\sin(4\pi f_{\mathrm{sub}}t) \tag{11.9}$$

$$h_c(t)\sin(2\pi f_{\mathrm{sub}}t) = \tfrac{1}{2}h_R(t)\sin(4\pi f_{\mathrm{sub}}t) + \tfrac{1}{2}h_B(t)[1 - \cos(4\pi f_{\mathrm{sub}}t)] \tag{11.10}$$

Coherent detectors require the locally generated components to be in phase with the suppressed chrominance sub-carriers. The maximum phase error that can be tolerated is in fact $\pm 5°$. The local oscillator thus requires synchronization which is accomplished by transmitting 10 cycles of sub-carrier tone on the

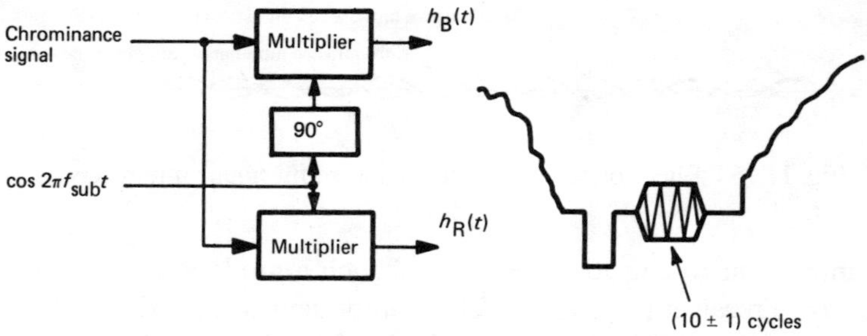

Fig. 11.17 Recovery of colour-difference signals.

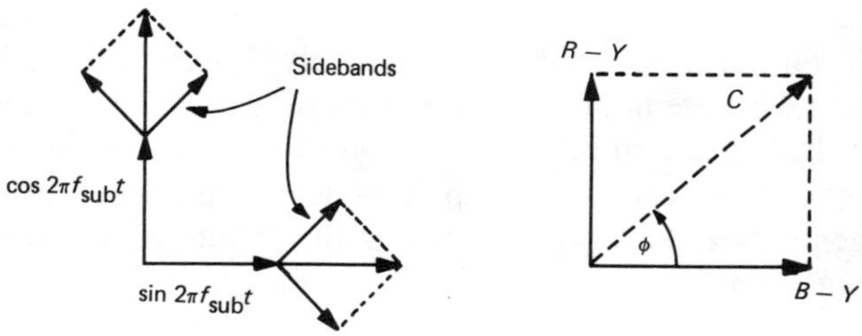

Fig. 11.18 Phasor representation of chrominance sub-carrier.

back porch of each line synchronizing pulse. This tone is used to synchronize the local oscillators. The detection process is illustrated in Fig. 11.17.

It is convenient to represent the chrominance sub-carrier in terms of a phasor diagram. This is derived in Fig. 11.18 from the phasor diagram of two DSB–AM carriers in phase quadrature. When each carrier is suppressed the result is two modulated components in phase quadrature, one representing the instantaneous amplitude of the red colour-difference signal and the other representing the amplitude of the blue colour-difference signal. These phasors do of course represent modulated components and are not the colour-difference signals themselves. The resultant is a modulated component that varies both in amplitude and phase. It is important at this point to consider the relevance of the phase and amplitude of this component. To do this we note that the amplitude of the resultant phasor C is proportional to

$$\{(E_R - E_Y)^2 + (E_B - E_Y)^2\}$$

and the phase ϕ is proportional to arctan $(E_R - E_Y)/(E_B - E_Y)$.

It is convenient to represent various hue/saturation combinations in tabular form. In this context the percentage saturation of any colour in terms of RGB primaries is defined as

$$S = \frac{\text{Maximum component} - \text{minimum component}}{\text{Maximum component}} \times 100\% \qquad (11.11)$$

Table 11.1 Amplitude and phase of sub-carrier

Colour	Saturation	E_R	E_G	E_B	E_Y	$E_R - E_Y$	$E_B - E_Y$	C	ϕ
Red	100	1.0	0	0	0.3	0.7	−0.3	0.76	113.19
Red	50	1.0	0.5	0.5	0.65	0.35	−0.15	0.38	113.19
Red	25	1.0	0.75	0.75	0.825	0.175	−0.075	0.19	113.19
Yellow	100	1.0	1.0	0	0.89	0.11	−0.89	0.896	172.95
Yellow	50	1.0	1.0	0.5	0.945	0.055	−0.445	0.448	172.95
Yellow	25	1.0	1.0	0.75	0.9725	0.0275	−0.2225	0.224	172.95
Magenta	100	1.0	0	1.0	0.41	0.59	0.59	0.834	45
Magenta	50	1.0	0.5	1.0	0.705	0.295	0.295	0.417	45
Magenta	25	1.0	0.75	1.0	0.825	0.1475	0.1475	0.208	45

Table 11.1 indicates that the phase of the sub-carrier represents the colour transmitted (i.e. the hue) and the amplitude of the sub-carrier represents the saturation. Any phase error that occurs during transmission will thus produce an incorrect hue. PAL is a development of the NTSC system which compensates for errors of this kind. The sub-carrier frequency used in the British system is 4.43361875 MHz, which is equal to (283.5 + 0.25)* line frequency + 25 Hz. This is, of course, different from the odd multiple of the half-line frequency specified earlier (which incidentally is the value used in the NTSC system). The reason for the difference is related to the effect of the PAL compensation for sub-carrier phase error. If the NTSC value were used in the PAL system an objectional dot pattern which 'crawls' across the CRT would be produced.

11.12 The Transmitted PAL Signal

The bandwidth of the luminance (i.e. monochrome) signal is fixed at approximately 5.5 MHz. To fit the sub-carrier into this bandwidth means that the sidebands must be restricted to a bandwidth of 1 MHz. The resolution of the chrominance signal is therefore considerably less than the luminance signal.

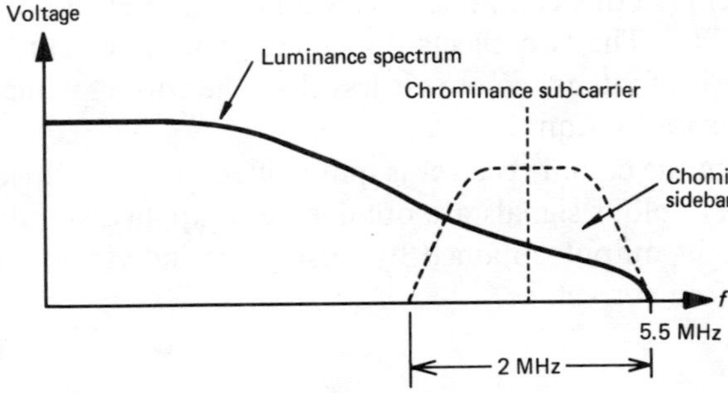

Fig. 11.19 Combined luminance and chrominance signals.

This is not usually detectable by the eye which is insensitive to high colour definition. The combined luminance and modulated chrominance signal as shown in Fig. 11.19 is then used to modulate the main vision carrier.

This combined signal is less than ideal for several reasons, e.g. in the monochrome receiver there is no provision for removing the chrominance sub-carrier and this signal will consequently be applied to the CRT along with the luminance signal. In the colour receiver it is not possible to separate the luminance and chrominance signals completely. Before discussing these effects in detail we must consider the significance of gamma correction in the colour television case.

11.13 Gamma Correction

Gamma correction is required to compensate for nonlinearities that exist (principally in the CRT) between the transducers at the transmitter and receiver. The gamma corrected luminance signal should be $(E_Y)^{1/\gamma}$ where $\gamma = 2.2$, i.e.

$$(E_Y)^{1/\gamma} = (0.3E_R + 0.59E_G + 0.11E_B)^{1/\gamma} \tag{11.12}$$

In the colour system each of the separate colour signals is individually gamma corrected. The resultant luminance signal is then

$$E'_Y = 0.3E_Y{}^{1/\gamma} + 0.59^{1/\gamma} + 0.11E_B{}^{1/\gamma} \tag{11.13}$$

The actual light output produced is obtained by raising the expressions in Eqns (11.12) and (11.13) to the power γ. Equation (11.13) only produces the correct luminance [i.e. the value given by raising the expression in Eqn (11.12) to the power γ] when $E_R = E_G = E_B$, which only occurs on black and white scenes. For coloured scenes the light output produced by Eqn (11.13) is less than the value produced by Eqn (11.12). This means that on a monochrome receiver intensely coloured parts of the picture will be reproduced with a much lower luminance than the correct value. This defect is reduced to some extent by the presence of the chrominance sub-carrier. On saturated coloured portions of the picture the amplitude of the chrominance sub-carrier will be large. Since there is no attempt to remove this component, it will also produce a non-linear light output on the CRT. The two effects do cancel to some extent, i.e. the total luminance in areas of intense colours is less than the correct value, but greater than the value given by Eqn (11.13).

The situation in the colour receiver is quite different. The individual gamma corrected primary colour signals are obtained and applied simultaneously to the CRT. The light output obtained by raising the individual signals to the power γ is therefore correct, i.e.

$$(E_R{}^{1/\gamma})^\gamma = E_R$$

This means that if the light output is correct in the colour receiver, then part of the luminance information in the gamma corrected signals must be contained in the chrominance signal which is not used in the monochrome receiver. In other words the gamma corrected signals do not really provide the required monochrome compatibility.

The signal used to modulate the vision carrier is E'_Y + chrominance sub-carrier (which at any instant can be considered as a sinusoidal component with a given amplitude and phase). In monochrome transmissions the maximum depth of modulation occurs on peak white. In colour transmissions maximum modulation occurs on bright yellow, the total video signal amplitude being 1.78 times the peak white value. The next peak occurs on bright cyan the total amplitude being 1.46 times the peak white value. Referring to Fig. 11.5 it is clear that serious overmodulation (up to 78%) of the vision carrier can result. This overmodulation is reduced to a practically acceptable value of 33% which only occurs rarely, e.g., for bright yellow, taking $\gamma = 2.2$,

$$E'_Y = 0.92 \qquad ({E_R}^{1/\gamma} - E'_Y) = 0.08 \qquad ({E_B}^{1/\gamma} - E'_Y) = -0.67$$

for bright cyan

$$E'_Y = 0.78, ({E_R}^{1/\gamma} - E'_Y) = -0.53, ({E_B}^{1/\gamma} - E'_Y) = 0.22$$

Choosing multiplying factors m and n for the gamma corrected colour difference signals and restricting the total amplitude to 1.33 gives

$$1.33 = E'_Y + [n^2({E_R}^{1/\gamma} - E'_Y)^2 + m^2({E_B}^{1/\gamma} - E'_Y)^2]^{1/2} \tag{11.14}$$

The equations for yellow and cyan are

$$1.33 = 0.92 + [(0.08n)^2 + (0.67m)^2]^{1/2}$$

$$1.33 = 0.78 + [(0.53n)^2 + (0.22m)^2]^{1/2}$$

Solving these equations for m and n gives $m = 0.877$ and $n = 0.493$. The colour-difference signals (γ corrected) transmitted in the PAL system are called the U and V signals and are given by

$$V = 0.877[{E_R}^{1/\gamma} - {E_Y}^{1/\gamma}]$$

$$U = 0.493[{E_R}^{1/\gamma} - {E_Y}^{1/\gamma}] \tag{11.15}$$

The original components ${E_R}^{1/\gamma}$, ${E_G}^{1/\gamma}$ and ${E_B}^{1/\gamma}$ are reproduced at the receiver.

11.14 The NTSC Chrominance Signal

This signal is significantly different in several respects to the PAL signal. The red and blue colour-difference signals are not transmitted in the NTSC system but I and Q signals are transmitted instead. The I and Q signals and their relationship to the colour-difference signals are shown in Fig. 11.20. This takes advantage of the fact that the human eye is most sensitive to colour detail in the orange and cyan hues and least sensitive to colour detail in green and magenta hues. This means that the Q signal can be transmitted in a relatively

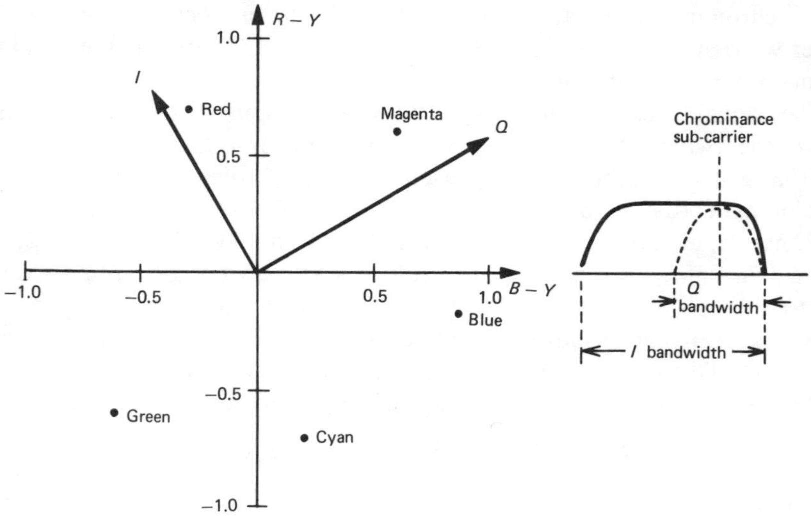

Fig. 11.20 NTSC chrominance signals.

narrow bandwidth DBS transmission being employed. The I signal on the other hand uses a much wider bandwidth with VSB transmission. The two signals can be fully separated at the receiver, over the Q bandwidth both sidebands are present and normal coherent quadrature detection separates the signals. Compensation must be introduced into the detection of the I signal for the portion present in only one sideband, as is normal for VSB transmission. Restricting the bandwidth of the Q signal in this way is important because it allows the use of the highest possible chrominance sub-carrier frequency. The higher this frequency the finer will be the dot pattern produced in the monochrome receiver. In the colour receiver, a high chrominance sub-carrier frequency means that spurious colour effects due to adjacent luminance components are minimized because the luminance energy is reduced as the frequency is increased. This is also true for PAL (see Fig. 11.19).

The I and Q signals can be derived from the red and blue colour-difference signals according to the relationship specified by Eqns (11.16). Evidently the colour difference signals can be reproduced at the receiver from the I and Q signals.

$$I = 0.74(E_R - E_Y) - 0.27(E_B - E_Y)$$

$$Q = 0.48(E_R - E_Y) + 0.41(E_B - E_Y) \tag{11.16}$$

11.15 The PAL Receiver

The PAL system of transmission is a development of the NTSC system, in which special provision is made to correct any phase errors that may occur

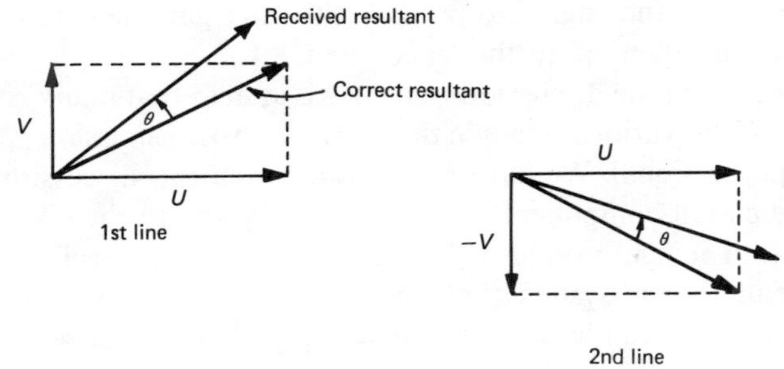

Fig. 11.21 Alternate line inversion of the V signal.

in the chrominance sub-carrier. Such phase errors are mainly due to incorrect synchronization of the local oscillator and sub-carrier generator in the television receiver or differential phase distortion in the transmission system. As we have seen from Table 11.1, the phase of the chrominance sub-carrier is related to the hue of the image and any phase error will produce incorrect colours. The PAL system compensates for this by reversing the phase of the V signal on alternate lines (PAL in fact means "phase alternation, line"). In the receiver the correct phase of the V signal on alternate lines is restored, the overall effect being to average the phase error over a period of two lines.

A sub-carrier that has a phase error of θ is illustrated in Fig. 11.21. When the V signal is inverted on alternate lines by the receiver the phase error is also inverted (i.e. a positive phase shift of θ on the received signal becomes a negative phase shift of θ after the phase of the V signal is inverted). If the chrominance information on two adjacent lines is the same, the average phase error over this period is zero. In fact the chrominance signal will not necessarily be identical on adjacent lines so that total cancellation will not result. The averaging is a carried out in most modern receivers by adding and subtracting the chrominance signal from its value on the previous line. This requires a delay of 64 μs. This length of delay is easily produced using a charge coupled delay line based on MOS transistors. A block diagram is shown in Fig. 11.22. The matrix unit combines the U and V signals in the required ratios to produce the three gamma corrected colour-difference signals. These signals are then

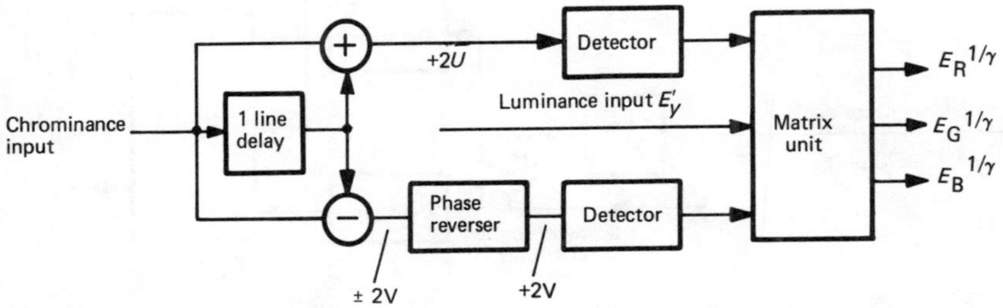

Fig. 11.22 Demodulation of PAL chrominance sub-carrier.

added to the luminance signal to produce the three gamma corrected primary signals which are then fed to the tri-colour CRT.

It is evident from the discussion of the PAL system that many compromises have been made at various stages in the production, transmission and detection of the composite signal. We have not considered in this discussion the effects of non-ideal circuit components which invariably add to the distortion of the eventual colour image produced. The PAL system is an excellent example of a telecommunications system that has been constrained very much by the existence of a set of standards that were developed for an entirely different (i.e. monochrome) broadcast service.

Before briefly considering the principles of operation of modern colour cameras and display devices we shall describe the essential features of an alternative colour transmission system which is used in parts of Europe. This system was also constrained by established monochrome standards and, as with PAL, produces an acceptable but less than perfect colour image.

11.16 The SECAM System

Unlike PAL and NTSC, which both transmit two colour-difference signals simultaneously, SECAM transmits two colour-difference signals sequentially. The red and blue colour-difference signals are transmitted separately on alternate lines using FM of the chrominance sub-carrier. There is no local oscillator required in this system, and hence phase distortion is not a problem. However both colour-difference signals are required at the receiver at the same time. SECAM meets this requirement by delaying each colour-difference signal by 1 line interval and using the delayed signal on the next line. Hence each transmitted colour-difference signal is used on two adjacent lines. This obviously halves the colour definition, but this is virtually undetected since the luminance signal transmits the full definition. A block diagram of the SECAM action is given in Fig. 11.23.

The inherent simplicity of the SECAM system suggested by the block

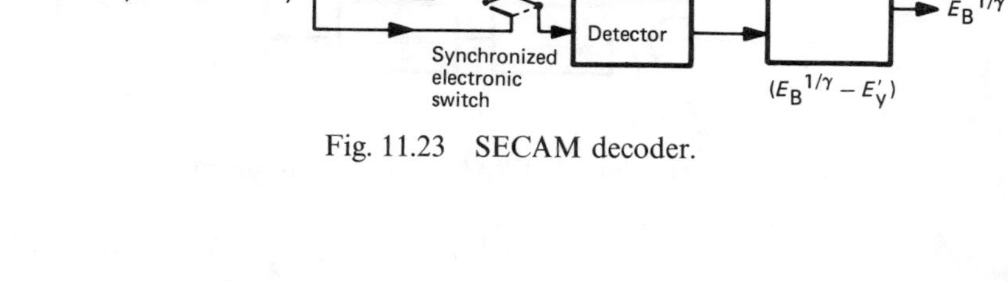

Fig. 11.23 SECAM decoder.

diagram of Fig. 11.23 is not in fact realized in practice. The FM chrominance sub-carrier produces a visible dot pattern on monochrome and colour receivers which varies with the frequency deviation. Several signal-processing techniques are used to reduce this effect, which consequently increases the complexity of the receiver. Discussion of these techniques is beyond the scope of this text, and the reader is referred to Sims[2] for further information.

11.17 Display Devices and Television Cameras

For picture reproduction the display tube must be able to produce three separate pictures in the primary colours. The eye averages these separate pictures to form a single subjective colour image. A colour picture tube essentially contains three electron guns producing three electrons beams, one for each of the primary colours. These three beams are deflected magnetically to form the usual scan, but each of the three beams is arranged to strike only the screen phosphor corresponding to its colour.

Early display tubes used a delta formation of electron guns in the neck of the tube and a shadow mask that had many holes designed to ensure each beam would strike a phosphor dot (arranged in triads) corresponding to its colour. There were very significant problems with the scanning arrangements of this tube in order to ensure registration of the three colour images at the centre and edges of the picture. In recent years the shadow mask tube has been replaced by the precision in-line tube. In this device the three electron guns have an in-line formation in the neck of the tube. The dot triads are replaced by phosphor strips and the holes of the shadow mask are replaced by vertically elongated slots. The construction of the precision in-line display tube is shown in Fig. 11.24. The in-line gun formation used in these tubes greatly simplifies the convergence problems. Horizontal convergence is achieved by a simple horizontal shift. The magnetic deflection system used in the colour tube is similar to the system discussed for the monochrome case.

The television camera performs the inverse function of the display tube, i.e. it transforms incident light into the three electrical signals E_R, E_G, E_B. It must also of course provide a luminance signal for monochrome compatibility. There are several ways of achieving these requirements. Some cameras use three tubes, one for red, green, blue, and then form the luminance signal electrically. Other cameras use four tubes, the fourth tube producing the luminance signal directly. The camera tubes themselves may be broadly classified as photoemissive or photoconductive. We shall describe briefly the principles of operation of a photoconductive tube known as the vidicon.

Incident light is focused on to a target disc of photosensitive semi-conductor material. The resistance of this material is inversely proportional to the intensity of incident illumination. The construction of the vidicon tube is shown in Fig. 11.25. The photoconductive target is scanned by a low-velocity electron beam

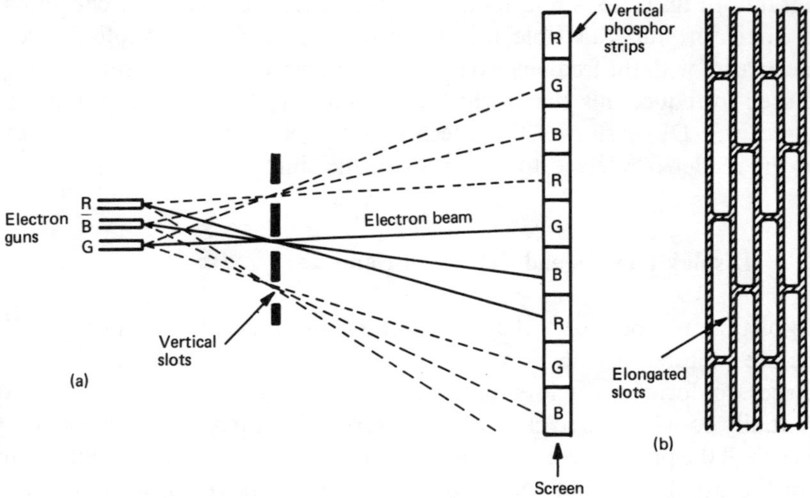

Fig. 11.24 Precision in-line display tube: (a) plan view of tube layout; (b) front view of mask.

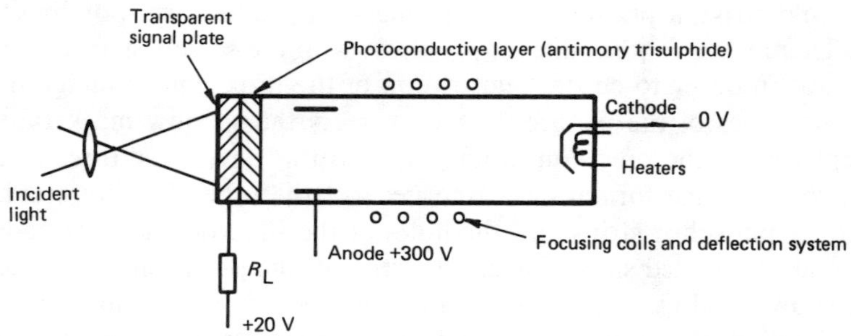

Fig. 11.25 The vidicon tube.

that stabilizes its rear surface at approximately cathode potential. The front surface or signal plate is held at about 20 V. Hence there will be current flowing through the load R_L. When there is no illumination this current will be small and is known as the dark current. When an image is focused on to the target disc the conductivity rises in areas of high illumination thus allowing charge to build up on the rear surface. The rear surface thus gradually becomes positive with respect to the cathode during the interval between scans. The scanning beam deposits sufficient electrons to neutralize this charge, and in so doing generates a varying current in the load. Since charge is conducted to the rear face of the target during the whole interval, the vidicon is very sensitive. It does, however, suffer from long persistence at low illumination levels, and for this reason is normally used in high illumination conditions.

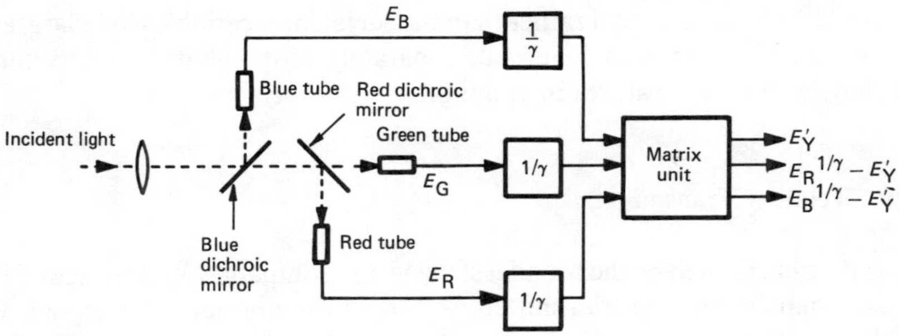

Fig. 11.26 Three-tube colour camera.

The colour television system requires three separate signals E_Y, $E_R - E_Y$ and $E_B - E_Y$. There are several ways of obtaining these signals, and colour cameras have been designed using from one to four different tubes. A typical three-tube layout is shown in Fig. 11.26. As we noted earlier, the response of each tube is adjusted electronically to produce an equal output for equal energy white light. The light is split into its primary colours using dichroic mirrors. These mirrors reflect light of one colour and transmit all other colours.

A single-tube vidicon is available in which the target photoconductive disc is split up into many electrically insulated vertical strips. Immediately in front of each strip is a light filter composed of vertical strips of red, green, blue sections. Each section of the target will then be exposed to only red or green or blue light, and in this way the individual colour signals can be generated.

We have concentrated up to this point on the use of television for the production of broadcast entertainment programmes. The remainder of this chapter will be devoted to the role of television as an information (i.e. data) transmission system.

11.18 Teledata

During its relatively short history, television has had a profound effect on society. This effect has now entered a new dimension with the emergence of the domestic television receiver as a household data terminal. The data service is known generally as teledata, and in Great Britain there are two distinct sources of textual information. The broadcast television companies transmit data signals over the same medium as normal entertainment programmes, and the service is known as teletext. British Telecom transmits data signals over the public telephone network, and this service is known as viewdata, the commercial name being Prestel. The viewdata service is potentially the most important, because it allows user interaction via special purpose interface units between the television receiver and telephone network. In this book we are concerned with the technical rather than the social implications of teledata, and we shall consider teletext and viewdata separately, although it should be noted that they have many features in common.

11.19 Teletext Transmission

This is the system used by the broadcast television companies, and consequently the data signals must be transmitted as part of the normal video signal. We noted in Section 11.4 that between each field of a television picture 25 lines are blanked out to allow for field flyback to return the scan to the top picture. Some of these blanked lines are used to transmit the field synchronizing and equalization pulses and receiver circuits are adjusted so that the remaining blank lines do not appear on the screen. The lines are therefore available for the transmission of data pulses that can be transmitted above black level because these lines will not be displayed.

In the present service, data signals are transmitted on lines 17 and 18 of the even fields and lines 330 and 331 of the odd fields. The complete specification allows for a future expansion to transmit data on lines 7 to 22 of the even fields and lines 320 to 335 of the odd fields. The data signals are undetected on a standard receiver in normal adjustment, but can be separated from the normal video signal in special-purpose receivers.

We do not propose to discuss the design of teletext receivers in detail, but we will consider the standards used and, where appropriate, include block diagrams of some special-purpose circuits. The information transmitted in the teletext system is divided into a number of pages containing 24 rows with 40 characters per row. Each character is represented by a 7 bit international standard code with an odd parity check. The odd parity has an additional receiver synchronization function when all seven bits of the standard code have the same value. Unlike the asynchronous data transmission described in Chapter 4, the teletext transmission is synchronous and is therefore more efficient because start and stop bits are not required. Synchronizing information is required, however, and this is transmitted during the first five binary words (bytes) of each row. This means that each row contains a total of 45 eight-bit bytes.

The format of each line is shown in Fig. 11.27. The bits in each byte are identified by sampling the data at the centre of each bit interval, the sampler being driven by a locally generated clock. This means that each row contains 360 bits and these bits are transmitted during 1 line scan interval using NRZ pulses. The line-scan duration in the 625 line system is 64 μs with 12 μs for line synchronization. The bit rate is thus 360 bits per 52 μs or 6.9375 Mb/s. The television receiver has a local clock running at this frequency. The clock is synchronized with the incoming data stream by the first two bytes of each line that contain alternate 1s and 0s at the clock frequency. This

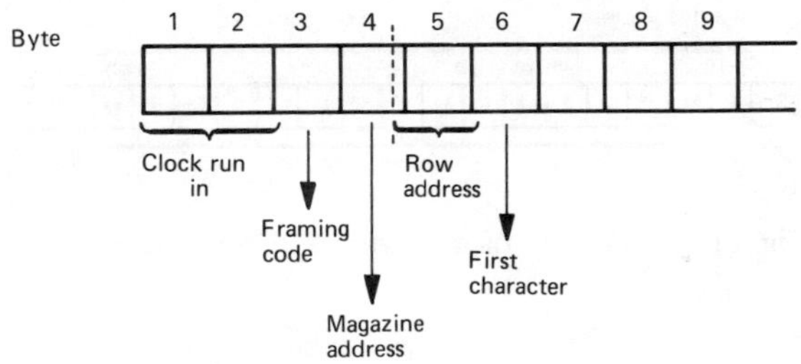

Fig. 11.27 Format of teletext row.

is known as clock run in. Once the local clock is synchronized it is necessary to detect the beginning of each individual 8 bit byte in order that correct decoding can take place. This is achieved by transmitting a special framing code 11 100100 as byte 3.

The data stream is used as a serial input to an 8 bit shift register. When the register contains the framing code a flag is set indicating that the next bit will be the first bit of byte 4, i.e., byte synchronization is achieved. For the moment we will skip bytes 4 and 5 and consider byte 6, which is the first character byte. The first step in decoding a character byte is to check the parity. A circuit that will produce an 8 bit parity check is shown in Fig. 5.9 and is composed of seven exclusive-OR gates. If a parity check fails (i.e. if the output of this circuit is binary 0) the character is not decoded, but is replaced by the code for a blank.

Each page of text is composed of 24 rows. Hence it is necessary for the decoder to be able to select individual pages and once a page has been selected, to assemble the rows of the page in the correct order. The rows within a page are recognized by a row address code that is transmitted in bytes 4 and 5 of each row. Since there are 24 rows in each page, a minimum of 5 bits is required to specify the number of row addresses. It is evident that the occasional error in a character byte can be tolerated as the character is simply omitted when this occurs. However an error in the bytes containing the row address is much more serious, as this can cause a row to be wrongly placed within a page. To reduce the probability of such errors occurring, the page and row address are error-protected using a Hamming code. Hamming codes and error detection and correction are discussed in detail in Chapter 5. The code used for row addressing has a Hamming distance of 4. This means that it can correct a single error and can detect a double error.

To provide a Hamming distance of 4, each address bit is accompanied by a parity check bit, hence a total of 10 bits (address + parity) is required. Eight of these bits are transmitted in byte 5 of each row and the other two in byte 4. The remainder of byte 4 is used for a Hamming coded magazine address which forms part of the page identification. Both row and magazine address codes are transmitted least significant bit first. The format is shown in Fig. 11.28.

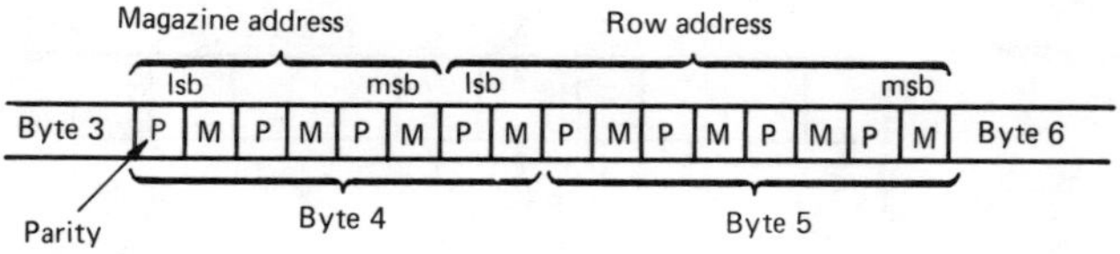

Fig. 11.28 Format of magazine and row address bits.

The function of the row address is to direct the following 40 bytes of text data to the appropriate locations in the page memory. Before this is done however the required page must be selected. The top row of each page is called the header row and has the row address 00000. This row only contains 32 text characters instead of the usual 40. The first 8 bytes of text data in the header row carry a page number code, a time code and a control code. These 8 bytes are Hamming coded in the same way as bytes 4 and 5 of the other lines. The text in the header row with the exception of the page number display, is the same for each page.

The page address code (tens and units) is transmitted as two 4 bit BCD numbers with the appropriate Hamming error coding. BCD code is used because this can be compared directly with the page selection number entered by the viewer from his decimal keyboard. The page address code occupies bytes 6 (units) and 7 (tens) of the header row. The next 4 bytes are used for transmitting the minutes (units and tens) and hours (units and tens) of the time code, the Hamming coded BCD format being used for these bytes also. The next two bytes are used to transmit control information.

Page Selection

The page is selected by keying in the required 3 digit number on the viewer's keypad. Each page is identified by a combination of the page code in the header row and the magazine address code that is transmitted as part of byte 4 of every row. The magazine row contains three digits that are Hamming coded and can thus have any value between 0 and 7. This code is used as the "hundreds" of the page number identification. The page code is selected from the header row and the magazine code is selected from each transmitted row of text. When a complete match occurs, the following data is written into the page memory. The row address codes are used at this stage to select the appropriate locations within the page memory. When another header row is detected (row address 00000) the page address code are again compared. If there is no match the following data is ignored. In this way only the data from the requested page is transferred to memory. A block diagram of the page selection circuit is shown in Fig. 11.29. If an uncorrected error occurs in either

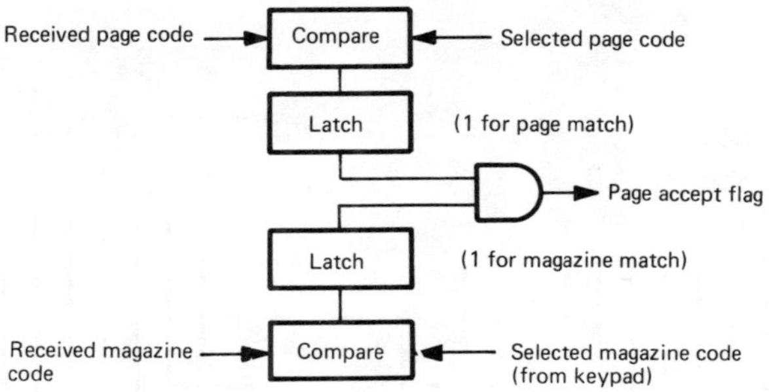

Fig. 11.29 Page-selection hardware.

the page or row address codes the following data is ignored until the requested page header is repeated.

Normally all header rows are displayed until the requested page header is received. This avoids a blank screen that could occur for up to 30 seconds in an average sized magazine. In this case the only part of the header that changes on the display is the page number and the time. Full pages are transmitted at the rate of approximately four per second, which means that 25 seconds are required to cycle through 100 pages. This can be inefficient when several blank rows within a page exist. To increase the efficiency of transmission, the page memory is completely filled with the code for "space" when a new page code is keyed in. Blank lines within a page can then be omitted, i.e., gaps can occur in the row address. Filling the page memory with blanks prevents the row of a previous page from being displayed when the requested page has rows omitted. This technique increases the speed of transmission by up to 25%.

If there are several pages dealing with a common subject, they are sent out in sequence using the same page number. Each new page is transmitted after a delay of about 1 min, which is sufficient time for the viewer to read the displayed text. The whole sequence of "self changing" pages is then repeated continuously.

The Page Memory

In order to produce the illusion of a fixed image the data representing each page must be scanned 50 times per second. We have seen that even when there are no more than 100 pages any one page is only transmitted once in every 25 seconds. Hence there is a requirement for a complete page memory that can then be scanned sequentially at the required rate. Since each page has 24 rows of 40 characters and each character has a 7-bit code the minimum storage requirement is 6720 bits per page. This may be readily achieved by use of a 1024 × 8 bit random access memory (RAM) as shown in Fig. 11.30.

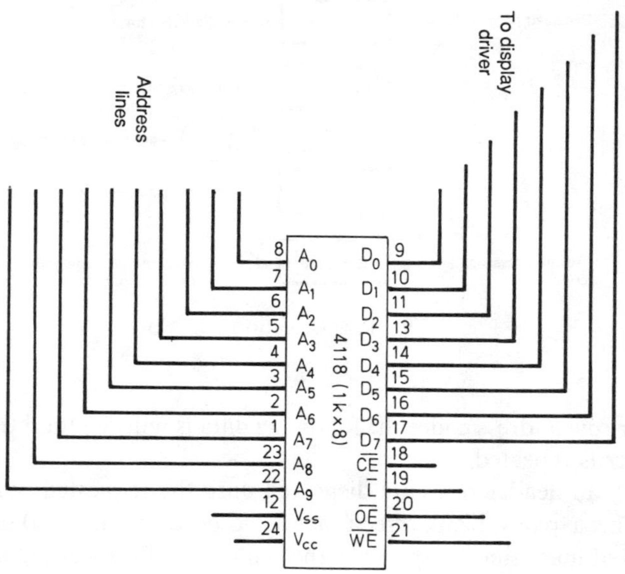

Fig. 11.30 Page memory using 1 k × 8 bit RAM.

Character Display

The character set in a teletext display is produced using a 6 × 10 dot matrix. The format allows both upper and lower case letters and special-purpose symbols to be displayed. The vertical dot resolution is made equal to the line spacing, which means that ten scanned lines are required for one row of text. A typical example is shown in Fig. 11.31.

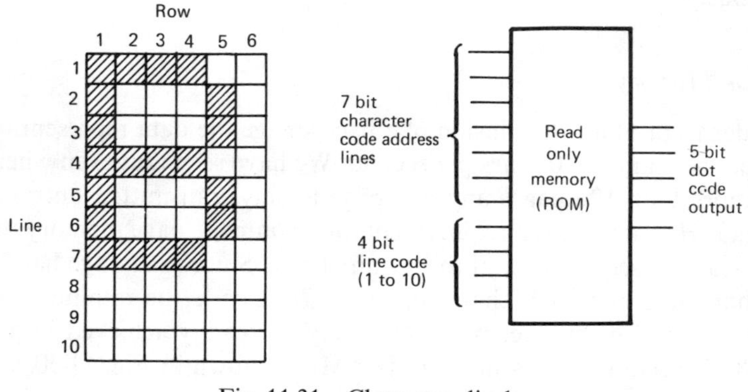

Fig. 11.31 Character display.

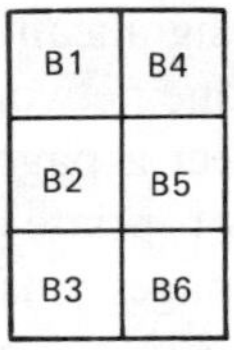

Fig. 11.32 Graphics dot matrix.

Each character is in fact represented by a 5×9 dot matrix, column 6 being reserved for character separation and line 10 being used for row separation. The character patterns are stored permanently in a read only memory (ROM). The scan proceeds on a line-by-line basis. During each line scan any character will have a corresponding 5 dot code. Hence the ROM must store a series of 5 bit numbers corresponding to the data code. The ROM is addressed by a combination of the character code (seven bits) and the line number (four bits are required for one to ten lines), the appropriate 5 dot code is then stored at each 11 bit address.

As each line is scanned, the character codes in the particular row of text being displayed are placed upon the address lines of the ROM in sequence together with the 4 digit line scan code. This produces the correct sequence of 5 dot codes on the data lines of the ROM. Smoother characters than the one illustrated can be generated by producing slightly different dot codes on alternate scans. This is known as character rounding.[3]

Graphics

The teletext display is not restricted to text but can also operate in a graphics mode. This mode is used for the display of simple diagrams and extra large characters. The display in the graphics mode is also divided into a dot matrix, but in this case a 2×6 rather than 6×10 matrix is used. The graphics mode is indicated by a control code and when in this mode each 7-bit character is interpreted directly as a graphics symbol. The graphics dot matrix and the corresponding bit number is shown in Fig. 11.32. The total number of graphics symbols which can be defined is $2^6 = 64$ symbols.

There are many other features of teletext, such as mixing of text and graphics, colour of display, super position of teletext on normal programme pictures, etc. that it is not possible to cover in this volume; the interested reader is referred to Money.[3]

11.20 Viewdata

This is an information service operated by British Telecom. The display format uses the same standard as teletext and thus a specially adapted receiver can be used for both services. The significant difference is that viewdata is transmitted over the public telephone network, and since two-way communication is possible the user can select a page of information from an almost limitless number by keying the appropriate digits. This is not possible in the teletext system as all the available pages must be transmitted in sequence and the total number of pages is limited by the acceptable delay. In the viewdata system only one page is transmitted in response to the page number keyed in by the user. The arrangement of the British national viewdata network is shown in Fig. 11.33. Each of the local centres contains a computer with about 5×10^4 pages of information. Most of the information accessed by the average user will be stored at the local centre, hence the pattern of telephone usage will be on a local call basis. The regional centre is the next stage in the hierarchy. If the requested information is not stored at the local centre it will be automatically transferred from the regional centre over a high-speed data link. The regional centres are connected by high-speed links to a national centre. By this means any user is able to access local information from any one of the regional or local centres. The capabilities of the system in terms of the amount of information are vast, and can, of course, be extended to an international centre at a future date.

A second difference between teletext and viewdata is that the viewdata system uses a cursor instead of page and row address. The position of the cursor is in effect controlled by the contents of two counters. One counter represents the position of a character within a row, the other represents the position of a row within a page. The viewdata system uses a similar 7-bit code for character transmission plus a parity check bit. The viewdata transmission uses an even parity in contrast to teletext which uses odd parity. There are thus $2^7 = 128$ possible characters that can be transmitted. Several of these characters are used for cursor control. The code 0001100 (form feed) causes the cursor to return to the top left-hand corner of the screen (i.e. character and row counters are set to zero) and the page memory is filled with blanks. As symbols are received the cursor is moved along the top line (i.e. the character counter is incremented). When the cursor moves to the end of the line, the character counter is reset and the row counter is incremented by 1. The location

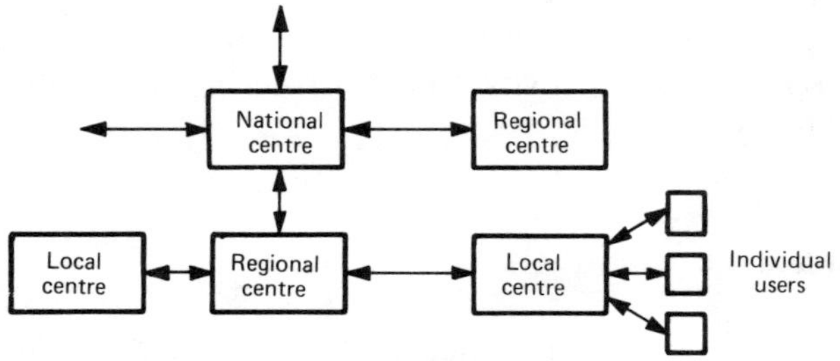

Fig. 11.33 British national viewdata network.

address in the page memory RAM is formed from a combination of character and row counter. The cursor can be placed at any point on the screen by using cursor controls, this avoids transmitting blank rows and blanks within a row. For example the code 0001101 (carriage return) causes the cursor to return to the beginning of the current line (resets character counter to zero) and the code 0001010 (line feed) causes the cursor to move to the next line (increments the row counter).

Line signals

Data is transmitted to the subscriber using frequency shift keying with asynchronous transmission as described in Section 3.8. The data rate used is 1200 baud with binary 1 being transmitted as a 1300 Hz tone and binary 0 as a 2100 Hz tone. This means that a modem is required in the television receiver to interface the FSK signals. This modem also allows transmission of the page-selection code in the reverse direction again, using FSK. The transmission speed in this case is 75 baud, a 390 Hz tone being used for binary 1 and 450 Hz for binary 0. The modem in this case is also required to drive the line and field

Table 11.2 Television channels and frequencies

Band	Channel	Frequencies (MHz)	Band	Channel	Frequencies (MHz)
IV	21	470–478	V	46	670–678
	22	478–486		47	678–686
	23	486–494		48	686–694
	24	494–502		49	694–702
	25	502–510		50	702–710
	26	510–518		51	710–718
	27	518–526		52	718–726
	28	526–534		53	726–734
	29	534–542		54	734–742
	30	542–550		55	742–750
	31	550–558		56	750–758
	32	558–566		57	758–766
	33	566–574		58	766–774
	34	574–582		59	774–782
V	39	614–622		60	782–790
	40	622–630		61	790–798
	41	630–638		62	798–806
	42	638–646		63	806–814
	43	646–654		64	814–822
	44	654–662		65	822–830
	45	662–670		66	830–838
				67	838–846
				68	846–854

scan circuits of the CRT in the receiver. This means that the viewdata service, unlike the teletext service, does not require the provision of broadcast line and field synchronization. It is thus available outwith normal broadcast periods.

11.21 Conclusion

In this chapter we have outlined the fundamental engineering principles of broadcast television systems and have shown that the television receiver is not just a home entertainment medium. The potential of viewdata, in particular, is enormous; for example it is currently possible to make airline reservations directly from a viewdata terminal. This is but one small step in the revolution in all commercial activities that is going to take place in the coming years.

References

1. Patchet, G.N., *Television Servicing*, Vol. 2, Norman Price, 1971.
2. Sims, H.V., *Principles of PAL Colour Television*, Newnes-Butterworth, London, 1974.
3. Money, S., *Teletext and Viewdata*, Newnes-Butterworth, London, 1979.

Problems

11.1 The transmission standards for television in the USA are as follows:

number of lines/picture = 525
number of fields/picture = 2
number of pictures/second = 30
field blanking = 14 lines
line blanking = 14 μs
displayed aspect ratio = 1.33:1

Differentiate between the transmitted and displayed aspect ratio and calculate a value for the former figure. What is the theoretical bandwidth of the transmitted video waveform?

Answer: 1.545:1, 6.37 MHz.

11.2 If the standards of the previous question were modified to accommodate 3 fields/picture what is the theoretical bandwidth of the transmitted waveform? Increasing the number of fields/picture produces a corresponding reduction in signal bandwidth. Comment on this statement and suggest why in practice there is a limit to the number of fields/picture.

Answer: 4.25 MHz.

11.3 The UHF television allocation for channel 50 is 702 MHz to 710 MHz. A television receiver with an IF of 39.5 MHz is tuned to this channel. Calculate the frequency of the local oscillator in this receiver.

Using the channel allocation table (Table 11.2) determine which channel forms the image frequencies for channel 50.

Answer: 743.25 MHz, channel 60.

11.4 Three primary light sources R G B are designed to produce an output light power that is linearly proportional to a control voltage. When these three sources are used to match equal energy white light of intensity 1 lumen, the required voltages are $R = 6.9$ V, $G = 3.8$ V, $B = 8.6$ V. The same light sources are used to match an unknown colour, the corresponding voltages being $R = 5.6$ V, $G = 2.2$ V, $R = 1.5$ V. Find the trichromatic coefficients of the unknown colour and its luminance. What are the chromacity coefficients of this colour on the colour triangle?

Answer: 0.81 T, 0.58 T, 0.17 T, 0.6 lumen, 0.52, 0.37.

11.5 The output voltages from a three-tube colour camera when viewing equal energy white light are adjusted such that $R = G = B = 1.0$ V. When the camera is directed towards an object of uniform colour the output voltages are $R = 0.56$ V, $G = 0.21$ V and $B = 0.75$ V. Calculate the amplitude of the resulting luminance signal when:

(a) gamma correction is applied after the formation of the luminance signal;
(b) gamma correction is applied to each colour signal separately.

What is the percentage difference in the luminance produced when these two signals are applied to a monochrome CRT? Assume the overall gamma is 2.2.

Answer: 0.639 V; 0.617 V; 7.42%.

11.6 When the output voltages of the camera of the previous question are $R = 0.5$ V, $G = 0.9$ V and $B = 0.8$ V, find the percentage saturation for this colour. What would be the values of the output voltages for a fully saturated colour of the same hue?

Answer: 44%, 0 V, 0.4 V, 0.3 V.

11.7 The row addressing data in a teletext transmission is coded to have a Hamming distance of 4. If the probability of a single digit error is 0.0001, find the probability that an undetected error will occur in the coded address information. State all assumptions made.

Answer: 4.33×10^{-4}.

12 OPTICAL FIBRE COMMUNICATIONS

Optical fibre communications present the most exciting, and probably the most challenging, aspects of modern systems. Fibres are exciting because they seem to offer so many benefits – low cost, enormous bandwidth, very small attenuation, low weight and size, and very good security against external interference. Physically, fibres occupy very little space, and they are so flexible that they can be used in places that would not be accessible to conventional cable.

Until recently, the high cost of development, which has produced some four orders of magnitude reduction in attenuation, has been reflected in the cost of the fibre. However, the large-scale installation being undertaken by many telephone administrations will require the production of substantial quantities of cable, and hence reduce the cost.

The challenges referred to are those of connecting fibres together, and of producing optical systems that will allow processing of signals to be carried out in the optical mode, rather than requiring a down-conversion in frequency first. Considerable progress has been made, particularly in the field of integrated optics and, eventually, optical systems will become much more versatile.

The optical components of a fibre communications system are, in simple terms, a light emitter, which initiates the optical signal, a fibre which transmits it, and a detector which receives it and converts it into an electrical equivalent. If several fibres need to be joined, end to end, the couplers must ensure that the fibres are correctly aligned and butted, to reduce any joining losses to a minimum.

Each of these components has essential ancillary parts; the detector and emitter are driven by stabilized voltages, and mounted in such a way that maximum transfer of light between them and the fibre is achieved. The fibre itself must be clad in some sort of protective coating and made up into a cable that will withstand the rigours of installation over long distance. However, although these factors are essential, they are, in a sense, of secondary importance, and we shall not be considering them further here. Our interest is confined to the operating principles of the basic devices, and, for further information on system details, reference should be made to the literature listed at the end of this chapter.

Before considering some aspects of a system, we will examine the way in which the fibres, detectors and emitters work.

12.1 Optical Fibre

An optical fibre is, in essence, a dielectric waveguide. It has been known for a long time that high-frequency electromagnetic energy can be transmitted along a glass or plastic rod and, indeed, observation shows that short rods are translucent to light. However, two factors prevented that knowledge from being used to produce useful light guides:

(i) energy leaked from the outside of the dielectric to the surrounding air, and
(ii) the attenuation was so large that worthwhile lengths could not be achieved.

The first difficulty, though virtually insurmountable at microwave frequencies, can be overcome in the optical and infrared parts of the spectrum by enclosing the guide in a cladding of similar material, but which has a slightly smaller refractive index. The boundary between the cladding and the core acts as a reflecting surface to the transmitted light. The second problem, that of high attenuation, could be reduced only by refining the methods of producing and drawing the glass so that the impurities and irregularities were reduced to a minimum. The attenuation now achievable in the laboratory is almost as low as possible, at about 0.2 dB/km.

Fibres of varying quality are used for communications, but when distances are significant, care is taken to ensure that the lowest attenuation possible is achieved. This involves choosing the best operating frequency for the particular fibre material, and ensuring that any contaminating elements are removed from the glass during manufacture. Before considering the loss mechanisms inherent in any fibre, we will look at the different fibres used, and examine, with the help of ray theory, the way in which light propagates along an optical waveguide.

12.2 Stepped-index Fibre

As we noted earlier, light can be made to propagate down a fibre waveguide consisting of an inner dielectric, of refractive index n_1, and an outer cladding of refractive index n_2, if n_1 is slightly larger than n_2. In Fig. 12.1 we see the path taken by a beam incident on the end face of the fibre at an angle of incidence θ_0. Outside the fibre, the air is assumed to have a refractive index n_0.

By the laws of refraction,

$$n_0 \sin\theta_0 = n_1 \sin\theta_1 = n_1 \cos\theta_2 \qquad (12.1)$$

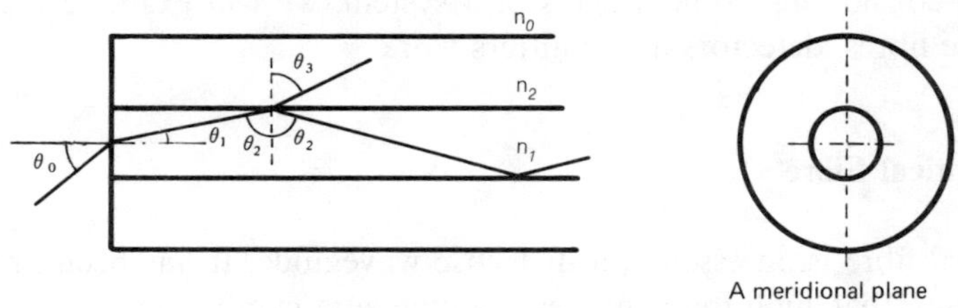

Fig. 12.1 Path of meridional ray in fibre.

At the core–cladding boundary

$$n_1 \sin \theta_2 = n_2 \sin \theta_3$$

Since $n_2 < n_1$, total reflection will occur at the boundary for all values of θ_2 such that

$$\sin \theta_2 \leqslant n_2/n_1 \tag{12.2}$$

From Eqn (12.1), the maximum value of θ_2 that will produce a reflected ray at the core–cladding interface is therefore given by

$$n_0 \sin \theta_{0_{\max}} = n_1 \cos \theta_2$$

where, from Eqn (12.2)

$$n_1 \cos \theta_2 = n_1 [1 - \sin^2 \theta_2]^{1/2} = n_1 \left[1 - \left(\frac{n_2}{n_1} \right)^2 \right]^{1/2}$$

Hence

$$\sin \theta_{0_{\max}} = \frac{1}{n_0} [{n_1}^2 - {n_2}^2]^{1/2} \tag{12.3}$$

The value of $\theta_{0_{\max}}$ given by Eqn (12.3) is known as the maximum acceptance angle for the fibre. From that equation we can see that it is determined by the refractive indices of the core and the cladding. The relationship

$$\text{NA} = [{n_1}^2 - {n_2}^2]^{1/2}$$

is known as the numerical aperture of the fibre. It is a useful indicator of the launching efficiency of the guide, and we will consider it later in relation to light sources.

When the surrounding medium is air, $n_0 = 1$, and

$$\theta_{0_{\max}} = \sin^{-1} (\text{NA}) \tag{12.4}$$

If θ_0 exceeds $\theta_{0_{\max}}$, θ_2 exceeds the value for which total reflection takes place at the core–cladding interface, and some of the energy will be refracted out into the cladding itself, where it is absorbed.

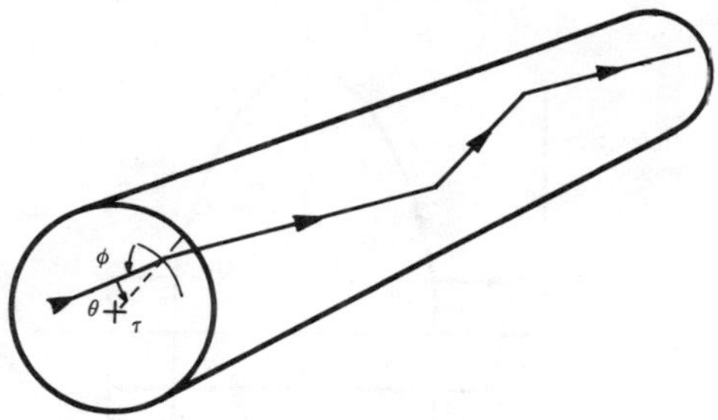

Fig. 12.2 Path of skew ray.

12.3 Skew Rays

The ray we have just discussed, with the help of Fig. 12.1, travels in a plane through a diameter of the fibre, and it is called a meridional ray. However, rays may also be launched into the guide in other directions, and these are known as skew rays. Provided the angle of incidence of a skew ray falls within $\theta_{0_{\max}}$ it will propagate along the fibre by total internal reflection, but, rather than following a path back and forth across a diameter, it will travel in a helix, as shown in Fig. 12.2.

12.4 Modes

The ray-theory approach to the mechanism of propagation along a fibre has the attraction of being simple, and so gives a good insight into fibre transmission. It does have some disadvantages, however. It assumes that the wavelength of the signal is extremely small, and therefore that the light can travel at any angle of incidence to the cladding, provided that the maximum acceptance angle is not exceeded. In fact that is not the case. As with all guides used to transmit electromagnetic waves, the boundary imposes conditions that restrict propagation to a series of modes. The details of these modes can be found by solving Maxwell's equations for the particular guide being studied. In the case of optical fibres, the solution is complicated by both the dielectric nature of the material, and its cylindrical geometry. We will not discuss these wave-theory solutions here, and for details of how they are obtained, reference should be made to one of the many specialist texts available.[1] We can note that as a result of solving Maxwell's equations, the various propagating modes in stepped-index fibre can be shown to have different wave velocities, a property that is called dispersion. As a result of this mode dispersion, a pulse of light launched on to the guide will, because it consists of energy in several modes, gradually broaden as it travels. This dispersion places a limit on the bandwidth of the fibre, a limit that is a function of the stepped index profile, and not of the quality of the fibre material.

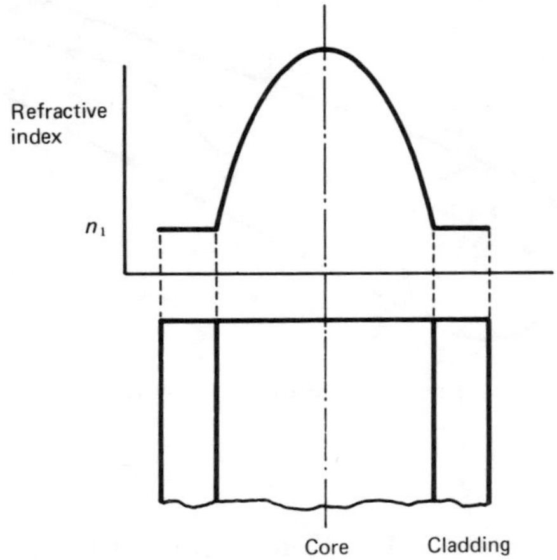

Fig. 12.3 Ideal graded profile to minimize mode dispersion.

12.5 Graded-index Fibre

The effect of mode dispersion can be minimized, and in theory eliminated,[2] by using a graded profile of the shape shown in Fig. 12.3. The difference between theory and practice is that it is very difficult to produce the exact profile required, and the effect of a slight deviation from it is marked. Not only is it difficult to produce the fibre to conform with theory, but there is bound to be some variation of profile with length, and this will allow dispersion to occur, with a consequent reduction in bandwidth. The ideal profile is almost a square law, i.e. the index of refraction of the core is given by

$$n^2(r) = n_1^{\,2}\left[1 - \frac{2[n_1^{\,2} - n^2(a)]}{2n_1^{\,2}}\left(\frac{r}{a}\right)^g\right] \tag{12.5}$$

where $n(r)$ is the refractive index at radius r, $n_1 = n(0)$ where the exponent $g \simeq 2$.

The graded-index changes the paths followed by the waves travelling down the fibre. In terms of the ray-theory of Fig. 12.1, the path is sinusoidal in shape, with the light travelling more quickly at the core–cladding boundary than in the centre of the guide, where the refractive index is at a maximum.

Skew waves will still follow a helical path, with their velocity increasing with distance from the centre of the fibre.

The effectiveness of a particular profile on reducing the mode dispersion is also a function of wavelength, so it is possible to reduce the dispersion in a fibre by operating it at its most favourable wavelength.

We can see the effect of varying the profile index on the mode dispersion from Fig. 12.4. The sharp minimum shows why great importance is attached to producing fibre with the optimal value of g.

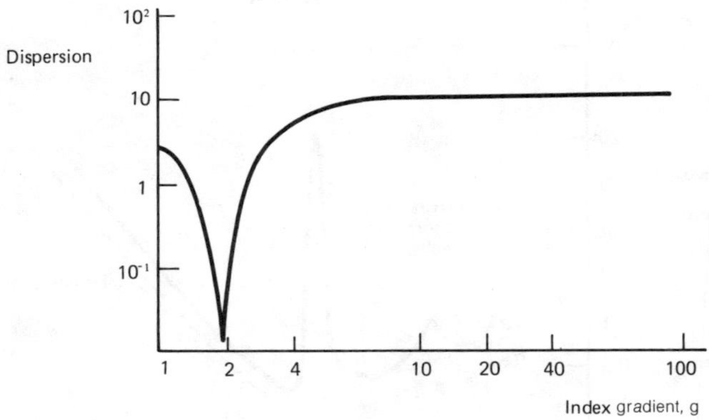

Fig. 12.4 Effect of profile index on mode dispersion.

12.6 Loss Mechanisms

Attenuation of the signal as it travels along the guide is caused by several factors. The main ones are absorption, scatter and radiation; each is important, but as fibre-production methods have improved, and the limits on the operating conditions of the fibre have been optimized, the attenuation is eventually determined by the scattering loss.

12.7 Absorption

The contaminating elements in glass, particularly the transition metals (V, Cr, Mn, Fe, Co, Ni), have electron energy levels that will absorb energy from incident light. The amount of energy absorbed depends on the proportion of impurity atoms present. With care in the manufacture of the glass, and in the drawing of the fibre, these impurities can be reduced to about 1 part per billion, and at this level the loss is negligible.

The absorption produced by the transition metals is not very dependent on frequency, and therefore is not easily avoided by choosing a suitable wavelength for the light source.

The other main absorption mechanism is that due to the presence of water. The hydroxyl ion, CH^-, absorbs readily at 2.8 μm, and there are less strong, but still pronounced, absorption peaks at wavelengths which are almost direct harmonics 1.4 μm, 0.97 μm, and 0.75 μm. The position of the peaks can be seen clearly in an attenuation–frequency curve, such as that shown in fig. 12.5. Some reduction in this absorption is obtained by pre-drying the powders used to form the glass. The strong frequency dependence of this mechanism allows an operating wavelength to be chosen that will avoid the severest levels of loss.

The two absorption mechanisms discussed are, in reality, complicated factors

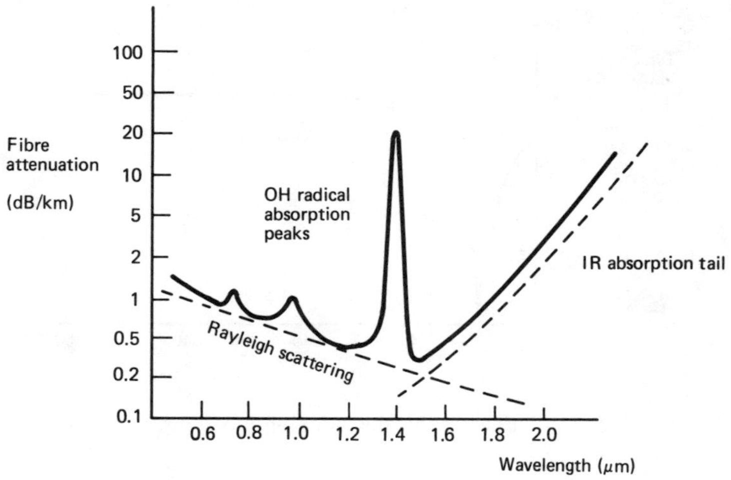

Fig. 12.5 Typical attenuation characteristic of fibres.

which depend on the composition of the glass and the method of fibre production. A deeper discussion is to be found in[1], and there are many specialist papers on the subject.

Optical fibres are sensitive to far-infrared radiation, and the loss peak in that region has a tail that reaches to wavelengths as low as 1.6 μm, as shown in Fig. 12.5.

12.8 Scatter

The fine variations in atomic shape, occurring over distances very small compared with a wavelength, scatter the light by the classical Rayleigh mechanism, which in the atmosphere gives the blue colour to the sky. The loss coefficient is proportional to $1/\lambda^4$, and since it is unavoidable, provides a bound to the lowest attenuation which can be achieved. The Rayleigh loss curve is also shown in Fig. 12.5. The value of the loss is a function of the fibre and its constituent glass:

$$\alpha = \frac{8\pi^3}{3\lambda^4}[n^2 - 1]kT\beta \tag{12.6}$$

where T is the transition temperature at which any imperfections in the glass structure are "frozen in", and β is the isothermal compressibility of the glass.

Other types of scatter can occur. In particular, Mie scatter, which is due to variations in the structure of the glass that occur at about one wavelength intervals, and waveguide scatter, which is caused by fluctuations in the geometry of the core along its length. These losses can be made insignificant if the fibres are produced with care.

12.9 Radiation from Bends

Energy is radiated at a bend in the fibre. The amount of radiation is usually small, but it is determined by the bend radius. As the radius is decreased there is a very rapid change, at a critical radius, R_c, from very little loss to very high loss. The value of R_c depends on the numerical aperture of the guide, and the radius of its core. Increasing NA and reducing the core size will help to decrease the minimum bending radius.

The mechanism that causes radiation results from the fact that the field pattern of the guided wave penetrates into the cladding. At a bend, the field at the outside of the bend has to travel faster than that at the inside, to maintain the phase relationship in the mode. At some distance from the fibre the velocity of the wave will reach the velocity of light; therefore the energy at a larger radius must be radiated. As the bend radius decreases, the radius at which radiation occurs also decreases, moving closer to the fibre. At the critical radius, half the light into the bend is lost. For many materials, the critical bend will be very small, and sometimes less than the limit imposed by the mechanical stress due to bending.

12.10 Single-mode Fibre

As we have seen above, the principal reasons for attenuation and bandwidth limitations in multi-mode optical fibre are the various losses described, and mode dispersion. Multi-mode fibre is used because of its larger size compared with mono-mode fibre. The core diameter is usually 50–60 μm, with a cladding of 100–150 μm, and hence coupling into source or detector devices, and jointing fibres can be done efficiently, and easily.

Single-mode fibre has the attraction that there is no mode dispersion, and therefore the bandwidth available can be very high indeed. There is a limitation caused by dispersion, not of modes on the fibre, but of the various frequencies emanating from the source. The spread of frequencies generated by the emitter will depend on the type of device used, but there will be some, and each frequency will propagate at a slightly different velocity. This dispersion is called chromatic, and will decrease as the line-width of a source is reduced.

The size of a single-mode guide depends on the free space wavelength and the refractive indices of the core and cladding. For a particular waveguide, operating at a given wavelength λ_0, the cut-off radius for modes higher than the fundamental is a, where a is given by

$$a \leqslant \frac{2.405\lambda_0}{2\pi(n_1{}^2 - n_2{}^2)^{1/2}} \tag{12.7}$$

and n_1 and n_2 are the refractive indices of the core and cladding, respectively. Typically, single-mode fibre has a core diameter of about 5 μm. Consequently, although it is very attractive for high bandwidth, long distance communications, multi-mode is usually preferred. It is doubtful however that single-mode fibre will be confined to laboratory study for much longer, as mentioned in Section 12.4.

12.11 Detectors

There are many devices available for detecting optical frequency energy. Some are vacuum tube, others solid state, either semi-conductor or not. There are also several photosensitive effects employed to convert incident light into a proportional electrical signal.

In communications systems the power in the incident beam may be extremely small, of the order of 10^{-14} W, and this places additional requirements on suitable detectors. Although several devices have been used, and there is extensive research and development on new techniques, we will limit our attention to the PIN semiconductor, and its associated device, the avalanche photodiode (APD).

To be suitable for use in communications, the detector must be very sensitive, have high efficiency of conversion between light and electrical energy, respond very rapidly for high bandwidth, have low noise power, and good light-collecting properties. In addition, it should operate at low voltage, be easy to use, be robust and insensitive to changes in ambient conditions, have a long life, good reliability, and be inexpensive. Such an ideal specification is, of course, unattainable; in any real device compromises are necessary, and priorities must be settled between the various parameters for any application. Here we cannot look at the devices on the market; may companies produce a range of detectors, and the choice of which to use must be made by the system designer. However, we can examine some of the general features of photodiodes and the materials from which they are made.

In photodetection, light is incident on a semi-conductor material that, because of the energy gap between the valence and conduction bands, will convert the light energy into electrical energy. On average, the photocurrent is related to the incident light energy by

$$I = \frac{\eta q}{h\nu} P \tag{12.8}$$

where q is the electron charge, $h\nu$ is the photon energy, P the optical power into the device and η is the quantum efficiency. η is the fraction of the incident power producing electron–hole pairs.

There is a statistical nature in the whole photodetection process that causes us to use mean values for the various parameters. These mean values, and the associated rms quantities, give us a long-term view of the device behaviour.

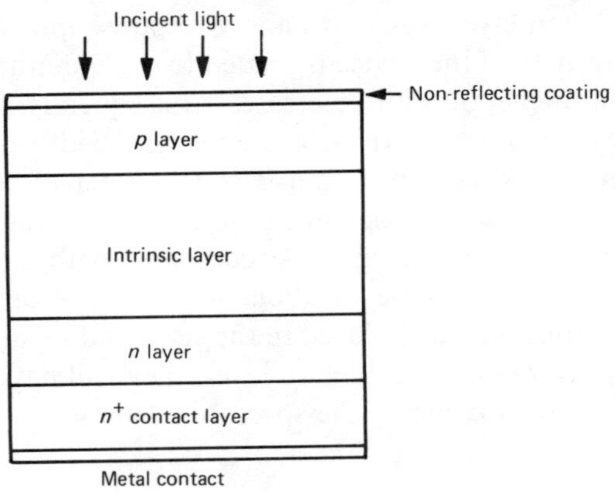

Fig. 12.6 Schematic of PIN diode.

12.12 PIN Diode

The diode consists of heavily doped p and n sections separated by an intrinsic layer, as shown in Fig. 12.6. The reverse bias across the device sweeps the carriers from the I region, leaving a depletion layer. Light falling on the depletion layer creates electron–hole pairs, and these carriers drift across the region to the terminals, where they leave as light-induced current. The device therefore appears to be a current source, across the depletion layer capacitance.

12.13 APD Device

In an APD, the applied voltage is much higher, and there is a region where the electric field is in excess of 10^5 V/cm. At such a high level, photoinduced electrons are accelerated to a velocity at which they can ionize atoms in the intrinsic region, and produce an avalanche effect. In this way the device has an internal gain, M, which is the mean number of electrons at the output per photoinduced electron. The APD is therefore more sensitive than the PIN diode, but because of the stochastic nature of the avalanche gain, its noise level is higher.

12.14 Speed of Response and Quantum Efficiency

When discussing the PIN diode, we noted that the photoelectrons move across the depletion layer at the drift velocity produced by the applied voltage. This transit time limits the speed of response, and hence the bandwidth, of the device. If the depletion layer is narrow, a low response time can be achieved. However, the sensitivity of the device is related to the quantum efficiency, and this increases as the depletion width increases. Consequently, there is a conflict between these two parameters. In Si the depletion width is approximately 50 μm and the minimum response time is of the order of 50 ns, whereas in GaAs the depletion layer can be much narrower, because of the higher absorption coefficient of photons in GaAs compared with Si, and therefore it has a much faster response time of about 40 ps. If the depletion width is narrow, photoelectrons may be induced in the surrounding diffusion area and before being swept through the depletion layer, they will have to diffuse from the diffusion region, thus reducing the speed of response.

12.15 Reliability

Si detectors have been developed over many years for use in the 0.8–0.9 μm range and improvements to the configuration and production techniques used have made available devices with projected lifetimes of many years, and high reliability is therefore readily available. For GaAs or other III–V devices, which are used at the longer wavelength of about 1.3 μm, experience is more limited, but long lifetimes are again anticipated. The reason for the limitation of Si to the shorter wavelength region is that its energy gap is 1.1 eV, and this cannot be achieved by photons of wavelength greater than about 1.1 μm. At the longer 1.3 μm wavelength, which is becoming increasingly attractive because of the development of very low-loss guides at that wavelength, Ge has high sensitivity but, because the ionization coefficients for electrons and holes are about the same, the noise level is high compared with that which can be achieved with GaAs, InP or some more complex compound such as InGaAs.

12.16 Noise

The sources of noise in PIN and APD devices are well known. In PIN diodes noise is produced by:

(i) random fluctuations in the photo-current itself,
(ii) background radiation from other than the intended light sources, and
(iii) current generated in the device when there is no light present, i.e. dark current.

In Si devices, noise produced by (ii) and (iii) can be made extremely small, leaving the shot noise of (i), and this provides a lower bound on the sensitivity of the device.

In APDs another source of noise exists – that due to the randomness of the gain process. It magnifies the shot noise by a factor $F \times M$ where M is the mean avalanche gain, as in Section 12.13, and F is a quantity called the excess noise factor. If the value of F is unity, the avalanche process is not itself noisy. In practice F is much greater than that. Smith[3] states that F is related to M, and a factor k, by the approximate relationship

$$F \simeq 2(1-k) + kM \tag{12.9}$$

where k is the ratio of the smallest to the largest ionization coefficient in devices in which the avalanche is initiated by the carrier with the highest coefficient. In Si, for example, the avalanche is initiated by electrons that have a much higher ionization coefficient than holes, giving $k < 0.1$. By contrast, Ge is intrinsically more noisy because the electron and hole ionization coefficients are equal.

The choice between PIN and APD will depend on the priorities given to sensitivity and SNR, as well as on operating constraints. The APD can be much more sensitive than the PIN if the internal mean gain, M, is high. However, the shot noise is also a function of M, and, for some devices, this quantity will increase more rapidly than the gain as M is increased. An optimal value of M may exist at which, for a given SNR, the sensitivity is maximum. In practice, for applications in which good SNR is required, PIN devices are used, but if sensitivity is at a premium, an APD would be more appropriate. Compared with PIN diodes, avalanche devices have less attractive operating conditions. They require a much larger voltage, possibly of the order of several hundred volts, and they are sensitive to temperature variations. This means that more complicated driving circuitry is required to provide compensation if the temperature changes.

12.17 Optical Sources

The special requirements of optical communications make many demands on optical sources, some of which may not be necessary for other applications. Apart from supplying sufficient power, at reasonable cost, over a long, reliable, lifetime, and with high efficiency, the source should have an emission wavelength that coincides with a loss minimum of the fibre. It should also show a linear output power versus drive current characteristic, emit over a narrow bandwidth, and be capable of transmitting high bit rates.

These constraints and requirements limit the types of suitable emitter to three:

(i) light emitting diode (LED)
(ii) semiconductor laser, and
(iii) solid-state laser.

Here we will discuss only (i) and (ii), since they are the principal devices presently in use. The solid-state laser is likely to become an important source in the future.

Before considering the LED and the semiconductor laser separately, we will examine some general properties of these optical sources.

LED emitters are simple in their construction, and do not require complex drive circuitry. They are particularly suited to relatively short distance links in which the bit-rate requirement is modest and the channel capacity low.

Alternatively, semiconductor lasers, which require higher drive currents and more complex circuitry, can produce higher power than LEDs, and are capable of high bit rates. They can therefore be used for high-speed, long-distance systems.

For both types of device, operating in the 0.8–0.9 μm range, the most commonly used material is a doped GaAs, usually GaAlAs. Several III–V compounds have been used successfully. In general, to be suitable a material must have a direct energy gap in the region of 2 eV so that, when electron–hole recombination takes place, a photon is released. The material must also be one in which it is possible to form a p–n junction very easily. This requirement rules out high melting point compounds. Ideally, the emission wavelength should be adjustable by changing the composition of the material. This requirement is usually satisfied by varying the mole fraction. For example, if x is varied in $Ga_{1-x}AsAl_x$, the emitted wavelength can be varied between that corresponding to GaAs and to AlAs.

The basic mechanism in both LEDs and lasers is that minority electrons in the conduction band recombine with valence band holes and release energy in the form of radiation, i.e. if the direct gap between the valence and conduction bands is Eg, the radiation produced by recombination is

$$\nu h = Eg \tag{12.10}$$

where h is Planck's constant and ν is the radiation frequency of the photon emitted.

Although p–n junctions in III–V materials are excellent sources of light, the basic compound will absorb light easily as it passes through, and therefore the separation between the active area and the outer surface of the diode must be very small, otherwise the semiconductor will absorb some of the energy produced in the active region, and thereby reduce overall efficiency.

Here we will discuss the performance features of sources, not the physical and material aspects of their operation. Information on the choice of compounds, and the physical mechanisms involved in the devices, is given in Section 16.2 of Miller and Chynoweth,[4] and Chapters 4 and 5 of Sandbank.[5]

12.18 Light Emitting Diodes (LEDs)

In its basic, homojunction form, an LED consists of a forward-biased p–n junction. Carriers are injected, under the influence of the electric field, from the majority to the minority side where they create an excess of minority carriers. As they diffuse away from the junction, recombination occurs, and if the energy gap Eg is approximately 2 eV, optical radiation will be emitted at a frequency given by Eqn (12.10). The optical energy radiates from the device, and is launched on to a fibre guide.

For efficient performance, the recombination process must have a very high proportion of radiating events, and the optical power must not be absorbed in its path from the junction to the surface of the device, or lost between the device and the fibre.

The quantum efficiency, defined as the proportion of radiating carrier recombinations, is a measure of the internal conversion between electrical and optical power. Values of about 80% can be achieved. However, all of the generated light cannot be used. Some will be absorbed by the semiconductor material itself, as we mentioned earlier. In addition, the semiconductor–air interface will not transmit all the energy incident on it, only that which falls within the critical angle, which is about 15° for III–V materials (see Fig. 12.7). In practice this means that only about 1% of the light generated will be useful. Some improvement can be achieved in this external quantum efficiency, η_c, by using a reflecting back plane.

The purpose of the diode is to provide a light source for fibre transmission, and one of the loss mechanisms occurs between the device and the fibre-end. The acceptance angle of the fibre is $\theta = \sin^{-1}(\mathrm{NA})$, where NA is defined in Section 12.2. For sources of area less than that of the fibre-end, the efficiency is approximately equal to $\sin^2\theta = (\mathrm{NA})^2$. This results in a fibre-launching efficiency of about 5%. The total efficiency of coupling, given as the proportion of diode input power that is launched into the fibre as light, is therefore about 1% of 5%, or 0.05%.

Clearly, considerable effort is involved in improving this figure. Some mechanical changes to the diode can be made. The simplest diode structure has a planar geometry, as shown in Fig. 12.7. By forming a well in the upper

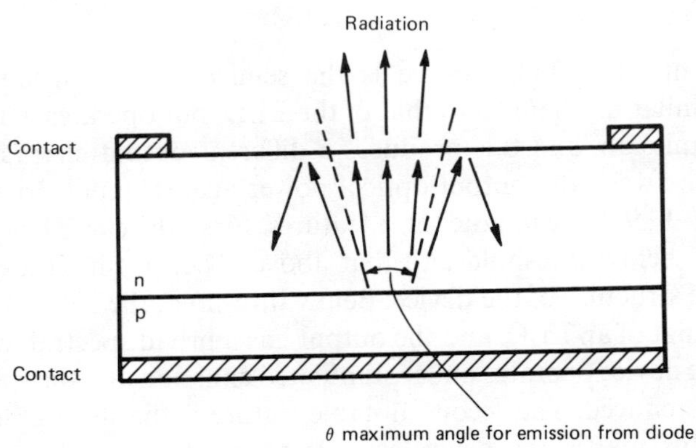

Fig. 12.7 Exit of light from an LED with planar geometry.

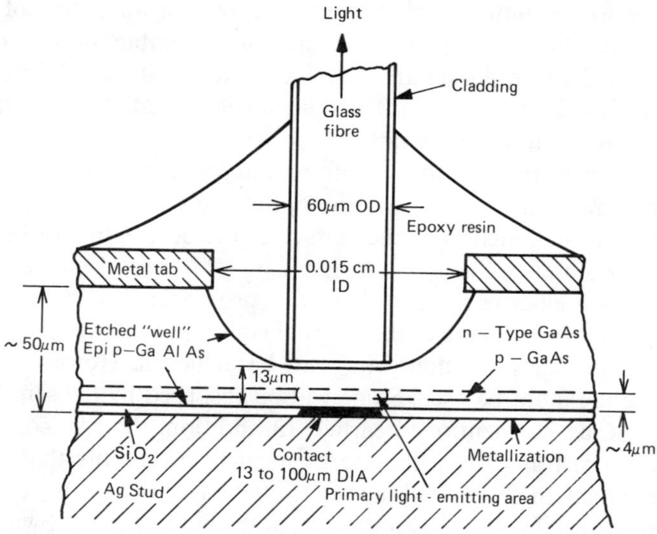

Fig. 12.8 Schematic of Burrus Light emitting diode. Reproduced from Dawson and Burrus, *Applied Optics* vol. 10, pp. 2367–2369, Oct. 1971.

material, the fibre end can be placed near to the active junction, thus reducing significantly the absorption of light within the semiconductor material. If, in addition, the light-generating area is smaller than that of the diode, an epoxy resin layer between the device and the fibre will increase the launching efficiency. An example of this type of device, known as the Burrus diode, is shown in Fig. 12.8. Although it has a high launching efficiency, it has the disadvantage of a relatively large spectral width. In general, the range of spectral widths produced by LEDs is from about 100 to 400 nm.

12.19 Lasers

The other important light source is the semiconductor injection laser. It employs a similar mechanism to that of the LED, but operates with laser-type stimulated emission, and this produces a different output characteristic. The relationship between the output optical power against input drive current, is shown in Fig. 12.9. We can note three features. First, the curve has two distinct sections, that below threshold and that above. The threshold level is closely related to the structure of the device. Below threshold, the action of the device is similar to that of an LED, and the output has a broad spectral width. Above threshold, the device operates under stimulated emission and the spectral width is very much reduced. The second notable feature is the steep gradient above threshold. This makes the device very fast, and this speed is exploited in

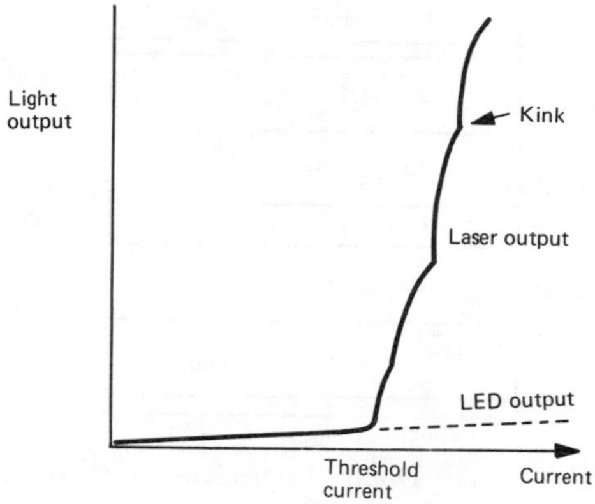

Fig. 12.9 Relationship between output optical power and input drive current for laser and LED.

high-speed digital systems. Finally, the output characteristic above threshold has nonlinear features, called kinks. These kinks are thought to result from slight changes in current paths through the active region of the device, and they can be eliminated by using a stripe configuration.

Several structures have been used for injection lasers. In its simplest form, the device consists of an active p–n junction in GaAs, with planar terminal plates parallel to the junction, to which the supply voltage is applied. Two of the opposite faces, perpendicular to the junction plane, are cut to provide reflecting surfaces, between which the essential gain of the carriers across the junction can be achieved. Through one of these faces (or facets, as they are called) light is coupled to the fibre. The opposite, rear facet is sometimes used to monitor the performance of the device. Figure 12.10 shows such a simple laser. The active region has a slightly higher refractive index than the surrounding layers, and this causes a weak waveguiding mechanism, allowing the generated light to be constrained within a narrow region.

This waveguiding is a very useful feature because it allows the device to produce an intense light source that is more suitable for launching on to fibres than that from an LED, which produces a less directional output. To improve the containment of the light into the active region, a double-heterojunction

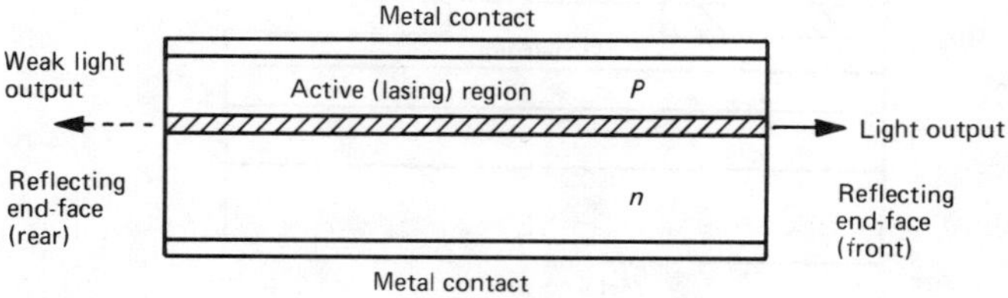

Fig. 12.10 Schematic of laser.

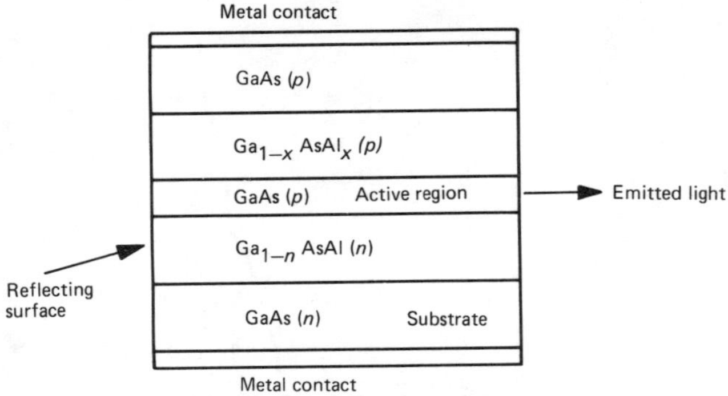

Fig. 12.11 Double heterodyne laser structure.

(DH) structure is used. Figure 12.11 shows the main features. The heterojunction is an interface between GaAs and GaAlAs, and in the DH device GaAlAs is placed at each side of the GaAs active region. Apart from providing a stronger waveguiding action, the DH structure causes the threshold current to be considerably lower than that of the homojunction arrangement shown in Fig. 12.12. This reduction in threshold current is sufficient to allow the device to operate in a continuous mode, whereas in the homojunction structure the current density required to sustain the lasing action can only be achieved in a pulse mode.

In Fig. 12.12 we can see the stripe referred to earlier. Several experiments on the best width of stripe to use have been performed, and that most commonly available is about 20 μm across. The effect of the stripe is to confine the active region of the device to an area below the stripe, thus producing a two-dimensional waveguiding mechanism, and improving the stability and directivity of the generated light.

The DH GaAlAs semiconductor injection laser is only suitable for the lower, 0.8–0·9 μm, wavelength range. For the longer wavelength of about 1.3 μm, a different base material has to be developed. Present research suggests that the

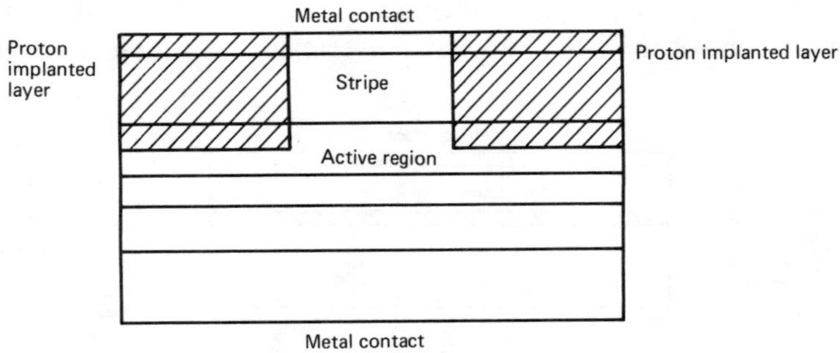

Fig. 12.12 Stripe laser.

most promising is InP, with doped layers of GaInPAs forming the heterojunctions.

From a systems point of view, the injection laser offers the advantages of higher intensity, more directive output, with a faster response time and narrower spectral width than those of the LED. These advantages are gained at the expense of higher operating voltages and more complex drive circuitry, which is necessary to counteract the variation in threshold current produced by temperature changes and ageing in the injection laser.

12.20 Communications Systems

Of the many applications of optical fibre systems, the one that will have, and indeed has had, the most significant effect on equipment development is long-distance telephony. Telephone operating companies around the world, stimulated by the paper of Kao and Hockham,[6] have investigated optical fibre systems in the hope that they could be used as a future trunk transmission system, with significant economic advantages over any alternatives. In the mid-1960s, it was clear that coaxial cable would not be the best medium to cater for the anticipated growth in telephone and television traffic, and considerable effort was directed to developing what seemed to be the most attractive alternative, overmoded circular waveguide. The special feature of an inherent reduction in attenuation with increase in frequency of the fundamental TE_{01} mode, made the circular waveguide approach very appealing, in spite of the essential expense of producing high-precision, helix-lined guides. At that time the major telephone administration research laboratories around the world were performing test trials on waveguide, before bringing it into service. Optical fibres were also interesting, but their high loss made them unable to challenge coaxial cable. Gradually, over the next five years, the considerable amount of research into the material and equipment aspects of fibre production began to be successful and the attenuation fell rapidly way from 20 dB/km, or more, to levels that were more reasonable. A comprehensive understanding of the mechanism of transmission over fibre guide was being established, and it was becoming clear that there was no inherent loss mechanism to prevent transmission over very long distances. Gradually, the relative importance given to research into circular waveguide and optical fibre systems changed, and the big decision to commit the future to optical transmission was taken. The reasons why are, from our viewpoint, clear. The demand was for a large-capacity system that was cost effective, secure, comparatively easy to instal, and low-loss. Circular waveguide could be made low-loss, but otherwise it could not compete with fibre, and once the potential very low attenuation of fibre began to be realized, its other outstanding attributes made it the obvious choice. It is now highly possible that the transmission length between repeaters will be at least twenty times that which could be achieved with coaxial cable, and the small size and weight of fibre cable makes the installation a very straightforward operation. Rather than provoke additional difficulties, whose cost could be offset against its advantages, optical fibre is so light weight, flexible and slim that it can be threaded into ducts that had previously been thought to be full, and thus the lifetime of the trunk system can be extended. The cost of the fibre was, in the early days, very high. This was a reflection of the very high investment in development made by both the glass companies and the cable manufacturers, but cost is no longer a penalty. Once the large telephone administrations had decided to use fibre for all new trunk transmission, the per unit cost had to fall dramatically, to make it the cheapest medium on a capacity basis. The spin-off will be seen in other systems, as the technical properties of cheap fibres are exploited.

The system designer has several choices to make, once the decision to use fibre has been taken. The fibre type, refractive index profile, and operating frequency must be specified, and the types of emitter and detector must be chosen. The type of fibre to be chosen depends on the bandwidth required, the length of run, the future development of the system (called upgradability in[7]), and the cost. There is a good case for installing as good a quality of fibre as possible, so that the system can be upgraded without having to replace the fibre. If high bandwidth is required, the choice lies between graded index multi-mode or single-mode guides. As we saw earlier, the core diameter of single-mode fibre is about a tenth that of multi-mode, which is therefore easier to launch light on to, and easier to join. However, advanced designs of injection laser diodes, and microprocessor-controlled coupling jigs, have reduced the difficulties of single-mode working such that the comparison with multi-mode is much more balanced. Several administrations have plans for 500 Mbit/s single-mode experimental trunk routes, and this development suggests that the attractive features of single-mode will outweigh its disadvantages. Another aspect of future systems that makes the choice of single-mode guide more appealing is the use of integrated processing devices, known as integrated optics. Several research laboratories have been working on integrated processing devices, such as switches, modulators and filters, for over a decade. Much of the effort has been in producing satisfactory materials, and making simple devices in a reproducible way. Gradually we are moving towards optical integrated circuits that will be used in mono-mode systems. When such devices are available, many additional communication applications will be possible.

As we have seen already, the choice of wavelength is governed by the behaviour of the fibre. A combination of low attenuation and low dispersion is desirable. Until 1982 the most commonly used wavelengths were 0.83 μm and 0.9 μm, or thereabouts. After that time the lower dispersion available at longer wavelengths began to be exploited with silica based fibres, operating at 1.3 μm.

This change of frequency required new emitters and detectors. Silicon detectors had been clearly the best performers at the shorter wavelength, but at 1.3 μm it is necessary to use one of the III–V compounds, usually based on GaAs. The structure of the detector should offer a large area for light-collection and the device should be sensitive, without being too noisy, as we discussed in Section 12.16.

The type of emitter used will depend on the application. The almost linear output versus drive current relationship of an LED makes it particularly attractive. It is also the obvious choice for low-speed, low-cost digital systems, particularly those using step-index fibre, which has a large value of NA. However, for launching on to graded-index fibre for high-speed systems, the injection laser might be preferred, and for single-mode systems it is essential. The small chromatic dispersion of the laser makes it necessary for long-distance communication systems. Semiconductor lasers are more temperamental than LEDs. Their performance characteristics change with temperature, and with age. To counteract these effects, some form of feedback is used, taking a monitoring signal from the rear facet of the laser, behind which is situated a simple photodiode. Two other problems with lasers are (i) rise-time delay and (ii) overshoot spikes, in response to a pulse input current (see Fig. 12.13). The dc bias used can affect the delay, and by biasing the device near its threshold level the delay is minimized. Here a compromise is required, for the spectral width increases as the bias increases, and the life-time falls. In Section 12.19 we discussed the use of stripe conductors on the laser to stabilize its performance, and one of the benefits of this geometry is that the output oscillatory spikes are reduced.

The length of an optical fibre system will not be limited to that which can be achieved without intermediate amplification. In digital systems, regenerators will be required at intervals. The spacing will depend on the quality of the system, but it will be somewhere in the range 10–60 km. A block diagram of a regenerator is shown in Fig. 12.14. Essentially, it detects the incoming light

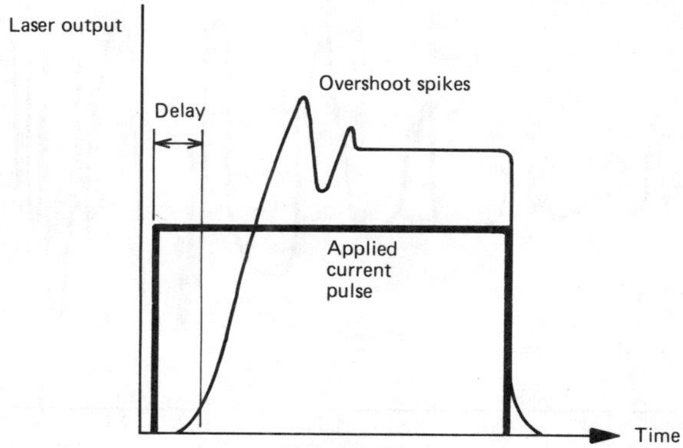

Fig. 12.13 Rise-time delay and output overshoot in laser response to pulse drive current.

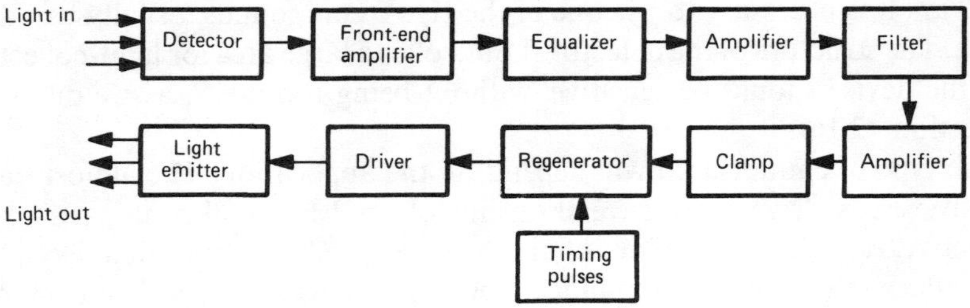

Fig. 12.14 Block diagram of optical regenerator.

signal, re-forms and re-times the pulses, and modulates an emitter for the next stage. Regenerators introduce additional problems for the designer. Supply voltages must be provided, and since regenerators are often placed in remote locations, the supply has to be fed along the optical cable. Monitoring, to allow remote fault finding from a control point at the end of a link, is usually included in the design. These features increase the cost of the system, so the gradual extension of the distance that can be covered without regenerators is a welcome result of fibre research.

Unlike systems that operate at lower radio frequencies, little attention has been paid to the use of free-space optical communications. This results from the heavy attenuation in the atmosphere due to moisture. Precipitation causes very large losses, but even under dry conditions there are attenuation peaks at certain frequencies (Fig. 12.15). In addition, the movement of air caused by localized heating produces refractive index variations that deflect the path of the light beam. Free-space links are limited to a maximum of about 2 km.

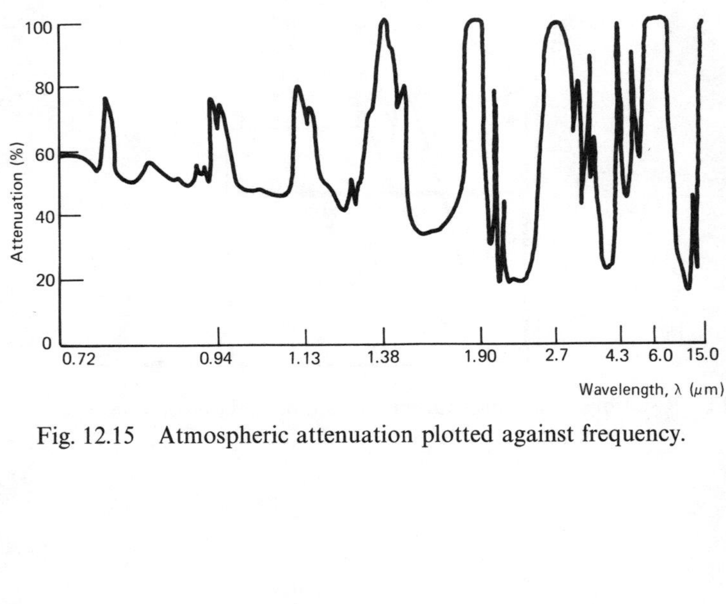

Fig. 12.15 Atmospheric attenuation plotted against frequency.

12.21 Conclusion

We have touched on some advantages that optical fibre systems possess. Many are attractive from the operations and management point of view, giving trouble-free performance. Some, such as no earth loops, and immunity to sparking or electrical interference, make communications links possible in environments that were previously very difficult, or even impossible. Apart from using fibres in such situations, we can expect them to form an integral part of the systems that will be required to satisfy the enormous increase in the quantity and type of traffic that will be generated as the revolution in office, computer and home communications takes place. Already there are designs for two-way, low-speed, data terminals in homes, and eventually all our domestic communications – telephony, television, automatic transactions, teletext, facsimile and systems not yet devised – will be transmitted over a single digital line – the optical fibre. Plans are being developed in the UK for national optical fibre cable television and fibre-based high capacity data systems. Similar systems are being developed in other countries. Coupled with digital representation of signals and integrated circuits of enormous density and complexity, optical fibre systems are transforming the aspirations of communications engineers, and encouraging a tremendous upsurge of innovation and enterprise. In fact, at present it seems that, for the first time in the history of telecommunications, we have more capacity and flexibility than we can use for any systems that presently come to mind.

References

1. Midwinter, T.E., *Optical Fibres for Transmission*, Wiley, Chichester, 1979.
2. Kaminow, I.P., Marcuse, D. and Presby, H.M., "Multimode fiber bandwidth: theory and practice", *Proc. I.E.E.E.*, **68** (10), 1209–1213 (Oct 1980).
3. Smith, R.G., Photodetectors for fiber transmission system", *Proc. I.E.E.E.*, **68** (10), 1247–1253 (Oct 1980).
4. Miller, S.E. and Chynoweth, A.G., *Optical Fiber Communications*, Academic Press, New York, 1979.
5. Sandbank, C.P., *Optical Fiber Communication Systems*, Wiley, Chichester, 1980.
6. Wilson, T. and Hawkes, T.F.B., *Optoelectronics: An Introduction*, Prentice-Hall, London, 1983.
7. Kao, K.C. and Hockham, G.A., "Dielectric – fibre surface waveguides for optical frequencies", *Proc. I.E.E.*, **113** (7), 1151–1158 (July, 1966).
8. Elion, G.R. and Elion, H.A. *Fiber Optics in Communications Systems*, Dekker, New York, 1978.
9. Wolf, H.F., *Handbook of Fibre Optics*, Granada, London, 1979.
10. Optical-fibre Communications, Special Issue, *Proc. I.E.E.E.*, **68** (10), (Oct 1980).

13 PACKET SWITCHED NETWORKS

13.1 Introduction

This chapter deals with the topic of packet switched technology which is now in common use on both wide area networks (WANs) and local area networks (LANs). The most intensive use of packet transmission at the present time (1989) is on local area networks, and a substantial portion of the chapter is devoted to this topic. Packet switching was, however, initially developed for connecting computers separated by long distances, and some consideration of WANs will also be given. Recent years have seen the development of both WANs and LANs to carry integrated digital services including data, voice and video traffic.

The main difference between wide area and local area networks is that the former cover very large distances and have limited bandwidth and substantial propagation delay. Local area networks, on the other hand, cover distances up to a few kilometres and have relatively high bandwidths and low propagation delays. This means that the packet transmission techniques for the two types of network have some important differences. It is worth pointing out here that although WANs and LANs can be categorized separately, many LANs have gateways to WANs so that global communications often make use of combinations of both type of network.

A packet is essentially a block of data which is transmitted from a source to a destination. A typical packet format for Ethernet is shown in Fig. 13.1. The data is preceded by a header which contains the destination and source address and is followed by a cyclic redundancy check. The length of the data section (for Ethernet) can be anywhere between 46 and 1500 bytes (octets). The format of the packet will be discussed in detail later in this chapter; it suffices at this point to indicate that the channel between transmitter and receiver is occupied only during the actual transmission of the packet.

Preamble	Destination address	Source address	Type field	Data field	CRC
64 bits	48 bits	48 bits	16 bits	8 N bits	32 bits

Fig. 13.1 Ethernet packet format.

The actual transmission rate is much higher than the data rate of the source, which means that packet switching may be considered as a form of time division multiplexing on demand, i.e. transmission is only requested when sufficient data to fill a packet is available at the transmitter. At other times the transmission medium may be used to transmit packets between other sources and destinations. For this reason packetized systems are often referred to as asynchronous time division multiplexed (ATDM) or statistically multiplexed transmission systems.

This contrasts with the public switched telephone network which allocates a circuit between caller and called party for the duration of the call. This is known as circuit switching and differs from packet switching in that the circuit remains allocated throughout any periods of inactivity between the communicating parties. Packet switched systems are thus generally more efficient than circuit switched systems. The analysis of packet switched networks is often very complex and, as a result, powerful software simulation tools have been developed to aid the design process. Discussion of these tools is outside the scope of a text of this nature; instead a simplified treatment will be given, where appropriate, and the results of simulations will be quoted.

13.2 Wide Area Networks

The characteristic feature of wide area networks is that distances, and hence propagation delays, tend to be large. A typical topology is shown in Fig. 13.2 and it is usually the case that more than one path exists between a source and a destination node. Each node accepts data from computers or terminals connected to it, in a series of bursts, and assembles the data into packets with the appropriate address information.

The nodes then determine the routing, provide buffering and error control, and return acknowledgement information to the sender when the packet reaches the final destination; an early example of such a network is ARPANET[1]. There will clearly be several hops between intermediate nodes and inter-node acknowledgements are given when a neighbouring node accepts a packet. Should an acknowledgement not be received within a specified interval (e.g.

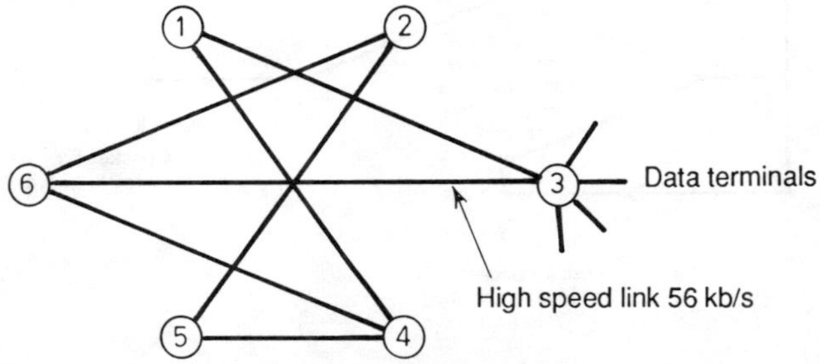

Fig. 13.2 Mesh connected wide area network topology.

125 ms) the packet is retransmitted. The end to end delay for short messages varies between 50 and 250 ms, depending on the packet length used. The delay can increase substantially under heavy network loads, when lengthy queues can form at intermediate nodes.

The topology of Fig. 13.2 is an example of a store and forward packet switched network as each node stores data until a complete packet is received

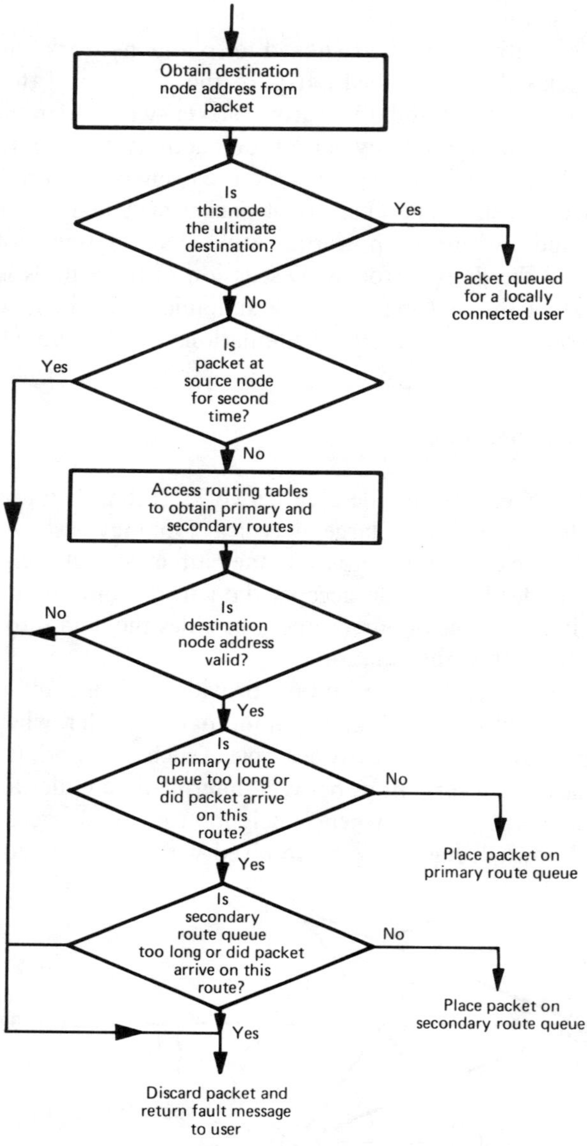

Fig. 13.3 Network routing algorithm.

and then queues the packet for forwarding to a neighbouring node. Clearly physical links must be provided between the individual nodes and there are essentially two packet transmission techniques which can be identified. A *virtual circuit* transmission system is one in which a path is set up between source and destination and all packets follow the same path and arrive in the sequence in which they were transmitted. A *datagram* transmission is one in which individual packets are transmitted by whatever route is available when they are presented for transmission. No fixed route is set up and each intermediate node decides on the appropriate path, according to some routing algorithm. In the case of a datagram the packets do not necessarily arrive in the order in which they are transmitted, and each packet must identify its position in the transmitted sequence.

The main feature of the datagram type of packet switched network is that the data links between nodes are used at near to their full capacity. Further, packets are scheduled for transmission so that there is no contention. Essentially each node queues packets for transmission and, since there is more than one outlet, each node can select an alternative route if the queue for the direct route becomes too long. There are many possible routing algorithms one of which is shown in Fig. 13.3.

In this example several priority ordered choices are stored in a routing table at each node. Considering Fig. 13.2, the first choice of a route between a source connected to node 1 and a destination connected to node 4 would be the direct route, i.e. 1 to 4. The second choice might then be via node 3. Once it has received an error-free packet from node 1, node 3 then selects the appropriate route for node 4. All of the nodes in a system of this type are referred to as packet switches.

In the topology of Fig. 13.2 the individual links are quite separate and routing algorithms are required. An alternative topology is one where each individual node is connected to a common broadcast channel and has to compete in some way for access, with all other nodes. This principle is used in most local area networks and in some wide area networks also. Wide area networks based on a single communications channel use either radio, to connect many geographically separate nodes, or satellite communications. This approach to wide area networks can be very cost-effective because it replaces the very expensive long distance dedicated links with a single wideband channel. A typical statellite system is shown in Fig. 13.4 in which each ground station transmits on a frequency f_{up} and receives on a frequency f_{down}. In such a system each ground station can monitor its own output for error-free transmission.

However, a single bus does introduce the problem of contention, i.e. several nodes requiring access to the channel at the same time. There is a number of techniques which have been developed to resolve such contention, and they can be categorized either as scheduled access or random access. The properties of these *access techniques* will be considered in some detail in this chapter. This will illustrate the suitability of specific access mechanisms for wide or

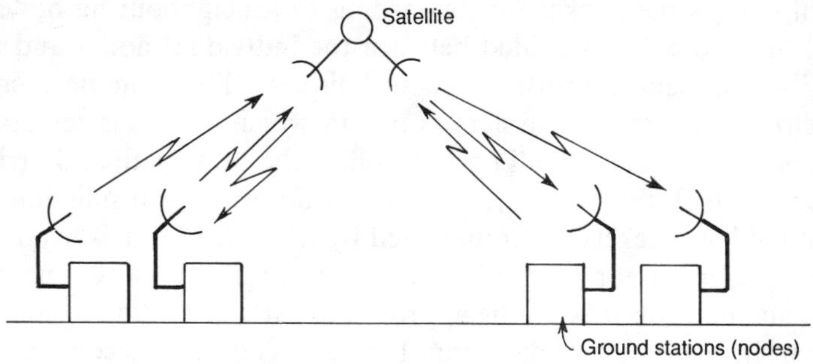

Fig. 13.4 Satellite packet broadcast system.

local area application. Of particular interest will be the relationship between throughput, offered traffic and delay.

13.3 The Poisson Distribution

Before any progress can be made in considering packet switched networks it is necessary to describe the statistics of packet arrivals on the network in any specified time interval T. To achieve this the time interval is assumed to be divided up into M intervals each of duration t seconds. As $t \to 0$ the probability of a packet actually arriving in the time interval t will be proportional to the actual value of t. Hence the probability of a packet arriving in t seconds is λt and the probability of no packet arriving in this interval is $(1-\lambda t)$. A fundamental assumption is that the probability of a packet arriving in one interval t is completely independent of arrivals in all other similar intervals. The system is thus said to be memoryless and the probability that there will be no arrivals in the entire interval T is then

$$P(0) = (1-\lambda t)^M = (1-\lambda t)^{T/t} \tag{13.1}$$

As $t \to 0$ then $M \to \infty$ and $P(0) = e^{-\lambda T}$ since $\lim_{x \to 0} (l - nx)^{1/x} = e^{-n}$

The probability of k arrivals in time T is the joint probability of one arrival occurring in k of the intervals t and no arrivals occurring in the remaining $M-k$ intervals, this is

$$P(k) = \frac{M!}{k!(M-k)!}(\lambda t)^k (1-\lambda t)^{M-k} \tag{13.2}$$

When $M \gg 1$ it may be shown that

$$M! \approx \sqrt{(2\pi)} e^{-M} M^{M+0.5} \tag{13.3}$$

This may be verified by replacing M by any integer (greater than 2) in Equation

13.3. Similarly if k is fixed

$$(M-k)! \approx \sqrt{(2\pi)}e^{-(M-k)}(M-k)^{M-k+0.5}$$

Substituting into Eqn (13.2)

$$k!P(k) = e^{-k}(1-k/M)^{-(M+0.5)}(1-k/M)^{k}(M\lambda t)^{k}(1-\lambda t)^{M-k} \tag{13.4}$$

But $\lim_{M\to\infty} (1-k/M)^{-(M+0.5)} = e^{k}$ and if $M \gg k$, then $(1-k/M)^{k} \approx 1$

i.e.

$$k!P(k) = (\lambda t)^{k}\frac{(1-\lambda t)^{T/t}}{(1-\lambda t)^{k}}$$

as

$$t \to 0\ (1-\lambda t)^{k} \to 1 \text{ and } (1-\lambda t)^{T/t} \to e^{-\lambda T}$$

Hence

$$P(k) = \frac{(\lambda t)^{k}e^{-lT}}{k!} \tag{13.5}$$

This is known as the Poisson distribution and gives the probability of k arrivals in the interval T. In this expression λ may be interpreted as the average rate of arrivals per second.

13.4 Scheduled Access Techniques

To illustrate the principles of scheduled access, consideration will be given to a single bus system of the type shown in Fig. 13.5 although it should be realized that the same arguments apply to ring topologies and the shared broadcast radio channel.

Scheduled access can be under centralized control or can be completely distributed. In the former case access is determined by a designated controller, and this will be considered first. The performance of such systems is usually

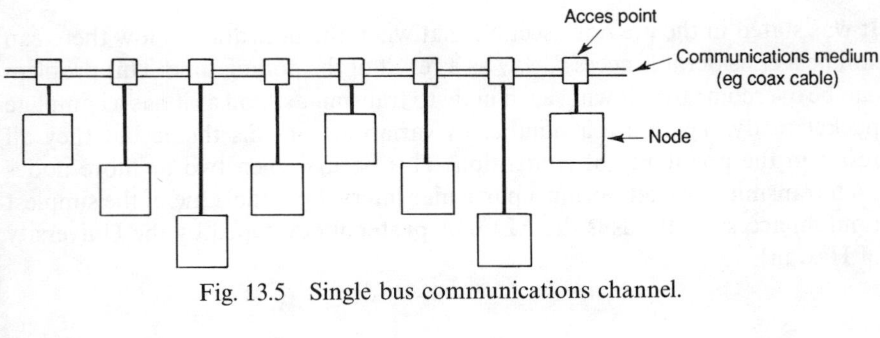

Fig. 13.5 Single bus communications channel.

measured in terms of access delay. This is the time which elapses between a packet being available for transmission at a node and commencement of transmission.

In order to avoid contention on the bus the controller poles each node in turn. If a node has a packet for transmission the transmission is commenced as soon as the polling signal is received. If a node has no data for transmission the controller is made aware and then poles the next node in the system. When the controller sends out a polling signal there will be a finite time which elapses before the polled node is able to reply (polling interval), followed by an interval required for data transmission (transmission interval).

The polling interval p_i will be a combination of the propagation delay on the network and the time required for the node to synchronize with the polling signal. The transmission interval t_i will depend on the amount of data a particular node has for transmission when polled. Both p_i and t_i are random variables, but it is clear that when the network extends over several hundred (or thousand) kilometres the polling delay p_i can greatly exceed the value of t_i. The efficiency of such a system is thus relatively poor when the propagation delay is very high.

A significant improvement can be achieved if central control is replaced with distributed control. When a node completes its transmission it essentially poles the next node on the network. The polling time is now, on average, significantly shorter than with centralized control, because the mean distance between a node and its immediate neighbour is much less than the mean distance between a designated controller and all other nodes. This type of distributed control is essentially the same as the control used on token passing LANs. The analysis for the token bus and token ring LAN is covered in Sections 13.8 and 13.9 respectively. Similar results are obtained for wide area token passing communications systems and the analysis will not be repeated here. It should be borne in mind, however, that propagation delay is usually insignificant in local area networks but this is not the case with long distance networks.

13.5 Random Access Techniques

It was stated in the previous section that when the utilization is low there can still be a considerable access delay as a result of the polling time. This problem can be overcome by allowing each node to transmit as soon as it has a complete packet ready. There are a number of variations on this theme but they all result in the possibility of contention. This occurs when two or more nodes each transmit a packet during a particular interval of time. One of the simplest random access methods is the ALOHA protocol developed by the University of Hawaii[1].

The ALOHA and Slotted ALOHA Access Protocol

The ALOHA protocol was developed for radio transmission and allows each user to transmit a fixed length packet as soon as it is formed. It relies on a positive acknowledgement to indicate that the packet was received without error. In the case of the single bus system of Fig. 13.5 the receiving node would be required to send an acknowledgement. In the satellite system of Fig. 13.4 the transmitting node can monitor the packet retransmitted by the satellite and this produces an automatic acknowledgement. Assuming that each packet requires a transmission time of P seconds then if such an acknowledgement is not received within a time $P+2t_p$, where t_p is the maximum end to end propagation delay on a bus or the propagation delay between transmitting node and satellite, the transmitter retransmits its packet.

A packet will be successfully transmitted at a particular instant t only if no other packet is transmitted P seconds before or after t (i.e. the vulnerable period is $2P$); this is illustrated in Fig. 13.6.

The normalized channel throughput S (the average number of successful transmissions per interval P) is related to the normalized traffic offered G (the number of attempted transmissions per interval P, including new and retransmitted packets) by

$$S = Gp_0 \tag{13.6}$$

Where p_0 is the probability that no additional packet transmissions are attempted in the vulnerable interval $2P$.

If it is further assumed that packet arrival times are independent and exponentially distributed with a mean arrival rate of λ per second, the probability of k arrivals in an interval of duration t is then a Poisson process given by $P_k(t)$ where

$$P(k) = (\lambda t)^k e^{-lt}/k!$$

The probability of no arrivals in time t is $P(0) = e^{-\lambda t}$. In P seconds there will be $\lambda P = G$ arrivals, hence $\lambda = G/P$. Letting $t = 2P$ gives the channel throughput as

$$S = Ge^{-2G} \tag{13.7}$$

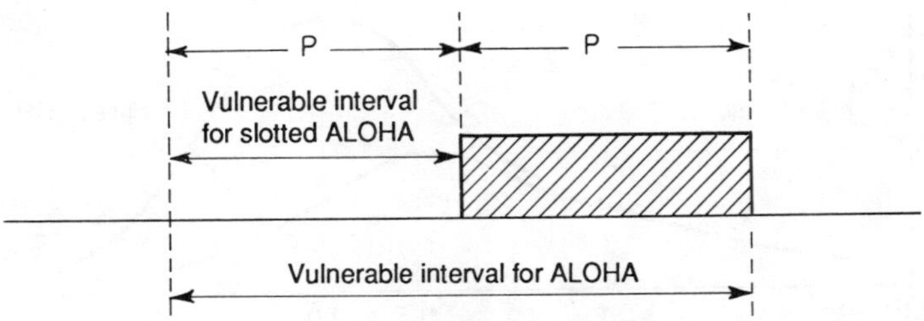

Fig. 13.6 Vulnerable period for ALOHA and slotted ALOHA.

Equation (13.7) is valid for satellite systems where all nodes are on the earth's surface and are approximately the same distance from the satellite. For earth based packet radio the satellite will be replaced by a base station and the expression must be modified to account for the fact that some nodes will be nearer the base station than others. In such cases there will be a maximum *difference* in propagation delay of t_{dm} between near and far nodes which has a normalized value of $a = t_{dm}/P$. In such circumstances equation (13.7) becomes

$$S = Ge^{-2(1+a)G}$$

note that for satellite systems the normalized difference in propagation delay $a = 0$.

The maximum throughput for a pure ALOHA channel is found by differentiating Eqn (13.7) with respect to G and occurs at $G = 0.5$

i.e.

$$S_{\max} = 1/2e = 0.184 \qquad (13.8)$$

The throughput characteristic for pure ALOHA is shown in Fig. 13.7.

Figure 13.7 shows that when the offered load G is low there are very few collisions and virtually all transmissions are successful and $S \approx G$. At higher values of offered traffic the number of collisions increases, which increases the number of retransmissions causing still more collisions, and so on. The characteristic is thus unstable and S actually begins to drop. The maximum throughput of the ALOHA protocol is thus limited to only 18% of the system capacity. However, this is still adequate for many purposes, especially when very bursty traffic, such as that produced by a human user at a remote terminal. If the system bit rate is, say, 1 Mb/s the useful capacity, including packet overheads, will be 180 kHz.

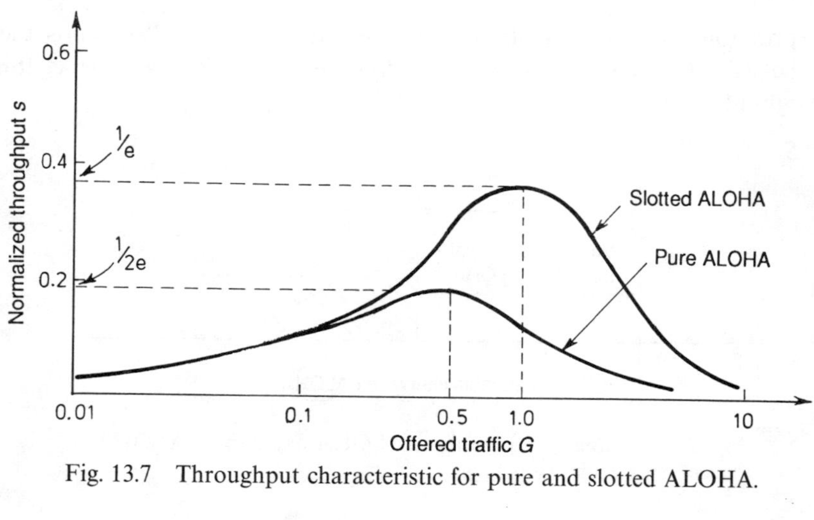

Fig. 13.7 Throughput characteristic for pure and slotted ALOHA.

In order to calculate the average time required to successfully transmit a packet, some knowledge of the retransmission procedure is required. A retransmission will be necessary if a node does not receive an acknowledgement within an interval $(P + 2t_p)$ after the first transmission. If a node retransmits immediately after this interval then the probability of a further collision will be unity since any other node involved in the collision will use the same retransmission procedure. To avoid a certain collision on retransmission some randomness should be introduced into the retransmission timing. One possibility is to define a retransmission interval $m \cdot P$ and to retransmit at time $n \cdot P$ where n is uniformly distributed between 1 and m. The average delay before a retransmission is attempted is then $P(m+1)/2$. The time required to confirm a successful retransmission is thus $2t_p + P(m+1)/2$.

If there is a total of E retransmission attempts the mean access delay is

$$t_a = P + 2t_p + E[2t_p + P(m-1)/2] \quad (13.9)$$

The relationship between throughput and offered traffic is

$$\frac{S}{G} = \frac{1}{1+E}$$

If all packet arrivals are independent (which will be true only for large values of m) then eliminating S from Eqn (13.7)

$$E = e^{2G} - 1 \quad (13.10)$$

The number of retransmissions can now be substituted into Eqn (13.9) to give the mean access delay. As an example consider the satellite system of Fig. 13.4 with a transmission rate of 1 Mb/s and a packet length of 8000 bits. The round trip delay for a synchronous satellite is approximately $2t_p = 270$ ms and the packet transmission time is 8 ms (note that the packet transmission time is much less than round trip delay, this is typical of WANs). If the normalized offered traffic is $G = 0.1$ and $m = 5$ then $E = 0.22$ and $t_a = 337.8$ ms. The major contribution to this access delay is the round trip delay of 270 ms. A typical round trip delay for a LAN would be of the order of 50 μs, and the corresponding access time would be $t_a = 8.413$ ms.

The throughput characteristic of pure ALOHA can be improved significantly if all transmissions are synchronized. The time axis is divided into intervals of time (or slots) of duration P and each user may transmit only at the start of a slot. Under these circumstances, if collisions do occur packets will overlap completely and the vulnerable period is reduced to P. The throughput of slotted ALOHA is then given by

$$S = Ge^{-G}$$

This has a maximum value when $G = 1$ given by

$$S_{\max} = 1/e = 0.368.$$

The maximum throughput of slotted ALOHA is thus increased to 36.8% of the system capacity. The characteristic for slotted ALOHA is also shown in Fig. 13.7.

It is clear from Fig. 13.7 that slotted ALOHA has a similar instability when the offered load exceeds the value at which the throughput is a maximum. The access delay is developed in a similar way to that for pure ALOHA. There is, however, one important difference: a packet may be transmitted only at the beginning of a slot. This means that on average (assuming packet arrivals are independent) a packet will have to wait $P/2$ seconds until the beginning of the next slot. Hence, on average, the time required to ensure a successful transmission is $1.5P + 2t_p$. The access delay for slotted ALOHA is thus

$$t_a = 1.5P + 2t_p + E[2t_p + 0.75P(m - 1)] \tag{13.11}$$

Hence the improved throughput of slotted ALOHA is achieved at the expense of a slight increase in access delay.

Carrier Sense Multiple Access with Collision Detection

This access protocol is one of a group of protocols known as carrier sense multiple access, which differ from ALOHA in that a node listens to the channel and transmits a packet only if the channel is idle. Determination of whether a channel is idle or busy presents particular problems for radio systems because not all users within a radio network can hear all others, giving rise to 'hidden users'. There are methods of overcoming this problem, but as far as this text is concerned consideration of CSMA will be restricted to the bus system of Fig. 13.5. A coverage of several variants of CSMA is given by Kleinrock[2]. In this chapter we limit our attention to CSMA/CD which is the most widely used variant.

The CSMA/CD access protocol requires each node to continuously monitor the transmission medium, and as a result each node is able to detect a collision. Once a collision is detected all nodes stop transmitting immediately. Retransmission is attempted at some later interval. CSMA/CD is more efficient than ALOHA since transmission is interrupted as soon as a collision is detected. The vulnerable period is also reduced to twice the end to end propagation delay on the network. This can be shown by considering the topology of Fig. 13.5. Assume a node at one end of the bus senses the bus idle and transmits a packet. A node at the other end of the bus will not be aware of this transmission until an interval t_p, equal to the end to end propagation delay, has elapsed. If it is supposed that the second node transmits a packet just prior to becoming aware of the first transmission then clearly a collision will occur. However the first node will not be aware of the collision until a further period t_p elapses; thus the minimum time required to detect a collision is $t_{sl} = 2t_p$. This argument assumes that each node will stop transmitting immediately upon detecting a collision. In practice there will be an additional period known as the jamming period. This will be ignored in the following analysis, but will be considered further in the description of Ethernet.

The characteristics of the CSMA/CD access technique are determined by assuming that a large number of nodes have a packet ready for transmission and that the length of each packet is the same (the restriction on packet length is imposed only to produce a tractable analysis and would actually be a disadvantage in practice). It is assumed that there are Q nodes with a packet ready for transmission and that the probability that any of the ready nodes will transmit within the interval t_{sl} is p. In order for this transmission to be successful one node only must transmit. The probability of a successful transmission is thus

$$P_r = Qp(1-p)^{-1} \tag{13.12}$$

The maximum value of P_r occurs when $dP_r/dQ = 0$, which gives $p = Q^{-1}$. In essence each node must determine the value of Q and it is assumed in the following analysis that this has been done (an approximate method of achieving this is considered in the Section 13.7 on Ethernet).

The probability of a successful transmission is thus

$$P_r = (1 - Q^{-1})^{Q-1} \tag{13.13}$$

As $Q \to \infty$, $P_r = e^{-1}$.

Thus when the network is sensed idle each node will transmit a packet with probability Q^{-1}. The probability that anode will experience a collision will be $(1 - P_r)$. The probability that the node will be delayed by J time slots (t_{sl}) is

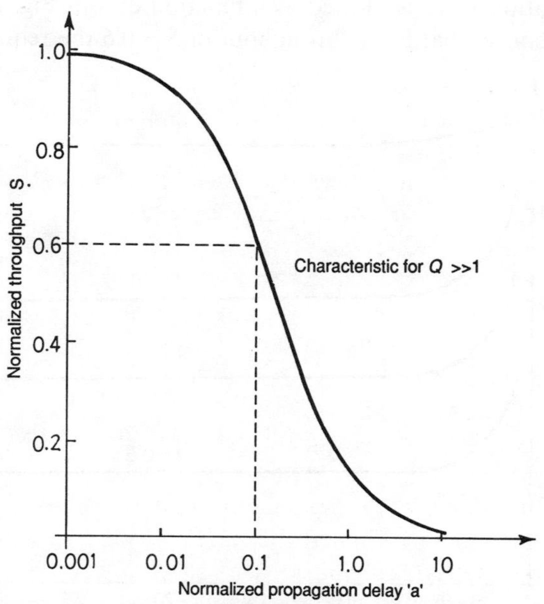

Fig. 13.8 Throughout characteristic for CMSA/CD as a formation of the normalized propagation time '*a*'.

thus the probability of $(J-1)$ collisions followed by a successful transmission i.e.

$$P_J = P_r(1-P_r)^{J-1}$$

The average number of collisions is thus

$$J_{av} = \sum_{J=0}^{\infty} J \cdot P_J = 1/P_r$$

The mean access delay is thus $t_a = 2t_p \cdot J_{av}$ and the average time required to deliver a packet is

$$t_d = P + t_p + 2t_p \cdot J_{av}$$

i.e.

$$t_d = P\{1 + a(1 + 2/P_r)\} \quad (13.14)$$

where $a = t_p/P$ is the ratio of propagation delay to packet transmission time.

If t_d is the average time to deliver a packet then the average number of packets delivered/second is $1/t_d$. The normalized throughput (the average number of packets delivered per interval P) is then

$$S = \frac{1}{\{1 + a(1 + 2/P_r)\}} \quad (13.15)$$

The limiting case occurs when the number of nodes Q tends to ∞, in which case $1/P_r = e$. This characteristic does not collapse to zero, as is the case with ALOHA. Equation 13.15 is plotted as a function of a in Fig. 13.8.

Figure 13.8 shows that for a throughput of $S = 0.6$ the required value of a

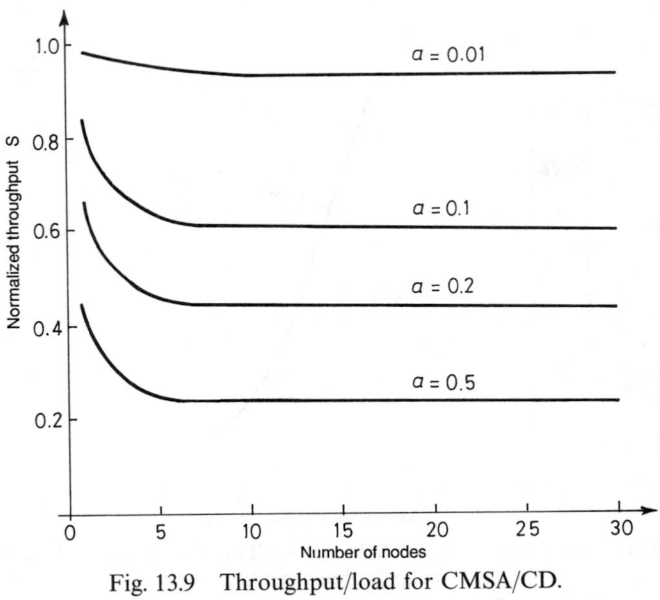

Fig. 13.9 Throughput/load for CMSA/CD.

is 0.1. This means that the propagation delay t_p is restricted to 10% of the packet transmission time. It may be concluded from this that CSMA/CD is not suited to long distance communication, with corresponding large propagation delays, unless the packets are excessively long or the data rate is low. An alternative way of presenting the characteristic is shown in Fig. 13.9 where S is plotted as a function of the number of users Q.

The significance of the parameter a is once again evident. If the packet size is equal to the smallest that will allow collision detection then $P = t_{sl}$ (i.e. $a = 0.5$) and the asymptotic value of S becomes 0.24. In typical applications with much longer packets and moderate loading the throughput is of the order of 80% and this increases to approximately 98% when relatively few users generate longer packets[3].

The delay characteristic associated with CSMA/CD is very much influenced by the method of determining the transmission probability Q^{-1} and this is covered in Section 13.7 which is devoted to Ethernet.

13.6 Local Area Networks

In this section we shall consider networks developed specifically for short distance communication where the propagation delay is low and the bit rate is high. The three most common LAN topologies are shown in Fig. 13.10.

The star connected network is a simple topology in which the central node performs the routing function for communication between nodes attached to it. For this region the central node must have sufficient capacity to handle all simultaneous traffic and is a potential bottleneck. In particular a very high reliability of the central node is demanded, as a failure at this point completely destroys the network. In this chapter we shall be concentrating on the bus and ring structures and the access protocols which they employ.

It is useful at this point to consider some of the properties of bus and ring topologies as this will provide some insight to the reasons for their use in local area networks.

(i) A bus network is generally more reliable than a ring network as it tends to be purely passive. This means that failure of one or more nodes does not cause the failure of the network in total. In a ring network each node effectively repeats any transmission not destined for itself to the next node. This means that failure of any node in the ring can be catastrophic. One method of reducing the risk of network failure is to provide each node with bypass circuitry which is brought into action automatically on a node failure.

(ii) Additional nodes can be incorporated in bus networks without disrupting network operation. Ring networks require the insertion of new cable segments which requires the network to be taken out of service.

(iii) A bus network is limited (currently) to twisted pair or coaxial cable because of the need to transmit signals in two directions, bi-directional

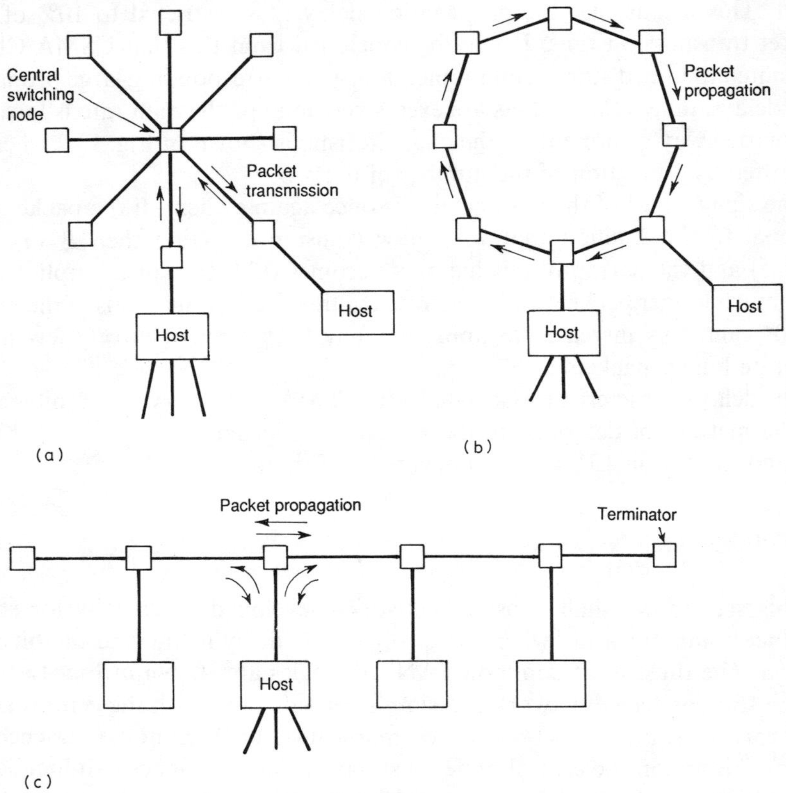

Fig. 13.10 LAN topologies, (a) star topology, (b) ring topology, (c) bus topology.

optical couplers would be required for optical systems. Ring networks tend to transmit signals in one direction only and lend themselves to optical fibre operation with all the inherent advantages of this medium.

(iv) Bus networks are prone to reflections and care is required to avoid impedance mismatch at tapping points.

(v) Fault isolation is very difficult in bus networks but is relatively straightforward in a segmented ring structure.

Ring networks tend to support some form of token passing for network access because of their sequential nature. Bus networks, on the other hand, are suitable for both random access and sequential (or polling) access. We shall consider two common examples of bus based LANs and compare these with the alternative ring structure.

13.7 The Ethernet Local Area Network

Ethernet is based on a bus network operating at 10 Mb/s baseband using the CSMA/CD access protocol. This local area network was developed for

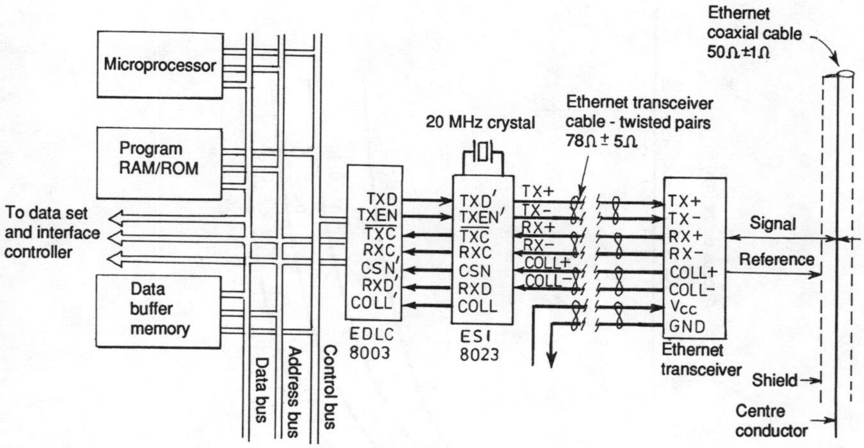

Fig. 13.11 CSMA/CD Ethernet node hardware.

computer communications, but there has been significant interest in recent times for using Ethernet for packetized voice communications[4,5] also. The basic CSMA/CD protocol was covered in the previous section; however, there are some further points which should be considered with reference to Ethernet and its very close derivative IEEE 802.3. The hardware configuration of a typical Ethernet node is shown in Fig. 13.11. This consists of a microprocessor, buffer memory, an Ethernet data link controller (EDLC) and its associated Ethernet serial interface (ESI) and transceiver unit. This hardware performs both transmission and reception of Ethernet packets.

Any node with a packet ready for transmission monitors the medium for ongoing transmission. If there is an ongoing transmission the ready node waits for an interval of 9.6 μs (the interpacket gap) and then transmits immediately. This differs from the CSMA/CD discussed previously where the node transmitted with a probability Q^{-1}. Collisions can occur during the interval $t_{sl} = 2t_p$ and this 'slot time' is set equal to 51.2 μs which limits the effective maximum length of the Ethernet bus to 2.5 km. (An Ethernet system is actually composed of a number of segments, see Fig. 13.13 for details.) If a collision is detected all nodes involved abort the current transmission attempt after transmitting a 32 bit jam pattern (010101...). This jam pattern ensures that all other nodes are aware of the collision. To prevent repeated collisions with contending nodes each node involved in a collision then generates a backoff interval given by

$$T_r = \mathrm{RND}[0, 2^r - 1]t_{sl}$$

where RND[ ] is a uniformly distributed random number between 0 and $2^r - 1$ and r is the number of successive collisions that a particular packet has experienced. The value of r is held at 10 should there be more than 10 but fewer than 16 collisions. On the 16th collision r is reset to 0 and a higher level collision warning is generated. The backoff algorithm is known as the

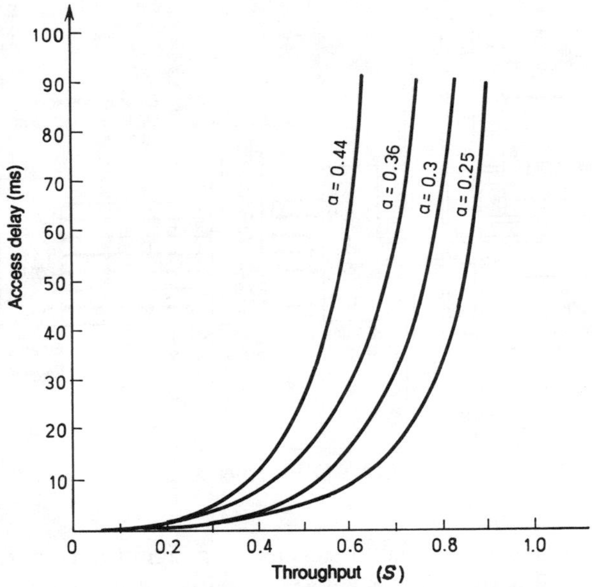

Fig. 13.12 Delay/throughput characteristic for Ethernet.

(truncated) binary exponential backoff algorithm and is intended to facilitate the calculation of the transmission probability Q^{-1} which maximizes the chance of a collision-free despatch. In effect the binary backoff algorithm sets the probability of transmission as $1/2^r$, i.e. the number of ready users is estimated as $Q = 2^r$. For a highly loaded Ethernet with many collisions r will reach the value 10, which estimates the number of ready users as 1024. This is the maximum number of nodes specified for Ethernet. Clearly the optimum transmission probability is not implemented as when a new packet is available a transmission is attempted 9.6 μs after the network becomes idle. Further, the Ethernet packet may have a data segment between 46 and 1500 bytes long, hence the constant packet assumption is also not valid.

In reality the performance of CSMA/CD with a binary exponential backoff is virtually impossible to analyse mathematically without making gross approximations. Hence performance measures are usually obtained by computer simulation. However it should be stated that Ethernet does exhibit a stable throughput for very high loads and this is greatly influenced by the parameter 'a' considered earlier. In the case of Ethernet the packet length can vary between 576 bits and 12 208 bits (including the overhead of 208 bits). With a transmission rate of 10 Mb/s and $t_p = 25.6\ \mu$s the parameter 'a' can vary between 0.002 and 0.44. Of particular importance, especially for speech transmission, is the access delay which increases as the traffic on the network increases. A typical access delay/throughput characteristic for speech packets derived from computer simulation is shown in Fig. 13.12.

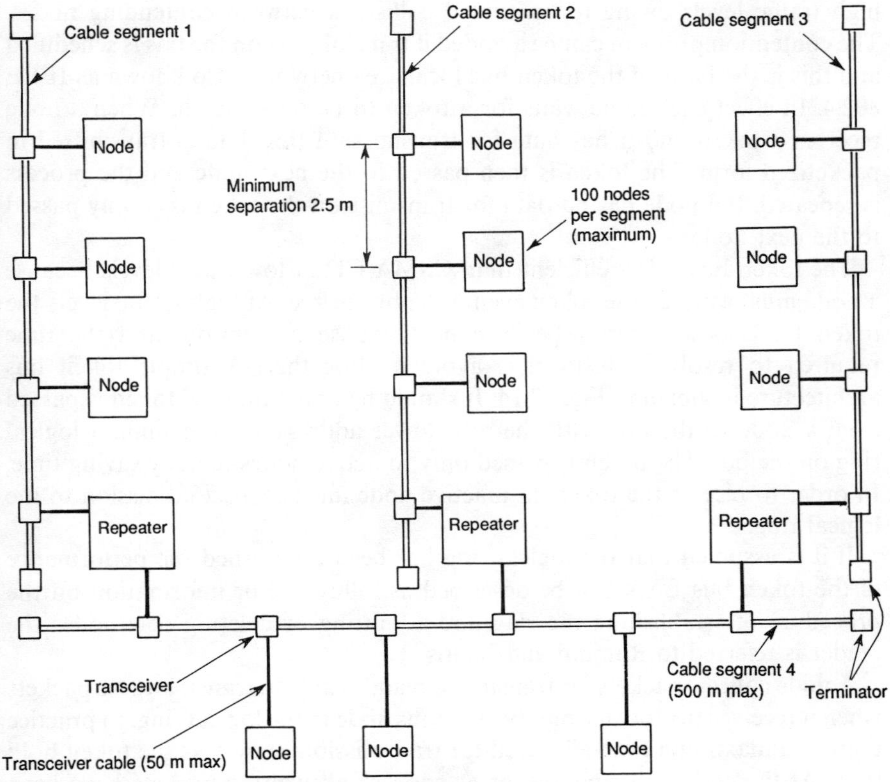

Fig. 13.13 Large scale Ethernet installation.

The reader's attention is drawn to the very rapid increase in delay which occurs at the higher network loads. For example when $a = 0.44$ the delay reaches 60 ms at a throughput of about 60%. The total delay in the case of speech is an important parameter. This will be composed of the access delay plus the packetization delay (the time required to collect sufficient speech samples to fill a packet).

A typical Ethernet network is shown in Fig. 13.13. This network is made up of several segments connected by repeaters or bridges. The maximum length of each segment is specified as 500 m and the maximum number of nodes per segment is 100. The maximum round trip delay for the whole system is specified as 51.2 μsec.

13.8 The Token Bus

The random access nature of CSMA/CD is very attractive when the traffic levels are low because it is possible to transmit a packet almost as soon as it arrives. However, Fig. 13.12 illustrates that delays can become very large at high traffic levels owing to continued collisions between contending nodes. The contention problem can be avoided if transmission on the bus is scheduled and this is the basis of the token bus local area network, also known as IEEE 802.4. In effect each node waits for a token to be passed to it. When a node receives a token and it has data for transmission this data is transmitted in packetized form. The token is then passed to the next node and the process is repeated. If a node has no data for transmission the token is simply passed to the next node.

The token bus is less efficient than CSMA/CD at low traffic levels because a node must wait for the token even if the bus is free. At high traffic levels the token bus has a superior performance because contention (and the time required to resolve collisions) is avoided altogether. A simple token bus architecture is shown in Fig. 13.14. It should be noted that the token is passed from a node to the one with the next lower address, thus forming a logical ring on the bus. The token is passed only to active nodes thereby saving time. In order to receive the token an inactive node must request admission to the logical ring.

If it is assumed that the logical ring has been established the performance of the token bus LAN can be described as follows. (For information on the procedure of establishing the ring and admitting or deleting new nodes the reader is referred to Ramimi and Jelatis[6].)

A node collects packets for transmission and is able to transmit these packets when it receives the token from the previous node in the logical ring. In practice there is a maximum time allocated for transmission known as the token hold time. At the end of this period, or as soon as all waiting packets have been transmitted, the node passes the token to the next node on the logical ring. The token rotation time t_r is defined as the interval between successive arrivals of the token at a specific node. Evidently the token bus network will never be

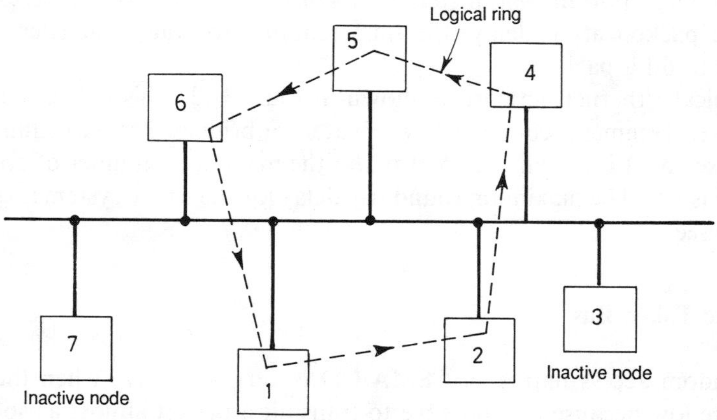

Fig. 13.14 Token passing local area bus network.

silent, as is possible with CSMA/CD, but will always be transmitting either message packets or a token.

In order to determine the delay characteristics of the token bus it will be assumed that token hold time is not limited, i.e. when a node receives the token it is able to transmit all its accumulated data. The token rotation time will thus be composed of the time to transmit N tokens (Nt_t) and the time required to transmit the total number of packets which accumulate during the token rotation time. If the mean packet arrival rate at each node is λ per second the total number of packet arrivals during the interval t_r is $\lambda N t_r$. If the bus can transmit packets at a rate of μ/second the token rotation time is

$$t_r = Nt_t + \lambda N/\mu \tag{13.16}$$

Defining the bus utilization as $U = \lambda N/\mu$, gives

$$t_r = Nt_t/(1 - U) \tag{13.17}$$

The mean access delay will thus be $t_a = t_r/2$. Note: in order to ensure stable operation $U < 1$. The minimum token rotation time will occur when $U = 0$. In this case none of the N nodes has packets for transmission, when the token arrives, and the bus traffic consists merely of token transmissions. The bus utilization U is effectively the percentage of time that the bus is transmitting data packets. The factor $(1 - U)$ is then the percentage of the time that the bus is transmitting tokens. The normalized access delay (for $Nt_t = 2$) is plotted in Fig. 13.15.

As the utilization approaches 1 the access delay approaches ∞. It may also be noted from Eqn (13.17) that the access delay is related directly to the time

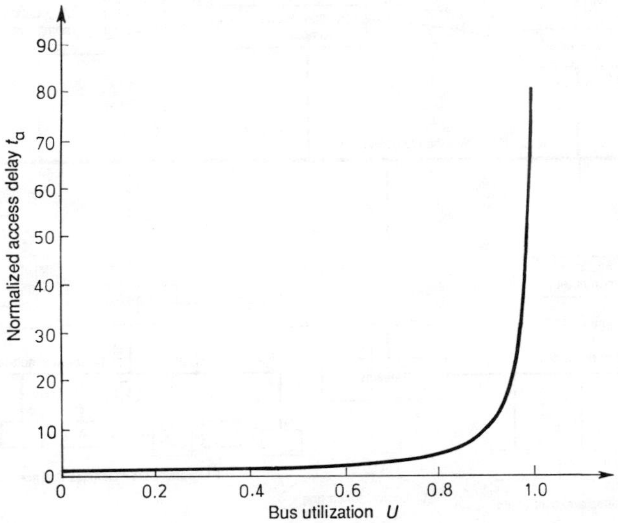

Fig. 13.15 Normalized access delay characteristic for token bus.

required to transmit the token t_t which in turn is proportional to the end to end propagation delay on the bus and the token length, and is inversely proportional to the data rate on the bus.

If the propagation time is large (as would be the case for a wide area network) the access delay can be reduced by increasing the data rate. Hence scheduled access is far more suited to long distance communication than is CSMA/CD which (according to Eqn (13.15)) is limited to low data rates in order to minimize the value of the parameter 'a'. In Eqn (13.15) the characteristic is given in terms of the throughput S which is defined as the average number of successful packet transmissions in a specified interval. The throughput S is related to the utilization U, i.e.

$$U = \frac{\text{mean packet arrival rate}}{\text{mean packet transmission rate}} = \frac{\text{mean arrival rate}}{S}$$

The IEEE 802.4 specification has been adopted for industrial use in the manufacturing environment. In this context a number of token bus networks operating at 1 Mb/s (known as carrier band) are connected via a bridge to a broadband cable operating at 10 Mb/s. These integrated systems operate under what is known as the Manufacturing Automation Protocol (MAP). A typical factory communications layout is shown in Fig. 13.16.

Each of the token bus subnets may be regarded as individual token bus

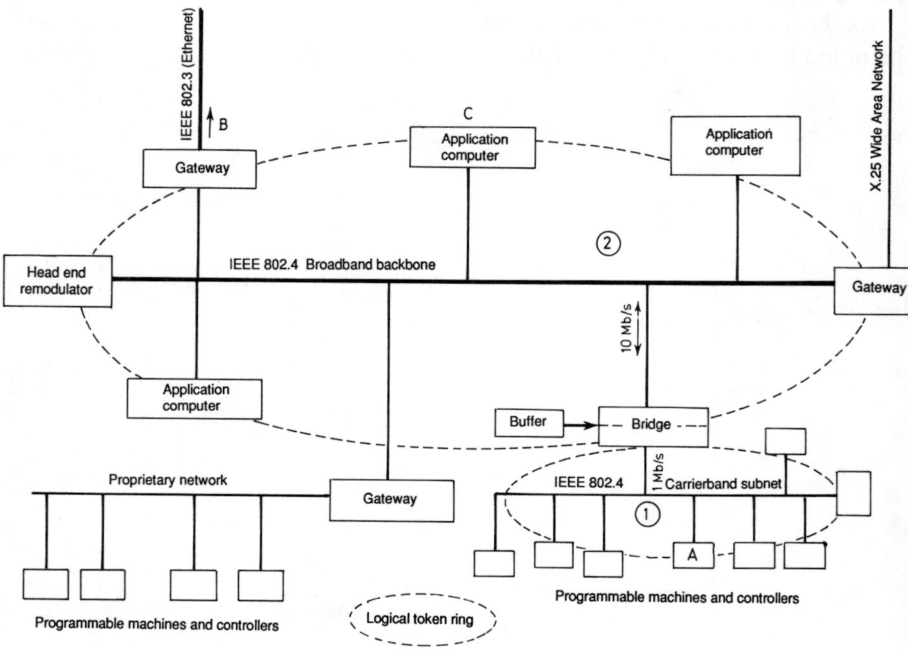

Fig. 13.16 Implementation of MAP in a factory environment.

networks in which the bridge to the broadband cable simply acts as an additional node. Tokens are not passed across the bridge so that the token passing on the subnets is independent from the token passing on the broadband cable. The subnets use phase coherent frequency shift keying with a logic low being represented by a carrier of 5 MHz and a logic high being represented as a carrier of frequency 10 MHz, the data transmission is omnidirectional. The broadband cable uses frequency division multiplex transmission, one frequency being allocated for transmission and a second frequency allocated for reception. In this case all transmissions are unidirectional, a head end repeater being used for translating data between transmit and receive frequencies. The broadband cable is also able to handle other independent transmissions on a frequency multiplexed basis.

13.9 The Token Ring

Ring networks are fundamentally different from bus networks in that the nodes on a ring network tend to act as repeaters and are therefore active devices. In the token ring access to the channel is controlled by passing a permission token around the ring. When the ring is initialized a designated node generates a free 8-bit token (11111111) which travels around the ring in one direction only until a node with packets ready to transmit changes the token to busy (11111110) and then transmits all its packets. The sending node is responsible for converting the token back to free, after it has travelled around the ring, and for removing its own packets. A further difference in ring structures is that each node adds delay to the circulating token and data. This known as the node latency and is an essential requirement because each node must be able to store a bit in order to decide whether to change the last bit in a token (i.e. from free to busy or vice versa). Thus each node contributes a 1-bit delay and, if the number of nodes in a ring is large, this delay can become very considerable.

The token is usually eight bits long which puts a lower limit on the number of nodes in a ring. If there are fewer than eight nodes on the ring the first bit of the token would return to the sending node before it had transmitted the last bit. Thus it is necessary to have sufficient delay on the ring to ensure that the complete token can be accommodated.

The mean access delay of the token ring may be obtained by assuming that there are N nodes on the ring and that each node has on average q packets ready for transmission. It is further assumed that packet arrivals have a Poisson distribution with a mean arrival rate, for all nodes, of l packets/second. The mean arrival rate for the complete ring is thus $N\lambda$ packets/second. The time required for one bit to go all the way around the ring is termed the walk time t_w and is given by

$$t_w = N(\text{latency} + \text{propagation time between nodes}) \tag{13.18}$$

The scan time t_s is the mean time between free token arrivals at a given node (this is equivalent to the token rotation time specified for the token bus LAN). This will consist of the walk time and the time required to service each of the packets ready for transmission. If the mean number of packets/second the ring can transmit is μ, the scan time is

$$t_s = t_w + N\lambda t_s/\mu \tag{13.19}$$

Defining ring utilization $U = N\lambda/\mu$

$$t_s = \frac{t_w}{(1-U)} \tag{13.20}$$

The mean acquisition delay is then $t_s/2$ and this will approach ∞ as U approaches 1. This is similar to the characteristic for the token bus, which is not surprising as the token bus is, in effect, a logical ring. The token bus has the advantage that each individual connection to the bus is essentially passive.

13.10 Open Systems Interconnection Reference Model

Packet switched networks were designed initially for computer communications. In such circumstances the computers themselves became an integral part of the communications system, performing tasks such as channel demand, addressing, routing, error checking buffering, etc. Clearly the proliferation of various computer architectures could result in the creation of barriers to efficient communication. The early realization of this problem resulted in the adoption of an international standard for communication, in 1983, known as the Open Systems Interconnection reference model[7]. This model divides up the various aspects of communication into seven layers as shown in Fig. 13.17.

The OSI model was designed to facilitate global communications, primarily for computers, irrespective of the characteristics of the particular networks of which they are part. The model does not specify how systems are implemented but rather how they communicate. This means that many different networks, using products of different manufacturers, may be coupled together by mapping the OSI model onto the complete communications path A brief description of each layer will now be given.

(i) The Physical Layer (layer 1)
This layer deals with transmitting individual bits over a communications system. It deals with the way in which the bits are represented, e.g. voltage levels, bit duration, etc. on the transmission medium and whether the transmission may proceed simultaneously in two directions.

(ii) The Data Link Layer (layer 2)
The data link layer makes use of the network layer to ensure error-free communication between the network layer at each end of the communi-

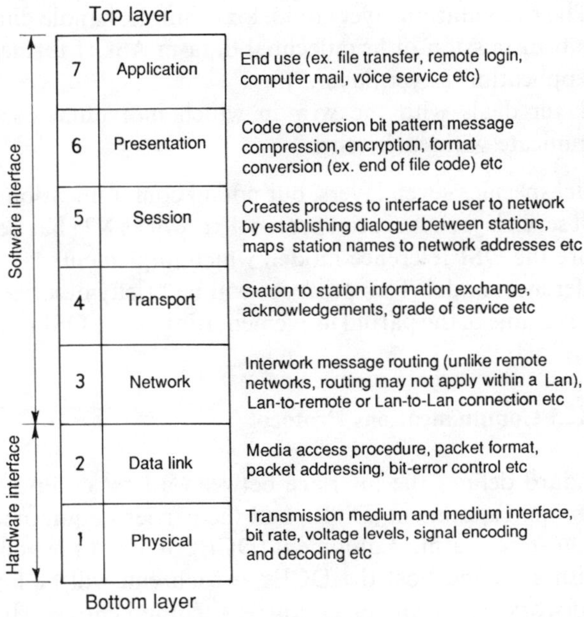

Fig. 13.17 Layered architecture of the OSI reference model.

cations link. In Ethernet, for example, the data link layer is responsible for transforming data into the packet format of Fig. 13.1. Layer 2 also handles acknowledgements from the receiver and responds to errors.

(iii) The Network Layer (layer 3)
This layer essentially deals with routing of packets over the communications link and ensures that the data is assembled in the correct order for presentation at the receiver. This layer would be concerned also with providing some form of error control.

(iv) The Transport Layer (layer 4)
This layer provides an end to end communication between host machines. It is not concerned with the hardware constraints of providing the actual connection, as are the lower layers, but is involved with such functions as how to create the most efficient use of the communications media for a particular throughput. The transport layer handles flow control, which is essential if a high speed transmitting host is communicating with a low speed receiving host.

(v) The Session Layer (layer 5)
This is the layer which deals with setting up the connection between users. This layer would handle, for example, the procedure for logging in on a remote machine and initiating a file transfer.

(vi) The Presentation Layer (layer 6)
This layer deals with the way in which the data is presented for transmission and is usually performed by subroutines called by the

user. The presentation layer could, for example, handle data enscription or the transmission of hexadecimal data in ASCII format.

(vii) The Application Layer (layer 7)
This layer deals with the way in which individual users' programs communicate with each other.

The OSI model specifies seven layers, but not all communications connections will involve all seven layers. A special protocol known as X25 has been developed (actually before the OSI reference model) which implements layers 1, 2 and 3 of the OSI reference model. This protocol will be briefly described in the next section, as an example of the partial implementation of the OSI reference model.

13.11 The X25 Communications Protocol

The X25 standard defines the interface between a host computer, known as the Data Terminal Equipment (DTE), and the carrier's equipment, known as the Data Circuit-terminating Equipment (DCE). When a host communicates, using X25, with a second host the DCEs at each end will be linked over an unspecified network by means of various switching centres. The connection between the two DCEs is given the general label of Data Switching Exchange (DSE) and is transparent to the X25 protocol. The relationship between X25 and the OSI reference model is shown in Fig. 13.18.

X25 defines three layers of communication, the physical layer (equivalent to the physical layer of OSI), the frame layer (equivalent to the data link layer of OSI) and the packet layer (equivalent to the network layer of OSI). The physical layer of X25 deals with the electrical representation of binary 1 and 0, timing, etc. and is described by a standard known as X21. This standard specifies how the DTE exchanges signals with the DCE to set up and clear calls, etc. and is illustrated in Fig. 13.19.

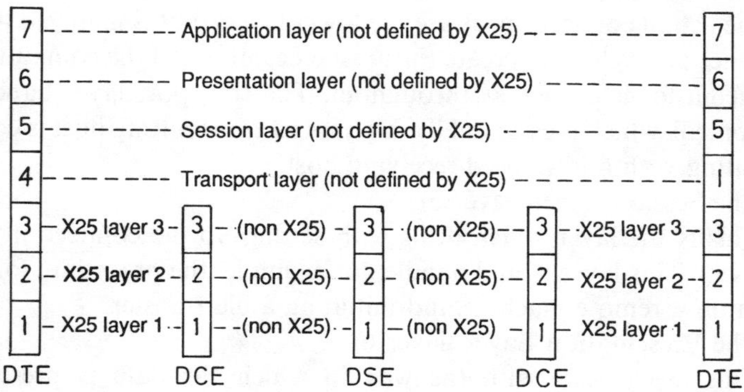

Fig. 13.18 Relationship between OSI reference model and X25.

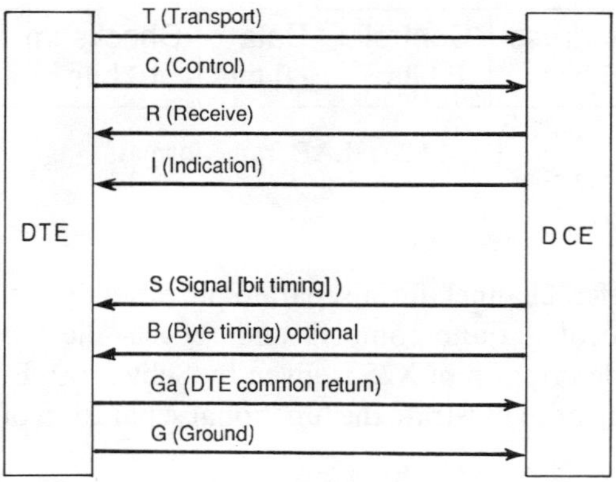

Fig. 13.19 Signal lines specified in X21.

In many installations the DTE will be connected to the DCE over a standard analogue telephone line which may not support the full X21 specification.

The frame layer ensures reliable communication between the DTE and DCE. In effect the frame layer adds additional bits to the data stream produced by the packet layer to make the communication reliable. The additional bits and the data produced by the packet layer are combined into a frame to produce what is known as the Link Access Procedure (LAP) protocol, a typical LAP format is shown in Fig. 13.20.

The LAP frame structure will not be discussed in detail in this text but it should be noted that there are three kinds of frame known as Information, Supervisory and Unnumbered. The reader is referred to Tanenbaum[8] for further information.

The third layer in X25 deals with the meaning of the various bits in the data field of the frame. The data within the frame is itself known as a packet hence X25 packets are embedded in a LAP frame. There are actually 15 different types of X25 packets dealing with various aspects of call set up and monitoring. As an example, when a DTE wishes to communicate with a second DTE a virtual circuit must be set up between them. The initiating DTE sends a 'call request packet' to its DCE which delivers this packet to the destination DCE which then passes the packet to the destination DTE. The calling DTE defines the logical channel on which communication is to take place by means of a channel field in the call request packet. If the distant DTE wishes to accept the call it returns a call accepted packet to the originating DTE and communication commences until a clear request packet is sent to terminate the call.

There are many facilities available to X25, for example if the logical channel requested by the initiating DTE is in use at the far end the called DCE will substitute an unused channel. X25 is able to cope with collisions which could occur when a logical channel request from an initiating DTE coincides with

Start byte	Address	Control	Data	Checksum	End byte
01111110	8 bits	8 bits	⩾ 0 bits	16 bits	01111110

Fig. 13.20 LAP frame format.

an allocation of that channel (for a separate purpose) by the called DTE. The actual X25 protocol is quite complex and reflects the number of facilities available. A full description of X25 is given by Schwartz[9]. For the purpose of this text it is sufficient to illustrate the functional separation of the three defined layers of X25.

It is common practice to use X25 links for devices, such as simple terminals, which do not themselves have X25 capability. This has resulted in the widespread use of devices known as Packet Assemblers/Disassemblers or PADs. A PAD is essentially a multiplexer or concentrator which collects data from a number of asynchronous terminals and periodically outputs a correctly formatted X25 packet made up of this data. The X25 packets may then be passed to a DCE in the normal fashion. The inverse operation is also required to convert X25 packets, from a host, into the asynchronous format for the individual terminals. The PAD contains significant intelligence and may itself be considered as an X25 DTE host.

13.12 Conclusions

This chapter has concentrated on the description of packet switched networks with a detailed consideration of local area networks. Some reference has been made to the integration of speech into local area networks. Packetization of speech and other real time signals is a topic of intense interest both for wide area and local area communications. It is envisaged that all analogue telephone communications will eventually be replaced by integrated services digital networks. Major research projects currently being funded in Europe are directed towards replacing the fixed telephone by a completely mobile cellular digital radio telephone system. Initially this system is likely to be a PAN European system where any mobile user (which will eventually be all citizens) may communicate with any other mobile user without a prior knowledge of that user's physical location.

Such a system will be entirely digital and probably packet switched. The concept of a mobile digital cellular radio system is shown in Fig. 13.21. A given area is divided up into a number of cells, each cell being covered by a 'base station'. The mobiles move around the cells and communicate by radio with the base station of whichever cell the mobile happens to be in at that instant. There will be a fixed number of radio channels available within each

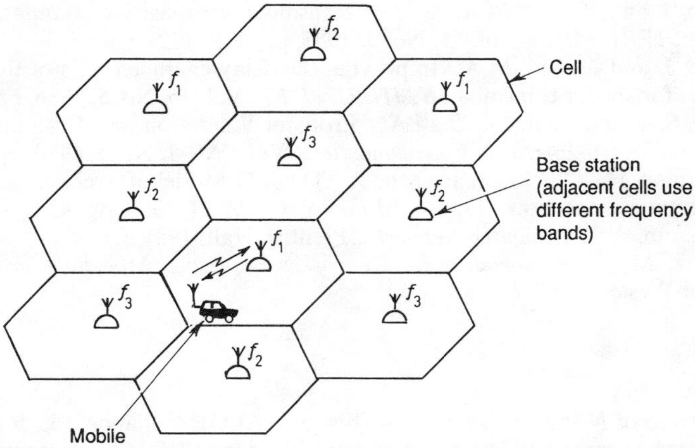

Fig. 13.21 Future digital mobile cellular radio system.

cell and hence the problems of accessing one of these cells is very similar to the access problems for local area networks.

There are, however, many problems for digital cellular radio which do not exist in LANs. The mobile can move freely between cells and hence the communications system must be able to determine the mobile's position in order to switch in the appropriate base station. There will clearly be a limit on the amount of radio spectrum available for this service (200 MHz has been suggested) and to accommodate the very large numbers of users the cells must be very small (e.g. 100 m radius). This means that radio channels can be re-used in other cells far enough away to avoid mutual interference. Interference between adjacent cells is avoided by using different radio frequencies. The problems that would arise from a mobile telephone in a fast vehicle traversing such a micro-cellular system would be significant to say the least. These problems can and will be solved in the fairly near future.

The technology of packet switched systems is very much in its infancy and it is likely that the development of this technology in the areas mentioned in the previous paragraphs will have an even more profound effect on civilization than telecommunications engineering has had up to the present time.

References

1. Abrahamson, N., 'The ALOHA System – Another Alternative for Computer Communications', *AFIPS Conference Proceedings*, 1970 Fall Joint Computer Conference, Vol. 37, pp. 281–285.
2. Kleinrock, L., *Queuing Systems Vol. II, Computer Applications*, Ch. 5, John Wiley, 1976.
3. Shoch, J. and Hupp, J.A., 'Measured Performance of an Ethernet Local Network', *Communications of the ACM*, Vol. 23, No. 12. 1980, pp. 711–721.

4. Dunlop, J. and Rashid, M.A., 'Speech transmission capacity of standard Ethernet system', *JIERE (UK)*, Vol. 55, No. 3, 1985, pp. 119–122.
5. Dunlop, J. and Rashid, M.A., 'Improving the delay characteristics of the standard Ethernet for speech transmission', *JIERE (UK)*, Vol. 56, No. 5, 1986, pp. 184–186.
6. Rahimi, S.K. and Jelatis, G.D., 'LAN Protocol Validation and Evaluation', *IEEE Journal on Selected Areas in Communications*, Vol. SAC-1, No. 5, 1983, pp. 790–802.
7. Zimmerman, H., 'OSI Reference Model – The ISO Model of Architecture for Open Systems Interconnection', *Trans. IEEE*, Vol. COM-28, 1980, pp. 425–432.
8. Tanenbaum, A.S., *Computer Networks*, Prentice-Hall, 1981, Ch. 4.
9. Schwartz, M., *Telecommunication Networks: Protocols, Modelling and Analysis*, Addison-Wesley, 1987, Ch. 5.

Problems

13.1 A group of N stations share a 64 kb/s pure ALOHA channel. Each station, on average, outputs a 1024 bit packet every 50 seconds. What is the maximum value of N which may be accommodated?

Answer: 575

13.2 A large number of nodes using an ALOHA system generate 50 requests/second; this figure includes new and re-scheduled packets. The time axis is slotted in units of 40 ms.

(i) What is the probability of a successful packet transmission at the first attempt?
(ii) What is the probability of exactly four collisions followed by a successful transmission?
(iii) What is the average number of transmission attempts per packet?

Answer: 0.157, 0.079, 6.39

13.3 An Ethernet cable operating at 10 Mb/s has a length of 1.5 km and a propagation speed of 2.2×10^8 m/second. Packets are 848 bits long which includes an overhead of 208 bits. The first bit slot after a successful transmission is reserved for the destination node to access the channel to send an acknowledgement packet (data field empty). What is the effective (i.e. useful) data rate assuming no collisions? Assume an inter-packet gap of 9.6 μs.

Answer: 5.33 Mb/s

13.4 One hundred nodes are distributed over a CSMA/CD bus of length 4 km. The transmission rate is 5 Mb/s and each packet is 1000 bits long including the standard overhead of 208 bits. Calculate the maximum number of packets/second that each node can generate. Assuming an inter-packet gap of 9.6 μs and a propagation delay of 5 μs/km, calculate also the useful data rate of the system.

Answer: 45.45, 3.61 Mb/s

13.5 The bit rate on the system of the previous question is increased to 10 Mb/s. Determine the effect of this increase on the number of packets/second that each node can generate and the useful data rate of the system. Repeat the calculation if the packet length is increased to 10 000 bits. Comment on the results.

Answer: 83.61, 6.62 Mb/s, 980.8, 9.6 Mb/s

13.6 A coaxial cable of length 15 km supports the token bus protocol, the propagation speed being 2×10^8 metres/second and the data rate being 10 Mb/s. There are 512 nodes connected to the bus each producing packets 12 kbits long, with an overhead of 96 bits. The token, which is a standard packet with an empty data segment, is passed sequentially to each node. Assuming a latency of 8 bits at each node and that each node generates 6 packets per 10 seconds on average calculate the average token rotation time. What is the maximum rate of transmission of packets at each node?

Answer: 17.9 ms, 1.55 packets/s

14 / SATELLITE COMMUNICATIONS

14.1 Introduction

For many years, the notion of satellite communications was a fantasy produced by the fertile mind of Arthur C. Clarke;[1] a brilliant idea but rather impractical. As with so many creative ideas, technology eventually caught up and now satellites are commonplace. For the communication engineer however, they represent a challenging and stimulating field of work. Within satellite communication systems we find the whole gamut of technologies operating in a strange and demanding environment. Coupled with the essential requirement of demanding as little energy to run as possible, the system must be capable of withstanding the arduous journey from earth to orbit. Consequently a careful balance must be struck between the mechanical, structural, electronic, electrical and electromagnetic engineering requirements of the system.[2] However, in this chapter we will limit our attention to system considerations.

Satellite communications provide opportunities, and pose problems, in communication methods. Their large area of access (footprint) allows a single transmission to cover an enormous number of receivers, thus allowing broadcast signals to be transmitted simultaneously to large numbers of people. However, this feature itself creates difficulties; partly political and partly economic. National boundaries are no barrier whatsoever, and the charging mechanism required to allow the satellite operator to recover the cost of development and provide continuous support requires a novel solution.

Satellites can travel in a variety of orbits, basically eliptical in shape, but we will limit our attention to geostationary systems, in which the orbit is fixed, and essentially circular. A geostationary satellite maintains a fixed position relative to points on earth, and this allows a cheap receiving antenna to be set-up once and then fixed in position.

The enormous attraction of satellites for a whole range of communication objectives has meant that there is considerable danger of overcrowding. More frequency bands will have to be allocated, and the satellites themselves will have to be pushed closer together so that more can be placed in the geostationary orbit. Recently there has been a compression from 3° to 2° separation.

In a sense, there is no new communications work in this chapter. What we have is an unusual application which produces unusual operating conditions and therefore requires novel developments of existing knowledge. In many of the following sections we will refer to earlier chapters in this book for explanations of principles, and we will limit our attention here to those features of communication systems which are peculiar to satellite applications.

14.2 Geostationary Orbits

The analysis of the satellite orbit is complex and rather beyond our scope. Excellent treatments are given elsewhere[2,3]. For our purpose however, we only need to know that it is possible to place a satellite in a geostationary orbit at a height of approximately 36 800 km above the earth, and that the station can be maintained, provided it has the provision for slight adjustments of orientation and position by way of small rockets. Although we will concentrate on geostationary satellite communications, it should be realized that there are areas near the poles which can be reached only by satellites in other orbits. For example, satellites in tundra orbit are not geostationary, but they provide regular coverage for the extreme latitudes. Fig. 14.1.

For a geostationary orbit the angular velocity of the satellite is constant and therefore, since the angle swept per unit time (Kepler's third law) is also constant, the orbit must be circular. There is a simple relationship between the period and the radius of the orbit

$$a \propto (T^{2/3}) \tag{14.1}$$

and the equality is provided by Kepler's constant μ:
i.e.

$$a = \left(\frac{T^2\mu}{4\pi^2}\right)^{1/3} \tag{14.2}$$

Assuming that the period T is the same as a normal (sidereal) day = 24.60.60 sec

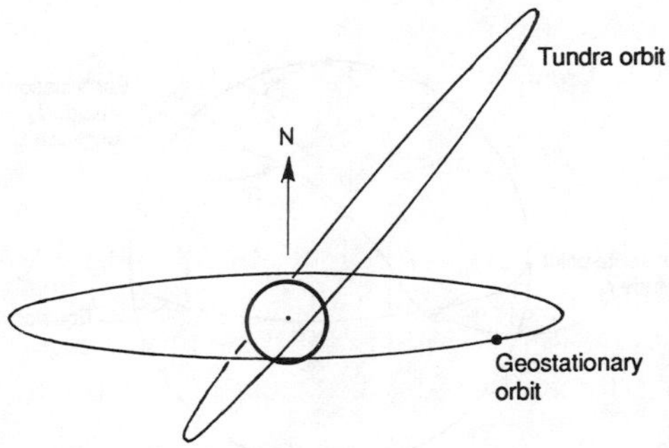

Fig. 14.1 Communication satellite orbits.

and taking $\mu = 3.9861352.10^5\,\text{km}^3/\text{s}^2$, the above equation gives the orbital radius to be

$$a = 42\,241.558\,\text{km} \tag{14.3}$$

14.3 Angle from Satellite to Earth Station

Again, the case of a geostationary satellite is a much simplified version of the general case. The point at which a line between the satellite and the centre of the earth intersects the earth's surface is called the sub-satellite point. There are two angles of interest between the earth station and the satellite: the elevation and the azimuth. The elevation is the angle of the satellite above the horizon, from the earth station, and the azimuth is the angle between the line of longitude through the earth and the direction of the sub-statellite point. These are shown diagrammatically in Figs. 14.2 and 14.3.

From Fig. 14.2 the angle γ at the centre of the earth is given by

$$\cos\gamma = \cos L_e \cos(l_s - l_e) \tag{14.4}$$

where L_e = latitude of the earth station

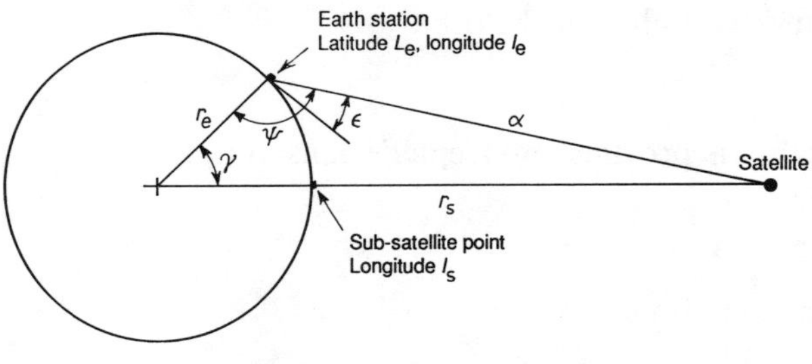

Fig. 14.2 Satellite position relative to Earth.

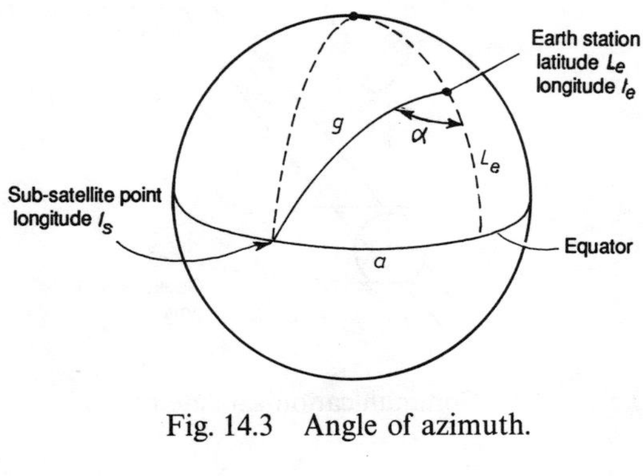

Fig. 14.3 Angle of azimuth.

$l_s - l_e$ = difference in latitude between the earth station and the sub-satellite point.

Applying the cosine rule

$$d_s^2 = r_e^2 + d_s^2 - 2r_e d_s \cos\gamma \tag{14.5}$$

From the sine rule

$$\frac{r_s}{\sin\psi} = \frac{d}{\sin\gamma} \tag{14.6}$$

giving

$$\sin\psi = \frac{r_s \sin\gamma}{d}$$

$$= \frac{r_s \sin\gamma}{[r_e^2 + r_s^2 - 2r_e r_s \cos\gamma]^{1/2}}$$

$$= \frac{\sin\gamma}{\left[\left(\frac{r_e}{r_s}\right)^2 + 1 - 2\left(\frac{r_e}{r_s}\right)\cos\gamma\right]^{1/2}} \tag{14.7}$$

and since, from Fig. 14.2

$$\varepsilon = \psi - 90$$

$$\cos\varepsilon = \frac{\sin\gamma}{\left[1 + \left(\frac{r_e}{r_s}\right)^2 - 2\left(\frac{r_e}{r_s}\right)\cos\gamma\right]^{1/2}} \tag{14.8}$$

Using $r_e = 6370$ km and $r_s = 42\,242$ km in Eqn (14.8) gives

$$\cos\varepsilon = \frac{\sin\gamma}{[1.02274 - 0.30159\cos\gamma]^{1/2}} \tag{14.9}$$

Clearly the angle of elevation will depend on the latitude of the earth station via angle γ. ε will be large towards the equator and small towards the poles. A minimum value of about 5° limits the highest latitude. The azimuth angle $180 + \alpha$ in Fig. 14.3, is related to the latitude and longitude of the earth station, L_e and l_e and the longitude of the sub-satellite point l_s by

$$\tan^2\frac{\alpha}{2} = \frac{\sin(S-\gamma)\sin(S-L_e)}{\sin S \sin(S+L_e)} \tag{14.10}$$

where

$$S = \tfrac{1}{2}(a + L_e + g) \qquad \text{and} \qquad a = (l_s - l_e)$$

The elevation and azimuth angles define the 'boresight' of the receiving antenna for a given receiver location. The azimuth angle uses the line from earth station to north pole as a reference.

It was noted earlier that towards the poles there is a restriction on the minimum angle of elevation considered suitable because the noise increases as $\varepsilon \to 0$. Another consideration is the additional attenuation produced by the increased distance to the satellite (slant height) with latitude.

From Eqn (14.5) the slant height is

$$d = [r_s^2 + r_e^2 - 2r_s r_e \cos \gamma]^{1/2}$$

14.4 Satellite Links

In satellite communications the direction of transmission is indicated by the term used: the downlink is transmission from satellite to earth station and the uplink is transmission in the opposite direction.

As we have mentioned before, communications by satellite impose enormous problems on the systems designer. In the uplink we have the availability of a high intensity transmission beam, because the transmitter is ground based and therefore power is not a problem, coupled with a sensitive receiver on the satellite, while in the downlink we have a low power source in the satellite and a highly sensitive receiver in the earth station. However, this presupposes that the earth station cost is not a constraint. For public utility applications such as telephony, cost will not be a limiting factor; the high cost of an expensive receiving station can be spread over a very large number of users. In direct broadcast television however, the story is different. A single user has to pay for the receiver and therefore its cost must be low. This means that the earth station will be 'basic'. The dish will be as small as possible and the amplifier as cheap as possible, consistent with obtaining an adequate signal for most of the year. Outage will be high, compared with that for a better quality receiver, but must still be at a reasonable level.

The preoccupation in satellite link design is the power budget; how much power can be obtained from the transmitter, how much of that power can be directed towards the receiver, how much power is lost over the link, and at the transmitter and receiver, and how much is left at the detector? What margin is necessary above the minimum detectable signal for the detector being used? Essentially the problem is to ensure that the signal level at the detector is large enough to produce a satisfactory output for a large part of the year.

Since the satellite distance is fixed, the attenuation (except for the effect of rain) is also fixed, and it is high. It cannot be avoided and so any improvement to be made must occur at the transmitter or receiver, and will be marginal, but very important. Fig. 14.4 shows the basic features of a satellite link.

We will discuss first the link transmission loss and the effect of antenna design.

Let P_t be the power from the transmitter.

G_t be the transmitter antenna gain

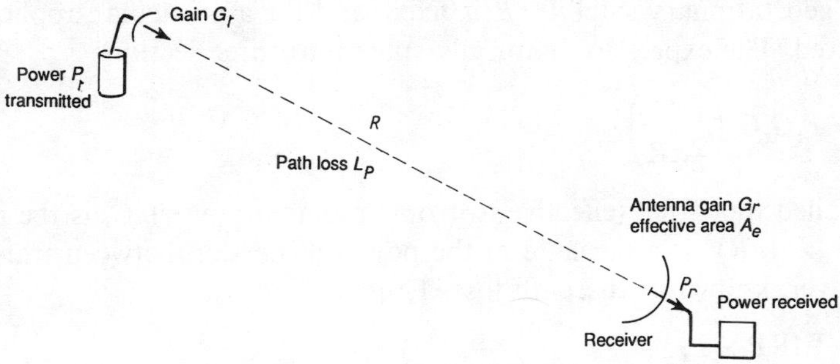

Fig. 14.4 Transmission link.

R the range from transmitter to receiver
P_r the received power
G_r the receiver antenna gain

We know from Chapter 7 that the power density, P, at a distance R from an isotropic source radiating total power P_t is

$$P = \frac{P_t}{4\pi R^2} \text{ W/m}^2 \tag{14.11}$$

The transmitting antenna of gain G_t will therefore concentrate the power in the direction of its maximum lobe to produce a flux density

$$P = \frac{P_t G_t}{4\pi R^2} \text{ W/m}^2 \tag{14.12}$$

The receiver antenna, when properly lined up with the transmitter, intercepts power of density P. The effective area of the receiving antenna is $A_e = \eta A_r$ where A_r is the antenna aperture area and η is known as the antenna efficiency.

The power received = power density at the receiver × effective antenna area

$$= \frac{P_t G_t}{4\pi R^2} \cdot A_e$$

$$= \eta \frac{P_t G_t A_r}{4\pi R^2} \tag{14.13}$$

There is a simple relationship between the effective area and the antenna gain which is

$$G_r = \frac{4\pi A_e}{\lambda^2}$$

Hence

$$P_r = \frac{P_t G_t G_r \lambda^2}{(4\pi R)^2} \tag{14.14}$$

For a geostationary satellite, R is fixed, and for a particular application λ is also fixed. The expression naturally splits into three sections

$$P_r = P_t G_t G_r \left(\frac{\lambda}{4\pi R}\right)^2 \tag{14.15}$$

$P_t G_t$ is called the EIRP (effective isotropic radiated power), G_r is the receiver gain and $(\lambda/4\pi R)^2$ is a measure of the power dispersion between transmitter and receiver, known as the path loss. Thus

$$P_r = \frac{\text{EIRP} \times P_r}{L_p} \tag{14.16}$$

where L_p = path loss
More simply, this can be expressed in dB as

$$P_r = (\text{EIRP} + P_r - L_p) \quad \text{dBW} \tag{14.17}$$

where

$$\text{EIRP} = 10 \log_{10} P_t G_t \quad \text{dBW}$$

$$G_r = 10 \log_{10} \frac{4\pi A_e}{\lambda^2} \quad \text{dB}$$

$$L_p = 20 \log_{10} \frac{4\pi R}{\lambda} \quad \text{dB}$$

The resultant power received must be sufficient to overcome losses not included in the above equation. As mentioned earlier, there will be precipitation losses, and there will be small losses at the transmitter and receiver which will need to be taken into account. For example, it has been assumed that all the power produced at the transmitter will be radiated from the transmitting antenna; that will not be so since there will be small losses in the transmitter circuits.

In Eqn (14.15) the received power was related to other system parameters by

$$P_r = P_t \cdot G_t \cdot G_r \cdot \left(\frac{\lambda}{4\pi R}\right)^2$$

This is known as the carrier power C. Of considerable interest is the ratio of this power to the noise power N at the detector. From the discussion in Chapter 4 the noise power at a receiver is

$$P_n = \text{k} T_s B G$$

where T_s system noise temperature

B receiver bandwidth
G receiver gain
k Boltzman's constant

and the power received due to the carrier is

$$P_r G$$

$\therefore$ the ratio $\frac{C}{N}$ is given by

$$\frac{C}{N} = \frac{P_t G_t G_r}{\mathrm{k} T_s B}\left(\frac{\lambda}{4\pi R}\right)^2 \qquad (14.18)$$

assuming that the receiver has been replaced by an equivalent system characterised by T_s, and B

or

$$\frac{C}{N} = \frac{P_t G_t}{\mathrm{k} B}\left(\frac{\lambda}{4\pi R}\right)^2 \frac{G_r}{T_s}$$

or, assuming the subscripts to be implicit, the carrier/noise ratio at a particular earty station is

$$\frac{C}{N} = \frac{P_t G_t}{\mathrm{k} B}\left(\frac{\lambda}{4\pi R}\right)^2 \frac{G}{T} \qquad (14.19)$$

where G/T is a measure of the goodness of the earth station. This quantity is widely used as a figure of merit to provide a means of comparison between different systems.

14.5 Antennas

For satellite communications the ideal antenna performance is to have high gain, high efficiency (low loss), low sidelobe levels, and narrow beam width. The beam width and sidelobe levels are important in determining the inteference from adjacent satellites and hence have a strong impact on the proximity which adjacent satellites can have in orbit.

The efficiency of the antennas used in satellite communications is extremely important and much care is taken to restrict any losses. Factors which reduce efficiency are wide ranging. The design of the antenna and its feed are obvious elements, but equally significant is the design of the waveguide and other components from, in the case of the receiver, the antenna feed to the first stage of the amplifier.

The degree to which clever antenna and feed design can be used to improve antenna efficiency will depend on the cost budget for a system. If the antenna forms part of an earth station carrying heavy traffic then a high cost can be tolerated, and sophisticated design techniques will be justified. However, for a simple broadcast television receiving station low cost and ease of manufacture are paramount, consequently antenna loss will be comparatively high.

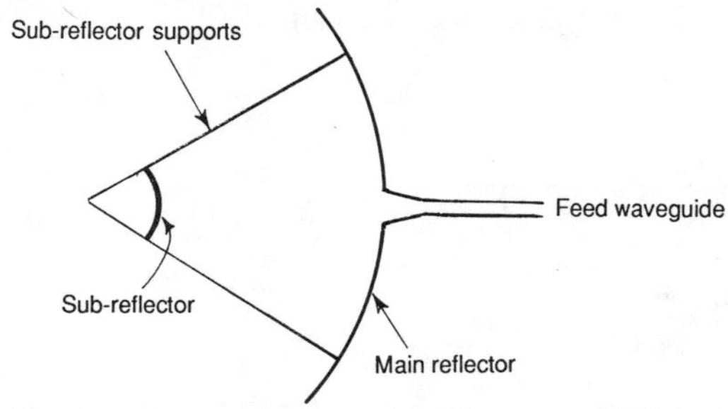

Fig. 14.5 Cassegrain dish antenna with direct feed.

The simplest antenna system, Fig. 14.5, is a straightforward dish paraboloid with a waveguide feed at its focus. The sources of loss are:

(i) poor illumination profile over the dish;
(ii) blocking by the feed;
(iii) long waveguide run in the feed;
(iv) high side lobe level.

As far as the antenna itself is concerned the paraboloid, in principle, allows a parallel beam of radiation from a distant source to be reflected to a focal point (section 7.18). However the ray theory approach on which that is based assumes that the wavelength is very small compared to the diameter of the antenna, but the wavelengths used in practice in satellite communications do not satisfy that condition. As a consequence there are defraction effects which have a significant bearing on the performance of the system. Ideally, a uniform distribution of E-field amplitude is desirable at the antenna aperture, but this can only be achieved for an antenna of infinite diameter. Usually some compromise is required between cost and performance. To reduce loss and limit cost, the direct feed arrangement is replaced by an indirect feed system, and in practice nearly all 'A' standard antennas for use in the INTELSAT system use this approach. Two methods are favoured, both based on the same type of principle and known after the 16th century inventors of the equivalent optical telescope. They are Cassegrain and Gregorian antennas. In both, the primary feed is at the centre of the paraboloid. This feed illuminates a secondary reflector; in the case of the Cassegrain of hyperbolic cross-section and in the case of the Gregorian of eliptical cross-section. These sub-reflectors in turn illuminate the main parabolic reflector. The sub-reflector and its mounting, block some of the energy from the paraboloid. To reduce this effect the sub-reflector can be offset from the main axis as shown in Fig. 14.6.

The compromise between low sidelobe level and high directivity (narrow beam width) is achieved by adjusting the illumination used across the aperture. To obtain the required distribution the primary feed, sub-reflector shape and dish shape can all be modified.[4]

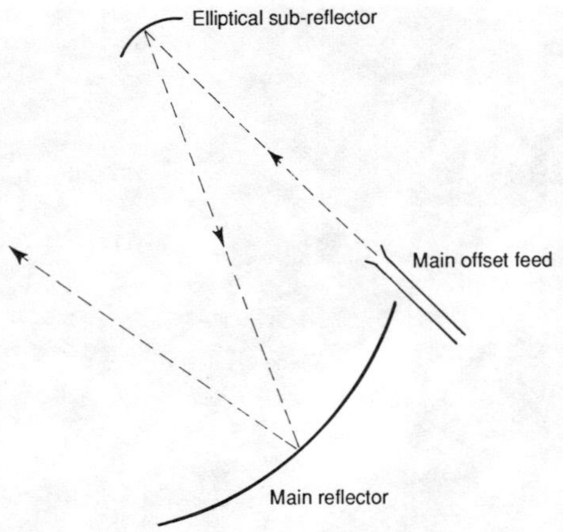

Fig. 14.6 Schematic of offset Gregorian reflector.

Apart from the losses associated with defraction effects and limited antenna aperture, there are other sources of antenna loss. In receiver mode the signal is transmitted from the primary feed into the receiver. For practical reasons the receiver is preferably placed away from the antenna, with the consequent waveguide run adding to the system loss. This loss can be reduced to a small value by using a beam waveguide, which is almost a free-space system enclosed in a large envelope, as shown in Fig. 14.7. This system is employed in a number

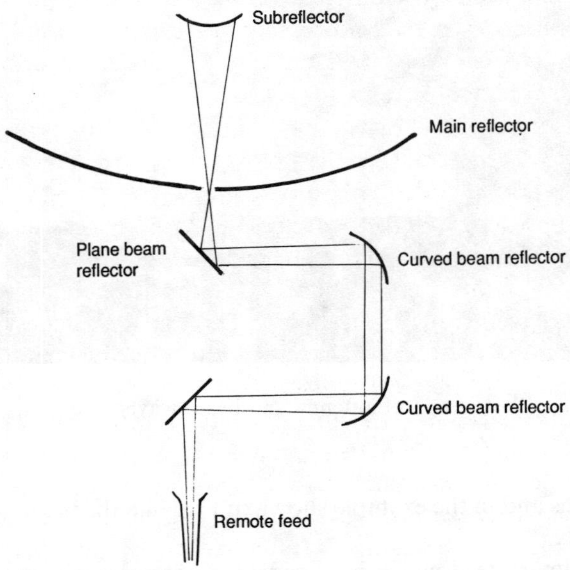

Fig. 14.7 Schematic of beam waveguide feed.

Fig. 14.8 Photograph of antenna and beam waveguide feed.

of practical systems and in the example shown in Fig. 14.8 the beam waveguide can be seen clearly.

One type of primary feed which is of considerable interest is a corrugated pyramidal horn. The corrugations can be broadbanded and it has the advantage that it provides excellent cross polarization rejection. In many systems the channel utilization can be increased significantly by using two directions of polarization to separate channels.

The function of spacecraft antennas is to direct the beam to those areas of the earth which have appropriate receivers. Although this may appear to be an obvious statement, it does emphasize the difficulties for the designer. Wasting power by sending a signal into areas with no receivers is undesirable, but to achieve coverage over specified areas only is also difficult.

With a standard paraboloidal dish the coverage on the earth's surface is circular. Since the angle subtended by the earth at the satellite is 17°, such a beam width will provide coverage over almost one hemisphere. To reduce the coverage area significantly requires a much larger antenna, and to provide a contoured (non-circular) coverage pattern involves the use of complex feed arrangements.

Gradually, linear array feeds are being introduced. They give the designer considerable flexibility in the range of patterns which can be obtained, at the cost of increased complexity. However, there are other benefits which may arise from these feeds. By making them electronically fed, the beam can be reconfigured to suit new users of the satellite who may require different coverage patterns.

Even further variability, and therefore options, can be achieved in antennas using linear arrays to illuminate a sub-reflector. The reflecting surfaces of both the sub-reflector and the main reflector can be distorted to modify the shape of the ground coverage profile.

There is a limit to the size of antenna array a satellite can accommodate. By using collapsible lightweight structures which can be opened out once the satellite is in orbit gives scope for large systems.

Offset feeds are often used, both to eliminate any blocking of the beam and to provide good mechanical support for the feeds. Several feeds, for different frequencies, may be mounted on the same support.

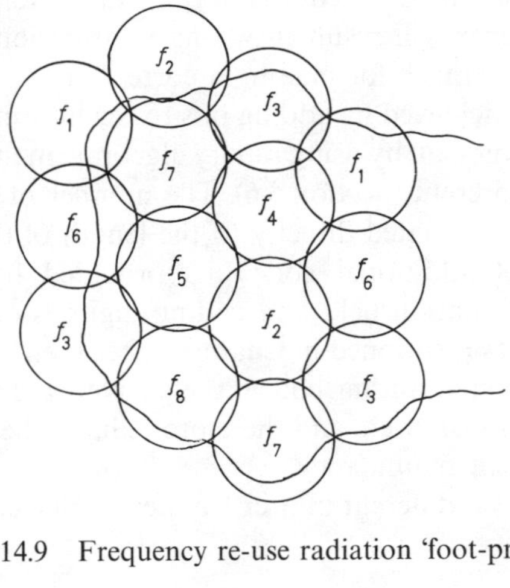

Fig. 14.9 Frequency re-use radiation 'foot-prints'.

Earlier we discussed polarization of the beam to extend the utilization of the channel. This can be achieved by relying on the polarization from the feed horn being maintained, or by using a polarizer in the form of a grid of appropriately orientated wires in front of the reflector.

A future development in spacecraft antenna design, planned for business satellites in particular, is to provide space switching. Earlier we noted that additional capacity we can obtained by using polarization. Space switching will also provide more capacity over a given area by allowing frequency re-use on a spatial basis. Beams will be very narrow and once the area of potential interference has been exceeded the same frequencies can be used again (Fig. 14.9). This is the same basic idea as that used in cellular radio (Section 13.12).

14.6 Error Coding

It is in the nature of transmission channels that from time to time interference with the transmitted signal will produce errors. In digital systems these errors will result in bits being incorrectly received. Coding is a methodology that provides some protection against such errors.

Often the provision of error detection and correction is the responsibility of the satellite user; the operator merely supplies the link to meet some specification, but does not provide the means of ensuring accurate transmission. One reason for this division of functions between the operator and the user is that the operator is not concerned with the type of information being transmitted, and therefore cannot predict when designing the system the degree of correction the user will require.

Normally speech channels are not error corrected; there is sufficient redundancy in speech to ensure that it is made intelligible in the presence of such errors (Section 1.13). In transmitting data however error detection is essential. There may be no redundancy in the information stream and consequently any error will result in wrong information being received and there will be no mechanism for making a correction.

Error detection is achieved by adding bits to the information packet in such a way that the receiver can, by using binary algebraic manipulation, determine whether errors are present (Section 5.6). The number of errors which can be detected in this way is related directly to the length of the information field, and the number of additional code bits provided by the encoder. The combination of information field and coding digits is called the code-word. There is a trade off, as mentioned in Chapter 5, between error detection power and transmitted capacity. Adding code bits effectively reduces the rate at which real information is transmitted, and therefore reduces the capacity of the link; the code bits represent redundancy.

Error correction is a different matter from error detection. It is possible to

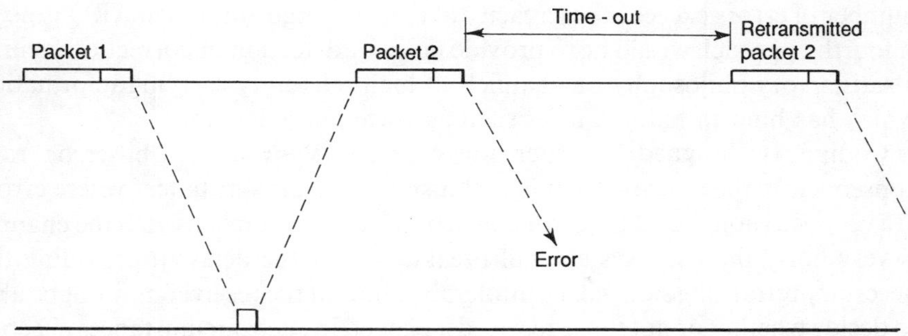

Fig. 14.10 Sample stop and wait protocol.

provide codes which will allow the receiver to detect and then correct a number of errors in a bit stream, but correction is more difficult to achieve than detection and therefore more additional bits are required. In practice this means that codes can correct a much smaller number of errors than they can detect. However, this type of correction, called forward error correction (FEC) has a significant role in satellite links. The very long transmission time between earth station and satellite discourages the use of retransmission for error correction, and therefore forward error correction in which the error is detected and corrected by the receiver is highly desirable for this application.

A simple method for correcting errors on short terrestrial links is to use a stop and wait protocol with automatic request for transmission (ARQ). The transmitter sends one packet at a time; when it is received correctly an acknowledgement signal (Ack) is returned. If the receiver detects an error no Ack is returned and the packet is retransmitted after a suitable time-out. The time-out is slightly longer than the expected time for the return of Ack., Fig. 14.10. Clearly the main drawback with this protocol is that the time between transmission of the packet and reception of the acknowledgement is wasted. For satellite applications, where that delay is about a quarter of a second, such wasted time cannot be tolerated. Instead a duplex link is used and packets are transmitted continuously. When an error occurs a retransmission mechanism causes those packets, and probably a few following, to be retransmitted.[5]

Clearly a number of mechanisms exist for handling the problem of error correction and detection. One is to provide such a strong link that the excess power over the minimum required for normal circumstances is sufficient to cope with nearly all the occurrences of poor transmission conditions. This is a very expensive and crude approach, and does not attack the problem of error correction directly. The second method is to arrange for all the incorrect packets to be retransmitted automatically (using ARQ). The long round-trip delay discourages complete reliance on this method so a third scheme is used in which a measure of forward error correction is provided at the receiver and if the number of errors exceeds the capacity of the FEC algorithm then ARQ is used. A fourth approach would be to provide no error detection or correction facility; a satisfactory philosophy on channels of high reliability carrying information which has built-in redundancies such as voice and television.

Coding is designed to cater for channels with a low bit-error rate, consequently the improvement is best displayed in circumstances where errors are very occasional, and the error correction facility is rarely used. If the channel is very noisy the FEC system will breakdown, or the delay in providing the necessary retransmission will be intolerable. Indeed the receiver may not be able to decide whether or not it has received a packet. In such circumstances effective transmission breaks down. Practical systems are designed to ensure that breakdown occurs only within some specified limit, measured in terms of minutes per month.

Types of error

The incidence of errors is classified as random or bursty. Random errors occur as isolated erroneous bits in a stream of good bits; the probability of an error is the same for all bits, regardless of whether the previous bits were correct or not. Bursty errors are somewhat different, and in a sense more typical of some types of interference. As the name suggests bursty errors turn up in blocks; a sequence of correct bits is followed by a sequence in which the incidence of errors is very high, followed by a stream of good bits. During an error burst the probability of bit error is high; during a sequence of good bits the probability of error is low.

Codes can also be classified into several types, but the principal division is between block codes and convolutional codes.

In a block coder the information is presented in a block of k digits, and this block is coded into a longer word of n digits; known as a (n, k) block code. The additional k–n digits provide the redundancy required to enable error detection and correction to take place. Successive coded blocks are independent of each other.

Convolutional codes consist similarly of (n, k) blocks, but in addition there is a dependency between the codes used for successive incoming frames. The memory order of codes is denoted by m, the number of previous frames having an influence on the coder used for the present frame.

A feature of the decision making process in the detector is to determine if an error has occurred and that could be based on a threshold related decision (hard decision) or on a probability related decision (soft decision). In the hard decision system there is no allowance for any doubt; either a bit has or has not been received. However in soft decision decoders the doubt can be included as a probability and taken into account in the algorithm for decoding the received code word. Block code decoders do not allow soft decisions whereas in convolutional decoding it can be part of the decision making process.

14.7 Interference and Noise

Background noise on the downlink has a significant effect on receiver performance since it reduces the C/N factor. This particular form of noise is represented as a noise temperature; and for an antenna that temperature is variable and depends strongly on the elevation of the boresight and the frequency of the channel. The quiescent noise temperature will be less than 70 K.

However, the direction of the antenna boresight is determined by the relative position of the earth station and the satellite. Consequently the antenna may pick up atmospheric noise from time to time well in excess of this value. Most noticeably, excess temperature occurs when the sun is behind the satellite, in which case the noise temperature will rise to over 10^4 K at frequencies up to about 10 GHz.

Interference increases the effective noise of the satellite channel as spurious signals from nearby links, or from terrestrial sources, are detected. However there is also a reciprocal effect in which our channel interferes with other communication systems, both satellite and terrestrial.

Atmospheric attenuation

The link path between satellite and earth is not homogeneous to radio wave propagation, nor is it immune from interference from nearby radio systems. There is always some attenuation due to the excitation of electrons in the path of the radio wave, but at satellite frequencies it is a minor factor, accounting for about 0.5 dB loss. Much more important are those effects which are transitory in nature, and consequently unpredictable. On a 'clear day' there are none of these loss mechanisms, but at other times some or all of them may occur. We can consider such mechanisms in several categories: here we choose three; attenuation, depolarization and interference. The first two are properties of the radio link itself. The third results from other sources of radiation in the proximity of the link.

Attenuation At satellite frequencies the most important variable attenuation mechanisms are those associated with water droplets, hydrometeorites, in the form of rain, ice, snow, sleet and combinations of these. Rain is the most important type of precipitation from this point of view. Snow produces relatively little attenuation, but wet ice can impair the signal significantly.

There are several parameters associated with rainfall, but in this context the most important is rainfall rate; measured in millimetres per hour, it has a direct effect on attenuation.

The attenuation A can be related to the rainfall rate R and the effective path length, L by

$$A = kR^{a}L \text{ dB} \qquad (14.20)$$

This simple relationship assumes that the rainfall rate R is constant over the whole of the path length L. Several very difficult problems arise in utilizing this relationship. The constants k and a are highly dependent on frequency and on rain drop size, as well as depending to a small degree on temperature. The relationships between k, a, frequency and drop size that are used in practice are empirical and have been tabulated by a number of organisations such as CCIR. Resort must therefore be made to such tables[6] to find values of A. However, knowing k and a is only part of the solution. R is impossible to determine precisely. The rainfall rate will vary along the length of the path in a statistical way, and even estimating the value of R at a particular point by using a ground-based measure of quantity of water versus time may be a poor approximation to the rate at which, over the same spot, raindrops are falling through the link path. The turbulence in the atmosphere precludes a straight fall of rain and hence complicates the assessment of fall rate.

To provide designers with a procedure to calculate A, CCIR have produced an algorithm based on tabulated values for k and a, and cumulative distributions for R. This gives, for example, a figure $A_{0.01}$ based on $R_{0.01}$, where the suffix indicates the percentage of the average year for which the attenuation will exceed the value $A_{0.01}$ given that the mean rainfall rate exceeded for 0.01% of the average year is $R_{0.01}$. In this way the designer can dimension the system to meet satisfactory performance specification for all but 0.01% of the year.

The ionosphere can produce absorption and other effects. At satellite frequencies the absorption is negligible unless there is electron activity in the ionosphere in the form of clouds of free electrons and the earth's magnetic field, producing not only absorption but also depolarization due to the rotational effect on the wave, and a slight misalignment caused by variations in refractive index. If the variations are rapid they are called scintillations and these can produce variations of up to 10 dB in the received signal amplitude.

Depolarization Previously, we noted that better use of the links can be achieved by sending out two orthogonally polarized signals. The success of this approach depends on the receiver being aligned to the correct polarization. Unfortunately there are several mechanisms which make this difficult to achieve consistently. The effect of changing the direction of polarization from that expected is called depolarization and that has two results. First, the intensity of the intended signal is reduced because the polarization of the incoming signal and that of the receiver are not matched. Second, the receiver can pick up some of the signal from the other polarization, hence effectively increasing the channel noise.

The main cause of depolarization is Faraday rotation, caused by action of the magnetic field and the ionospheric free electrons on the signal. The degree of rotation depends on both the electron density, and the length of path over which the electron cloud or clouds exist. The ionosphere does not exhibit the same electrical properties in all directions; it is said to be anisotropic. This lack of homogeneity itself results in depolarization.

As estimate of the rotation, φ, produced by the Faraday effect can be obtained from the formula

$$\varphi = \int k_1 \frac{NL}{f^2} B \cos\theta \, \mathrm{d}l \qquad (14.21)$$

where k_1 is a constant
N is the ionospheric plasma electron density
B is the magnetic flux density
L is the path length
f is the channel frequency
and θ is the angle between the magnetic field vector and the link

The integration is over the length of the path in the ionosphere. In practice N will not be constant and some suitable model will have to be chosen to determine its value as a function of L.

14.8 Multi-access

As the number of earth stations increases, and satellite communications are more widely used, the multi-access techniques used become more significant. In principle the function of multi-access is to provide the many earth stations wishing to use a particular transponder with channel access, at the required bandwidth.

In many respects the problems of multi-access in satellite systems are similar to those in local area networks, and the same mechanisms can be used in each. In particular, the traffic to be transmitted is a combination of voice and data and the same problems, and solutions, as discussed in Chapter 13 occur. There are however some additional issues in satellite systems, and they will be mentioned here.

FDMA

In FDMA the bandwidth can be divided on a multi-carrier or single carrier basis. In multi-carrier the transponder bandwidth is divided into sub-bands, each sub-band being assigned to an earth station. The sub-bands need not be all of the same width; the number of channels in each will depend on the earth station's requirements and the allocation they have been made. This arrangement gives the earth station some flexibility in the allocation of bandwidth within its own sub-band. The main disadvantage of this type of access is that the power output from the transponder must be reduced compared with the output that could be achieved with a single carrier arrangement. Except for a few satellites using solid-state amplifiers, the output device at the transponder is normally a travelling wave tube. It has a non-linear input-output characteristic, and this leads to serious inter-modulation products in the output. This effect

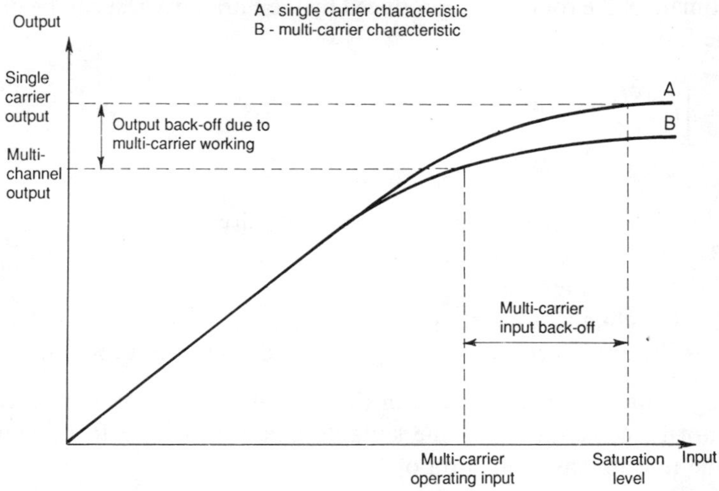

Fig. 14.11 Effect of multi-carrier working.

becomes more serious as the number of carriers increases. To avoid the non-linearities from producing serious degradation on the output signal, it is necessary to impose an input-back-off. This back-off has to be large enough to ensure that the output back-off brings the operating point to a position where the inter-modulating products are insignificant. The non-linear characteristic means, of course, that the output back-off will be smaller than the input back-off, Fig. 14.11. The amount of back-off required increases with the number of carriers in the system.

A single carrier FDMA arrangement, in which the available 36 MHz bandwidth is divided into uniform channels, allows more efficient use of the travelling wave tube output. If QPSK is used, and each channel is spaced by 45 kHz, there is room for 800 channels to avoid cross channel interference, and some back-off is necessary because of the inter-modulation products between channels. Most satellites currently use FDMA access methods of one sort or another.

TDMA

Throughout communications systems design, the trend for many years has been to use digital representation of signals, and digital transmission. Its ease of transmission, and robustness in noise, makes it very appealing, and the same arguments in its favour in terrestrial systems also apply to satellite networks. As far as the TWT operating conditions are concerned, in digital transmission there is no reason why the full output should not be used: there is only one carrier, and since the transmission is digital, inter-modulation products are not a problem.

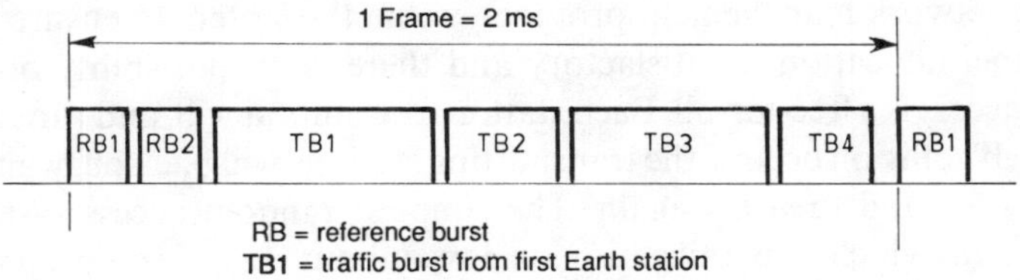

Fig. 14.12 Typical Intelsat frame.

The discussion on TDMA in local area networks, in Chapter 13, is highly relevant here. The different characteristics of voice and data traffic, their different delay and error-correction requirements and the methods of accommodating them are all explained there.

Two issues are however of especial interest in satellite operations. The frame size, which has to be greater than 125 μs if we assume that normal pcm is used for speech, can have any value up to about 25 ms; the limit imposed by the maximum tolerable delay to speech traffic. In the *INTELSAT* system, for example, a frame size of 2 ms is used, with the frame structure shown in Fig. 14.12. The reference bursts are for synchronization and each traffic burst may be from a different earth station. The traffic bursts contains time multiplexed speech channels.

The other issue is synchronization. The up-link packets arriving at the transponder from different earth stations must not overlap in time. Consequently the whole network related to one satellite has to be in synchronism. This can be achieved in different ways in detail, but it depends on the transponder sending out synchronizing bits to indicate the start of each frame, and to provide reference times to which earth stations relate their own transmission. The variable nature of the satellite position will produce some changes in this synchronization timing, and guard intervals are used to avoid overlap due to this cause. The guard interval adds to the transmission overhead, leaving less time available for revenue earning traffic. Other overheads which are included concern addresses, acknowledgement for data signals and network management.

Random Access

The bandwidth allocated to each channel depends on the method of access employed. Fixed assignment means that at the outset the designer determines the bandwidth allocation, and there is no flexibility for the operator to adjust the system in use. Demand assignment does give the operator flexibility. Channel bandwidths can be adjusted to cater for the traffic mix, which may vary from time to time. For such demand assignment to be feasible, an overhead in terms of network management information is imposed.

An alternative approach is to use a random access mechanism. Instead of having a network management protocol within the system to ensure that both the channel allocation is satisfactory and there is no possibility of collision, random access is a free-for-all. Each station transmits at will, and runs the risk of collision. If collision occurs, the transmitting stations will back off, wait for some random time and then try again. The simplest random access system is the ALOHA, in which the utilization is low (about 18%); some considerable improvement is achieved with the slotted ALOHA (Section 13.5). It is worth remembering that the more sophisticated the access mechanism, the higher is the overhead and the more complex is the management protocol, so that there is a trade-off to be made.

Speech interpolation

One of the early developments in transatlantic submarine communications, when the demand immediately saturated the circuits provided, was to utilize the silent intervals in normal speech to allow channels to be shared. In normal speech about 45% of the time is not used, in the sense that it is occupied by silence intervals. Speech interpolation techniques allow these intervals to be utilized, and this same method can be applied to satellite links. In packet based transmission systems, since there is no packet to be transmitted during silence intervals, speech interpolation is automatic.

References

1. Clarke, A.C., 'Extra-terrestrial Relay', *Wireless World*, Oct. 1945.
2. Pratt, T. and Bostian, C.W., *Satellite Communications*, John Wiley, 1986.
3. Spilker, J.J., *Digital Communication by Satellite*, Prentice–Hall, 1977.
4. Evans, B.G. (ed.), *Satellite Communication Systems*, I.E.E. London, 1987, Chapter 12.
5. Tanenbaum, A.S., *Computer Networks*, Prentice–Hall, 1981, Section 4.2.2.
6. Propagation in non-ionised Media. *CCIR, Recommendations and Reports*, Vol. V, 1982, I.T.U., Geneva.

APPENDIX 1 FOUR-FIGURE BESSEL FUNCTIONS

β	$J_0(\beta)$	$J_1(\beta)$	$J_2(\beta)$	$J_3(\beta)$	$J_4(\beta)$	$J_5(\beta)$	$J_6(\beta)$	$J_7(\beta)$	$J_8(\beta)$	$J_9(\beta)$	$J_{10}(\beta)$	$J_{11}(\beta)$	$J_{12}(\beta)$	$J_{13}(\beta)$	$J_{14}(\beta)$	$J_{15}(\beta)$	$J_{16}(\beta)$	$J_{17}(\beta)$	$J_{18}(\beta)$	$J_{19}(\beta)$	$J_{20}(\beta)$	$J_{21}(\beta)$	$J_{22}(\beta)$	$J_{23}(\beta)$	$J_{24}(\beta)$	$J_{25}(\beta)$
0.2	0.990	0.995	0.0050	0.0002																						
0.4	0.9604	0.1960	0.0197	0.0013																						
0.6	0.9120	0.2867	0.0437	0.0044	0.0003																					
0.8	0.8463	0.3688	0.0758	0.0102	0.0010																					
1.0	0.7652	0.4401	0.1149	0.0196	0.0025	0.0002																				
1.25	0.0659	0.5106	0.1711	0.0369	0.0059	0.0007																				
1.5	0.5118	0.5579	0.2321	0.0610	0.0118	0.0618	0.0002																			
1.75	0.3690	0.5802	0.2940	0.0918	0.0209	0.0038	0.0006	0.0001																		
2.0	0.2239	0.5767	0.3528	0.1289	0.0340	0.0070	0.0012	0.0002																		
2.5	−.0484	0.4971	0.4461	0.2166	0.0738	0.0195	0.0042	0.0008	0.0001																	
3	−.2601	0.3391	0.4861	0.3091	0.1320	0.0430	0.0114	0.0025	0.0005																	
4	−.3971	−.0660	0.3641	0.4302	0.2811	0.1321	0.0491	0.0152	0.0040	0.0009	0.0002															
5	−.1776	−.3276	0.0466	0.3648	0.3912	0.2611	0.1310	0.0534	0.0184	0.0055	0.0015	0.0004														
6	0.1506	−.2767	−.2429	0.1148	0.3576	0.3621	0.2458	0.1296	0.0565	0.0212	0.0070	0.0020	0.0005	0.0001												
7	0.3001	−.0047	−.3014	−.1676	0.1578	0.3479	0.3392	0.2336	0.1280	0.0589	0.0235	0.0083	0.0027	0.0008	0.0002											
8	0.1717	0.2346	−.1130	−.2911	−.1054	0.1858	0.3376	0.3206	0.2235	0.1263	0.0608	0.0256	0.0096	0.0033	0.0010	0.0003										
9	−.0903	0.2453	0.1448	−.1809	−.2655	−.0550	0.2043	0.3275	0.3051	0.2149	0.1247	0.0622	0.0274	0.0108	0.0039	0.0013	0.0004	0.0001								
10	−.2459	0.0435	0.2546	0.0584	−.2196	−.2341	−.0145	0.2167	0.3179	0.2919	0.2075	0.1231	0.0634	0.0290	0.0120	0.0045	0.0016	0.0005	0.0002							
11	−.1712	−.1768	0.1390	0.2273	−.0150	−.2383	−.2016	0.0184	0.2250	0.3089	0.2804	0.2010	0.1216	0.0643	0.0304	0.0130	0.0051	0.0019	0.0006	0.0002						
12	0.0477	−.2234	−.0849	0.1951	0.1825	−.0735	−.3437	−.1703	0.0451	0.2304	0.3004	0.2704	0.1953	0.1201	0.0650	0.0316	0.0140	0.0057	0.0022	0.0008	0.0003					
13	0.2069	−.0703	−.2177	0.0033	0.2193	0.1316	−.1180	−.2406	−.1410	0.0670	0.2338	0.2927	0.2615	0.1901	0.1118	0.0656	0.0327	0.0149	0.0063	0.0025	0.0009	0.0003	0.0001			
14	0.1711	0.1334	−.1520	−.1768	0.0762	0.2204	0.0812	−.1508	−.2320	−.1143	0.0850	0.2357	0.2855	0.2536	0.1855	0.1174	0.0661	0.0337	0.0158	0.0068	0.0028	0.0010	0.0004	0.0001		
15	−.0142	0.2051	0.0418	−.1940	−.1192	0.1305	0.2061	0.0345	−.1740	−.2200	−.0901	0.1000	0.2367	0.2787	0.2464	0.1813	0.1162	0.0665	0.0346	0.0166	0.0074	0.0031	0.0012	0.0004		
16	−.1749	0.0904	0.1862	−.0438	−.2026	−.0575	0.1667	0.1823	−.0070	−.1895	−.2062	−.0682	0.1124	0.2368	0.2724	0.2399	0.1775	0.1150	0.0668	0.0354	0.0173	0.0079	0.0034	0.0013	0.0005	0.0002
17	−.1699	−.0977	0.1584	−.1349	−.1107	−.1870	0.0007	0.1875	0.1537	−.0429	−.1991	−.1914	−.0486	0.1228	0.2364	0.2666	0.2340	0.1739	0.1138	0.0671	0.0362	0.0180	0.0084	0.0037	0.0015	0.0006
18	−.0134	−.1880	−.0075	0.1863	0.0696	−.1554	−.1560	0.0514	0.1959	0.1228	−.0732	−.2041	−.1762	−.0309	0.1316	0.2356	0.2611	0.2286	0.1706	0.1127	0.0673	0.0369	0.0187	0.0089	0.0039	0.0017
19	0.1466	−.1057	−.1578	0.0725	0.1806	0.0036	−.1788	−.1165	0.0929	0.1947	0.0916	−.0984	−.2055	−.1612	−.0151	0.1389	0.2345	0.2559	0.2235	0.1676	0.1116	0.0675	0.0375	0.0193	0.0093	0.0042
20	0.1670	0.0665	−.1603	−.0989	0.1307	0.1512	−.0551	−.1842	−.0739	0.1251	0.1865	0.0614	−.1190	−.2041	−.1464	−.0008	0.1452	0.2331	0.2511	0.2189	0.1647	0.1106	0.0676	0.0381	0.0199	0.0098

APPENDIX 2

USEFUL TRIGONOMETRIC IDENTITIES

$$\sin(A+B) = \sin A\cos B + \cos A\sin B$$
$$\sin(A-B) = \sin A\cos B - \cos A\sin B$$
$$\cos(A+B) = \cos A\cos B - \sin A\sin B$$
$$\cos(A-B) = \cos A\cos B + \sin A\sin B$$
$$\sin 2A = 2\sin A\cos A$$
$$\cos 2A = \cos^2 A - \sin^2 A$$

$$\sin A + \sin B = 2\sin\left(\frac{A+B}{2}\right)\cos\left(\frac{A-B}{2}\right)$$

$$\sin A - \sin B = 2\cos\left(\frac{A+B}{2}\right)\sin\left(\frac{A-B}{2}\right)$$

$$\cos A + \cos B = 2\cos\left(\frac{A+B}{2}\right)\cos\left(\frac{A-B}{2}\right)$$

$$\cos A - \cos B = 2\sin\left(\frac{A+B}{2}\right)\sin\left(\frac{B-A}{2}\right)$$

$$\sin A\sin B = \tfrac{1}{2}[\cos(A-B) - \cos(A+B)]$$
$$\sin A\cos B = \tfrac{1}{2}[\sin(A+B) + \sin(A-B)]$$
$$\cos A\cos B = \tfrac{1}{2}[\cos(A+B) + \cos(A-B)]$$
$$\cos A\sin B = \tfrac{1}{2}[\sin(A+B) - \sin(A-B)]$$

$$\sin^2 A = \tfrac{1}{2}(1 - \cos 2A)$$
$$\cos^2 A = \tfrac{1}{2}(1 + \cos 2A)$$

$$\exp(j\omega t) = \cos\omega t + j\sin\omega t$$
$$\exp(-j\omega t) = \cos\omega t - j\sin\omega t$$

$$2\cos\omega t = \exp(j\omega t) + \exp(-j\omega t)$$
$$2\sin\omega t = \frac{1}{j}[\exp(j\omega t) - \exp(-j\omega t)]$$

APPENDIX 3

NORMAL ERROR FUNCTION TABLES

$$\operatorname{erf} x = \frac{2}{\sqrt{\pi}} \int_0^x \exp(-y^2)\,\mathrm{d}y$$

x	erf x	x	erf x	x	erf x	x	erf x	x	erf x	x	erf x
0.00	0.000 000	0.40	0.428 392	0.80	0.742 101	1.20	0.910 314	1.60	0.976 348	2.00	0.995 322
0.01	0.011 283	0.41	0.437 969	0.81	0.748 003	1.21	0.912 956	1.61	0.977 207	2.01	0.995 525
0.02	0.022 565	0.42	0.447 468	0.82	0.753 811	1.22	0.915 534	1.62	0.978 038	2.02	0.995 719
0.03	0.033 841	0.43	0.456 887	0.83	0.759 524	1.23	0.918 050	1.63	0.978 843	2.03	0.995 906
0.04	0.045 111	0.44	0.466 255	0.84	0.765 143	1.24	0.920 505	1.64	0.979 622	2.04	0.996 086
0.05	0.056 372	0.45	0.475 482	0.85	0.770 668	1.25	0.922 900	1.65	0.980 376	2.05	0.996 258
0.06	0.067 622	0.46	0.484 655	0.86	0.776 100	1.26	0.925 236	1.66	0.981 105	2.06	0.996 423
0.07	0.078 858	0.47	0.493 745	0.87	0.781 440	1.27	0.927 514	1.67	0.981 810	2.07	0.996 582
0.08	0.090 078	0.48	0.502 750	0.88	0.786 687	1.28	0.929 734	1.68	0.982 493	2.08	0.996 734
0.09	0.101 281	0.49	0.511 668	0.89	0.791 843	1.29	0.931 899	1.69	0.983 153	2.09	0.996 880
1.10	0.112 463	0.50	0.520 500	0.90	0.796 908	1.30	0.934 008	1.70	0.983 790	2.10	0.997 021
0.11	0.123 623	0.51	0.529 244	0.91	0.801 883	1.31	0.936 063	1.71	0.984 407	2.11	0.997 155
0.12	0.134 758	0.52	0.537 899	0.92	0.806 768	1.32	0.938 065	1.72	0.985 003	2.12	0.997 284
0.13	0.145 867	0.53	0.546 939	0.93	0.811 564	1.33	0.940 015	1.73	0.985 578	2.13	0.997 407
0.14	0.156 947	0.54	0.554 939	0.94	0.816 271	1.34	0.941 914	1.74	0.986 135	2.14	0.997 525
0.15	0.167 996	0.55	0.563 323	0.95	0.820 891	1.35	0.943 762	1.75	0.986 672	2.15	0.997 639
0.16	0.179 012	0.56	0.571 616	0.96	0.825 424	1.36	0.945 561	1.76	0.987 190	2.16	0.997 747
0.17	0.189 992	0.57	0.579 816	0.97	0.829 870	1.37	0.947 312	1.77	0.987 691	2.17	0.997 851
0.18	0.200 936	0.58	0.587 923	0.98	0.834 232	1.38	0.949 016	1.78	0.988 174	2.18	0.997 951
0.19	0.211 840	0.59	0.595 936	0.99	0.838 508	1.39	0.950 673	1.79	0.988 641	2.19	0.998 046
0.20	0.222 703	0.60	0.603 856	1.00	0.842 701	1.40	0.952 285	1.80	0.989 091	2.20	0.998 137
0.21	0.233 522	0.61	0.611 681	1.01	0.846 810	1.41	0.953 852	1.81	0.989 525	2.21	0.998 224
0.22	0.244 296	0.62	0.619 411	1.02	0.850 838	1.42	0.955 376	1.82	0.989 943	2.22	0.998 308
0.23	0.255 023	0.63	0.627 046	1.03	0.854 784	1.43	0.956 857	1.83	0.990 347	2.23	0.998 388
0.24	0.265 700	0.64	0.634 586	1.04	0.858 650	1.44	0.958 297	1.84	0.990 736	2.24	0.998 464
0.25	0.276 326	0.65	0.642 029	1.05	0.862 436	1.45	0.959 695	1.85	0.991 111	2.25	0.998 537
0.26	0.286 900	0.66	0.649 377	1.06	0.866 144	1.46	0.961 054	1.86	0.991 472	2.26	0.998 607
0.27	0.297 418	0.67	0.656 628	1.07	0.869 773	1.47	0.962 373	1.87	0.991 821	2.27	0.998 674
0.28	0.307 880	0.68	0.663 782	1.08	0.873 326	1.48	0.963 654	1.88	0.992 156	2.28	0.998 738
0.29	0.318 283	0.69	0.670 840	1.09	0.876 803	1.49	0.964 898	1.89	0.992 479	2.29	0.998 799
0.30	0.328 627	0.70	0.677 801	1.10	0.880 205	1.50	0.966 106	1.90	0.992 790	2.30	0.998 857
0.31	0.338 908	0.71	0.684 666	1.11	0.883 533	1.51	0.967 277	1.91	0.993 090	2.31	0.998 912
0.32	0.349 126	0.72	0.691 433	1.12	0.886 788	1.52	0.968 413	1.92	0.993 378	2.32	0.998 966
0.33	0.359 279	0.73	0.698 104	1.13	0.889 971	1.53	0.969 516	1.93	0.993 656	2.33	0.999 016
0.34	0.369 365	0.74	0.704 678	1.44	0.893 082	1.54	0.970 568	1.94	0.993 923	2.34	0.999 065
0.35	0.379 382	0.75	0.711 156	1.15	0.896 124	1.55	0.971 623	1.95	0.994 179	2.35	0.999 111
0.36	0.389 330	0.76	0.717 537	1.16	0.899 096	1.56	0.972 628	1.96	0.994 426	2.36	0.999 155
0.37	0.399 206	0.77	0.723 822	1.17	0.902 000	1.57	0.973 603	1.97	0.994 664	2.37	0.999 197
0.38	0.409 009	0.78	0.730 010	1.18	0.904 837	1.58	0.974 547	1.98	0.994 892	2.38	0.999 237
0.39	0.418 739	0.79	0.736 103	1.19	0.907 608	1.59	0.975 462	1.99	0.995 111	2.39	0.999 275

x	erf x	x	erf x	x	erf x	x	erf x	x	erf x	x	erf x
2.40	0.999 311	2.67	0.999 481	2.94	0.999 968	3.21	0.999 994 36	3.48	0.999 999 141	3.75	0.999 999 886
2.41	0.999 346	2.68	0.999 849	2.95	0.990 970	3.22	0.999 994 73	3.49	0.999 999 201	3.76	0.999 999 895
2.42	0.999 379	2.69	0.999 858	2.96	0.999 972	3.23	0.999 995 07	3.50	0.999 999 257	3.77	0.999 999 903
2.43	0.999 411	2.70	0.999 866	2.97	0.999 973	3.24	0.999 995 40	3.51	0.999 999 309	3.78	0.999 999 910
2.44	0.999 441	2.71	0.999 873	2.98	0.999 975	3.25	0.999 995 70	3.52	0.999 999 358	3.79	0.999 999 917
2.45	0.999 469	2.72	0.999 880	2.99	0.999 997	3.26	0.999 995 98	3.53	0.999 999 403	3.80	0.999 999 923
2.46	0.999 497	2.73	0.999 887	3.00	0.999 977 91	3.27	0.999 996 24	3.54	0.999 999 445	3.81	0.999 999 929
2.47	0.999 523	2.74	0.999 893	3.01	0.999 979 26	3.28	0.999 996 49	3.55	0.999 999 485	3.82	0.999 999 934
2.48	0.999 547	2.75	0.999 899	3.02	0.999 980 53	3.29	0.999 996 72	3.56	0.999 999 521	3.83	0.999 999 939
2.49	0.999 571	2.76	0.999 905	3.03	0.999 981 73	3.30	0.999 996 94	3.57	0.999 999 555	3.84	0.999 999 944
2.50	0.999 593	2.77	0.999 910	3.04	0.999 982 86	3.31	0.999 997 15	3.58	0.999 999 587	3.85	0.999 999 948
2.51	0.999 614	2.78	0.999 916	3.05	0.999 983 92	3.32	0.999 997 34	3.59	0.999 999 617	3.86	0.999 999 952
2.52	0.999 635	2.79	0.999 920	3.06	0.999 984 92	3.33	0.999 997 51	3.60	0.999 999 644	3.87	0.999 999 956
2.53	0.999 654	2.80	0.999 925	3.07	0.999 985 86	3.34	0.999 997 68	3.61	0.999 999 670	3.88	0.999 999 959
2.54	0.999 672	2.81	0.999 929	3.08	0.999 986 74	3.35	0.999 997 838	3.62	0.999 999 694	3.89	0.999 999 962
2.55	0.999 689	2.82	0.999 933	3.09	0.999 987 57	3.36	0.999 997 983	3.63	0.999 999 716	3.90	0.999 999 965
2.56	0.999 706	2.83	0.999 937	3.10	0.999 988 35	3.37	0.999 998 120	3.64	0.999 999 736	3.91	0.999 999 968
2.57	0.999 722	2.84	0.999 941	3.11	0.999 989 08	3.38	0.999 998 247	3.65	0.999 999 756	3.92	0.999 999 970
2.58	0.999 736	2.85	0.999 944	3.12	0.999 989 77	3.39	0.999 998 367	3.66	0.999 999 773	3.93	0.999 999 973
2.59	0.999 751	2.86	0.999 948	3.13	0.999 990 42	3.40	0.999 998 478	3.67	0.999 999 790	3.94	0.999 999 975
2.60	0.999 764	2.87	0.999 951	3.14	0.999 991 03	3.41	0.999 998 582	3.68	0.999 999 805	3.95	0.999 999 977
2.61	0.999 777	2.88	0.999 954	3.15	0.999 991 60	3.42	0.999 998 679	3.69	0.999 999 820	3.96	0.999 999 979
2.62	0.999 789	2.89	0.999 956	3.16	0.999 992 14	3.43	0.999 998 770	3.70	0.999 999 833	3.97	0.999 999 980
2.63	0.999 800	2.90	0.999 959	3.17	0.999 992 64	3.44	0.999 998 855	3.71	0.999 999 845	3.98	0.999 999 982
2.64	0.999 811	2.91	0.999 961	3.18	0.999 993 11	3.45	0.999 998 934	3.72	0.999 999 857	3.99	0.999 999 983
2.65	0.999 822	2.92	0.999 964	3.19	0.999 993 56	2.46	0.999 999 008	3.73	0.999 999 867		
2.66	0.999 831	2.93	0.999 966	3.20	0.999 993 97	3.47	0.999 999 077	2.74	0.999 999 877		

INDEX